Clostridium botulinum

FOOD SCIENCE AND TECHNOLOGY

A Series of Monographs, Textbooks, and Reference Books

1. *Flavor Research: Principles and Techniques,* R. Teranishi, I. Hornstein, P. Issenberg, and E. L. Wick
2. *Principles of Enzymology for the Food Sciences,* John R. Whitaker
3. *Low-Temperature Preservation of Foods and Living Matter,* Owen R. Fennema, William D. Powrie, and Elmer H. Marth
4. *Principles of Food Science*
   Part I: Food Chemistry, edited by Owen R. Fennema
   Part II: Physical Methods of Food Preservation, Marcus Karel, Owen R. Fennema, and Daryl B. Lund
5. *Food Emulsions,* edited by Stig E. Friberg
6. *Nutritional and Safety Aspects of Food Processing,* edited by Steven R. Tannenbaum
7. *Flavor Research: Recent Advances,* edited by R. Teranishi, Robert A. Flath, and Hiroshi Sugisawa
8. *Computer-Aided Techniques in Food Technology,* edited by Israel Saguy
9. *Handbook of Tropical Foods,* edited by Harvey T. Chan
10. *Antimicrobials in Foods,* edited by Alfred Larry Branen and P. Michael Davidson
11. *Food Constituents and Food Residues: Their Chromatographic Determination,* edited by James F. Lawrence
12. *Aspartame: Physiology and Biochemistry,* edited by Lewis D. Stegink and L. J. Filer, Jr.
13. *Handbook of Vitamins: Nutritional, Biochemical, and Clinical Aspects,* edited by Lawrence J. Machlin
14. *Starch Conversion Technology,* edited by G. M. A. van Beynum and J. A. Roels

15. *Food Chemistry: Second Edition, Revised and Expanded,* edited by Owen R. Fennema
16. *Sensory Evaluation of Food: Statistical Methods and Procedures,* Michael O'Mahony
17. *Alternative Sweetners,* edited by Lyn O'Brien Nabors and Robert C. Gelardi
18. *Citrus Fruits and Their Products: Analysis and Technology,* S. V. Ting and Russell L. Rouseff
19. *Engineering Properties of Foods,* edited by M. A. Rao and S. S. H. Rizvi
20. *Umami: A Basic Taste,* edited by Yojiro Kawamura and Morley R. Kare
21. *Food Biotechnology,* edited by Dietrich Knorr
22. *Food Texture: Instrumental and Sensory Measurement,* edited by Howard R. Moskowitz
23. *Seafoods and Fish Oils in Human Health and Disease,* John E. Kinsella
24. *Postharvest Physiology of Vegetables,* edited by J. Weichmann
25. *Handbook of Dietary Fiber: An Applied Approach,* Mark L. Dreher
26. *Food Toxicology, Parts A and B,* by Jose M. Concon
27. *Modern Carbohydrate Chemistry,* Roger W. Binkley
28. *Trace Minerals in Foods,* edited by Kenneth T. Smith
29. *Protein Quality and the Effects of Processing,* edited by R. Dixon Phillips and John W. Finley
30. *Adulteration of Fruit Juice Beverages,* edited by Steven Nagy, John A. Attaway, and Martha E. Rhodes
31. *Foodborne Bacterial Pathogens,* edited by Michael P. Doyle
32. *Legumes: Chemistry, Technology, and Human Nutrition,* edited by Ruth H. Matthews
33. *Industrialization of Indigenous Fermented Foods,* edited by Keith H. Steinkraus
34. *International Food Regulation Handbook: Policy · Science · Law,* edited by Roger D. Middlekauff and Philippe Shubik
35. *Food Additives,* edited by A. Larry Branen, P. Michael Davidson, and Seppo Salminen
36. *Safety of Irradiated Foods,* J. F. Diehl
37. *Omega-3 Fatty Acids in Health and Disease,* edited by Robert S. Lees and Marcus Karel
38. *Food Emulsions: Second Edition, Revised and Expanded,* edited by Kåre Larsson and Stig E. Friberg
39. *Seafood: Effects of Technology on Nutrition,* George M. Pigott and Barbee W. Tucker
40. *Handbook of Vitamins: Second Edition, Revised and Expanded,* edited by Lawrence J. Machlin

41. *Handbook of Cereal Science and Technology,* Klaus J. Lorenz and Karel Kulp
42. *Food Processing Operations and Scale-Up,* Kenneth J. Valentas, Leon Levine, and J. Peter Clark
43. *Fish Quality Control by Computer Vision,* edited by L. F. Pau and R. Olafsson
44. *Volatile Compounds in Foods and Beverages,* edited by Henk Maarse
45. *Instrumental Methods for Quality Assurance in Foods,* edited by Daniel Y. C. Fung and Richard F. Matthews
46. Listeria, *Listeriosis, and Food Safety,* Elliot T. Ryser and Elmer H. Marth
47. *Acesulfame-K,* edited by D. G. Mayer and F. H. Kemper
48. *Alternative Sweeteners: Second Edition, Revised and Expanded,* edited by Lyn O'Brien Nabors and Robert C. Gelardi
49. *Food Extrusion Science and Technology,* edited by Jozef L. Kokini, Chi-Tang Ho, and Mukund V. Karwe
50. *Surimi Technology,* edited by Tyre C. Lanier and Chong M. Lee
51. *Handbook of Food Engineering,* edited by Dennis R. Heldman and Daryl B. Lund
52. *Food Analysis by HPLC,* edited by Leo M. L. Nollet
53. *Fatty Acids in Foods and Their Health Implications,* edited by Ching Kuang Chow
54. Clostridium botulinum: *Ecology and Control in Foods,* edited by A. H. W. Hauschild and Karen L. Dodds

Additional Volumes in Preparation

Low-Calorie Foods Handbook, edited by Aaron M. Altschul

Cereals in Breadmaking: A Molecular Colloidal Approach, Anne-Charlotte Eliasson and Kåre Larsson

Antimicrobials in Foods: Second Edition, Revised and Expanded, edited by P. Michael Davidson and Alfred Larry Branen

Clostridium botulinum
Ecology and Control in Foods

edited by
Andreas H. W. Hauschild
Karen L. Dodds
Health and Welfare Canada
Ottawa, Ontario, Canada

CRC Press is an imprint of the
Taylor & Francis Group, an **informa** business

CRC Press
Taylor & Francis Group
6000 Broken Sound Parkway NW, Suite 300
Boca Raton, FL 33487-2742

First issued in paperback 2019

ISBN-13: 978-0-8247-8748-6 (hbk)
ISBN-13: 978-0-367-40262-4 (pbk)

Visit the Taylor & Francis Web site at
http://www.taylorandfrancis.com

and the CRC Press Web site at
http://www.crcpress.com

Preface

When we were asked to write or edit a comprehensive book on *Clostridium botulinum*, we felt that a general review of the subject was not needed at the time. Smith and Sugiyama (1988) had only recently published such a review, Eklund and Dowell (1987) had edited a book entitled *Avian Botulism*, and Simpson's book *Botulinum Neurotoxin and Tetanus Toxin* (1989) was in press. However, while writing a chapter for Doyle's *Foodborne Bacterial Pathogens* (1989), it occurred to us that two aspects of the microorganism—its ecology and its control in foods—had been somewhat neglected. These are addressed in this book.

This book is intended to serve as a reference for professionals and graduate students in the fields of bacterial pathogens and microbial ecology, as well as for food microbiologists and food scientists. We are confident that the book will indeed be a useful tool, thanks to the wide-ranging expertise and enthusiasm of our coauthors. The chapters are arranged according to subjects, and the chapter formats have been left largely to the discretion of the authors.

In recent years, strains of recognized clostridial species other than *C. botulinum* have been found to be capable of producing neurotoxins and of causing botulism. These discoveries have fundamentally altered our concept of the group of microorganisms associated with botulism. Therefore, as an introduction to this book, and in particular to its ecological aspects, we asked Charles Hatheway to write on currently known clostridia producing botulinum neurotoxins.

A short chapter on infant botulism (Chapter 5) has been included because ecological aspects such as *C. botulinum* in the environment and in foods and its competitive development in the infant's intestine are significant factors in the etiology of infant botulism. Wound botulism and "unclassified" botulism have been dealt with briefly in Chapter 1.

In conclusion, we are much indebted to our coauthors for their excellent contributions, and we wish to thank them for doing their utmost to finish their work within the given time frame.

ANDREAS H. W. HAUSCHILD
KAREN L. DODDS

REFERENCES

Doyle, M. P. (1989). *Foodborne Bacterial Pathogens*, Marcel Dekker, New York.

Eklund, M. W., and Dowell, V. R. (1987). *Avian Botulism, an International Perspective*, Charles C Thomas, Springfield, IL.

Simpson, L. L. (1989). *Botulinum Neurotoxin and Tetanus Toxin*, Academic Press Inc., Toronto, Ontario, Canada.

Smith, L. DS., and Sugiyama, H. (1988). *Botulism: The Organism, Its Toxins, the Disease*, 2nd ed. Charles C Thomas, Springfield, IL.

Contents

Contributors

David A. Baker Nestec Research Center, Lausanne, Switzerland

David L. Collins-Thompson Bioscience Group, Westreco, Inc., New Milford, Connecticut

Karen L. Dodds Bureau of Microbial Hazards, Food Directorate, Health Protection Branch, Health and Welfare Canada, Tunney's Pasture, Ottawa, Ontario, Canada

Mel W. Eklund Utilization Research Division, Northwest Fisheries Science Center, Seattle, Washington

Peggy M. Foegeding Departments of Food Science and Microbiology, North Carolina State University, Raleigh, North Carolina

Constantin Genigeorgis Veterinary School, Aristotelian University, Thessaloniki, Greece

Charles L. Hatheway Division of Bacterial and Mycotic Diseases, Center for Infectious Diseases, Centers for Disease Control, Atlanta, Georgia

Andreas H. W. Hauschild Bureau of Microbial Hazards, Food Directorate, Health Protection Branch, Health and Welfare Canada, Tunney's Pasture, Ottawa, Ontario, Canada

Jungho Kim Department of Agricultural Chemistry, Sunchon National University, Sunchon, Republic of Korea

Friedrich-Karl Lücke Microbiology Laboratory, FB Haushalt & Ernährung, Fachhochschule, Fulda, Germany

Barbara M. Lund Genetics and Microbiology, Agricultural and Food Research Council Institute of Food Research, Norwich Laboratory, Colney, Norwich, United Kingdom

S. H. W. Notermans Laboratory for Water and Food Microbiology, National Institute of Public Health and Environmental Protection, Bilthoven, The Netherlands

Terry A. Roberts Department of Microbiology, Agricultural and Food Research Council Institute of Food Research, Reading Laboratory, Shinfield, Reading, Berkshire, United Kingdom

Lynn S. Siegel Department of Toxinology, Pathology Division, U.S. Army Medical Research Institute of Infectious Diseases, Fort Detrick, Frederick, Maryland

Robert B. Wainwright Arctic Investigations Program, Centers for Disease Control, Anchorage, Alaska

Diane S. Wood Department of Food Science, University of Guelph, Guelph, Ontario, Canada

I

ECOLOGY

1

Clostridium botulinum and Other Clostridia that Produce Botulinum Neurotoxin

Charles L. Hatheway

Centers for Disease Control, Atlanta, Georgia

I. INTRODUCTION

Van Ermengem (1897) first established that botulism is caused by a toxin consumed in a food and then isolated the organism responsible for the toxin from the food. He named the organism *Bacillus botulinus*. As subsequent incidents of botulism were investigated, it was found that sometimes the toxins had different serological properties and that the causative organisms had varying physiological characteristics. The organisms were always anaerobic and formed spores, and they were later placed into the genus *Clostridium* (Bengtson, 1924); the genus *Bacillus* was restricted to aerobic spore-forming rods.

The genus *Clostridium* includes all anaerobic spore-forming rods except for those classified as *Desulfotomaculum* (Cato et al., 1986). Clostridia are considered gram-positive, although many strains stain positive only in very young cultures and may appear as gram-negative cells in more mature cultures. Clostridia can be distinguished from non–spore-forming organisms by visualizing the spores on a stained smear or wet mount or by demonstrating heat resistance. The morphology of a typical strain of *Clostridium botulinum* is shown in the photomicrographs in Figure 1. Most species of *Clostridium* will not grow on agar surfaces in the presence of oxygen; some, however, are aerotolerant and could be judged to be aerobic organisms, i.e., *Bacillus* species, but they can be distinguished from these by their failure to produce catalase.

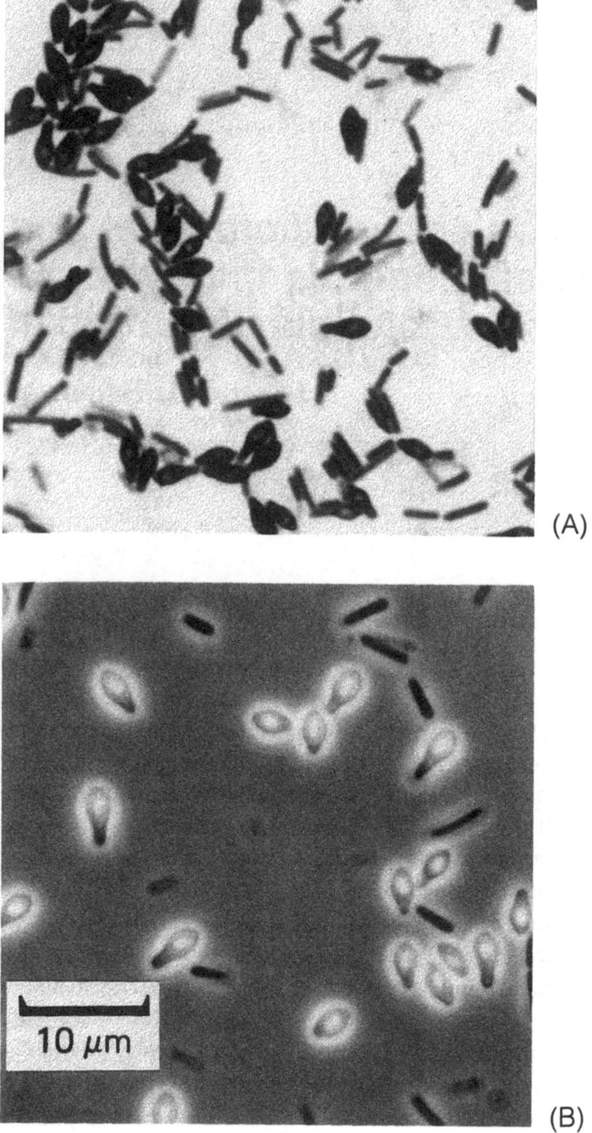

Figure 1 Photomicrographs of egg yolk agar culture of a typical strain of *Clostridium botulinum* type A (CDC strain 5356). (A) Gram-stained smear from 3-day culture. (B) Phase-contrast view of wet-mount suspension from 4-day culture. Scale for both photomicrographs is indicated by bar in B.

C. botulinum was defined in 1953 as the species designation for all organisms known to produce botulinum neurotoxin and thereby having the capability of causing botulism in humans or animals (Prévot, 1953). This nomenclature eliminated the use of the name *C. parabotulinum* to distinguish the proteolytic group of toxigenic organisms from the nonproteolytic group. There had been a problem of other ambiguous uses of that species name. The unifying feature of the organisms known as *C. botulinum*, then, has been their toxigenicity. It will be seen that *C. botulinum*, as presently constituted, can be separated into four different species. In the investigation of some recent botulism incidents, the causative organisms have been found to be other established species of *Clostridium* that possess the ability to produce botulinum neurotoxin and to cause botulism. This illustrates the difficulty inherent in a nomenclature rigidly based on the ability to produce a toxin (Hatheway and McCroskey, 1989).

Thus, botulinum toxin can be produced by a variety of clostridia. The neurotoxins produced by various organisms can be differentiated serologically by specific neutralization into seven toxin types: A, B, C, D, E, F, and G. A given toxin type is not necessarily restricted to one phenotypic group of organisms. Regardless of serological type, the toxins are structurally and pharmacologically similar, causing the same flaccid paralysis of botulism in susceptible animal species.

II. TOXIN TYPES

Nine years after van Ermengem (1897) established that botulism was caused by a bacterial toxin, a second outbreak due to bean salad was investigated and confirmed in a similar manner by Landmann (1904) in Darmstadt, Germany. Leuchs (1910) compared the organisms isolated from the Belgian and the German outbreaks and, using the method of Kempner (1897) for preparing antitoxins against botulinum toxin, found that the toxins were serologically distinct, each being neutralized only by its homologous antiserum. Letter designations were not assigned to toxin types until Burke (1919) reported the results of studies on the characteristics and toxins of strains recovered from botulism outbreaks in the United States. Toxins produced by these organisms were identified as either type A or type B on the basis of neutralization with rabbit antitoxins. Although the van Ermengem and the Landmann strains were no longer available for testing with the typing reagents, it was apparent from the descriptions of the cultural characteristics of the organisms in the earlier literature that the former would correspond to type B and the latter to type A. Additional types were established as other organisms were recovered from incidents of botulism in animals and humans that produced toxins that did not correspond to existing types.

The toxin produced by organisms causing botulism in chickens in the United States (Bengtson, 1922) and in cattle in Australia (Seddon, 1922) was designated as type C. Upon observing that antitoxin produced against the Bengtson strain neutralized the toxin of either strain, while antitoxin against the Seddon strain neutralized only toxin from the homologous strain, the former was designated C_α, and the latter strain, C_β (Gunninson and Meyer, 1929). Another outbreak in cattle in South Africa was caused by an organism subsequently designated type D because its toxin was serologically distinct from types A, B, and C (Meyer and Gunnison, 1928–29). Type E was established to designate the toxin type of organisms causing botulism in the United States and in the Soviet Union caused by consumption of fish (Gunnison et al., 1936–37). An outbreak of botulism caused by a liver paste in Denmark involved a sixth type of toxin, known as F (Moller and Scheibel, 1960). Finally, the toxin produced by an organism isolated from soil in Argentina was designated type G (Giménez and Ciccarelli, 1970b).

The distinction between the C_α and the C_β toxin types is a source of confusion. C_α strains produce C_1, a neurotoxin, and C_2, which is not a neurotoxin but is lethal in mice, ducks, and geese; congestion and hemorrhage in the lung and dripping of the nares are observed in birds injected with C_2 toxin, but paralytic signs of botulism are absent (Jensen and Duncan, 1980; Ohishi and DasGupta, 1987). C_β strains produce only C_2 toxin.

Thus, there are seven serologically differentiable types of botulinum neurotoxin. Although, in general, each type is serologically distinct, slight partial cross-neutralizations of the neurotoxins produced by types C and D strains occur because type C strains produce a minor amount of type D neurotoxin, and type D strains produce a trace of type C neurotoxin (Jansen, 1971a, b, 1987). Oguma et al. (1984) have also found evidence for common antigenic sites on types C and D toxin molecules. The initial studies of *C. botulinum* type F showed that its toxin was neutralized by a large excess of type E antitoxin (Moller and Scheibel, 1960; Dolman and Murakami, 1961). A later study indicates that the cross-neutralization between type E and type F is reciprocal (Yang and Sugiyama, 1975). There are also subtle differences between toxins produced by strains of the same toxin type (Ciccarelli and Giménez, 1971; Giménez and Ciccarelli, 1972; Hatheway et al., 1981). Strains have been found that produce mixtures of two types of toxin, A + F (Giménez and Ciccarelli, 1970a), A + B (Poumeyrol et al., 1983), and B + F (Hatheway and McCroskey, 1989).

III. PHYSIOLOGICAL AND GENETIC DIFFERENCES AMONG ORGANISMS PRODUCING BOTULINUM NEUROTOXIN

A. Physiological Groups

Since botulism had always been associated with meat products, as denoted by its name (derived from *botulus*, the Latin word for sausage), it was surprising that the illness in Darmstadt (Landmann, 1904) due to consumption of bean salad was also confirmed as botulism. Leuchs (1910) noted that the Ellezelles strain (van Ermengem, 1897) and the Darmstadt strain were similar in their morphological and cultural characteristics but differed in their optimal temperatures for growth and toxin production; for the former it was 18–25°C, and for the latter, 35–37°C. Burke (1919) found that the strains of both toxin variants (type A and type B) of organisms isolated from botulism incidents in the United States resembled the Darmstadt strain, with 37°C as their optimum temperature for growth and toxin production. Bengtson (1924) found that all the American isolates of either toxin type were strongly proteolytic, i.e., they digested meat particles, casein, and coagulated egg albumin, while a European isolate that produced type B toxin clearly was not proteolytic. The latter corresponded very closely to the organism isolated and described by van Ermengem (1897) in all of its cultural characteristics.

In examining an international collection of strains, Gunnison and Meyer (1929) confirmed the findings of Bengtson (1924). They found that two of three European toxin type B strains were nonproteolytic, as well as all strains of toxin type C and one strain (the only one tested) of toxin type D. Although the type C and D strains are nonproteolytic (or weakly so according to some researchers), they are distinguishable from the nonproteolytic type B strains by their high optimum growth temperature of 40°C (Smith and Sugiyama, 1988).

Type E strains resemble European strains of toxin type B in that they are clearly nonproteolytic (Gunnison et al., 1936–37). Although they may grow well at 37°C, toxigenicity is more constant at room temperature, and they are capable of growing and producing toxin at lower temperatures than the proteolytic strains. Nonproteolytic type B strains have been shown to grow and produce toxin at temperatures as low as 3.3°C (Eklund et al., 1967).

The initial isolate of *C. botulinum* type F (Moller and Scheibel, 1960) was proteolytic and resembled the American strains of toxin types A and B (Burke, 1919; Bengtson, 1924). A strain isolated from soil in Argentina also belongs to the proteolytic group (Giménez and Ciccarelli, 1968). Toxin type F strains isolated from marine sediments from the northwest coast of the United States (Eklund et al., 1967) and from the second outbreak of human type F botulism (Midura et al., 1972) are nonproteolytic, resembling the European type B strains (van Ermengem, 1897; Bengtson, 1924; Gunnison and Meyer, 1929) and the type E strains (Gunnison et al., 1936–37).

The type G organism isolated from soil in Argentina (Giménez and Ciccarelli, 1970b) is considerably different from the organisms discovered earlier that produce toxins of types A, B, C, D, E, and F. The latter all ferment glucose and may ferment several other carbohydrates; they also produce lipase, which is detectable on egg yolk agar (Cato et al., 1986). The type G organism is asaccharolytic and shows no reaction on egg yolk agar.

The various neurotoxigenic organisms described above are designated as *C. botulinum* and are divided, according to *Bergey's Manual* (Cato et al., 1986), into four groups. The distinguishing features are listed in Table 1.

Table 1 Characteristics of Organisms Capable of Producing Botulinum Neurotoxin[a]

	Groups					
	I	II	III	IV[b]	*C. butyricum*	*C. baratii*
Toxin types	A,B,F	B,E,F	C,D	G	(E)	(F)
Proteolysis	+	−	−	+	−	−
Liquefaction of gelatin	+	+	+	+	−	−
Fermentation of						
Glucose	+	+	+	−	+	+
Fructose	±	+	±	−	+	+
Mannose	−	+	+	−	+	+
Maltose	±	+	±	−	+	+
Sucrose	−	+	−	−	+	+
Trehalose	−	+	−	−	+	−
Lipase	+	+	+	−	−	−
Lecithinase	−	−	−	−	−	+
Metabolic acids[c]	A,iB,B,iV PP	A,B	A,P,B	A,iB,B,iV PA	A,B	A,B
Optimal growth temperature	35–40°C	18–25°C	40°C	37°C	30–37°C	30–45°C
Minimum growth temperature	10+°C	3.3°C	15°C		10°C	
Spore heat resistance (temp./D-value)	112°C/ 1.23	80°C/ 0.6–1.25	104°C/ 0.1–0.9	104°C 0.8–1.12		
Similar atoxic organism	*C. sporogenes*	(no species name assigned)	*C. novyi*	*C. subterminale*	(all typical strains)	(all typical strains)

[a] For biochemical reactions: +, all strains are positive; −, all strains are negative; ±, some strains are positive and some are negative. For temperature values, none are listed if not readily available in the literature.

[b] The name *Clostridium argentinense* has been proposed for group IV (Suen et al., 1988a).

[c] Metabolic acids: A, acetic; P, propionic; B, butyric; iB, isobutyric; iV, isovaleric; PP, phenylpropionic (hydrocinnamic); PA, phenylacetic.

Group I. The organisms typified by toxin type A strains are proteolytic, produce isobutyric, isovaleric, and beta-phenylpropionic (hydrocinnamic) acids, have an optimal growth temperature of 37°C and spores with a high heat resistance, and produce toxin of either type A, B, or F. Types A and B strains commonly produce 10^6 mouse lethal doses (LD) of toxin per ml in cultures;

type F strains generally produce 10^3–10^4 LD/ml. *C. sporogenes* appears to be a nontoxigenic variant of organisms in this group.

Group II. The organisms in this group are nonproteolytic, grow optimally at 30°C or less, are able to grow at refrigeration temperatures, have spores with a low heat resistance, and produce toxin of either type B, E, or F. Toxicity of the cultures is enhanced by trypsinization, since the organisms lack the endogenous proteolytic enzymes for activation of the toxin. Cultures of type E strains generally attain toxin levels of 10^5 LD/ml after activation. Nontoxigenic organisms resembling *C. botulinum* type E are encountered, but there is no separate species name to designate them.

Group III. This group consists of organisms that produce either type C or type D toxin. The organisms are generally nonproteolytic, and do not produce the iso-acids that are typical of proteolytic clostridia; they produce propionic as well as acetic and butyric acids. Toxigenicity of these organisms is mediated by bacteriophages (Eklund et al., 1987), and isolation of organisms with stable toxigenicity may be difficult (Hariharan and Mitchell, 1977; Bengtson, 1922). Organisms in this group grow optimally at 40°C and have an intermediate spore resistance to heat. They are phenotypically similar to *C. novyi*.

Group IV. The organism that produces type G toxin is vastly different from members of the other three groups in that it is asaccharolytic and does not produce lipase. Therefore, a fourth group was required (Giménez and Ciccarelli, 1970b). Organisms of this group are proteolytic and produce the iso-acids typical of proteolytic clostridia; they also produce phenylacetic acid. Spores are rarely seen in cultures in the usual media, but their presence can be demonstrated by survival after heat treatment. Toxin levels in cultures are much lower than levels produced by other types. Approximately 50 LD/ml is attained in chopped meat broth; although the organism is proteolytic, the toxicity of the cultures may be increased 10-fold by trypsinization (Ciccarelli et al., 1977). Higher toxin levels (10^4–10^5 LD/ml) may be obtained in dialysis cultures. The optimum growth temperature is 37°C. Lynt et al. (1984) determined the spore heat resistance at 104°C as $D = 0.1$–0.9, which is similar to that of group III. Group IV strains are similar to other asaccharolytic, proteolytic clostridia such as *C. subterminale* and *C. hastiforme*.

Four cases of human botulism, three cases of infant botulism and one adult case apparently due to colonization, were caused by other established species of *Clostridium*. Two were the only cases of type E infant botulism on record; they occurred in Italy, and the causative organisms were *C. butyricum* (Aureli et al., 1986; McCroskey et al., 1986). The other infant case was the only known type F case, and it was caused by *C. baratii* (Hall et al., 1985; Hoffman et al., 1982). The adult case was also type F and due to *C. baratii* (Hatheway and McCroskey, 1989; McCroskey et al., 1991). The characteristics of *C. butyricum* and *C. baratii* are listed in Table 1 for comparison with those of the organisms generally recognized as producers of botulinum neurotoxin. Both species are strongly saccharolytic and nonproteolytic and do not liquefy gelatin. *C. baratii* shows lecithinase activity on egg yolk agar.

Toxin from one of the *C. butyricum* stains has been purified and found to be very similar to the toxin produced by a known strain of *C. botulinum* type E (Giménez and Sugiyama, 1988). Purification and characterization of the toxin from the neurotoxigenic *C. baratii* has also been reported recently (Giménez et al., 1992). Serologically, it is not typical of type F toxin produced by *C. botulinum* strains in that it requires more type F antitoxin to neutralize a given amount of toxin, and there is slightly more cross-neutralization by type E antitoxin (McCroskey et al., 1991).

B. Genetic Studies

Based on DNA homology, Lee and Riemann (1970a) found little genetic relatedness among groups I, II, and III of *C. botulinum*. They found a high degree of relatedness among strains of different toxin types within each group, and even between toxigenic and nontoxigenic variants of phenotypically similar organisms (Lee and Riemann, 1970a, b). Thus, strains of *C. sporogenes* showed a high degree of relatedness to group I *C. botulinum*. Nakamura et al. (1977) found two groups of *C. sporogenes*: one with a high degree of relatedness with *C. botulinum* and the other with intermediate relatedness.

Nakamura et al. (1983) compared the DNA sequence relationships among *C. novyi* types A and B, *C. haemolyticum* (*C. novyi* type D), and *C. botulinum* type C. All these organisms had at least a low to moderate degree of relatedness with each other. The organisms in the study were separated into three groups of highly related organisms: (1) *C. novyi* type A, (2) *C. novyi* type B, *C. haemolyticum*, and one strain of *C. botulinum* type C, and (3) the remaining four strains of *C. botulinum* type C.

Suen et al. (1988a) studied the relatedness of the organisms that produce type G botulinum toxin to Group I *C. botulinum* and to other phenotypically similar (asaccharolytic and proteolytic) nontoxigenic clostridia. The type G organisms were very closely related (90–100%) to each other and highly related to two strains identified as *C. subterminale* (85 and 86%) and to one strain identified as *C. hastiforme* (76%). They were intermediately related (53%) to two strains identified as *C. subterminale* and unrelated (<20%) to any other clostridia in the study, including the group I organisms. They proposed the name *Clostridium argentinense* for the type G toxin–producing organisms as well as for the closely related nontoxigenic organisms. This name change was proposed as an initial step in rectifying the difficulty in nomenclature. The remainder of the asaccharolytic clostridia fell into six other DNA-related groups, each distinguishable by reactions in biochemical tests.

Toxigenic strains of *C. baratii* and *C. butyricum* were compared with nontoxigenic reference strains of their respective species (Suen et al., 1988b). They were 86% and 78% related, respectively, to their type strains in DNA hybridization tests carried out at 50°C, while all other clostridial species in those tests were less than 20% related to those strains.

IV. TOXINS

A. Nature of the Neurotoxins

The active botulinum neurotoxins are proteins with molecular masses of 150–167 kDa, each consisting of a heavy chain (ca. 100 kDa) and a light chain (ca. 50 kDa) joined by a disulfide bridge. The details of their structure have been recently reviewed by DasGupta (1989). All toxin types are similar, and their similarity to tetanus neurotoxin (Matsuda, 1989) is striking. They are synthesized as a single peptide with relatively low biological activity, requiring cleavage by a proteolytic enzyme to form the highly active di-chain molecule. In culture, the neurotoxins exist as complexes with nontoxic proteins (Sakaguchi, 1983). The toxin in the complex is rather stable, especially under acidic conditions (pH 3.5–6.5), but the complex dissociates under slightly alkaline conditions and the biological activity is readily inactivated in this state. The neurotoxin can be separated from the nontoxic components and purified by ion-exchange chromatography.

Botulinum neurotoxin acts by blocking release of acetylcholine at the neuromuscular junction in a three-step process (Simpson, 1986): (1) the toxin molecule binds to receptors on the nerve ending; (2) the toxin molecule, or a portion of it, is internalized; and (3) within the nerve cell, the toxin interferes with the release of acetyl choline by an unknown (probably enzymatic) mechanism. Thus far, no enzymic activity of the neurotoxins has been elucidated.

B. Neurotoxin Genes

The neurotoxin genes for types C and D organisms are carried by bacteriophages (Eklund et al., 1987). Organisms cured of their prophages no longer produce neurotoxin, and reinfection of cured organisms with a phage reestablishes toxigenicity for either type C or type D neurotoxin, depending on the source of the phage (Eklund and Poysky, 1974). Cured type C organisms may continue to produce C2 toxin. This suggests that C_β strains are derived from C_α strains upon loss of their prophage (Jansen, 1987).

A strain of *C. botulinum* type C, cured of its phage and reinfected with phage NA1 from *C. novyi*, produced the α-toxin of *C. novyi* and could not be distinguished from *C. novyi* type A (Eklund et al., 1974). The same organism may be interconverted among *C. botulinum* toxin types C and D and *C. novyi* type A, depending upon which phage is inserted. The interspecies conversion is consistent with DNA relatedness findings for these organisms (Nakamura et al., 1983). Phage association with production of botulinum neurotoxin has only been established for types C and D.

The entire nucleotide sequences for type C (Hauser et al., 1990) and type D (Binz et al., 1990a) neurotoxin genes have been derived from cloned phage DNA fragments. The type C gene encodes for 1291 amino acids (M_r 148, 698), and the type D gene encodes for 1276 amino acids (M_r 146, 872).

Noting that the gene for tetanus toxin is found in a plasmid (Finn et al., 1984), studies of plasmid association with botulinum neurotoxin production were undertaken (Strom et al., 1984; Weickert et al., 1986). Evidence for plasmid-mediated production of type G neurotoxin has been reported (Eklund et al., 1988).

The genes for types A, B, E, and F neurotoxins are likely present on the bacterial chromosome. This has been confirmed for type A neurotoxin with the complete determination of the nucleotide sequence of the structural gene using fragments of chromosomal DNA (Binz et al., 1990b; Thompson et al., 1990). The gene encodes a polypeptide of 1296 amino acids, with M_r of approximately 149,500. A fragment of the structural gene for type E neurotoxin derived from chromosomal DNA has been sequenced (Binz et al., 1990b). It codes for the first 252 amino acids of the molecule, indicating that the gene for type E is also chromosomal.

C. Other Toxins

Among the clostridia, the term *toxin* has been used to describe a variety of antigenic proteins with demonstrable activities (Hatheway, 1990). The various organisms that have been shown to produce botulinum neurotoxin may also produce lipase, lecithinase, C2 toxin, or C3 ADP ribosyltransferase. The latter two unique products will be discussed here.

C2 Toxin

Jansen (1971a) established that toxin type C_α organisms produce multiple lethal factors, which he designated C1, C2, and D, while toxin type C_β organisms produce only the C2 factor. It was initially assumed that both C1 and C2 as well as D were neurotoxins. Jensen and Duncan (1980) demonstrated that the lethal activity of C2 toxin was not associated with any paralytic activity. Purification of C2 toxin was a problem because it consists of two separate, unlinked peptides, designated component I (50 kDa) and component II (104 kDa) (Ohishi and DasGupta, 1987). Both are required for expression of biological activity. The combination is known as a binary toxin. Together they exhibit mouse lethality, vascular permeability activity, and enterotoxicity.

C2 functions enzymatically as an ADP-ribosyltransferase. Component II (C2-II) recognizes a receptor on susceptible cells and, after binding to the receptor, it effects the internalization of C2-I, which ADP-ribosylates intracellular actin (Aktories et al., 1986). This interferes with cell structure formation and results in cell damage. Without the assistance of C2-II, C2-I cannot enter the cell.

C3 ADP-Ribosyltransferase

Researchers have sought an enzymatic function for the botulinum neurotoxins as a key to understanding the mechanism of their action. ADP-ribosyltransferase activity was reported for type C and type D neurotoxins (Matsuoka et al., 1987), but this activity was due to a contaminant in the neurotoxin preparations (Aktories et al., 1987; Roesener et al., 1987). The activity is due to a novel ADP-ribosyltransferase designated as exoenzyme C3, which is distinct from C2-I. It has a molecular mass of about 25 kDa and a different substrate (21–24 kDa proteins common to membranes in a variety of mammalian cells). The enzyme does not ADP-ribosylate actin.

C3 appears simultaneously with C1 or D and has been shown to be encoded by the same phage that encodes the neurotoxins. The gene for C3 is distinct from the neurotoxin gene within the phage genome. It consists of 654 nucleotides that encode a peptide of 218 amino acids (Popoff et al., 1990). Seven amino acids appear to be cleaved from the initially synthesized peptide, yielding an extracellular product of 211 amino acids with a molecular mass of 23,546 Da.

V. ORGANISMS CAUSING BOTULISM

A. Human Botulism

Human botulism is almost always of type A, B, or E. Only six incidents of type F botulism have been documented. One foodborne outbreak in France, apparently caused by consumption of a paté, was identified as type C on the basis of isolation of *C. botulinum* type C_β from a sample of the suspected food (Prévot et al., 1955). One other incident was reported as type C botulism in the United States by K. F. Meyer (cited by Prévot et al., 1955) on the basis of isolation of the organism (type C_α) from the deceased patient's stomach contents.

One outbreak of type D botulism in humans is reported to have occurred in Chad (Demarchi et al., 1958). Samples of the suspected food vehicle, salted ham, were found to contain type D botulinum toxin; the same toxin was also found in cultures of heated samples of the ham.

Although one laboratory has reported isolation of organisms capable of producing type G botulinum toxin from autopsy specimens (Sonnabend et al., 1981), it is still open to question as to whether type G botulism has ever occurred, either in humans or in animals outside of laboratory experiments. No clinical signs of illness suggestive of botulism were observed in any of the autopsied persons, and no other laboratory has reported isolation of type G organisms from clinical specimens. In another report of nine positive autopsies (Sonnabend et al., 1985), the evidence in seven was for botulism types rarely, if ever, documented in humans. The isolated organisms were type G in three cases, type C in one, and type F in three. The more probable types A and B were found in only one case each.

Nearly 2000 cases of human botulism have been confirmed in the United States since 1950. The cases are summarized by disease category and toxin type in Table 2. In foodborne botulism, all type A and B cases except three cases of type B were caused by group I organisms. Those type B cases were caused by group II strains. The organism that caused an outbreak of type F botulism involving three patients was of group II (Midura et al., 1972). In one other case of type F foodborne botulism, the organism was not isolated.

All of the type A, B, and Bf cases of infant botulism were caused by group I strains. The only type F infant case in the United States was caused by a *C. baratii* strain that produces type F neurotoxin. The only type E infant cases on record were caused by *C. butyricum* in Italy (Aureli et al., 1986).

All organisms isolated from wound botulism cases belong to group I. Unclassified incidents involve adult patients or older children with evidence suggesting that colonization of the gut by toxigenic organisms may be responsible for the illness (McCroskey and Hatheway, 1988). One of the type F cases was caused by *C. baratii*. In the other type F case, no organism was recovered; confirmation was made on the basis of detection of type F toxin in the patient's serum (Green et al., 1983).

Table 2 Human Botulism Cases Confirmed in the United States 1950–1989 by Disease Category and Toxin Type[a]

	Type A	*Type B*	*Type E*	*Type F*	*Other*	*Unknown*	*Total*
Foodborne	388	171	156	4	0	300	1019
Infant	431	451	0	1[b]	3[c]	8	894
Wound	24	9	0	0	1[d]	10	44
Unclassified	4	1	0	2[b]	0	4	12
Total	847	632	156	7	4	322	1969

[a] Unpublished data, Centers for Disease Control; one of the wound botulism cases was confirmed in 1943.

[b] The case of type F infant botulism and one of the cases of unclassified type F botulism were caused by toxigenic *Clostridium baratii*.

[c] Two infant cases were caused by one organism that produces a major amount of type B and a minor amount of type F neurotoxins (toxin type Bf); one infant case was caused by a combination of two strains—one type A and the other type B.

[d] One wound case was caused by a combination of two strains—one type A and the other type B.

No group III or IV organisms were implicated in human botulism in the United States during this period.

B. Botulism in Animals and Birds

Smith and Sugiyama (1988) present an overview of botulism in mammalian and avian species and in turtles and fish. Among domestic animals, botulism has been a problem in cattle, horses, dogs, and mink. Most often it is type C or D, due to consumption of carrion, either intentionally or accidentally, for example, when dead animals are present in forage feeds. Horses are very susceptible to type B toxin. ''Forage poisoning'' in horses, mules, and cattle has been found to be of type B (Divers et al., 1986; Johnston and Whitlock, 1987). Shaker foal syndrome is due to intestinal growth of *C. botulinum* type B (Swerczek, 1980).

Avian botulism among domestic flocks is caused mostly by type C and involves large numbers of wild ducks and geese (Smith and Sugiyama, 1988). Massive spread of such outbreaks has often been due to birds eating toxin-laden maggots feeding on the carcasses of other birds that have died of botulism. Some outbreaks among other species of waterbirds (gulls, loons, and grebes) have been caused by type E and were probably due to consumption of toxic fish.

VI. SUMMARY

Botulism can be caused by neurotoxins produced by any one of a variety of organisms in the genus *Clostridium*. The neurotoxigenic organisms that have been recognized and characterized can be classified into the equivalent of six species. Four of these have been designated groups I–IV of *C. botulinum*. Neurotoxigenic strains of *C. baratii* and *C. butyricum* have been implicated in four cases of human botulism. Taxonomically, retention of a single species name for the four phenotypically and genetically different groups is a problem. Likewise, exclusion of genetically similar organisms from a species because of failure to produce a toxin is also a problem. *C. argentinense* has been proposed as a species to include group IV organisms and genetically related nontoxigenic clostridia.

Botulinum neurotoxins are differentiated into seven toxin types based on differential neutralization with type-specific antitoxin reagents. There are subtle differences among toxins produced by strains of the same toxin type. Some minor cross-reactions also occur between toxin types, i.e., between types C and D, and between types E and F.

The neurotoxins are structurally very similar, consisting of a heavy and a light chain joined by a disulfide bridge. C2 is a binary toxin consisting of two peptide molecules which together exhibit lethal, permease, and enterotoxic activity. Types C and D organisms also produce an ADP-ribosyltransferase known as C3.

Genes for types C and D neurotoxins, as well as C3, are encoded by bacteriophages carried by neurotoxigenic organisms as prophages. The nucleotide sequences of the genes for these molecules have been determined. The entire gene sequence for type A toxin has been determined from fragments of chromosomal DNA. A portion of the type E toxin gene has been identified among chromosomal DNA, and the sequence for about 20% of the toxin molecule has been determined. Evidence for the presence of the type G toxin gene on a plasmid has been reported.

Human botulism is caused by group I and group II organisms, while botulism in animal and avian species is caused mainly by group III organisms. Although isolation of group IV organisms from autopsy specimens has been reported, there is insufficient evidence that botulism was the cause of death.

REFERENCES

Aktories, K., Baermann, M., Ohishi, I., Tsuyama, S., Jacobs, K. H., and Habermann, E. (1986). Botulinum C2 toxin ribosylates actin. *Nature 322*: 390.

Aktories, K., Weller, V., and Chhatwal, G. S. (1987). *Clostridium botulinum* type C produces a novel ADP-ribosyltransferase distinct from C2 toxin. *FEBS Lett. 212*: 109.

Aureli, P., Fenicia, L., Pasolini, B., Gianfranceschi, M., McCroskey, L. M., and Hatheway, C. L. (1986). Two cases of type E infant botulism caused by neurotoxigenic *Clostridium butyricum* in Italy. *J. Infect. Dis. 154*: 201.

Bengtson, I. A. (1922). Preliminary note on a toxin-producing anaerobe isolated from the larvae of *Lucilia caesar. Public Health Rep. 37*: 164.

Bengtson, I. A. (1924). Studies on organisms concerned as causative factors in botulism. *Hyg. Lab. Bull. 136*: 101.

Binz, T., Kurazono, H., Popoff, M. R., Eklund, M. W., Sakaguchi, G., Kozaki, S., Krieglstein, K., Henschen, A., Gill, D. M., and Niemann, H. (1990a). Nucleotide sequence of the gene encoding for *Clostridium botulinum* neurotoxin type D. *Nucleic Acids Res. 18*: 5556.

Binz, T., Kurazono, H., Wille, M., Frevert, J., Wernars, K., and Niemann, H. (1990b). The complete sequence of botulinum neurotoxin type A and comparison with other clostridial neurotoxins. *J. Biol. Chem. 265*: 9153.

Burke, G. S. (1919). Notes on *Bacillus botulinus. J. Bacteriol. 4*: 555.

Cato, E. P., George, W. L., and Finegold, S. M. (1986). Section 13, Genus Clostridium. In *Bergey's Manual of Systematic Bacteriology* (P. H. A. Sneath, N. S. Mair, M. E. Sharpe, J. G. Holt, eds.), Williams and Wilkins, Baltimore, pp. 1141–1200.

Ciccarelli, A. S., and Giménez, D. F. (1971). Estudio serologico de toxinas de *Clostridium botulinum* tipo A y Cepa 84. *Rev. Lat-Amer. Microbiol. 13*: 67.

Ciccarelli, A. S., Whaley, D. N., McCroskey, L. M., Giménez, D. F., Dowell, V. R. Jr., and Hatheway, C. L. (1977). Cultural and physiological characteristics of *Clostridium botulinum* type G and the susceptibility of certain animals to its toxin. *Appl. Environ. Microbiol. 34*: 843.

DasGupta, B. R. (1989). The structure of botulinum neurotoxin. In *Botulinum Neurotoxin and Tetanus Toxin* (L. L. Simpson, ed.), Academic Press, Inc., San Diego, CA, pp. 53–67.

Demarchi, J., Mourgues, C., Orio, J., and Prevot, A. R. (1958). Existence du botulisme humain de type D. *Bull. Acad. Nat. Med. 142*: 580.

Divers, T. J., Bartholomew, R. C., Messick, J. B., Whitlock, R. H., and Sweeney, R. W. (1986). *Clostridium botulinum* type B toxicosis in a herd of cattle and a group of mules. *J. Am. Vet. Med. Assoc. 188*: 382.

Dolman, C. E., and Murakami, L. (1961). *Clostridium botulinum* type F with recent observations on other types. *J. Infect. Dis. 109*: 107.

Eklund, M. W., and Poysky, F. T. (1974). Interconversion of type C and D strains of *Clostridium botulinum* by specific bacteriophages. *Appl. Microbiol. 27*: 251.

Eklund, M. W., Poysky, F. T., Meyers, J. A., and Pelroy, G. A. (1974). Interspecies conversion of *Clostridium novyi* type A by bacteriophage. *Science 186*: 456.

Eklund, M. W., Poysky, F. T., Mseitif, L. M., and Strom, M. S. (1988). Evidence for plasmid-mediated toxin and bacteriocin production in *Clostridium botulinum* type G. *Appl. Environ. Microbiol. 54*: 1405.

Eklund, M. W., Poysky, F., Oguma, K., Iida, H., and Inoue, K. (1987). Relationship of bacteriophages to toxin and hemagglutinin production and its significance in avian botulism outbreaks. In *Avian Botulism* (M. W. Eklund, V. R. Dowell, Jr., eds.), Charles C Thomas, Springfield, IL, pp. 191–222.

Eklund, M. W., Poysky, F. T., and Wieler, D. T. (1967). Characteristics of *C. botulinum* type F isolated from the Pacific Coast of the United States. *Appl. Microbiol. 15*: 1316.

Eklund, M. W., Wieler, D. I., and Poysky, F. T. (1967). Outgrowth and toxin production of non-proteolytic type B *Clostridium botulinum* at 3.3–5.6 deg. C. *J. Bacteriol. 93*: 1461.

Finn, C. W., Silver, R. P., Habig, W. H., Hardegree, M. C., Zon, G., and Garon, C. F. (1984). The structural gene for tetanus neurotoxin is on a plasmid. *Science 224*: 881.

Giménez, D. F., and Ciccarelli, A. S. (1968). *Clostridium botulinum* type F in the soil of Argentina. *Appl. Microbiol. 16*: 732.

Giménez, D. F., and Ciccarelli, A. S. (1970a). Studies on Strain 84 of *Clostridium botulinum. Zentralbl. Bakt. I, Abt. Orig. A 215*: 212.

Giménez, D. F., and Ciccarelli, A. S. (1970b). Another type of *Clostridium botulinum. Zentralbl. Bakt. I, Abt. Orig. A 215*: 215.

Giménez, D. F., and Ciccarelli, A. S. (1972). Variaciones antigenicas en toxinas botulinicas del tipo F; ensayo de definiciones para la tipificacion serologica y classificacion de *Clostridium botulinum. Medicina 32*: 596.

Giménez, J. A., Giménez, M. A., and DasGupta, B. R. (1992). Characterization of the neurotoxin isolated from a *Clostridium baratii* strain implicated in infant botulism. *Infect. Immun. 60*: 518.

Giménez, J. A., and Sugiyama, H. (1988). Comparison of toxins of *Clostridium butyricum* and *Clostridium botulinum* type E. *Infect. Immun. 56*: 926.

Green, J., Spear, H., and Brinson, R. R. (1983). Human botulism (type F)—a rare type. *Am. J. Med. 75*: 893.

Gunnison, J. B., Cummings, J. R., and Meyer, K. F. (1936–37). *Clostridium botulinum* type E. *Proc. Soc. Exp. Biol. Med. 35*: 278.

Gunnison, J. B., and Meyer, K. F. (1929). Cultural study of an international collection of *Clostridium botulinum* and *parabotulinum. J. Infect. Dis. 45*: 119.

Hall, J. D., McCroskey, L. M., Pincomb, B. J., and Hatheway, C. L. (1985). Isolation of an organism resembling *Clostridium barati* which produces type F botulinal toxin from an infant with botulism. *J. Clin. Microbiol. 21*: 654.

Hariharan, H, and Mitchell, W. R. (1977). Type C botulism: The agent, host spectrum and environment. *Vet. Rec. 47*: 95.

Hatheway, C. L. (1990). Toxigenic clostridia. *Clin. Microbiol. Rev. 3*: 67.

Hatheway, C. L., and McCroskey, L. M. (1989). Unusual neurotoxigenic clostridia recovered from human fecal specimens in the investigation of botulism. In *Recent Advances in Microbial Ecology* (T. Hattori, Y. Ishida, Y. Maruyama, R. Y. Morita, A. Uchida, eds.), Japan Scientific Societies Press, Tokyo, pp. 477–481.

Hatheway, C. L., McCroskey, L. M., Lombard, G. L., and Dowell, V. R. Jr. (1981). Atypical toxin variant of *Clostridium botulinum* type B associated with infant botulism. *J. Clin. Microbiol. 14*: 607.

Hauser, D., Eklund, M. W., Kurazono, H., Binz, T., Niemann, H., Gill, D. M., Boquet, P., and Popoff, M. R. (1990). Nucleotide sequence of *Clostridium botulinum* C1 neurotoxin. *Nucleic Acids Res. 18*: 4924.

Hoffman, R. E., Pincomb, B. J., and Skeels, M. R. (1982). Type F infant botulism. *Am. J. Dis. Child. 136*: 270.

Jansen, B. C. (1971a). The toxic antigenic factors produced by *Clostridium botulinum* types C and D. *Onderstepoort J. Vet. Res. 38*: 93.

Jansen, B. C. (1971b). The quantitative determination of the toxic factors produced by *Clostridium botulinum* (Van Ermengem, 1896) types C and D. *Onderstepoort J. Vet. Res. 38*: 177.

Jansen, B. C. (1987). *Clostridium botulinum* type C, its isolation and taxonomic position. In *Avian Botulism* (M. W. Eklund, V. R. Dowell, Jr., eds.), Charles C. Thomas, Springfield, IL, pp. 123–132.

Jensen, W. I., and Duncan, R. M. (1980). The susceptibility of the mallard duck (*Anas platyrhynchos*) to *Clostridium botulinum* C2 toxin. *Jpn. J. Med. Sci. Biol. 33*: 81.

Johnston, J., and Whitlock, R. H. (1987). Botulism. In *Current Therapy in Equine Medicine*, 2nd ed. (N. E. Robinson, ed.), W. B. Saunders, Philadelphia, pp. 367–370.

Kempner, W. (1897). Weiterer Beitrag zur Lehre von der Fleischvergiftung. Das Antitoxin des Botulismus. *Z. Hyg. Infektionskh. 26*: 481.

Landmann, G. (1904). Ueber die Ursache der Darmstaedter Bohnenvergiftung. *Hyg. Rundschau. 14*: 449.

Lee, W. H., and Riemann, H. (1970a). Correlation of toxic and nontoxic strains of *Clostridium botulinum* by DNA composition and homology. *J. Gen. Microbiol. 60*: 117.

Lee, W. H., and Riemann, H. (1970b). The genetic relatedness of proteolytic *Clostridium botulinum* strains. *J. Gen. Microbiol. 64*: 85.

Leuchs, J. (1910). Beitraege zur Kenntnis des Toxins und Antitoxins des *Bacillus botulinus*. *Z. Hyg. Infektskh. 65*: 55.

Lynt, R. K., Solomon, H. M., and Kautter, D. A. (1984). Heat resistance of *Clostridium botulinum* type G in phosphate buffer. *J. Food Protect. 47*: 463.

Matsuda, M. (1989). The structure of tetanus toxin. In *Botulinum Neurotoxin and Tetanus Toxin* (L. L. Simpson, ed.), Academic Press, Inc., San Diego, CA, pp. 69–92.

Matsuoka, I., Syuto, B., Kurihara, K., and Kubo, S. (1987). ADP-ribosylation of specific membrane proteins in pheochromocytoma and primary-cultured brain cells by botulinum neurotoxins type C and D. *FEBS Lett. 216*: 295.

McCroskey, L. M., and Hatheway, C. L. (1988). Laboratory findings in four cases of adult botulism suggest colonization of the intestinal tract. *J. Clin. Microbiol. 26*: 1052.

McCroskey, L. M., Hatheway, C. L., Fenicia, L., Pasolini, B., and Aureli, P. (1986). Characterization of an organism that produces type E botulinal toxin but which resembles *Clostridium butyricum* from the feces of an infant with type E botulism. *J. Clin. Microbiol. 23*: 201.

McCroskey, L. M., Hatheway, C. L., Woodruff, B. A., Greenberg, J. A., and Jurgenson, P. (1991). Type F botulism due to neurotoxigenic *Clostridium baratii* from an unknown source in an adult. *J. Clin. Microbiol. 29*: 2618.

Meyer, K. F., and Gunnison, J. B. (1928–29). *C. botulinum* type D, Sp. Nov. *Proc. Soc. Exp. Biol. Med. 26*: 88.

Midura, T. F., Nygaard, G. S. Wood, R. M., and Bodily, H. L. (1972). *Clostridium botulinum* type F: *Isolation from venison jerky. Appl. Microbiol. 24*: 165.

Moller, V., and Scheibel, I. (1960). Preliminary report on the isolation of an apparently new type of *Cl. botulinum. Acta Pathol. Microbiol. Scand. 48*: 80.

Nakamura, S., Kimura, I., Yamakawa, K., and Nishida, S. (1983). Taxonomic relationships among *Clostridium novyi* types A and B, *Clostridium haemolyticum* and *Clostridium botulinum* type C. *J. Gen. Microbiol. 129*: 1473.

Nakamura, S., Okado, I., Nakashio, S., and Nishida, S. (1977). *Clostridium sporogenes* isolates and their relationship to *C. botulinum* based on deoxyribonucleic acid reassociation. *J. Gen. Microbiol. 100*: 395.

Oguma, K., Murayama, S., Syuto, B., Iida, H., and Kubo, S. (1984). Analysis of antigenicity of *Clostridium botulinum* type C1 and D toxins by polyclonal and monoclonal antibodies. *Infect. Immun. 43*: 584.

Ohishi, I., and DasGupta, B. R. (1987). Molecular structure and biological activities of *Clostridium botulinum* C2 toxin. In *Avian botulism* (M. W. Eklund, V. R. Dowell, Jr., eds.), Charles C. Thomas, Springfield, IL, pp. 223–247.

Popoff, M., Boquet, P., Gill, D. M., and Eklund, M. W. (1990). DNA sequence of exoenzyme C3, an ADP-ribosyltransferase encoded by *Clostridium botulinum* C and D phages. *Nucleic Acids Res. 18*: 1291.

Poumeyrol, M., Billon, J., Delille, F., Haas, C., Marmonier, A., and Sebald, M. (1983). Intoxication botulique mortelle due à une souche de *Clostridium botulinum* de type AB. *Méd. Malad. Infect. 13*: 750.

Prévot, A. R. (1953). Rapport d'introduction du President du Sous-Comité *Clostridium* pour l'unification de la nomenclature des types toxigeniques de *C. botulinum. Int. Bull. Bacteriol. Nomenclature 3*: 120.

Prévot, A. R., Terrasse, J., Daumail, J., Cavaroc, M., Riol, J., and Sillioc, R. (1955). Existence en France du botulisme humain de type C. *Bull. Acad. Med. (Paris) 139*: 355.

Roesener, S., Chhatwal, G. S., and Aktories, K. (1987). Botulinum ADP-ribosyltransferase C3 but not botulinum neurotoxins C1 and D ADP-ribosylates low molecular mass GTP-binding proteins. *FEBS Lett. 224*: 38.

Sakaguchi, G. (1983). *Clostridium botulinum* toxins. *Pharmacol. Ther. 19*: 165.

Seddon, H. R. (1922). Bulbar paralysis in cattle due to the action of a toxicogenic bacillus with a discussion of the relationship of the condition to forage poisoning (botulism). *J. Comp. Pathol. Ther. 35*: 147.

Simpson, L. L. (1986). Molecular pharmacology of botulinum toxin and tetanus toxin. *Ann. Rev. Pharmacol. Toxicol. 26*: 427.

Smith, L., and Sugiyama, H. (1988). *Botulism: The Organism, Its Toxins, the Disease*. Charles C. Thomas, Springfield, IL.

Sonnabend, O., Sonnabend, W., Heinzle, R., Sigrist, T., Dirnhofer, R., and Krech, U. (1981). Isolation of *Clostridium botulinum* type G and identification of type G botulinal toxin in humans: Report of five sudden unexpected deaths. *J. Infect. Dis. 143*: 22.

Sonnabend, O. A., Sonnabend, W. F., Krech, U., Molz, G., and Sigrist, T. (1985). Continuous microbiological study of 70 sudden and unexpected infant deaths: Toxigenic intestinal *Clostridium botulinum* infection in 9 cases of sudden infant death syndrome. *Lancet ii*: 237.

Strom, M. S., Eklund, M. W., and Poysky, F. T. (1984). Plasmids in *Clostridium botulinum* and related Clostridium species. *Appl. Environ. Microbiol. 48*: 956.

Suen, J. C., Hatheway, C. L., Steigerwalt, A. G., and Brenner, D. J. (1988a). *Clostridium argentinense*, sp. nov: A genetically homogenous group composed of all strains of *Clostridium botulinum* toxin type G and some nontoxigenic strains previously identified as *Clostridium subterminale* or *Clostridium hastiforme*. *Int. J. Syst. Bacteriol. 38*: 375.

Suen, J. C., Hatheway, C. L., Steigerwalt, A. G., and Brenner, D. J. (1988b). Genetic confirmation of identities of neurotoxigenic *Clostridium baratii* and *Clostridium butyricum* implicated as agents of infant botulism. *J. Clin. Microbiol. 26*: 2191.

Swerczek, T. W. (1980). Toxicoinfectious botulism in foals and adult horses. *J. Am. Vet. Med. Assoc. 176*: 217.

Thompson, D. E., Brehm, J. K., Oultram, J. D., Swinfield, T. J., Shone, C. C., Atkinson, T., Melling, J., and Minton, N. P. (1990). The complete amino acid sequence of the *Clostridium botulinum* type A neurotoxin, deduced by nucleotide sequence analysis of the encoding gene. *Eur. J. Biochem. 189*: 73.

Van Ermengem, E. (1897). Ueber einen neuen anaeroben Bacillus and seine Beziehungen zum Botulismus. *Z. Hyg. Infektionskh. 26*: 1.

Weickert, M. J., Chambliss, G. H., and Sugiyama, H. (1986). Production of toxin by *Clostridium botulinum* type A strains cured of plasmids. *Appl. Environ. Microbiol. 51*: 52.

Yang, K. H., and Sugiyama, H. (1975). Purification and properties of *Clostridium botulinum* type F toxin. *Appl. Microbiol. 29*: 598.

2

Clostridium botulinum in the Environment

Karen L. Dodds

Health Protection Branch,
Health and Welfare Canada,
Ottawa, Ontario, Canada

I. INTRODUCTION

Fundamental to a knowledge of the potential botulism hazard from any food is a knowledge of the possibility of contamination of the food with *Clostridium botulinum*. This contamination depends largely on the distribution and incidence of *C. botulinum* in the environment. The serious nature of botulism has led many investigators worldwide to conduct surveys to establish the presence or absence of *C. botulinum* in different environments.

There are several limitations to these surveys. Perhaps foremost among these is the fact that there are often factors present in the environment, including bacteriophages, boticins (microbial metabolites active against *C. botulinum*), and competing microorganisms, which may inhibit *C. botulinum*, and allowance is seldom made for them. Several surveys, for example, those of Meyer and Dubovsky (1922a,b,c,) predate knowledge of the existence of types E, F, and G, and the methodology used in these surveys was not adequate to detect all types; for example, type E was likely destroyed by the intensive heat treatment. In many surveys the number of samples analyzed was small, and often the sample size was small as well. The type was not always determined, and a distinction between proteolytic and nonproteolytic type B was rarely made. Also, the surveys were seldom quantitative.

C. botulinum is commonly present in soil and is often described as a soil organism, ubiquitous in nature. This statement does not very accurately reflect its distribution in the environment. As the survey data show, spores of *C. botulinum* are widely spread in the environment, but the factors governing the geographical distribution of spores and of the different types are still not well understood.

The purpose of this chapter is to summarize environmental data on the distribution of *C. botulinum* and to expand on an earlier review covering this subject by Hauschild (1989). Data included in the tables in this chapter are generally from surveys where it was possible to calculate a most probable number (MPN) count according to Halvorson and Ziegler (1933). It should be noted that the MPN counts are a compromise, because the equation presumes even distribution of spores among samples, which was unlikely in most of the surveys. The tables are not meant to be exhaustive, but to give representative results for different sample types and geographical areas.

II. INCIDENCE IN SOILS AND SEDIMENTS

C. botulinum has been found widely distributed throughout the land and coastal waters of North America (Table 1). The spore load varies considerably from region to region, as does the predominating type.

A. Continental United States

Perhaps the earliest surveys for the presence of *C. botulinum* in the environment are those of Meyer and Dubovsky (1922a,b,c). These surveys were initiated in California as a result of the rather frequent occurrence of botulism following consumption of vegetables there. Initially, attention was focused on cultivated and manured garden, field, and pasture soils. However, the frequent demonstration of *C. botulinum* in supposed control samples of "uncontaminated" mountain soil led to the examination of soil from uninhabited, virgin territories as well. A total of 312 cultures from various soil samples were examined; 45 (15%) were positive and typable, 40 (13%) with type A spores, and 5 (2%) with type B. A further 36 (12%) cultures were toxic but untypable. Of 78 soils from virgin territory, 45 (58%) were positive for *C. botulinum*, with a definite predominance of type A. A much lower incidence was discovered in cultivated fields; 59 (26%) of 226 samples were positive for *C. botulinum*, but most were untypable. Type A was the predominant type, but B appeared more frequently than in the virgin soil samples.

Meyer and Dubovsky (1922b) continued their surveys with samples taken from 46 states across the continent and the District of Columbia. A total of 963 samples from virgin, cultivated, garden and pasture soil, and manure from animal yards were examined. Of these, 111 (11.5%) were positive for type A spores and 60 (6%) for type B spores; 61 (6%) were weakly toxic but untypable. No samples from New Mexico, South Dakota, or Texas were positive. The

Table 1 Quantitative Soil and Sediment Surveys for *C. botulinum* Spores in North America

Location	Sample size (g)	% Positive samples	MPN per kg	Type (%)[a]					Ref.[b]
				A	B	C/D	E	F	
Eastern U.S. soil	10	19	21	12	64	12	12	0	1
Western U.S. soil	10	29	33	62	16	14	8	0	1
Great Lakes area									
Misc. soil	1	2	24	0	0	0	100	0	2
L. Michigan shore soil	1	34	100	0	0	0	100	0	2
Green Bay shore soil	1	28	300	0	0	0	100	0	2
L. Michigan sediment	1	60	920	0	0	0	100	0	2
Green Bay sediment	1	77	1280	0	0	0	100	0	2
Drainage area soil	1	5	81	0	0	0	100	0	2
Gulf of Maine sediment	10	1	10	0	0	0	100	0	3
New York to Florida coast sediment	1	4	36	8	17	42	33	0	4
Alabama coast sediment	2	4	20	0	0	67	33	0	5
Florida to Texas coast sediment	1	6	62	12	4	38	46	0	6
Washington coast sediment	5	71	250	1	3	0	96	0	7
Washington inland sediment	5	91	480	0	0	0	100	0	7
Oregon coast sediment	5	35	86	31	3	0	56	10	7
California north of 36° sediment	5	16	29	5	5	0	90	0	7
California south of 36° sediment	5	10	20	44	50	0	0	6	7
Alaska northwest beaches soil	5	74	270	0	0	0	100	0	8
Alaska inland soil	1	6	63	0	0	0	100	0	9
Alaska coast soil	1	41	660	0	0	0	100	0	9
Gulf of St. Lawrence sediment	10	18	20	0	0	0	100	0	10
Nova Scotia coast sediment	10	21	22	0	7	61	32	0	11
Newfoundland coast sediment	10	10	10	0	0	0	100	0	11
British Columbia coast sediment	4–10	6	9	8	0	0	92	0	12

[a]As percentage of the types identified.
[b] 1) Smith, 1978; 2) Bott et al., 1968; 3) Nickerson et al., 1966; 4) Ward et al., 1967b; 5) Presnell et al., 1967; 6) Ward et al., 1967a; 7) Eklund and Poysky, 1967; 8) Miller et al., 1972; 9) Miller, 1975; 10) Laycock and Loring, 1972; 11) Laycock and Longard, 1972; 12) Dolman and Iida, 1963.
Source: Dodds and Hauschild, 1989; Hauschild, 1989.

incidence of spores was low in soil samples collected from the middle states, the Great Plains states and states bordering the Great Lakes and the Mississippi River and its tributaries. Of the samples collected in the Atlantic states, 10–30% were toxic. Soils of the Pacific Coast and Rocky Mountain states were heavily contaminated with *C. botulinum*. This study was the first to document the disparity in the distribution of types A and B across the United States. The western states, including the Great Plains states, harbored mainly type A, while the Mississippi Valley and Atlantic states contained mainly type B. The authors also noted the predominance of type B spores in cultivated soils and the rare occurrence of *C. botulinum* spores in soils contaminated with manure or animal excreta, and hence concluded that fertilization cannot be considered a significant factor contributing to its incidence in the soil.

In his first study, Smith (1975) found 38% of soil samples positive for *C. botulinum*. All type A strains were found in samples taken west of the Missouri River. All the samples in which types A or B were demonstrated were neutral or alkaline in reaction (average pH 7.9 compared to an overall average of 6.3) and low in organic matter. Type E spores were only demonstrated in one sample, an acid soil from the Olympic rain forest. In a more extensive survey, Smith (1978) found *C. botulinum* in 24% of 260 soil specimens taken in four transects across the continental United States. Almost all type A strains were from samples taken west of the rise of the Rocky Mountains in soil that was neutral to alkaline (average pH 7.5), with a lower than average organic content. Type B strains were demonstrated in slightly more acidic soil (average pH 6.25) than usual, with a higher level of organic matter, and primarily in soil taken from under grass or from cultivated fields. While the distribution was not as skewed as for type A spores, type B predominated in samples from the eastern United States, and no samples in the southernmost transect were positive from type B. Type C and D spores were very localized; only three samples of acid soil along the Gulf Coast contained type C, and five alkaline soil samples from the northwest contained type D. Type E spores were infrequently demonstrated, usually in damp to wet locations.

After several botulism outbreaks in the 1960s from fish caught in the Great Lakes, this region was intensively studied, and only type E spores were detected (Bott et al., 1968). High numbers were found in shoreline and soil samples from northern Lake Michigan, as well as from Green Bay. Even higher numbers were found in sediment samples; 77% of samples from the aquatic environment of Green Bay were positive. So were 22% of the bottom and shoreline samples of the rivers emptying into Green Bay, but the incidence decreased in the upper reaches of the rivers. In samples taken from the land mass around Green Bay, the incidence was only 5%. The spores in Green Bay could originate in the surrounding land mass, but if this were the case, the distribution would be expected to be more uniform. The much higher concentration in the bay suggests multiplication in the bay itself.

Graikoski et al. (1970) looked at the incidence of type E spores in bottom sediments from the Great Lakes area over 3 years. The highest occurrence was from samples from Saginaw Bay of Lake Huron (45%), followed by samples collected from Lake Michigan (34%) and Lake Superior (20%). None of the other lakes yielded positive samples. A few samples of sand and soil from Saginaw Bay, Lake Huron, were positive for type C spores, and one sample was positive for type A.

In general, the incidence in sediment samples from the Atlantic and Gulf of Mexico are low. A very low incidence, (1%), with only type E spores, was found in the Gulf of Maine (Nickerson et al., 1966). Only 12 of 335 sediment samples (4%) from Staten Island and Key Largo were positive, four with type E spores, three with type C, two each with types B and D, and one with type A (Ward et al., 1967b). Presnell et al. (1967) found only 4% of samples from the Alabama coastline positive, with types C, D, and E being identified. A similar survey from Key West, Florida, to Brownsville, Texas, found 24 of 400 samples (6%) positive (Ward et al., 1967a), most (46%) with type E, followed by types C (25%), D (13%), A (12%), and B (4%).

Avian botulism is an important disease of waterfowl in many areas of the world, including North America. Many marshes inhabited by birds have all the requirements for an outbreak: the presence of toxigenic strains of *C. botulinum*, usually type C, a substrate and environmental conditions suitable for bacterial growth and toxigenesis, and a population of susceptible birds. The carcasses of either vertebrates or invertebrates dying in marshes provide a good substrate with its own microenvironment for *C. botulinum*. Therefore, the occurrence of avian botulism in some marches are likely related to the prevalence of spores in the marshes.

Two outbreaks of avian botulism in settling ponds of phosphate mines in northern Florida led to a survey by Marion et al. (1983) for the prevalence of *C. botulinum* in the region. Type C was identified in 26 of 467 (5.6%) sediment samples but was detected only in samples taken during the summer months.

A high incidence of *C. botulinum*, particularly of type E, was found in sediments from coastal areas of Washington and Oregon (Eklund and Poysky, 1967). The incidence of spores decreased further south off the coast of California, and types A, B, and F, rather than E, were found south of latitude 36°N.

B. Alaska and Hawaii

The high morbidity rate from type E botulism in Alaska led to surveys of the incidence of *C. botulinum* in the environment. Only type E spores were found, with a very high incidence (74%) on beaches in northwestern Alaska (Miller et al., 1972). Interestingly, at the time the samples were collected, marine mammals were being butchered on two of the beaches. The skinning and butchering of animals on contaminated beaches provides the opportunity for contamination of the meat and blubber. Further surveys (Miller, 1975) found 41% of samples from coastal beaches positive, but only 6% along inland rivers, streams, and ponds. The origin of the spores is unknown, but offal discarded from butchered marine animals and effluents discharged from canneries contribute organic wastes and may seed the sediments. Water temperatures in the summer are such that some in situ multiplication of *C. botulinum* may occur.

The only survey of Hawaii dates back to the work of Schoenholz and Meyer (1922). Of 19 soil samples examined, five were positive for type B *C. botulinum*, and two for type A. The type B strains were all found in cultivated soil, whereas only virgin soil was positive for type A.

C. Canada

Fewer environmental surveys have been done in Canada. Dubovsky and Meyer (1922) examined 100 soil specimens from eight provinces, most from British Columbia. Of these, 27 were toxic; 15 with type A, four with type B; eight were not typable.

The incidence of *C. botulinum* on Canada's Atlantic seaboard appears to be relatively low. A survey of the St. Lawrence River and Gulf (Laycock and Loring, 1972) found 18% of the sediment samples positive, all with type E spores. The survey indicated a close correlation between patterns of terrigenous sedimentation and distribution of *C. botulinum* type E. Higher concentrations in relatively shallow and brackish waters suggested multiplication in situ. In the vicinity of Nova Scotia, the overall incidence of positive samples was 21% (Laycock and Longard, 1972). Type B and E spores were found in low concentrations in sediments on the Scotian Shelf, whereas localized areas of high concentration of type C were found inshore in the Halifax-Dartmouth area and offshore in the deeper waters and on the edge of the Continental Shelf. Type C in the inshore areas is likely derived from the adjacent marshy coastline, which is rich in wild fowl. Off the coast of Newfoundland, only type E spores were found in sediments, with an overall incidence of 10% (Laycock and Longard, 1972). A high incidence of the organism (17%) was found as far as 150 miles off the northeast coast. All but one of the positive samples were found in the area where most rivers in the province drain into the Atlantic. The extent of the positive samples suggests that currents can carry spores over a considerable distance.

An outbreak of equine botulism in Saskatchewan led to an investigation into the source of the toxin. Type C toxin was detected in sand and dirt at the bottom of a feedtrough (Heath et al., 1990).

A survey by Woebster et al. (1987) of wetland soil in Saskatchewan found 35.7% of samples collected in the spring and 40.5% of samples collected in July positive for Type C. No other types were identified. No clear relationship was evident between water depth and the proportion of samples producing toxin, but there was a very strong relationship between the prior history of avian botulism in a wetland and the proportion of soil samples producing toxin ($p < 0.0001$).

Dolman and Iida (1963) conducted extensive surveys of sediments off the coast of British Columbia and found an overall incidence of 6%. Of 183 samples taken from coastal inlets at depths up to several hundred meters, five were positive with type E spores and one with type A. A total of 27 samples taken from shallow depths along the northern British Columbia coast yielded seven type E strains. Type E was also demonstrated in 11 of 47 samples of sand from an inland lake. A total of 142 sediment samples taken from various locations of the ocean bottom at some distance from the coast were all negative.

D. Europe

The soils and sediments of Europe have also been extensively surveyed for *C. botulinum* (Table 2). Perhaps the earliest reported survey is that of Meyer and Dubovsky (1922c), who found samples of soil from Belgium, Denmark, England, the Netherlands, and Switzerland positive for type B. Isolation rates varied from a low of 6% in Denmark to a high of 67% in Belgium, but generally very few samples were analyzed. Type A spores were consistently absent.

Surveys of soils and sediments in the United Kingdom show some interesting differences. An early survey of English soils by Haines (1942) found only 5 of 106 (5%) samples definitely positive, with another 9 samples possibly being positive. Four of the positive samples contained type A spores, and one type B. A later survey (Smith and Young, 1980) again found a low incidence (6%) in soil from various parts of Britain, and only type B spores were detected. Three sites associated with animals for many years were not found to have a higher than usual incidence, whereas an earlier study (Smith and Milligan, 1979) found 15 of 60 soil samples (25%) from the site of the Metropolitan Cattle Market positive. Type B spores were most frequently demonstrated, followed by types C, D, and E. This localized high incidence suggests that, at least in some areas, animals can contribute to the incidence of *C. botulinum* in soils.

British coastal waters have a relatively low incidence (4%) of *C. botulinum* in bottom sediments (Cann et al., 1968). Only samples taken from the immediate coastal waters were positive, all with type B spores. The highest incidence was in a virtually landlocked area, suggesting that the presence of *C. botulinum* in some bottom sediments may be due to terrestrial contamination. In contrast, Smith and Moryson (1975) found 73% of sediment samples from London lakes and waterways positive. Most of the positive samples contained type B spores (59%), followed by types C (22%), E (17%), and D (2%). A small survey in Edinburgh, made for comparison, showed 4 of 7 sediment samples positive with type B spores and one with type C. A later study (Smith and Moryson, 1977)

Table 2 Quantitative Soil and Sediment Surveys for *C. botulinum* Spores in Europe

Location	Sample size (g)	% Positive samples	MPN per kg	Type (%)[a]					Ref.[b]
				A	B	C/D	E	G	
Britain									
Misc. soil	50	6	2	0	100	0	0	0	1
Former cattle grounds	50	25	6	0	56	38	6	0	2
Coast sediment	2	4	18	0	100	0	0	0	3
Inland sediment	50	73	26	0	59	24	17	0	4
Inland sediment	50	37	9	0	80	12	8	0	5
Norfolk Broads sediment	50	98	76	0	36	29	35	0	6
Ireland inland sediment	50	18	4	0	100	0	0	0	5
Camargue, France sediment	50	9	2	0	25	0	75	0	7
Baltic Sea sediment	6	100	>320	0	0	0	100	0	8
Skagerrak, Kattegat sediment	6	100	>780	0	0	0	100	0	8
Sweden coast soil	3–6	84	410	0	0	1	99	0	8
Sweden inland coast	6	47	110	0	4	0	96	0	8
Sweden misc. soil	6	29	30	0	5	0	95	0	8
Norway shores soil	5	7	7	0	0	0	100	0	8
Norway coast soil	5	30	71	0	0	0	100	0	8
The Sound sediment	2	100	>1750	0	0	0	100	0	9
Kattegat sediment	2	88	1050	0	0	0	100	0	9
Skagerrak and North Sea sediment	2	19	100	0	0	0	100	0	9
Danish coast sediment	10	67	110	0	0	0	100	0	10
Denmark inland sediment	10	63	100	0	4	0	96	0	10
Denmark misc. soil	10	14	15	0	93	7	0	0	10
Iceland inland sediment	10	5	5	50	0	0	50	0	10
Iceland soil	10	3	3	0	100	0	0	0	10
Iceland coast sediment	10	0	<5						10
Faroe Islands coast sediment	10	1	1	0	0	0	100	0	10
Faroe Islands inland sediment	10	1	1	0	0	0	100	0	10
Faroe Islands soil	10	0	<3						10
Greenland coast sediment	10	37	46	0	0	0	100	0	10
The Netherlands									
Areas with avian botulism soil		75		0	4	80	16	0	11
Other soils		31		0	8	21	71	0	11

Table 2 Quantitative Soil and Sediment Surveys for *C. botulinum* Spores in Europe

Location	Sample size (g)	% Positive samples	MPN per kg	Type (%)[a] A	B	C/D	E	G	Ref.[b]
Misc. sediment	0.5	94	2500	0	22	46	32	0	12
Poland Baltic coast soil	5	32	76	3	0	0	97	0	13
Czechoslovakia sediment		42		0	0	100	0	0	14
Switzerland soil	12	44	48	19	56	3	0[c]	19	15
Bari, Italy soil	5	1	1	0	100	0	0	0	16
Rome, Italy soil	7.5	1	2	86	14	0	0	0	17

[a]As percentage of the types identified.
[b] 1) Smith and Young, 1980; 2) Smith and Milligan, 1979; 3) Cann et al., 1968; 4) Smith and Moryson, 1975; 5) Smith et al., 1978; 6) Borland et al., 1977; 7) Smith et al., 1977; 8) Johannsen, 1963; 9) Cann et al., 1965; 10) Huss, 1980; 11) Haagsma, 1974; 12) Notermans et al., 1979; 13) Zaleski et al., 1973; 14) Neubauer et al., 1988; 15) Sonnabend et al., 1987; 16) Barbuti et al., 1980; 17) Creti et al., 1990.
[c] Type F (3%) also detected.
Source: Dodds and Hauschild, 1989; Hauschild, 1989.

analyzed 25 soil samples, taken mostly within close proximity of lakes with sediments known to be contaminated with *C. botulinum*. Only one sample was positive with type B. Smith et al. (1978) undertook a wider survey of sediments in the United Kingdom. Of 499 samples from Britain, 184 (37%) were positive; type B spores were most frequently encountered, followed by types C, E, and D. Of 55 samples from Ireland, 10 (18%) were positive, all with type B spores.

An unusually high number of outbreaks of avian botulism occurred in the United Kingdom in 1975, with particularly high mortality in Norfolk (Borland et al., 1977). Of 45 mud samples collected from 22 well-distributed aquatic sites, 44 (98%) were shown to contain *C. botulinum*. Type B and E spores were most prevalent, followed by type C. Over half the samples contained more than one type.

Thus, the number of *C. botulinum* in British soils is much lower than in sediments. Type A has not been isolated since the earliest surveys, although type A botulism has occurred in Britain. The Loch Maree outbreak of 1923 involving eight deaths and at least one case of infant botulism were caused by type A (Turner et al., 1978).

A survey of sediments in the Camargue, an area of southern France, was done because of the importance of this area for resident and migratory birds (Smith et al., 1977). The prevalence of *C. botulinum* appears to be low, with only 9% of samples positive. Type B and E spores were detected.

Johannsen's surveys (1962,1963) of sediments in and around Sweden have shown a very high prevalence of *C. botulinum* type E. All bottom samples taken from the Baltic Sea, the Kattegat, and Skagerrak were contaminated. Many shore samples were also contaminated, with the incidence highest on the Swedish coast (84%), followed by inland shores (47%) and miscellaneous soils (29%). Type B and C spores were also isolated, but at very low levels, while type A was never encountered. Samples from the west coast of Norway gave rise to only a few toxic cultures.

The high contamination level in marine sediments in Scandinavian waters was confirmed by a later study (Cann et al., 1965). The incidence was highest in the Sound, near Copenhagen, slightly lower in the Kattegat, lower in the Skagerrak, and zero in the North Sea, with an overall detection rate of 90%. Only type E spores were detected.

Huss (1980) also found very high concentrations of type E spores in the Danish aquatic environment. The overall incidence in Danish soil was considerably lower, and the distribution pattern was different from that in sediments. Type E was noticeably absent in soil samples from woodlands and cultivated farmland, where type B predominated. Huss (1980) also noticed a relationship between the level of contamination with type E and flooding of land with sea water. Samples from the salt marsh, a type of coastal land occasionally flooded by sea water, were heavily contaminated. Land reclaimed from the sea and left uncultivated still showed a fairly high level of contamination, but the level seemed to decrease with the length of time that the land was dry. After cultivation, the contamination level was much lower, and only type B spores were detected.

Huss (1980) also examined samples from Iceland, the Faroe Islands, and Greenland. In Iceland, the incidence was very low; type A and E spores were found in freshwater sediments, type B in one soil sample, but no spores were detected in marine sediments. Similarly, few contaminated samples were discovered in the Faroe Islands; type E spores were found in one marine sediment sample and one freshwater sample, but all soil samples were negative. The incidence was higher in Greenland, where 37% of marine sediment samples harbored type E.

The highest incidence of type E in the Danish aquatic environment contrasts sharply with its very rare occurrence in dry soil samples, in which only type B was demonstrated (Huss, 1980). This suggests that type E is a true aquatic organism and that multiplication occurs in situ. Multiplication may occur in shallow, warm waters where there is little wave action, and where the bottom sediments often become anaerobic, providing excellent conditions for multiplication of *C. botulinum*, particularly in carrion. These areas, found in all Danish fiords, may act as foci for production of type E spores. Similar conditions exist around the Baltic Sea. The main current from the Baltic Sea is likely to carry a considerable number of spores, which may explain the survey data showing high numbers in the Baltic, but diminishing numbers as the water passes through the

Sound to the North Sea. None of the main currents pass the shores of Great Britain, the Faroe Islands, or Iceland, where the incidence of type E spores is low.

Mass mortality of waterfowl due to avian botulism, recorded in the Netherlands for the first time in 1970, led to a survey of the distribution of *C. botulinum* (Haagsma, 1974). Type B, C, and E spores were found in soil, but the incidence depended on the location. In areas in which avian botulism had occurred, the incidence was high (75%), and most samples contained type C spores (80%), followed by E (16%), and B (4%). Birds, mammals, and fish from these areas were also contaminated, mostly with type C spores. In areas free from botulism and in random soil samples, the contamination level was much lower (31%), and type E spores predominated, followed by types C and B. Type A spores were isolated from only one sample. A separate study demonstrated high numbers of type E and B spores in soil and sediment samples from across the Netherlands (Notermans et al., 1979). Zaleski et al. (1973) detected *C. botulinum* in 32% of soil samples from the Polish Baltic seaboard, nearly all of type E.

Recently, Czechoslovakia has been experiencing mass deaths of waterfowl due to unknown causes (Neubauer et al., 1988). Mud samples from nine localities in southern Moravia had a fairly high incidence of contamination with type C spores, suggesting botulism as the cause of death.

A survey of Swiss soil (Sonnabend et al., 1987) found 44% of samples positive for *C. botulinum*. Type B spores were most common, and almost half (45%) of the type B strains isolated were nonproteolytic. Type A, G, C, and F spores were also found, though C and F in only one sample each. Type A was found most frequently in samples from grassy areas. Almost all samples containing type B alone came from the Alpine region. This is the first survey reporting a substantial number of *C. botulinum* type G. Type G is difficult to detect in mixed cultures because it generally produces only small amounts of toxin which may require trypsin treatment for detection, and lacks the characteristic lipase reaction (see Chapter 1).

Barbuti et al. (1980) found a low incidence level (1%) in cultivated soil samples around Bari, Italy. Only type B was identified. After two incidents of infant botulism in Rome caused by *C. butyricum*, an extensive survey was carried out by Creti et al. (1990) for botulinum toxin–producing clostridia in soils in the vicinity of Rome. A low level of contamination was detected. Toxigenic *C. botulinum* types A and B, all proteolytic, were responsible for the contamination.

E. Asia

Survey results from Asia, including all of the former U.S.S.R. are presented in Table 3. An extensive survey of soils throughout the U.S.S.R. was carried out by Kravchenko and Shishulina (1967), mainly during the spring-summer period. *C. botulinum* was found in all regions of the country, but the frequency of de-

Table 3 Quantitative Soil and Sediment Surveys for *C. botulinum* Spores in Asia

Location	Sample size (g)	% Positive samples	MPN per kg	Type (%)[a] A	B	C/D	E	F	Ref.[b]
U.S.S.R.									
Northeastern soil		2		0	14	0	86	0	1
Southeastern soil		13		8	17	0	73	0	1
Central soil		14		16	77	0	6	0	1
Northwestern soil		4		14	0	0	86	0	1
Far East soil		14		0	8	5	87	0	1
Countrywide soil		11		8	28	2	62	0	1
Aral Sea soil		35		<1	0	0	>99	0	2
Lake Balkhash soil	15–20	22	13	21	14	13	52	0	3
Iran, Caspian Sea sediment	2	19	93	0	8	0	92	0	4
Bangladesh									
Freshwater sediment	10	37	19	0	0	100	0	0	5
Coast sediment	10	50	10	0	0	100	0	0	5
Thailand									
Hua-Hin sediment	10	3	3	0	0	83	17	0	6
Other coast sediment	10	0	<2	0	0	0	0	0	6
Mekong River sediment	10	0	<2	0	0	0	0	0	6
China, Shinkiang soil	10	75	25000[c]	47	32	19	2	0	7
Taiwan soil	10	78	153	34	37	4	20	5	8
Japan									
Hokkaido soil	5–10	4	4	0	0	0	100	0	9
Hokkaido coast soil and sediment	50	13	3	0	0	0	100	0	10
Hokkaido inland soil and sediment	50	19	4	0	0	0	100	0	10
Aomori, L. Towada soil and sediment		12		0	0	0	97	3	11
Ishikawa soil and sediment	40–50	56	16	0	0	100	0	0	12
Ishikawa, San'in, Kyushu, islands soil	10	15	16	0	0	75	25	0	7
Seto Inland Sea sediment	50	12	2	0	0	100	0	0	13

[a]As percentage of the types identified.
[b] 1) Kravchenko and Shishulina, 1967; 2) Kazdobina et al., 1976; 3) Bulatova et al., 1969; 4) Rouhbakhsh-Khaleghdoust, 1975; 5) Huss, 1980; 6) Tanasugarn, 1979; 7) Yamakawa et al., 1988; 8) Tsai and Ting, 1977; 9) Nakamura et al., 1956; 10) Kanazawa et al., 1970; 11) Yamamoto et al., 1970; 12) Serikawa et al., 1977; 13) Venkateswaran et al., 1989.
[c]Actual number of type A spores as determined by authors.
Source: Dodds and Hauschild, 1989; Hauschild, 1989.

tection, and the types detected, varied from region to region. The northern areas of the European region had a very low level of contamination (2%), mostly (86%) with type E spores, the remainder with type B. The frequency of detection in the southern part of the European region was much higher (13%), and again, type E spores were predominant (73%), followed by types B (18%) and A (9%). Types C and D were also detected in this region, but in only very few samples. In central Asia, the contamination rate was approximately 14%. In this region, type B spores predominated (77%), followed by types A (16%) and E (6%). Considerably fewer samples (4%) from western Siberia were contaminated, mostly with type E spores. In the far eastern regions of the U.S.S.R., 14% of samples were positive, predominantly with type E spores (87%). Countrywide, 11% of samples were positive; spores of type E were the most widespread (62%), followed by types B (28%), A (8%), and C (2%). Type E was only detected in soils in the vicinity of aquatic environments, soils rich in organic substances with a high moisture content. The distribution of types A and B appeared to be influenced by a complex set of factors peculiar to each region. In the west, types A and B were found more often in heavily populated and contaminated areas, but not in the east. Type C was found only in the southern regions.

A relatively high contamination rate was demonstrated in the soil of the coast of the Aral Sea (Kazdobina et al., 1976). All but one of 484 positive samples (35%) contained type E spores; the single remainder contained type A. The incidence was highest in soil close to fish-processing plants and moorages, lower on the shores at the fishing sites and near settlements, and lowest in the soil of uninhabited districts. While the contamination rate in soils in the vicinity of Lake Balkhash, approximately 900 km east of the Aral Sea, was lower (22%), a wider variety of types was detected (Bulatova et al., 1969). Type E spores were found in 134 samples (52%), type A in 55 (21%), type B in 35 (14%), and type C in 34 (13%). Several samples contained more than one type. The frequency of contamination varied between sampling areas. The highest level (58%) was found near the fishing station of Kara-Chaga, where type A and E spores were the most common contaminants. In the area of Kara-Uzyak, the incidence was lower (28%), and mostly type C and E spores were detected. The incidence in the area of Kuigan, on the bank of the Ili River was slightly lower (22%), and consisted mostly of type E spores. All of these areas had sandy to silty soil. In areas with sandy soil, the contamination rate was considerably lower (5–6%) and limited to type E.

A survey of bottom sediments of the Caspian Sea, prompted by a relatively high local incidence of botulism associated with consumption of fish, was carried out by Rouhbakhsh-Khaleghdoust (1975). Type E spores were most prevalent (92%), but type B spores were also present (8%).

At least one limited survey for *C. botulinum* has been conducted in India (Pasricha and Panja, 1940). Eight samples of garden soil were collected from different areas of Calcutta, and four of them were positive for type A. In Bangladesh, Huss (1980) found 37% of freshwater sediment samples and 50% of brackish sediment samples positive, with only type C and D spores.

C. botulinum was detected in 12 of 475 sediments samples (3%) collected by Tanasugarn (1979) at Hua-Hin on the coast of Thailand (Table 3). Ten of the positive samples contained type D spores, and two contained type E. *C. botulinum* was not detected in other coastal areas of Thailand, nor in the Mekong River.

Spores of types A and B are readily demonstrated in soils from mainland China. This was first demonstrated by Schoenholz and Meyer (1922). Analyses of 52 soil samples from two coastal provinces showed 13 samples positive for type B spores, and one for type A. A recent study by Yamakawa et al. (1988) found soils of the Shinkiang district heavily contaminated (75%). Type A spores were most frequently detected (47%), followed by types B (32%), C (19%), and E (2%).

A high incidence of contamination (78%) was also found in cultivated fields of different areas of Taiwan (Tsai and Ting, 1977). The incidence was highest in samples taken from northern parts of the island. Type B spores were found most frequently (37%), followed by types A (34%), E (20%), F (5%), and C (4%). Interestingly, all soils containing type E were obtained from inland areas. The majority of samples (71%) contained more than one type. The authors suggested several possible reasons for the high incidence of *C. botulinum*. Geographic characteristics, together with a subtropical climate, may be favorable to the existence of types A, B, and E. Violent summer thunderstorms may result in flooding and spreading of types in the cultivated land. There is the possibility of man-made contamination of *C. botulinum* due to fertilizing with sewage and manure and a high level of invertebrate debris, especially insect carrion in the soil, which may provide a natural niche for certain *C. botulinum* type.

The incidence of type E spores is also high in some areas of northern Japan. In Hokkaido, they were isolated from soil of the shores of Lake Abashiri (Nakamura et al., 1956), but the incidence was low (4%). A slightly higher incidence (13%), all of type E, was found in the soil of the coast line of Hokkaido (Kanazawa et al., 1968). Generally, a higher incidence was found at the northern coast of the Sea of Japan and a lower incidence at the southern coast, except for a conspicuously high rate at Okushiri Island. The same study examined 900 mud samples collected at river or lakesides in Hokkaido and found 169 (19%) positive for type E spores. Examination of 260 soil samples collected at 26 inland reforestation locations failed to find any *C. botulinum*.

The prevalence of type E spores is also apparent in the northern prefecture of Aomori on the main island, Honshu (Yamamoto et al., 1970). Soil and sand samples were taken from the shore of Lake Towada and from along the Oirase River. Of 244 samples, 30 (12%) were positive, 29 for type E and one for type F, the first such isolation in Japan.

In other areas of Japan, the distribution of *C. botulinum* is quite different. Following several outbreaks of avian botulism, an ecological study on *C. botulinum* type C was done by Serikawa et al. (1977) in the area of five lakes in Ishikawa Prefecture (western Honshu). Overall, the contamination of soil samples collected from shores was high (56%), and only type C spores were detected. However, all samples from the shore of Lake Kahoku were positive, whereas all of the samples from the coast of the Sea of Japan were negative. Yamakawa et al. (1988) determined that the level of contamination of soils from areas of Japan that are closely located to or facing mainland Asia, including Ishikawa Prefecture, San'in and northern Kyushu districts and the four islands of Tsushima, Yonaguni, Ishigaki, and Miyako, is considerably lower (15%). The most prevalent toxin type was C (75%) followed by E (25%). Wakamatsu (1953) identified *C. botulinum* by cultural characteristics in soils taken from Kyushu and typed one isolate as A, a very rare type in Japan.

The increase in mariculture in Japan prompted a recent survey for pathogens in the water and sediments from the Seto Inland Sea (Venkateswaran et al., 1989). Of 26 sediment samples analyzed for *C. botulinum*, three (12%) were positive. Only one was typable, as type C. The positive samples were taken from shallow areas, or areas receiving a discharge of pollutants.

F. Central and South America

There have been considerably fewer surveys for the presence of *C. botulinum* in southern areas (Table 4). Type A, B, C, and F spores were all detected in sediments from various sites in the Brazilian state of Ceara at an overall rate of 16% (Ward et al., 1967c). A survey of soils from São Paulo, Brazil (Leitão and Delazari, 1983), found a high incidence in soil from commercial vegetable plantations (35%) but no contamination in samples from horse or cattle pastures. Among the positive samples, there was a prevalence of type A (57%), followed by types B (7%) and F (7%). The remaining positive samples (29%) were not typed, but were likely either type C or D. Further evidence for the presence of *C. botulinum* in the Brazilian environment was the clinical diagnosis of enzootic botulism of water buffaloes in the northern lowland region of the state of Maranhao (Langenegger and Dobereiner, 1988). Approximately 10% of buffaloes grazing in the area together with cattle and horses suffered from botulism in 1978 and 1979.

Yamakawa et al. (1990a) detected *C. botulinum* in four of 17 sites (24%) examined in Paraguay. The prevalent spore type was F, followed by A and C. An

Table 4 Quantitative Soil and Sediment Surveys for *C. botulinum* Spores in South America, Africa, Indonesia, Australia, and New Zealand

Location	Sample size (g)	% Positive samples	MPN per kg	Type (%)[a] A	B	C/D	E	F	Ref.[b]
Brazil									
Ceara sediment	1	16	214	20	40	20	0	20	1
São Paulo									
Cultivated land soil	5	35	86	57	7	29	0	7	2
Pasture land soil	5	0	<3						2
Paraguay soil	5	24	10	14	0	14	0	71	3
Argentina soil		34		67	20	0	0	5[c]	4
South Orkney Islands soil and sediment	50	0	<1						5
Falkland Islands sediment	50	33	8	0	100	0	0	0	5
Kenya soil	5	25	33	89	0	11	0	0	6
South Africa soil	30	3	1	0	100	0	0	0	7
South Africa lakes sediment	30	21	8	0	0	100	0	0	8
Indonesia soil	50	21	5	0	0	100	0	0	9
Western Indonesia sediment	5	2	4	18	36	46	0	0	10
Java coast sediment	10	13	14	0	33	67	0	0	11
Java inland sediment	10	10	11	0	0	80	20	0	11
Australia soil	50	2	<1	100	0	0	0	0	12
Auckland, New Zealand, sediment	20	55	40	0	0	100	0	0	13

[a]As percentage of the types identified.
[b] 1) Ward et al., 1967c; 2) Leitao and Delazari, 1983; 3) Yamakawa et al., 1990a; 4) Ciccarelli and Giménez, 1981; 5) Smith et al., 1987; 6) Yamakawa et al., 1990b; 7) Knock, 1952; 8) Mason, 1968; 9) Hayashi et al., 1981; 10) Suhadi et al., 1981; 11) Haq and Suhadi, 1981; 12) Eales and Gillespie, 1947; 13) Gill and Penney, 1982.
[c] Types AF (7%) and G (1%) were also detected.
Source: Dodds and Hauschild, 1989; Hauschild, 1989.

additional site examined from Brazil was positive for both types B and F. All positive samples were from grass fields.

A high incidence (34%) and a diversity of serological types was discovered in Argentine soils (Ciccarelli and Giménez, 1981). Spores of type A were most prevalent (67%), followed by B (20%), AF (7%), F (5%), and G (1%).

A recent search for *C. botulinum* in the soils and sediments of the South Orkney and Falkland Islands (Smith et al., 1987) found only one sediment sample from the Falkland Islands positive for type B.

G. Africa

African surveys are included in Table 4. An outbreak of type A botulism in nomads in Kenya prompted a soil survey in which 12 sites were sampled (Yamakawa et al., 1990b). Two were positive with type A spores, and one with type C. The soil of the sites where type A was found, one inland and one on the Indian Ocean, seemed to be heavily contaminated because three of five cultures were positive from the inland site, and all five cultures from the Indian Ocean site were positive. Knock (1952) conducted a soil survey in South Africa even though no authenticated cases of human botulism have been reported there. Only three of 102 samples of soil, all from uncultivated land, were positive, all with type B spores. Mason (1968) examined sediment samples from lakes of the Zululand Game Park in South Africa and found type D in seven of 34 samples.

H. Indonesia, Australia, and New Zealand

Type C and D spores are the most prevalent in Indonesia (Table 4). Hayashi et al. (1981) detected only type C and D spores in soil. A survey of sediments from the western part of Indonesian waters by Suhadi et al. (1981) found only 11 of 592 samples (2%) positive. Types A, B, C, and D were all identified. Haq and Suhadi's (1981) survey of sediments from Java reported a higher incidence; 13% of samples from coastal areas and 10% from inland areas were positive. Most positive samples contained type C spores, followed by types B, D, and E.

The low incidence of botulism in Australia and New Zealand appears to reflect the low environmental incidence of spores (Table 4). Eales and Gillespie (1947) found only four of 183 samples (2%) of virgin soil from Victoria, Australia, positive for type A *C. botulinum*. Environmental sampling associated with five cases of infant botulism in Australia indicated that samples from 16–20% of the rural sites examined were contaminated with either type A or B spores (Murrell and Stewart, 1983). An outbreak of type C botulism in wild waterbirds in southern Australia indicated the probable existence of type C spores in the Lake Lalbert district, although no environmental testing was done (Galvin et al., 1985). While only one incident of foodborne botulism has been reported in New Zealand, botulism has been reported in wild waterfowl (Gill and Penney, 1982), indicating the presence of spores in the environment. Bottom sediments from lakes and waterways in the Auckland area were positive for *C. botulinum* types C and D (55%), while samples from other areas of the North Island were negative.

III. INCIDENCE IN FISH AND AQUATIC ANIMALS

Surveys of fish and other aquatic organisms for the presence of *C. botulinum* are seldom quantitative and are difficult to compare because sampling methods are vastly different. Single or multiple whole animals, gills, intestines, and other organs have been used for analysis. Results of surveys where the samples were obtained in situ are reported here (Table 5). The incidence of *C. botulinum* in samples obtained at the marketplace are reported in Chapter 3.

A. North America

The incidence of type E spores in the intestinal contents of fish from the Great Lakes depends on the lake (Sugiyama et al., 1970); 2% of samples from Lake Erie and Superior were positive, 3% from Lake Huron, and 8% from Lake Michigan. Green Bay of Lake Michigan has a particularly high incidence: 56% overall and 82% in the southern part.

Type E was detected in two of 32 (6%) fish caught in Cayuga Lake in upper New York State (Chapman and Naylor, 1966). The incidence of type E in the intestines of fish from the Gulf of Maine was 5% (Nickerson et al., 1966). The highest contamination level indicated by MPN methodology was 0.1 per g.

During an investigation of the prevalence of *C. botulinum* type E in crabs in the mid-Atlantic region of the United States, two crabs from the York River in Virginia were found to be contaminated with type F (Williams-Walls, 1968). This was the first reported isolation of proteolytic type F in the United States.

Very few samples (<1%) of fish and other aquatic species from the U.S. Atlantic coast between Key Largo, Florida, and Staten Island, New York, were contaminated (Ward et al., 1967b). The spore types identified were B, D, and E. A second survey (Ward et al., 1967a) of samples from the Gulf of Mexico between Key West, Florida, and Brownsville, Texas, found very few contaminated samples of fish (4%). Most positive samples (24) were contaminated with type E spores, followed by types C, B, and D. The contamination level of shrimp was also low (1%), with only type E spores. Type D and E spores were detected in crabs and sea turtles. In oysters from Mobile Bay, Alabama, only two of 74 samples (3%), each consisting of at least 12 live shellfish, were positive, both with type E spores (Presnell et al., 1967). *C. botulinum* was not found by Lovell and Barkate (1969) in any of 66 samples of freshwater crayfish from Louisiana fishing areas.

The incidence of *C. botulinum* in fish from the Pacific Northwest was comparatively high; 5% of ocean salmon samples and 17% of other fish from the ocean were positive, as were 23% of salmon from rivers (Craig et al., 1968). The incidence depended on the type of fish; it was 8% for coho salmon, 17% for sole and cod, 23% for steelhead trout, and about 24% for sockeye salmon. Shellfish also showed high levels of contamination (24%). Again, the level depended

Table 5 Incidence of *C. botulinum* Spores in Fish and Aquatic Animals

Location	Type of sample	% Samples positive	Type (%)[a] A	B	C/D	E	F	Ref.[b]
Great Lakes								
Lake Erie	Fish	2	0	0	0	100	0	1
Lake Superior	Fish	2	0	0	0	100	0	1
Lake Huron	Fish	3	0	0	0	100	0	1
Lake Michigan	Fish	8	0	0	0	100	0	1
Green Bay	Fish	56	0	0	0	100	0	1
New York, Lake Cayuga	Fish	6	0	0	0	100	0	2
Gulf of Maine	Fish	5	0	0	0	100	0	3
New York to Florida coast	Fish and invertebrates	<1	0	33	33	33	0	4
Gulf coast	Fish	4	0	7	13	80	0	5
	Shrimp	1	0	0	0	100	0	5
	Other	2	0	0	50	50	0	5
Alabama coast	Oysters	3	0	0	0	100	0	6
Washington/Oregon coast	Salmon	5	0	0	0	100	0	7
	Other fish	17	5	5	0	90	0	7
Washington/Oregon rivers	Salmon	23	8	8	0	79	4	7
Oregon coast	Shellfish	24	3	6	0	91	0	7
Alaska coast	Salmon	5	0	0	0	100	0	8
Sweden								
the Sound	Fish	100	0	0	0	100	0	9
South Baltic	Fish	55	0	0	0	100	0	9
Kattegat	Fish	0	0	0	0	0	0	9
North Sea, Skagerrak	Fish	0	0	0	0	0	0	10
United Kingdom trout ponds	Trout	10	0	7	14	63	16	11
Scottish trout ponds	Trout	1	0	100	0	0	0	12
Danish trout ponds	Trout	64	<1	<1	0	99	0	13
	Invertebrates	76	0	0	0	100	0	13
Finnish trout ponds	Trout	7	0	0	0	100	0	14
U.S.S.R.	Fish and other aquatic animals	8	5	10	3	34	48	15
Caspian Sea	Fish	35	0	0	0	100	0	16
Japan, Hokkaido	Fish	<1	0	0	0	100	0	17
	Dead fish	92	0	0	0	100	0	17
Japan, Aomori	Fish	5	0	0	0	60	40	18
Thailand	Fish	<1	0	0	67	33	0	19

Table 5 *Continued*

Location	Type of sample	% Samples positive	Type (%)[a]					Ref.[b]
			A	B	C/D	E	F	
Nicaragua/Honduras Caribbean coast	Fish	2	0	0	100	0	0	20
Gulf of Venezeula, Darien	Fish and invertebrates	43	46	7	40	7	0	21
Brazil	Fish	22	6	5	0	78	11	22
	Oysters and shrimp	35	13	16	23	29	19	23
Indonesia	Fish and invertebrates	13	6	18	65	0	12	24
Java, marine	Fish	17	7	21	50	14	7	25
Java, freshwater	Fish	13	0	0	86	14	0	25

[a]As percentage of the types identified.
[b] 1) Sugiyama et al., 1970; 2) Chapman and Naylor, 1966; 3) Nickerson et al., 1966; 4) Ward et al., 1967b; 5) Ward et al., 1967a; 6) Presnell et al., 1967; 7) Craig et al., 1968; 8) Miller, 1975; 9) Johannsen, 1963; 10) Cann et al., 1965; 11) Cann et al., 1975; 12) Burns and Williams, 1975; 13) Huss et al., 1974a; 14) Ala-Huikku et al., 1977; 15) Chulkova et al., 1976; 16) Rouhbakhsh-Khaleghdoust, 1975; 17) Kodama, 1970; 18) Yamamoto et al., 1970; 19) Tanasugarn, 1979; 20) Ward et al., 1967c; 21) Carroll et al., 1966; 22) Delazari et al., 1982b; 23) Delazari et al., 1982a; 24) Mortojudo et al., 1973; 25) Haq and Suhadi, 1981.
Source: Dodds and Hauschild, 1989; Hauschild, 1989.

on the type of shellfish; 31% of oysters were contaminated, 23% of clams, and 18% of crabs. Most contaminants were type E, but types A, B, and F were also detected. The type F strain was later isolated and found to be nonproteolytic, with properties similar to those of type E (Craig and Pilcher, 1966).

C. botulinum type E has been demonstrated in the gills of salmon from the Alaskan coast (Miller, 1975). Only one of 44 colon cultures from marine mammals was contaminated. The presence of spores in the gills of fish, but not in flesh, viscera, or roe, indicates environmental contamination with *C. botulinum*, rather than ingestion of contaminated food.

Since 1979, botulism has been recognized as a major cause of fish mortality in the United States (Eklund et al., 1982, 1984). Outbreaks during 1979 and 1980 resulted in the loss of over 1.25 million juvenile coho salmon in Washington and Oregon state hatcheries. Type E toxin was demonstrated in the stomach, intestinal contents and flesh of morbid fish. Pond sediments from the hatcheries contained 7.5×10^7 type E spores per kg. In an effort to eradicate the disease, the sediments from one affected pond were removed, and the pond was relined with new gravel. However, within approximately 10 months, the population of *C. botulinum* type E was even higher than before. The highest numbers occurred where waste feed, fecal material, and dead fish accumulated in the sediments. Increases in type E populations in pond sediments did not appear to be correlated with the water temperature. While live fish may become contaminated with *C. botulinum* from the environment, type E does not multiply in or on the living fish (Sugiyama et al., 1968), as fish fed spores rapidly cleared them. However, if these fish die from any cause, type E spores can germinate, grow, and produce high titers of toxin.

B. Europe

Surveys of fish from the coastal waters of Sweden confirm the high incidence of type E spores found in sediments in that area (Johannsen, 1963). The incidence was 100% in fish caught in the Sound, corresponding to an MPN of >2.91 organisms per fish. In fish caught in the southern Baltic the incidence was 55%, while none of the fish caught in the Kattegat were contaminated. The decrease in incidence with distance from the Baltic Sea was confirmed by Cann et al. (1965), who failed to find contaminated fish in the Skagerrak or the North Sea.

Trout have been raised in fish ponds in Great Britain for some time, but only recently has the production increased to a commercial scale, causing concern regarding the possible hazard of *C. botulinum*. Cann et al. (1975) found 10% of 1400 trout from 17 farms contaminated. Fish from 13 of the 17 farms were contaminated, with contamination rates ranging from 3 to 100%. Type C spores were detected at most farms, followed by types B, E, and F. However, in terms of the number of positive fish, type E predominated. Demonstration of type F was confined to Scotland.

Burns and Williams (1975) carried out a similar study on three fish farms in Scotland. Of 69 fish examined, only one (1%) was positive, with nonproteolytic type B spores. The positive sample was from a mud-bottomed pond; the other two ponds were concrete.

Huss et al. (1974a) found a high incidence of *C. botulinum* in Danish trout farms; 64% of fish and 76% of aquatic invertebrates were positive. Four farms were each sampled once in the winter and once in the summer. The contamination rate of fish ranged from 5 to 100%, with more positive samples in the summer than in the winter. All positive samples contained type E spores, with the exception of one sample containing both types A and E, and another with type B. Detailed examination of trout from one farm yielded MPN counts ranging from 340 to 5300 spores per kg. Interestingly, the incidence of type E spores in bottom material from these ponds was considerably lower than in the fish, and spores could not be detected in the sediment from one pond. Huss et al. (1974b) then examined methods aimed at decreasing the level of contamination in both fish and ponds. Starvation of fish gave variable results and, at best, only lowered the level of contamination. Quick-liming the bottom of the pond eliminated contamination of bottom mud within 7 days, and was proposed as a possible control measure, even though the effect on fish was not examined.

Two trout farms studied in Finland were both found to be contaminated with type E by Ala-Huikku et al. (1977). While fish from both farms were contaminated, only bottom deposits from one farm were. The level of incidence was similar to that in Great Britain, but less than that in Denmark.

C. Asia

Chulkova et al. (1976) found 8% of fish and other aquatic animals in the U.S.S.R. contaminated with *C. botulinum*. Most positive samples (128) were initially identified as containing type E spores, but 75 of these were later identified as type F. Types A, B, and C were also found. Of 34 fresh fish from the Caspian Sea, 12 (35%) were positive with type E (Rouhbakhsh-Khaleghdoust, 1975). The incidence was higher in carp, pike, and sturgeon than in anchovy.

Kodama (1970) determined the incidence of *C. botulinum* in fish in Lake Hachiro, Hokkaido, Japan. The isolation rate from living fish was low (<1%), but the rate from floating dead fish was very high (92%). Of 110 lake trout from Lake Towada, in Aomori prefecture, only 5 (5%) were found contaminated with type E and F spores (Yamamoto et al., 1970).

A low level of contamination of fish from the Gulf of Thailand was found by Tanasugarn (1979). Less than 1% of composite samples derived from over 16,000 fish were positive, and only type E and D spores were detected.

D. Central and South America

Only two batfish out of 82 aquatic animal specimens (2%) taken from the Nicaraguan and Honduran Caribbean coastal waters were positive for *C. botulinum*, both type C (Ward et al., 1967c). Types A, B, C, and E *C. botulinum* were detected by Carroll et al. (1966) in samples of fish and shrimp from the Gulfs of Darien and Venezuela, just to the north of Colombia and Venezuela, respectively. One sample each of sand trout and snapper was contaminated with type A. Several samples of shrimp were positive for types A, B, C, and E. Type A predominated, followed by C.

A survey by Delazari et al. (1982b) off the coasts of São Paulo and Southern Brazil found an overall incidence in fish of 22%. The incidence in white fish and catfish, both 31%, was considerably higher than in sardines (4%). Type E spores were prevalent at 74%, followed by types F, A, and B. The incidence in oysters and shrimp (Delazari et al., 1982a) was also high (35%). Again, type E spores predominated, followed by types F, B, and A. Types C and D were also detected, but were not differentiated.

E. Indonesia

Surveys by Mortojudo et al. (1973) of fish and aquatic animals from Indonesian waters found 17 of 130 (13%) samples positive. Type C spores predominated (59% of positives), followed by types B, F, D, and A. Haq and Suhadi (1981) found similar levels of contamination in both marine (17%) and freshwater fish (13%). Most positive samples contained type C spores, followed by types D, B, E, A, and E.

IV. *C. BOTULINUM* IN ANIMALS

Not only is the occurrence of botulism in animals evidence for the presence of *C. botulinum* in the environment, but the carcasses of animals dying of botulism provide a site for multiplication of *C. botulinum*. As well, healthy animals may carry small numbers of botulinal spores in their intestines, and may spread them out over considerable distances.

Botulism in cattle, usually caused by ingestion of type C or D toxin, appears to be a worldwide problem. It has been reported from North America, South Africa, Australia (Smith and Sugiyama, 1988), Europe (Haagsma and Ter Laak, 1978; Abbitt et al., 1984; Smart et al., 1987; McLoughlin et al., 1988; Hogg et al., 1990), and Asia (Egyed et al., 1978). Contaminated feeds containing poultry litter or carcasses have been implicated in several outbreaks in Europe and Asia. Outbreaks also occur in areas where cattle develop a phosphorus deficiency and feed on carcasses of small animals which may be contaminated with botulinum toxin, usually type D (Smith and Sugiyama, 1988). This occurs in South Africa, Australia, and in the southwestern United States, indicating wide distribution of type D in the environment.

Botulism in sheep appears to be a problem only in Australia and South Africa. As with cattle in these areas, a dietary insufficiency seems to be involved, provoking the sheep to consume carcasses of small animals, which may be toxic (Smith and Sugiyama, 1988).

Equine botulism occurs sporadically worldwide and is caused most frequently by toxin types C and D, but occasionally by type B toxin (Smith and Sugiyama, 1988). While equine type C botulism is common in Europe, it is rare in North America. Outbreaks in both Florida and Saskatchewan have been caused by ingestion of soil or dirt contaminated with type C toxin (Heath et al., 1990). Horses have also ingested botulinum toxin in contaminated silage, spoiled oats, hay, and potatoes (Heath et al., 1990). Botulinum type B intoxication in horses is enzootic in the eastern United States (Anonymous, 1990) where type B spores predominate in soil. Botulism in foals is referred to as Shaker Foal syndrome (SFS) (Whitlock and Messick, 1988). In SFS, *C. botulinum* type B has been isolated repeatedly from the feces of affected foals, although toxin has very rarely been detected in serum samples.

Avian botulism, discussed earlier in this chapter for wild birds, is a problem worldwide. Botulism in broiler chickens and in farmed turkeys caused by type C toxin has been reported in both the United States and the United Kingdom (Smart and Roberts, 1977; Smart et al., 1983). In Hiroshima Prefecture in Japan, Kurazono et al. (1985) reported an outbreak of botulism among penned pheasants caused by type C toxin.

Botulism in zoos, gameparks, etc., has been reported (Lewis et al., 1990), including outbreaks in birds, monkeys and baboons in the United Kingdom, waterfowl in Belgium, Argentina, Brazil, Uruguay, and the United States, sea lions in the United States, and gibbons in Australia. Most, if not all, of these outbreaks have been caused by ingestion of type C toxin in spoiled food, contaminated commercial maggots, material from stagnant ponds in hot weather, or rotting carcasses of waterfowl. Botulism has also been diagnosed in various other animals, for example, in minks and in dogs (Smith and Sugiyama, 1988). This chronology of botulism in animals is further evidence of the widespread nature of *C. botulinum* spores in the environment.

V. CONCLUSIONS

Although the factors affecting the distribution of the different *C. botulinum* types in nature are little understood, some salient points have emerged. The prevalence of different types in the environment depends on the geographical location. Type A spores predominate in soils in the western United States, Brazil, Argentina, and China. Type B is the major type found in the eastern United States, Great Britain, and much of continental Europe. However, most American type B strains are proteolytic, while the European strains seem to be nonproteolytic. In addition to aquatic environments and their surroundings, type E is the predominant type in many northern regions—Alaska, northern Canada, much of Scandinavia and the Soviet Union, and Japan. Types C and D have been found more frequently in warmer environments, for example, in the soils of Indonesia. Following outbreaks of avian botulism, the soil of other areas can also be heavily contaminated with type C spores.

The reasons why a specific serotype is prevalent in given areas are not well understood. It has been suggested that type A is favored by neutral to alkaline soil with low organic content, which is consistent with the virtual absence of type A in the highly cultivated soils of Europe and the eastern United States. Obviously, the psychrotolerance of type E is one reason for its prevalence in the north. There is still debate regarding the origin of type E, i.e., terrestrial versus aquatic. Its presence in aquatic environments has been linked to drainage, which occurs from the lands bordering many of the major fishing grounds. Obviously, water runoff will deposit terrestrial spores in the aquatic environment, but *C. botulinum* type E is capable of developing in the aquatic environment as well. Concentrations of type E spores in areas of the Great Lakes, the Baltic Sea and the North American Pacific coast appear higher than simple drainage could account for. Probably the best evidence for the development of type E spores in the aquatic environment is the Danish situation (Huss, 1980), where type E is prevalent in aquatic environments, but is seldom found in soil samples. Low salinity of water is undoubtedly a major factor in the concentration of type E, and likely accounts for the relatively high concentrations in the freshwater of Green Bay and in the Baltic Sea. Type E is also favored by humus-rich soils of high water capacity. Dormant spores may survive and accumulate in habitats where conditions are normally unsuitable for growth.

REFERENCES

Abbitt, R., Murphy, M. J., Ray, A. C., Reagor, J. C., Eugster, A. K., Gayle, L. G., Whitford, H. W., Sutherland, R. J., Fiske, R. A., and Pusok, J. (1984). Catastrophic death losses in a dairy herd attributed to type D botulism. *J. Am. Vet. Med. Assoc. 185*: 798.

Ala-Huikku, K., Nurmi, E., Pajulahti, H., and Raevuori, M. (1977). The occurrence of *Clostridium botulinum* type E in Finnish trout farms and the prevention of toxin formation in fresh-salted vacuum-packed trout fillets. *Nord. Vet. Med. 29*: 386.

Anonymous (1990). Rare form of botulism in horses reported. *J. Am. Vet. Med. Assoc. 196*: 529.

Barbuti, S., Quarto, M., Ricciardi, G., and Armenise, E. (1980). Research on the presence of *C. botulinum* in soils cultivated for vegetables. *Igiene Moderna 73*: 3.

Borland, E. D., Moryson, C. J., and Smith, G. R. (1977). Avian botulism and the high prevalence of *Clostridium botulinum* in the Norfolk Broads. *Vet. Rec. 100*: 106.

Bott, T. L., Defner, J. S., McCoy, E., and Foster, E. M. (1966). *Clostridium botulinum* type E in fish from the Great Lakes. *J. Bacteriol. 91*: 919.

Bott, T. L., Deffner, J. S., and Foster, E. M. (1967). Occurrence of *Clostridium botulinum* type E in fish from the Great Lakes with special reference to certain large bays. In *Botulism 1966* (M. Ingram, and T. A. Roberts, eds.), Chapman and Hall, London, p. 21.

Bott, T. L., Johnson, J., Foster, E. M., and Sugiyama, H. (1968). Possible origin of the high incidence of *Clostridium botulinum* type E in an inland bay (Green Bay of Lake Michigan). *J. Bacteriol. 95*: 1542.

Bulatova, T. I., Matveev, K. I., and Kazdobina, I. S. (1969). *Cl. botulinum* contamination of soil on shores of Lake Balkash. *Hyg. Sanit. 34*: 317.

Burns, G. F., and Williams, H. (1975). *Clostridium botulinum* in Scottish fish farms and farmed trout. *J. Hyg., Camb. 74*: 1.

Cann, D. C., Taylor, L. Y., and Hobbs, G. (1975). The incidence of *Clostridium botulinum* in farmed trout raised in Great Britain. *J. Appl. Bacteriol. 39*: 331.

Cann, D. C., Wilson, B. B., Hobbs, G., Shewan, J. M., and Johannsen, A. (1965). The incidence of *Clostridium botulinum* type E in fish and bottom deposits in the North Sea and off the Coast of Scandinavia. *J. Appl. Bacteriol. 28*: 426.

Cann, D. C., Wilson, B. B., and Hobbs, G. (1968). Incidence of *Clostridium botulinum* in bottom deposits in British coastal waters. *J. Appl. Bacteriol. 31*: 511.

Carroll, B. J., Garrett, E. S., Reese, G. B., and Ward, B. Q. (1966). Presence of *Clostridium botulinum* in the Gulf of Venezuela and the Gulf of Darien. *Appl. Microbiol. 14*: 837.

Chapman, H. M., and Naylor, H. B. (1966). Isolation of *Clostridium botulinum* type E from Cayuga Lake fish. *Appl. Microbiol. 14*: 301.

Chulkova, I. E., Bulatova, T. I., and Anisimova, L. I. (1976). Contamination of fish with the causative agents of botulism, type F, and its differentiation with type E. *Zh. Mikrobiol. Epidemiol. Immunobiol. 1*: 74.

Ciccarelli, A. S., and Giménez, D. F. (1981). Clinical and epidemiological aspects of botulism in Argentina. In *Biomedical Aspects of Botulism* (G. E. Lewis, ed.), Academic Press, New York, p. 291.

Craig, J. M., Hayes, S., and Pilcher, K. S. (1968). Incidence of *Clostridium botulinum* type E in salmon and other marine fish in the Pacific Northwest. *Appl. Microbiol. 16*: 553.

Craig, J. M., and Pilcher, K. S. (1966) *Clostridium botulinum* type F: isolation from salmon from the Columbia River. *Science 153*: 311.

Creti, R., Fenicia, L., and Aureli, P. (1990). Occurrence of *Clostridium botulinum* in the soil of the vicinity of Rome. *Curr. Microbiol. 20*: 317.

Delazari, I., Camargo, R., Leitao, M. F. F., and Anderson, A. W. (1981/1982a). *Clostridium botulinum* in seafood off the coast of Sao Paulo, Brazil. II. Occurrence in oysters (*Crassostrea brasiliana*, Lamarck) and shrimps (*Xyphopenaeus kroyeri*, Heller), and the influence of the methodology in the isolation of the bacterium. *Col. Ital, Campinas 12*: 179.

Delazari, I., Camargo, R., Leitao, M. F. F., Santos, C. A., and Anderson, A. W. (1981/1982b). *Clostridium botulinum* in seafood off the coast of Sao Paulo, Brazil. I. Occurrence in white fish (*Macrodon ancylodon*), cat fish (*Netuna barba*) and sardines (*Sardinella brasiliensis*). *Col. Ital, Campinas 12*: 163.

Dodds, K. L., and Hauschild, A. H. (1989). Distribution of *Clostridium botulinum* in the environment and its significance in relation to botulism. In *Recent Advances in Microbial Ecology*, Proceedings, 5th International Symposium on Microbial Ecology, Kyoto, Japan, p. 472.

Dolman, C. E., and Iida, H. (1963). Type E botulism: Its epidemiology, prevention and specific treatment. *Can. J. Public Health 54*: 293.

Dubovsky, B. J., and Meyer, K. F. (1922). The distribution of the spores of *B. botulinus* in the territory of Alaska and the Dominion of Canada. V. *J. Infect. Dis. 31*: 595.

Eales, C. E., and Gillespie, J. M. (1947). The isolation of *Clostridium botulinum* type A from Victorian soils. *Austral. J. Sci. 10*: 20.

Egyed, M. N., Shlosberg, A., Klopfer, U., Nobel, T. A., and Mayer, E. (1978). Mass outbreaks of botulism in ruminants associated with ingestions of feed containing poultry waste. *Refuah. Vet. 35*: 93.

Eklund, M. W., Peterson, M E., Poysky, F. T., Peck, L. W. and Conrad, J. F. (1982). Botulism in juvenile Coho salmon (*Oncorhynchus kisutch*) in the United States. *Aquaculture 27*: 1.

Eklund, M. W., and Poysky, F. (1965). *Clostridium botulinum* type F from marine sediments. *Science 149*(3681): 306.

Eklund, M. W., and Poysky, F. T. (1967). Incidence of *Cl. botulinum* type E from the Pacific Coast of the United States. In *Botulism 1966* (M. Ingram and T. A. Roberts eds.), Chapman and Hall Ltd., London, p. 49.

Eklund, M. W., Poysky, F. T., Peterson, M. E., Peck, L. W., and Brunson, W. D. (1984). Type E botulism in salmonids and conditions contributing to outbreaks. *Aquaculture 41*: 293.

Galvin, J. W., Hollier, T. J., Bodinnar, K. D., and Bunn, C. M. (1985). An outbreak of botulism in wild waterbirds in southern Australia. *J. Wildlife Dis. 21*: 347.

Gill, C. O., and Penney, N. (1982). The occurrence of *Clostridium botulinum* at aquatic sites in and around Auckland and other urban areas of the North Island. *N.Z. Vet. J. 30*: 110.

Graikoski, J. T., Bowman, E. W., Robohm, R. A., and Koch, R. A. (1970). Distribution of *Clostridium botulinum* in the ecosystem of the Great Lakes. In *Toxic Microorganisms*, Proceedings, 1st U.S.-Japan Conference on Toxic Microorganisms, Honolulu, HI, p. 271.

Haagsma, J. (1974). Etiology and epidemiology of botulism in water-fowl in the Netherlands. *Tijdschr. Diergeneesk. 99*: 434.

Haagsma, J., and Ter Laak, E. A. (1978). Atypical cases of type B botulism in cattle caused by supplementary feeding of brewers' grains. *Tijdschr. Diergeneesk. 103*: 312.

Haines, R. B., (1942). The occurrence of toxigenic anaerobes, especially *Clostridium botulinum*, in some English soils. *J. Hyg. 42*: 323.

Halvorson, H. O., and Ziegler, N. R. (1933). Application of statistics to problems in bacteriology. *J. Bacteriol. 25*: 101.

Haq, I., and Suhadi, F. (1981). Incidence of *Clostridium botulinum* in coastal and inland areas of west Java. *Japan J. Med. Sci. Biol. 34*: 231.

Hauschild, A. H. (1989). *Clostridium botulinum*. In *Foodborne Bacterial Pathogens* (M. P. Doyle, ed.), Marcel Dekker, New York, p. 111.

Hayashi, K. I., Tokuchi, M., Teramoto, K., Fujita, N., Okuno, Y., Rahim, A., Sujudi, and Hotta, S. (1981). Distribution of *Clostridium botulinum* in Indonesian soil. *ICMR Ann. 1*: 187.

Heath, S. E., Bell, R. J., Chirino-Trejo, M., Schuh, J. C. L. and Harland, R. J. (1990). Feedtrough dirt as a source of *Clostridium botulinum* type C intoxication in a group of farm horses. *Can Vet. J. 31*: 13.

Hogg, R. A., White, V. J., and Smith, G. R. (1990). Suspected botulism in cattle associated with poultry litter. *Vet. Record 126*: 476.

Huss, H. H. (1980). Distribution of *Clostridium botulinum*. *Appl. Environ. Microbiol. 39*: 764.

Huss, H. H., Pedersen, A., and Cann, D C. (1974a). The incidence of *Clostridium botulinum* in Danish trout farms. I. Distribution in fish and their environment. *J. Food Technol. 9*: 445.

Huss, H. H., Pedersen, A., and Cann, D. C. (1974b). The incidence of *Clostridium botulinum* in Danish trout farms, II. Measures to reduce contamination of the fish. *J. Food Technol. 9*: 451.

Johannsen, A. (1962). Forekomst och utbredning av *Cl. botulinum* typ E med sarskild hansyn till Oresundsomradet. *Nord. Vet.-Med. 14*: 441.

Johannsen, A. (1963). *Clostridium botulinum* in Sweden and the adjacent waters. *J. Appl. Bacteriol. 26*: 43.

Kanazawa, K., Ono, T., Karashimada, T., and Iida, H. (1970). Distribution of *Clostridium botulinum* type E in Hokkaido, Japan. In *Toxic Microorganisms*. Proceedings, 1st U.S.-Japan Conference on Toxic Microorganisms, Honolulu, HI, p. 299.

Kazdobina, I. S., Matveev, K. K., and Bulatova, T. I. (1976). *Clostridium botulinum* in the soil of the coast of the Aral Sea. *Hyg. Sanitation 29*: 18.

Knock, G. G. (1952). Survey of soils for spores of *Clostridium botulinum* (Union of South Africa and South West Africa). *J. Sci. Food Agric. 3*: 86.

Kodama, E. (1970). Epidemiological observations on botulism in Japan, especially to the present status in the Akita Prefecture. In *Toxic Microorganisms*, Proceedings, 1st U.S.-Japan Conference on Toxic Microorganisms, Honolulu, HI, p. 309.

Kravchenko, A. T., and Shishulina, L. M. (1967). Distribution of *Cl. botulinum* in soil and water of the U.S.S.R. In *Botulism 1966* (M. Ingram and T. A. Roberts, eds.), Chapman and Hall Ltd., London, p. 13.

Kurazono, H., Shimozawa, K., Sakaguchi, G., Takahashi, M., Shimizu, T., and Kondo, H. (1985). Botulism among penned pheasants and protection by vaccination with C1 toxoid. *Res. Vet. Sci. 38*: 104.

Langenegger, J., and Dobereiner, J. (1988). Enzootic botulism of water buffaloes in Maranhao, Brazil. *Pesq. Vet. Bras. 8*: 37.

Laycock, R. A., and Longard, A. A. (1972). *Clostridium botulinum* in sediments from the Canadian Atlantic Seaboard. *J. Fish. Res. Bd. Can. 29*: 443.

Laycock, R. A., and Loring, D. H. (1972). Distribution of *Clostridium botulinum* type E in the Gulf of St. Lawrence in relation to the physical environment. *Can. J. Microbiol. 18*: 763.

Leitao, M. F. F., and Delazari, I. (1983). *Clostridium botulinum* em solos no estado de Sao Paulo. *Col. Ital. Campinas 13*: 75.

Lewis, J. C. M., Smith, G. R., and White, V. J. (1990). An outbreak of botulism in captive hamadryas baboons (*Papio hamadryas*). *Vet. Rec. 126*: 216.

Lovell, R. T., and Barkate J. A. (1969). Incidence and growth of some health-related bacteria in commercial fresh water crayfish (genus *Procambarus*). *J. Food Sci. 34*: 268.

Marion, W. R., O'Meara, T. E., Riddle, G. D., and Berkhoff, H. A. (1983). Prevalence of *Clostridium botulinum* type C in substrates of phosphate-mine settling ponds and implications for epizootics of avian botulism. *J. Wildlife Dis. 19*: 302.

Mason, J. H. (1968). *Clostridium botulinum* type D in mud of lakes of the Zululand game parks. *J. S. Afr. Vet. Med. Assoc. 39*: 37.

McLoughlin, M. F., McIlroy, S. G., and Neill, S. D. (1988). A major outbreak of botulism in cattle being fed ensiled poultry litter. *Vet. Rec. 122*: 579.

Meyer, K. F., and Dubovsky, B. J. (1922a). The distribution of the spores of *B. botulinus* in California. II. *J. Infect. Dis. 31*: 541.

Meyer, K. F., and Dubovsky, B. J. (1922b). The distribution of the spores of *B. botulinus* in the United States. IV. *J. Infect. Dis. 31*: 559.

Meyer, K. F., and Dubovsky, B. J. (1922c). The occurrence of the spores of *B. botulinus* in Belgium, Denmark, England, the Netherlands and Switzerland. VI. *J. Infect. Dis. 31*: 600.

Miller, L G. (1975). Observations on the distribution and ecology of *Clostridium botulinum* type E in Alaska. *Can. J. Microbiol. 21*: 920.

Miller, L. G., Clark, P. S., and Kunkle, G. A. (1972). Possible origin of *Clostridium botulinum* contamination of Eskimo foods in northwestern Alaska. *Appl. Microbiol. 23*: 427.

Mortojudo, J. W., Siagian, E. G., Suhadi, F., Ward, B. Q., and Ward, W. M. S. (1973). The presence of *Clostridium botulinum* in Indonesian waters. *J. Appl. Bacteriol. 36*: 437.

Murrell, W. G., and Stewart, B. J. (1983). Botulism in New South Wales, 1980-1981. *Med. J. Aust. 1*: 13.

Nakamura, Y., Iida, H., Saeki, K., Kanzawa, K., and Karashimada, T. (1956). Type E botulism in Hokkaido, Japan. *Jap. J. M. Sc. Biol. 9*: 45.

Neubauer, M., Hudec, K., and Pellantova, J. (1988). The occurrence of *Clostridium botulinum* type C bacterium and botulotoxin in an aquatic environment in southern Moravia. *Folia Zoologica 37*: 255.

Nickerson, J. T. R., Goldblith, S. A., DiGioia, G., and Bishop, W. W. (1966). The presence of *Clostridium botulinum* type E in fish and mud taken from the Gulf of Maine. In *Botulism 1966* (M. Ingram, and T. A. Roberts, eds.), Chapman and Hall, London, p. 25.

Notermans, S., Dufrenne, J., and van Schothorst, M. (1979). Recovery of *Clostridium botulinum* from mud samples incubated at different temperatures. *Europ. J. Appl. Microbiol. Biotechnol 6*: 403.

Pasricha, C. L., and Panja, G. (1940). *Clostridium botulinum* in samples of Calcutta soil. *Ind. J. Med. Res. 28*: 49.

Presnell, M. W., Miescier, J. J., and Hill, W. F. (1967). *Clostridium botulinum* in marine sediments and in the oyster (*Crassostrea virginica*) from Mobile Bay. *Appl. Microbiol. 15*: 668.

Rouhbakhsh-Khaleghdoust, A. (1975). The incidence of *Clostridium botulinum* type E in fish and bottom deposits in the Caspian Sea coastal waters. *Pahlavi Med. J. 6*: 550.

Schoenholz, P., and Meyer, K. F. (1922). The occurrence of the spores of *B. botulinus* in the Hawaiian Islands and China. VII. *J. Infect. Dis. 31*: 610.

Serikawa, T., Nakamura,S., and Nishida, S. (1977). Distribution of *Clostridium botulinum* type C in Ishikawa Prefecture, and applicability of agglutination to identification of nontoxigenic isolates of *C. botulinum* type C. *Microbiol. Immunol. 21*: 127.

Smart, J. L., Jones, T. O., Clegg, F. G., and McMurtry, M. J. (1987). Poultry waste associated type C botulism in cattle. *Epidem. Inf. 98*: 73.

Smart, J. L., Laing P. W., and Winkler C. E. (1983). Type C botulism in intensively farmed turkeys. *Vet. Rec. 113*: 198.

Smart, J. L., and Roberts, T. A. (1977). An outbreak of type C botulism in broiler chickens. *Vet. Rec. 100*: 378.

Smith, G. R., and Milligan, R. A. (1979). *Clostridium botulinum* in soil on the site of the former metropolitan (Caledonian) cattle market, London. *J. Hyg., Camb. 83*: 237.

Smith, G. R., Milligan, R. A., and Moryson, C. J. (1978). *Clostridium botulinum* in aquatic environments in Great Britain and Ireland. *J. Hyg., Camb. 80*: 431.

Smith, G. R., and Moryson, C. J. (1975). *Clostridium botulinum* in the lakes and waterways of London. *J. Hyg. Camb. 75*: 371.

Smith, G. R., and Moryson, C. J. (1977). A comparison of the distribution of *Clostridium botulinum* in soil and in lake mud. *J. Hyg., Camb. 78*: 39.

Smith, G. R., Moryson, C. J., and Walmsley, J. G. (1977). The low prevalence of *Clostridium botulinum* in the lakes, marshes and waterways of the Camargue. *J. Hyg., Camb. 78*: 33.

Smith, G. R., Turner, A. M., Wynn-Williams, D. D., Collett, G., Wright, D., and Keymer, I. F. (1987). Search for *Clostridium botulinum* in the South Orkney and Falkland Islands. *Vet. Rec. 121*: 404.

Smith, G. R., and Young, A. M. (1980). *Clostridium botulinum* in British soil. *J. Hyg., Camb. 85*: 271.

Smith, L. D. (1975) Common mesophilic anaerobes, including *Clostridium botulinum* and *Clostridium tetani*, in 21 soil specimens. *Appl. Microbiol. 29*: 590.

Smith, L. D. (1978). The occurrence of *Clostridium botulinum* and *Clostridium tetani* in the soil of the United States. *Health Lab. Sci. 15*: 74.

Smith, L. D., and Sugiyama, H. (1988). *Botulism*, Charles C. Thomas, New York.

Sonnabend, W. F., Sonnabend, U. P., and Krech, T. (1987). Isolation of *Clostridium botulinum* type G from Swiss soil specimens by using sequential steps in an identification scheme. *Appl. Environ. Microbiol. 53*: 1880.

Sugiyama, H., Bott, T. L., and Foster, E. M. (1970). *Clostridium botulinum* type E in an inland bay (Green Bay of Lake Michigan). In *Toxic Microorganisms*, Proceedings 1st U.S.-Japan Conf. on Toxic Microorganisms, Honolulu, HI p. 287.

Sugiyama, H., Bott, T. L., and Soucheck, J. A. (1972). *Clostridium botulinum* in the Great Lakes. In *Spores V* (H. O. Halverson, R. Hanson, and L. L. Campbell, eds.), American Society of Microbiology, Washington, DC, p. 314.

Suhadi, F., Thayib, S. S., and Sumarsono, N. (1981). Distribution of *Clostridium botulinum* around fishing areas of the western part of Indonesian waters. *Appl. Environ. Microbiol. 41*: 1468.

Tanasugarn, L. (1979). *Clostridium botulinum* in the Gulf of Thailand. *Appl. Environ. Microbiol. 37*: 194.

Turner, H. D., Brett, E. M., Gilbert, R. J., Ghosh, A. C., and Liebeschuetz, H. J. (1978). Infant botulism in England. *Lancet 1*: 1277.

Tsai, C. C., and Ting, S. C. (1977). Ecological studies of *Clostridium botulinum* in soils of Taiwan. *J. Formosan Med. Assoc. 76*: 563.

Venkateswaran, K., Nakano, H., Okabe, T., Takayama, K., Matsuda, O., and Hashimoto, H. (1989). Occurrence and distribution of *Vibrio* spp., *Listonella* spp., and

Clostridium botulinum in the Seto Inland Sea of Japan. *Appl. Environ. Microbiol. 55*: 559.

Wakamatsu, T. (1953). Ecological study of clostridia in Kyushu, especially its southern part. *Kitasato Arch. Exp. Med. 25*: 163.

Ward, B. Q., Carroll, B. J., Garrett, E. S., and Reese, G. B. (1967a). Survey of the U.S. Gulf Coast for the presence of *Clostridium botulinum. Appl. Microbiol. 15*: 629.

Ward, B. Q., Carroll, B. J., Garrett, E. S., and Reese, G. B. (1967b). Survey of the U.S. Atlantic Coast and estuaries from Key Largo to Staten Island for the presence of *Clostridium botulinum. Appl. Microbiol. 15*: 964.

Ward, B. Q., Garrett, E. S., and Reese, G. B. (1967c). Further indications of *Clostridium botulinum* in Latin American waters. *Appl. Microbiol. 15*: 1509.

Whitlock, R. H., and Messick, J. B. (1988). Foal botulism (shaker foal syndrome) clinical signs, diagnosis, treatment and prevention. *Proceedings of the Annual Convention of the American Assoc. of Equine Practitioners*, Manhattan, KS, pp. 359–366.

Williams-Walls, N. J. (1968). *Clostridium botulinum* type F: Isolation from crabs. *Science 162*: 375.

Wobeser, G., Marsden, S., and MacFarlane, R. J. (1987). Occurrence of toxigenic *Clostridium botulinum* type C in the soil of wetlands in Saskatchewan. *J. Wildlife Dis. 23*: 67.

Yamakawa, K., Hayashi, T., Kamiya, S., Yoshimura, K., and Nakamura S. (1990a). *Clostridium botulinum* in the soil of Paraguay. *Jpn. J. Med. Sci. Biol. 43*: 19.

Yamakawa, K., Kamiya, S., Nishida, S., Yoshimura, K., Yu, H., Lu, D., and Nakamura, S. (1988). Distribution of *Clostridium botulinum* in Japan and in Shinkiang district of China. *Microbiol. Immunol. 32*: 579.

Yamakawa, K., Kamiya, S., Yoshimura, K., Nakamura, S., and Ezaki, T. (1990b). *Clostridium botulinum* in the soil of Kenya. *Ann. Trop. Med. Parasitol. 84*: 201.

Yamamoto, K., Kudo, H., Asano, H., Seito, Y., Nabeya, S., Horiuchi, Y., Awasa, K., Sasaki, J., and Kimura K. (1970). Examination of *Clostridium botulinum* in samples taken from Lake Towada. *Hirosaki Med. J. 22*: 92.

Zaleski, S., Fik, A., and Daczkowska, E. (1973). *Clostridium botulinum* type E in the soil samples from the Polish Baltic seaboard. *Proceedings of 6th International Symp. World Assoc. Vet. Food Hygienists*, Elsinore, Denmark.

3

Clostridium botulinum in Foods

Karen L. Dodds

Health Protection Branch,
Health and Welfare Canada,
Ottawa, Ontario, Canada

I. INTRODUCTION

Worldwide, the incidence of foodborne botulism continues to be much higher than the incidence of any other type of botulism. It is well recognized that the highest potential hazard for botulism lies in our food supply. However, in general, the food industry is aware of the hazard that *C. botulinum* presents, and has taken many steps to ensure its control. The sequence of events resulting in foodborne botulism starts with the contamination of a food with *C. botulinum.* This often occurs while a product is growing or being harvested, and is therefore more likely when a product originates in an evironment where the incidence of spores is high. However, contamination can also occur during or after processing.

This chapter summarizes surveys carried out to determine the incidence of *C. botulinum* spores in a variety of foods, and expands an earlier review covering this subject (Hauschild, 1989). There have been considerably fewer surveys of foods for contamination with *C. botulinum* than environmental surveys (see Chapter 2). As Hauschild (1989) noted, food surveys have focused primarily on fish, meats, and infant foods, particularly honey. Data included in the tables in this chapter are generally from surveys where it was possible to calculate a most probable number (MPN) count according to Halvorson and Ziegler (1933). In a few instances, actual or MPN counts were determined by authors, and these numbers are so noted. The tables are not meant to be exhaustive, but to give representative results for each food type and geographical area surveyed.

Finally, there is often evidence of contamination of certain foods with *C. botulinum* due to their involvement in a foodborne outbreak. For example, even though survey results show the level of contamination in dairy products to be very low, botulism outbreaks have been associated with Liederkranz cheese (Meyer and Eddie, 1951), Brie cheese (Sebald et al., 1974), and a commercial cheese spread with onions (de Làgarde, 1974). These foods will be covered in Chapter 4.

II. INCIDENCE IN PREPARED FISH

Fish may carry spores of *C. botulinum* as a result of exposure to spores before harvest or due to contamination during processing and handling. Since the aquatic environment of fish is frequently contaminated with spores (Chapter 2), it is to be expected that fish will often be contaminated as well. A large number of surveys of fish and other aquatic animals have been carried out, mainly as an extension of surveys of aquatic sediments. The results of surveys of fish and aquatic animals taken in situ are reported in Chapter 2. Surveys of fish and other aquatic species taken at the marketplace or after any form of processing are reported here.

A. North America

The presence of *C. botulinum* in fish in North America is readily demonstrated (Table 1), although the numbers found are generally lower than in environmental surveys. In the Great Lakes, the frequency of contamination of freshly caught and eviscerated whitefish chubs depends on the processing stage. Pace et al. (1967) found that, while chubs sampled at the brining step showed the highest incidence of contamination (20%), smoking at 82.2°C for 30 minutes reduced the incidence to approximately 1%. Most positive samples were contaminated with type E spores. Similarly, Fantasia and Duran (1969) showed that smoking reduced the contamination of chubs from 50 to 14%. When commercially dressed chubs from Lake Michigan were tested, 31 of 500 (6.2%) were positive for type E spores. When noneviscerated chubs were "properly" dressed in the laboratory, the contamination rate fell to less than 1%.

Results from the Atlantic seaboard have generally indicated lower levels of contamination. A survey by Goldblith and Nickerson (1965) demonstrated high numbers of type E spores in commercially produced haddock fillets from five processing plants in Boston. Insalata et al. (1967) found a lower level of contamination, also with type E spores, in randomly collected samples of commercial vacuum-packed, frozen flounder harvested in fishing areas close to Nova Scotia. Very few specimens out of 400 frozen packaged fish purchased in Canada contained detectable *C. botulinum* (Thatcher et al., 1967). Only one sample

Table 1 Incidence of *C. botulinum* Spores in Prepared Fish in North America

Product	Origin	Sample size (g)	% positive samples	MPN per kg	Types identified	Ref.[a]
Whitefish chubs	Great Lakes					
Eviscerated		10	13	14	E, C	1
In brine tank		10	20	22	B, C, E	1
Brined, washed		10	14	15	E	1
Fresh smoked		10	1	1	B, E	1
Haddock fillets	Atlantic		24	170[b]	E	2
Frozen flounder, vacuum-packed	Atlantic	1.5	10	70	E	3
Frozen, packaged fish	Canada		<1		A, B, E	4
Filleted cod, whiting, flounder	New York City	100	<1	<1	E	5
Smoked fish	Pacific NW	5–10	5	9	E	6
Dressed rockfish	California	10	100[c]	2400	A, E	7
Fish and seafood	California	70	22	4	A, B, E, F	8
Salmon	Alaska		1		E	9
Salmon	Washington		8		E	9
Salmon	Oregon		6		E	9
Salmon	Alaska	24–36	100	190[b]	A	10

[a] 1) Pace et al., 1967; 2) Goldblith and Nickerson, 1965; 3) Insalata et al., 1967; 4) Thatcher et al., 1967; 5) Post et al., 1985; 6) Hayes et al., 1970; 7) Lindroth and Genigeorgis, 1986; 8) Baker et al., 1990; 9) Houghtby and Kaysner, 1969; 10) Garcia and Genigeorgis, 1987.
[b]MPN determined by authors.
[c]All lots examined were contaminated.
Source: Hauschild, 1989.

from the Atlantic coast was contaminated, containing both types A and B spores. Two samples from the Pacific coast were contaminated with type E spores. Similarly, the contamination level of whiting, cod, and flounder purchased in New York City during a 2-year study was very low (Post et al., 1985). Contamination was not detected in 1074 samples of commercial vacuum-packed fresh fish produced throughout the United States (Lilly and Kautter, 1990), giving a contamination rate between 0 and 0.34%.

Surveys have also examined products from the Pacific. Smoked fish products, including salmon, sturgeon, black cod, and others, produced by 28 small processors in the Pacific Northwest and examined for the presence of *C botulinum*, had a fairly low contamination level (5%) (Hayes et al., 1970). In contrast, fresh dressed rockfish caught off the California coast contained type A spores at levels estimated at 2400 spores per kg (Lindroth and Genigeorgis, 1986). One lot tested was contaminated with type A and E spores at a level of 90 per kg. Baker et al. (1990) examined the incidence of *C. botulinum* in fresh fish and other seafoods purchased in California. While the overall contamination level was 22%, it ranged from 0% for butterfish, kingfish, mackerel, Pacific oysters, prawns, shark, shrimp, smelt, and squid to 100% for composite samples each of rockfish and king salmon. Most samples contained type A spores (20%), followed by types B (2%), E, and F (<1% each). Seasonal prevalence rates determined separately were highest in the winter (87%), and declined through spring (78%) and summer (60%) to the lowest rate in the fall (55%). Types identified included A (33%), E (20%), proteolytic B (4%), nonproteolytic B (6%), and nonproteolytic F (2%).

The potential risk of botulism from commercial seafood cocktails in California was assessed by Lerke (1973). These seafoods are preserved primarily by the action of acetic acid and secondarily by refrigeration. *C. botulinum* type E was not detected in any of 35 samples from nine processors. Houghtby and Kaysner (1969) determined that the incidence of type E *C. botulinum* in salmon from routine salmon cannery inspections was highest in Washington, followed by Oregon and Alaska. Of 74 samples from Washington, 6 (8%) were positive; of 70 samples from Oregon, 4 (6%) were positive, and of 589 samples from Alaska, 6 (1%) were positive. Garcia and Genigeorgis (1987) estimated the numbers of type A *C. botulinum* in salmon tissue homogenates from fresh, gutted, dressed king salmon from Alaska at 30–440 spores per kg, averaging 190 per kg.

B. Europe and Asia

With the exception of Scandinavia and especially the Caspian Sea, the level of contamination of prepared fish in Europe and Asia appears to be much lower than that in the United States (Table 2). *C. botulinum* was not detected in over 200 samples of vacuum-packed fish purchased at random from shops in England, nor in a small number of samples packaged at the Torry Research Station or in fresh North Sea herring (Hobbs et al., 1965). Only five of 646 samples of vacuum-packed fish purchased at random from shops in England were contaminated with type E *C. botulinum* (Cann *et al.*, 1966). While herring caught in the North Sea showed no evidence of contamination, herring caught in the Norwegian Sea and in the Viking Bank were positive with type E spores at 44% and 42%, respectively. Estimated levels ranged from 5 to 63 spores per kg of fish. In Sweden, Johannsen (1965) detected type E *C. botulinum* in 18 of 144 samples of smoked herring and in one sample of smoked trout. Smoked Danish salmon bought on the open market had a low level of contamination, with only type B spores (Nielsen and Pedersen, 1967). Smoked Baltic eel processed at 55°C for 2

Table 2 Incidence of *C. botulinum* Spores in Prepared Fish in Europe and Asia

Product	Origin	Sample size (g)	% positive samples	MPN per kg	Types identified	Ref.[a]
Vacuum-packed fish	England	3	0			1
	England		<1		E	2
	North Sea		0			2
	Norwegian Sea		44		E	2
	Viking Bank		42	63[b]	E	2
Smoked herring	Sweden		13		E	3
Smoked salmon	Denmark	20	2	<1	B	4
Smoked eel	Baltic Sea	20	20	11	E	5
Salted carp	Caspian Sea	2	63	490	E	6
Smoked carp	Caspian Sea	2	5	28	E	6
Salted fish	Caspian Sea	2	0	<60	E	6
Smoked fish	Caspian Sea	2	0	<68	E	6
Fish	Indonesia	5	3	6	A, B, C D, F	7
Shellfish	Indonesia	5	1	2	A, C	7
Shrimp	Indonesia	5	2	4	B, C, F	7
Crab	Indonesia	5	7	13	A, C, D	7
Boiled fish	Indonesia	5	<1	<1	D	7
Fish and seafood	Osaka	30	8	3	C, D	8

[a] 1) Hobbs et al., 1965; 2) Cann et al., 1966; 3) Johannsen, 1965; 4) Nielsen and Pedersen, 1967; 5) Abrahamsson, 1967; 6) Rouhbakhsh-Khaleghdoust, 1975; 7) Suhadi et al., 1981; 8) Haq and Sakaguchi, 1980.
[b]MPN as determined by authors.
Source: Hauschild, 1989.

hours and at 60°C for a further 30 minutes was still contaminated with type E spores (Abrahamsson, 1967). Very high levels of contamination occur in both salted and smoked carp from the Caspian Sea (Rouhbakhsh-Khaleghdoust, 1975). However, none of 17 samples of salted and smoked fish of an unknown variety were contaminated. Environmental samples from this area were also shown to be heavily contaminated (see Chapter 2).

Suhadi et al. (1981) examined various fish and seafood samples purchased from local commercial fishermen and samples of pindang, boiled fish, from various markets in Indonesia. In fish, most of the positive samples contained either type C or D spores (27% each), but types A (20%), B (17%), and F (8%) were also detected. The incidence was similar in shellfish and shrimp, and slightly higher in crabs. Types A, B, C, D, and F spores were all present. Only one sample of boiled fish was contaminated, with type D *C. botulinum*.

The potential health hazard from fish and other seafoods sold at retail in Osaka was evaluated by Haq and Sakaguchi (1980). While only the large fish had been eviscerated and dressed, all fish were packed in small plastic trays for sale. Of 142 samples examined, 11 proved to be contaminated: six with type C spores, and one with type D. Thus, it is evident that not only are raw fish fairly frequently contaminated, but they can also serve as a source of contamination for equipment, surfaces and uncontaminated fish during processing, and perhaps as a source of postprocess contamination if handled in the same area as processed fish.

III. INCIDENCE IN MEATS AND POULTRY

In comparison to fish and fish products, there have been few surveys for *C. botulinum* in meat and meat products. The level of contamination of meats is generally low (Table 3). These products are less likely than fish to be contaminated with spores prior to slaughter of the animals, since contamination of the farm environment is lower than the aquatic environment (see Chapter 2).

Table 3 Incidence of *C. botulinum* Spores in Meat and Meat Products

Product	Origin	Sample size (g)	% positive samples	MPN per kg	Types identified	Ref[a]
Raw meat	Europe		36		E	1
Raw meat	North America	3	<1	0.1	C	2
Cured meat	North America	30	2	0.5	A, B	3
Raw pork	United Kingdom	30	0	<0.1	A	4
		30	2	0.7	A, B,	4
		30	14	5	A, B, C	4
		30	6	2	C	4
Vacuum-packed	United Kingdom	25	4	2	A, B	5
bacon		50	5	1	A, B	5
		175	73	7	A	5
Vacuum-packed bacon	Canada	75	<1	0.1	A or B[b]	6
Cured meat	Canada	75	2	0.2	A	6
Liver sausage	Canada	75	2	0.2	A	7

[a] 1) Klarmann, 1989; 2) Greenberg et al., 1966; 3) Abrahamsson and Riemann, 1971; 4) Roberts and Smart, 1977; 5) Roberts and Smart, 1976; 6) Hauschild and Hilsheimer, 1980; 7) Hauschild and Hilsheimer, 1983.
[b]Not specifically typed.
Source: Hauschild, 1989.

There is evidence that the animals themselves occasionally carry spores that may lead to internal contamination of meat and to contamination of the meat-processing environment. To determine the level of contamination in meat animals in Europe, Narayan (1967) examined various tissues in clinically healthy cattle, pigs, and poultry. Of 167 caecal contents, four were positive for *C. botulinum*: three with type B and one with type A. Of 203 cultures from cattle and pigs, 17 were positive, 10 with type A spores and seven with type B, of which two were nonproteolytic. Positive cultures were from mesenteric lymph gland, muscle, liver, kidneys, and spleen. Similarly, Klarmann (1989) detected *C. botulinum* types B, C, or E in seven of 12 fecal samples from cattle and swine in Europe. He also examined raw material and pulverized dehydrated meat taken from rendering plants in Europe. Four samples and one sample were contaminated, respectively, all with type E spores.

The incidence of mesophilic *Clostridium* spores in raw pork, beef, and chicken in processing plants in the United States and Canada was examined by Greenberg et al. (1966). Of 19,727 putrefactive anaerobic spores isolated from 2358 samples of raw meats, one was confirmed as *C. botulinum* type C. The positive sample was from a western Canada chicken sample.

Riemann (1963) reviewed the natural contamination of commercially canned cured meats with clostridial spores. The majority of clostridia isolated from canned cured meats belong to the putrefactive type—*Clostridium sporogenes* and similar species. The incidence of putrefactive anaerobic spores in meat has been studied because the higher the level of these organisms, the greater the probability that *C. botulinum* might be present. However, Riemann could find no evidence that *C. botulinum* had ever been isolated from such products. He indicated that the number of anaerobic spores in cured meats and meat trimmings decreased from an MPN of 0.52 spores/g of raw luncheon meat in 1944 to below 0.001/g after 1952.

Thatcher et al. (1967) were unable to demonstrate the presence of *C. botulinum* in 436 samples of vacuum-packed sliced processed meat purchased in Canada. In a study of the prevalence of *C. botulinum* in cured meat products such as bologna, smoked beef, turkey and chicken, liver sausage, luncheon loaf, salami, and pastrami, Abrahamsson and Riemann (1971) found only six of 372 samples positive. Five of 100 samples of cooked ham contained type A spores, and one of 41 samples of smoked turkey contained type B.

Roberts and Smart (1976,1983) examined the occurrence of clostridia in raw pork and in vacuum-packed bacon in the United Kingdom. The incidence in raw pork at one bacon factory varied considerably between sampling occasions. Twice, *C. botulinum* was not found in 140 samples. Twice, between 2 and 6% of samples were positive. On one occasion, 20 of 138 samples were positive, equivalent to about 5 spores/kg. Most positive samples (65%) contained type C spores, 15% contained type A, and the rest were untyped. The incidence also varied in the finished product. Eleven of 263 25-g samples were positive, 10 with type B spores, and one with type A. In 50-g samples, six of 110 were positive, three with type A spores, and three with type B. In unsliced collar bacon, 19 of 26 175-g samples contained type A spores. These results show a fairly infrequent occurrence of *C. botulinum* but relatively high levels at times, with spore loads reaching 7 per kg of sample.

Hauschild and Hilsheimer (1980,1983) found a lower incidence of *C. botulinum* in processed meats in Canada. Of 208 samples of commercial bacon, only one was positive. An MPN was estimated at 0.064 spores/kg. In cured meats at the retail level, type A *C. botulinum* was detected in two of 132 samples, giving an MPN of 0.2 per kg. In commercial liver sausage, five of 276 samples were positive, all with type A spores. The MPN was estimated between 0.15 and 0.24 per kg.

These studies indicate a very low incidence of *C. botulinum* spores in meat and meat products in North America, while the incidence in Europe appears to be higher. This is reflected in the relative number of botulism outbreaks associated with meats in the two areas (see Chapter 4), with many more cases in Europe than in North America.

IV. INCIDENCE IN FRUITS AND VEGETABLES

C. botulinum may be present on fruits and vegetables, particularly those harvested from the soil, as a result of environmental contamination (Table 4). Different agricultural practices, such as the use of manure for fertilizer, may also affect the level of contamination.

The first report on the presence of *C. botulinum* on fruits and vegetables in the United States is that of Meyer and Dubovsky (1922a). As a result of the relatively frequent occurrence of botulism following consumption of vegetables in California, they undertook a very extensive survey for *C. botulinum*. Of 189 samples of fruits and vegetables collected at markets or from gardens and orchards across the state, 29 were positive, including asparagus, beans, carrots, celery, corn, lima beans, olives, potatoes, turnips, apricots, cherries, and peaches. Most of the isolates were type A. Meyer and Dubovsky (1922b) then extended their survey across the United States. Of 431 samples of fruits and vegetables examined, 54 were positive. Types A and B spores were detected in approximately the same number of samples. String beans gave the highest percentage of positive cultures.

Later, Rozanova et al. (1972) detected types A and B *C. botulinum* in samples of cabbage, carrots, onions, potatoes, tomatoes, horseradish leaves, dill, and parsley in the U.S.S.R. Prokhorovich et al. (1975) detected *C. botulinum* in the wash water from carrots, red peppers, and apricots, indicating its presence on the raw product. However, Vergieva and Incze (1979) failed to detect *C. botulinum* in any of 27 samples of potatoes or 18 samples of carrots.

Table 4 Incidence of *C. botulinum* Spores in Fruits and Vegetables

Product	Origin	% positive samples	Types identified	Ref.[a]
Fruits and vegetables	California	15	A, B	1
Fruits and vegetables	United States	13	A, B	2
Vegetables, herbs	U.S.S.R.	43	A, B	3
Wash water from fruits and vegetables	U.S.S.R.	7		4
Potatoes, carrots	U.S.S.R.	0		5
Vegetables	Italy	4	B	6
Potatoes, cooked, vacuum-packed	The Netherlands	0 (0.63)[b]		7
Potatoes, raw and cooked and vacuum-packed	Germany	0		8
Mushrooms	Canada	(2100)[b]	B	9
Mushrooms	United States	0		10
Mushrooms	The Netherlands	0		11

[a] 1) Meyer and Dubovsky, 1922a; 2) Meyer and Dubovsky, 1922b; 3) Rosanova et al., 1972; 4) Prokhorovich et al., 1975; 5) Vergieva and Incze, 1979; 6) Quarto et al., 1983; 7) Notermans et al., 1985; 8) Baumgart, 1987; 9) Hauschild et al., 1975; 10) Kautter et al., 1978; 11) Notermans et al., 1989.
[b]MPN per kg as estimated by authors.

Quarto et al. (1983) detected type B spores in 13 of 296 Italian vegetables. While *C. botulinum* was not detected on vacuum-packed, cooked potatoes in The Netherlands, the number was estimated at 0.6 per kg. The estimate was made by determining the ratio of the number of anaerobic sporeformers to the number of *C. botulinum* spores in soils from different potato fields, then determining the number of anaerobic sporeformers in vacuum-packed, cooked potatoes, and calculating the number of *C. botulinum* spores, assuming that the ratio of anaerobic sporeformers to *C. botulinum* would be constant (Notermans et al., 1985). Baumgart (1987) did not detect *C. botulinum* in 72 samples of raw and 48 samples of cooked vacuum-packed potatoes in Germany, but several other species of clostridia were identified.

Cultivated mushrooms are a product which has been of concern with regards to the presence of *C. botulinum*. In a survey of Canadian mushrooms, Hauschild et al. (1975) found up to 2100 type B spores per kg. In a survey done in the Washington, D.C. area (Kautter et al., 1978), all 1078 packages of mushrooms were free of botulinum toxin at the end of their normal shelf life, but the spore numbers were not determined. Notermans et al. (1989) did not detect *C. botulinum* in a survey of mushrooms produced in the Netherlands, but estimated the incidence at between 0.8 and 1.6 spores/kg in the same fashion as he estimated the numbers in vacuum-packed, cooked potatoes.

V. INCIDENCE IN DAIRY PRODUCTS

There has been very little work done to determine the level of contamination of dairy products with *C. botulinum*. Taclindo et al. (1967) did not detect *C. botulinum* in any samples of vacuum-packed cheese analyzed. Similarly, Insalata et al. (1969) found no evidence of *C. botulinum* in any of 40 samples of Edam and cheddar, or 10 samples of cheese spreads examined. Recent work (see Chapter 10) indicates that the contamination level of raw milk is less than 1 spore/liter.

VI. INCIDENCE IN HONEY AND INFANT FOODS

The presence of botulinum spores in most foods is of little concern unless the spores are able to germinate, grow, and produce toxin. However, spores in honey and other infant foods pose a greater hazard because, in some infants, the spores are able to colonize the infant's intestines, produce toxin, and cause infant botulism. One of the earliest studies of infant botulism (Midura and Arnon, 1976) examined numerous food items consumed by three patients and found one sample of honey contaminated with *C. botulinum*. By 1984, honey had been implicated as the likely source of botulinum spores in 20 cases of infant botulism in California (Anonymous, 1984).

As a result, honey has been extensively studied for the presence of *C. botulinum* spores (Table 5). A survey of honey from 32 U.S. states estimated the maximum contamination level as 8–28 spores per kg (Sugiyama et al., 1978; H. Sugiyama, personal communication, 1982). Types A and B were both detected. While Midura et al. (1979) detected few positive samples in honey purchased at random in California, the estimated contamination in honey incriminated in infant botulism was up to 8×10^4 spores per kg. Samples of honey from Illinois, New Jersey, Pennsylvania, South Dakota, and the Washington, DC, area were also contaminated (Huhtanen et al., 1981; Kautter et al., 1982). However, a survey of 48 samples of honey purchased in New York City did not find any with *C. botulinum* (Guilfoyle and Yager, 1983). Both Hartgen (1980) and Flemmig and Stojanowic (1980) found 301 commercial samples of honey in West Germany free of contamination. Similarly, no evidence of *C. botulinum* could be found in samples of honey produced in Italy (Aureli et al., 1983; Pastoni et al., 1986), Norway (Hetland, 1986), or the United Kingdom (Berry et al., 1987). A survey in Canada (Hauschild et al., 1988) failed to detect *C. botulinum* in samples of honey other than in one lot incriminated in infant botulism. The concentration of type A spores in the incriminated lot was 10^3–10^4 per kg. A recent

Table 5 Incidence of *C. botulinum* Spores in Honey

Country	Sample size (g)	% positive samples	MPN per kg	Types identified	Assoc. with illness[a]	Ref.[b]
United States	25–75	7	8–28	A, B	no	1,2
	30	1	0.4	A, B	no	3
	30	100	8×10^{4}[c]	A, B	yes	3
	30	7	2.5	A, B	no	4
	25	2	0.8	A	no	5
	25	0	<0.8		no	6
Germany	1	0	<5		no	7
	25	0	<0.4		no	8
Italy	10	0	<0.9		no	9
	30	0	<0.5		no	10
Norway	25	0	<0.3		no	11
United Kingdom	20	0	<0.4		no	12
Canada	75	0	<0.2		no	13
	75	100	8×10^{3}[c]	A	yes	13
Japan	20	14	7	A, C	no	14
China	20	15	8	A, B, C	no	14
Hungary	20	17	9	A, C	no	14
Spain	20	50	35	A	no	14
Mexico	20	50	35	C	no	14
Argentina	20	9	5	A	no	14
	30	50	23	C/D	no	10

[a]Honey was implicated in a case of infant botulism.
[b] 1) Sugiyama et al., 1978; 2) Sugiyama, personal communication, 1982; 3) Midura et al., 1979; 4) Huhtanen et al., 1981; 5) Kautter et al., 1982; 6) Guilfoyle and Yager, 1983; 7) Hartgen, 1980; 8) Flemmig and Stojanowic, 1980; 9) Aureli et al., 1983; 10) Pastoni et al., 1986; 11) Hetland, 1986; 12) Berry et al., 1987; 13) Hauschild et al., 1988; 14) Sakaguchi, 1990.
[c]MPN calculated by authors.

survey in Japan (Sakaguchi, 1989) showed that samples of honey from Japan, China, Hungary, Spain, Mexico, and Argentina were all contaminated with *C. botulinum*. Type A spores were detected most frequently, but types B and C were also found. Type C or D spores have also been detected in honey from Argentina (Pastoni et al., 1986).

This work suggests that in random samples of honey the botulinum spore level is low, in the order of 1–10 per kg. However, in honey samples associated with infant botulism the level is much higher, approximately 10^4 spores per kg. These relatively high numbers and the presence of only one serotype in honey associated with infant botulism suggest that the contamination of this honey is not a haphazard event, but the result of an external source, or multiplication of *C. botulinum* in the beehive (Hauschild et al., 1988).

A variety of other infant foods have also been examined. Kautter et al. (1982) discovered *C. botulinum* in 13 of 1001 samples of corn syrup and determined an MPN of approximately 50 spores/kg. All samples of cereals, formula, dry milk, whole milk, canned fruit, canned fruit juice, sugar, and carrots were negative. Similarly, Guilfoyle and Yager (1983) found no evidence of *C. botulinum* in samples of dry cereal, nonfat dry milk, evaporated milk, canned formula, and canned baby food purchased in New York City. Hauschild et al. (1988) examined oat, barley, rice, and mixed dry infant cereals, as well as various syrups, and found one sample of rice cereal with type B *C. botulinum*, but the contamination level was likely very low because only one of six replicates was positive.

Exposure of infants to botulinum spores via foods other than honey seems to be minimal. As detailed above, corn syrup and rice cereal have been shown to occasionally contain spores. However, the spore levels have been low, and they are unlikely to multiply during manufacture and storage.

VII. INCIDENCE IN OTHER FOODS

Taclindo et al. (1967) examined a variety of foods sold from refrigerated cases that require little or no heating prior to consumption, with an emphasis on vacuum-packed products. These products are considered vulnerable with respect to *C. botulinum*. Foods examined included luncheon meats, sausages, cheese, shredded vegetable mix, unshucked oysters, smoked salmon, smoked herring, barbecued cod, kippers, anchovy paste, and a variety of oriental foods. Of 113 samples, one sample of luncheon meat was positive with type B spores, and one sample of unshucked oysters was positive with type E.

In a similar study, Insalata et al. (1969) examined the incidence of *C. botulinum* in convenience foods. One hundred samples in each of the following categories were studied: boil-in-bag foods, vacuum-packed foods, pressurized foods, and dehydrated and freeze-dried foods. Processed vegetables, meats, seafood, cheeses, and other foods were included in the survey. Only one sample of vacuum-packed frankfurters was found positive with type B spores. These studies show a very low incidence of *C. botulinum* in these products, but the few positive samples detected emphasize the requirement for proper refrigeration and/or use of other controls to prevent a possible risk to the consumer.

VIII. SUMMARY

The results of food surveys indicate that fish have the highest level of contamination with *C. botulinum*. This is not unexpected, because surveys of aquatic environments often show fairly high levels of contamination (see Chapter 2). This is also in agreement with data on foodborne botulism, which show a large number of outbreaks associated with fish (see Chapter 4). The level of contamination of meats is generally low, with the incidence lower in North America than in Europe, again, in agreement with foodborne disease data. While fewer surveys have been done of fruits and vegetables, these products are occasionally contaminated with *C. botulinum*. More evidence for the contamination of vegetables with *C. botulinum* is provided by the data on foodborne botulism attributable to vegetables (see Chapter 4). Both survey results and foodborne botulism data support the conclusion that the level of contamination of dairy products with *C. botulinum* is very low. While some batches of honey, particularly those implicated in cases of infant botulism, appear to be contaminated with fairly high numbers of *C. botulinum*, the contamination of honey in general is low. Other infant foods do not seem to pose a risk. The level of contamination of convenience foods and so-called vulnerable foods with *C. botulinum* appears to be very low. This is reassuring because the market for these foods is increasing rapidly.

REFERENCES

Abrahamsson, K. (1967). Occurrence of type E *Cl. botulinum* in smoked eel. In *Botulism 1966*, Proceedings of the Fifth International Symposium on Food Microbiology, Moscow, July 1966 (M. Ingram and T. A. Roberts, eds.), Chapman and Hall Ltd., London, p. 73.

Abrahamsson, K., and Riemann, H. (1971). Prevalence of *Clostridium botulinum* in semipreserved meat products. *Appl. Microbiol. 21*: 543.

Anonymous. (1984). Honey, corn syrups and infant botulism. *Calif. Morbid. Weekly Rep. 14*.

Aureli, P., Ferrini, A. M., and Negri, S. (1983). Ricerca delle spore di *Cl. botulinum* nel miele. *Riv. Soc. Ital. Sci. Alimen. 12*: 457.

Baker, D. A., Genigeorgis, C., and Garcia, G. (1990). Prevalence of *Clostridium botulinum* in seafood and significance of multiple incubation temperatures for determination of its presence and type in fresh retail fish. *J. Food Prot. 53*: 668.

Baumgart, J. (1987). Presence and growth of *Clostridium botulinum* in vacuum-packed raw and cooked potatoes. *Chem. Mikrobiol. Technol. Lebensm. 11*: 74.

Berry, P. R., Gilbert, R. J., Oliver, R. W. A. and Gibson, A. A. M. (1987). Some preliminary studies on low incidence of infant botulism in the United Kingdom. *J. Clin. Pathol. 40*: 121.

Bott, T. L., Defner, J. S., McCoy, E., and Foster, E. M. (1966). *Clostridium botulinum* type E in fish from the Great Lakes. *J. Bacteriol. 91*: 919.

Cann, D. C., Wilson, B. B., Shewan, J. M., and Hobbs, G. (1966). Incidence of *Clostridium botulinum* type E in fish products in the United Kingdom. *Nature 211*: 205.

de Làgarde, E. A. (1974). *Boletin informativo del Centro Paramericano de Zoonosis*, Vol. 1. Centro Paramericano de Zoonosis, Buenos Aires, Argentina.

Fantasia, L. D., and Duran, A. P. (1969). Incidence of *Clostridium botulinum* in commercially and laboratory dressed whitefish chubs. *Food Technol. 23*: 85.

Flemmig, R., and Stojanowic, V. (1980). Untersuchungen von Bienenhonig auf *Clostridium botulinum* Sporen. *Arch. Lebensmittelhyg. 31*: 179.

Garcia, G., and Genigeorgis, C. (1987). Quantitative evaluation of *Clostridium botulinum* nonproteolytic types B, E, and F growth risk in fresh salmon tissue homogenates stored under modified atmospheres. *J. Food Prot. 50*: 390.

Goldblith, S. A., and Nickerson, J. T. R. (1965). The effect of gamma rays on haddock and clams inoculated with *Clostridium botulinum* type E. U.S. Atomic Energy Commission Report No. 22, Massachusetts Institute of Technology. *MIT-3325*.

Greenberg, R. A., Tompkin, R. B., Bladel, B. O., Kittaka, R. S., and Anellis, A. (1966). Incidence of mesophilic *Clostridium* spores in raw pork, beef, and chicken in processing plants in the United States and Canada. *Appl. Microbiol. 14*: 789.

Guilfoyle, D. E., and Yager, J. F. (1983). Survey of infant foods for *Clostridium botulinum* spores. *J. Assoc. Off. Anal. Chem. 66*: 1302.

Halvorson, H. O., and Ziegler, N. R. (1933). Application of statistics to problems in bacteriology. *J. Bacteriol. 25*: 101.

Haq, I., and Sakaguchi, G. (1980). Prevalence of *Clostridium botulinum* in fishes from markets in Osaka. *Japan. J. Med. Sci. Biol. 33*: 1.

Hartgen, V. H. (1980). Untersuchungen von honigproben auf botulinustoxin. *Arch. Lebensmittelhyg. 31*: 178.

Hauschild, A. H. W. (1989). *Clostridium botulinum*. In *Foodborne Bacterial Pathogens* (M. P. Doyle, ed.), Marcel Dekker, New York, p. 111.

Hauschild, A. H. W., Aris, B. J., and Hilsheimer, R. (1975). *Clostridium botulinum* in marinated products. *Can. Inst. Food Sci. Technol. J. 8*: 84.

Hauschild, A. H. W., and Hilsheimer, R. (1980). Incidence of *Clostridium botulinum* in commercial bacon. *J. Food Prot. 43*: 564.

Hauschild, A. H. W., and Hilsheimer, R. (1983). Prevalence of *Clostridium botulinum* in commercial liver sausage. *J. Food Prot. 46*: 242.

Hauschild, A. H. W., Hilsheimer, R., Weiss, K. F., and Burke, R. B. (1988). *Clostridium botulinum* in honey, syrups and dry infant cereals. *J. Food Prot. 51*: 892.

Hayes, S., Craig, J. M., and Pilcher, K. S. (1970). The detection of *Clostridium botulinum* type E in smoked fish products in the Pacific Northwest. *Can. J. Microbiol. 16*: 207.

Hetland, A. (1986). Absence of *Clostridium botulinum* spores in Norwegian-produced honey. *Norsk. Veterinaertidsskrift 98*: 725.

Hobbs, G., Cann, D. C., Wilson, B. B., and Shewan, J. M. (1965). The incidence of organisms of the genus *Clostridium* in vacuum packed fish in the United Kingdom. *J. Appl. Bacteriol. 28*: 265.

Houghtby, G. A., and Kaysner, C. A. (1969). Incidence of *Clostridium botulinum* type E in Alaskan salmon. *Appl. Microbiol. 18*: 950.

Huhtanen, C. N., Knox, D., and Shimanuki, H. (1981). Incidence and origin of *Clostridium botulinum* spores in honey. *J. Food Prot. 44*: 812.

Insalata, N. F., Fredericks, G. J., Berman, J. H., and Borker, E. (1967). *Clostridium botulinum* type E in frozen vacuum-packed fish. *Food Technol. 21*: 296.

Insalata, N. F., Witzeman, S. F., Fredericks, G. J., and Sunga, F. C. A. (1969). Incidence study of spores of *Clostridium botulinum* in convenience foods. *Appl. Microbiol. 17*: 542.

Johannsen, A. (1965). *Clostridium botulinum* type E. 1. Smoked fish products. *Sartryck Sv. Vet. 5*: 1.

Kautter, D. A., Lilly, T., and Lynt, R. (1978). Evaluation of the botulism hazard in fresh mushrooms wrapped in commercial polyvinylchloride film. *J. Food Prot. 41*: 120.

Kautter, D. A., Lilly, T., Solomon, H. M., and Lynt, R. K. (1982). *Clostridium botulinum* spores in infant foods: A survey. *J. Food Prot. 45*: 1028.

Klarman, D. (1989). *Clostridium botulinum* in faecal samples of cattle and swine and samples of raw material and pulverized dehydrated meat of different rendering plants. *Berl. Munch. Tierarztl. Wschr. 102*: 84.

Lerke, P. (1973). Evaluation of potential risk of botulism from seafood cocktails. *Appl. Microbiol. 25*: 807.

Lindroth, S. E., and Genigeorgis, C. A. (1986). Probability of growth and toxin production by nonproteolytic *Clostridium botulinum* in rockfish stored under modified atmospheres. *Int. J. Food Microbiol. 3*: 167.

Lilly, T., and Kautter, D. A. (1990). Outgrowth of naturally occurring *Clostridium botulinum* in vacuum-packaged fresh fish. *J. Assoc. Off. Anal. Chem. 73*: 211.

Meyer, K. F., and Dubovsky, B. J. (1922a). The distribution of the spores of *B. botulinus* in California. II. *J. Infect. Dis. 31*: 541.

Meyer, K. F., and Dubovsky, B. J. (1922b). The distribution of the spores of *B. botulinus* in the United States. IV. *J. Infect. Dis. 31*: 559.

Meyer, K. F., and Eddie, B. (1951). Perspectives concerning botulism. *Z. Hyg. Infektionskr. 133*: 255.

Midura, T. F., and Arnon, S. S. (1976). Infant botulism: Identification of *Clostridium botulinum* and its toxins in faeces. *Lancet 2*: 934.

Midura, T. F., Snowden, S., Wood, R. M., and Arnon, S. S. (1979). Isolation of *Clostridium botulinum* from honey. *J. Clin. Microbiol. 9*: 282.

Narayan, K. G. (1967). Incidence of the food poisoning clostridia in meat animals. *Zentralbl. Bakteriol. I. Orig. 204*: 265.

Nielsen, S. F., and Pedersen, H. O. (1967). Studies on the occurrence and germination of *Cl. botulinum* in smoked salmon. In *Botulism 1966*, Proceedings of the 5th International Symposium on Food Microbiology (M. Ingram and T. A. Roberts, eds.), Chapman and Hall Ltd., London, p. 66.

Notermans, S., Dufrenne, J., and Gerrits, J. P. G. (1989). Natural occurrence of *Clostridium botulinum* on fresh mushrooms (*Agaricus bisporus*). *J. Food Prot.52*: 733.

Pace, P. J., Krumbiegel, E. R., Angelotti, R., and Wisniewski, H. J. (1967). Demonstration and isolation of *Clostridium botulinum* types from whitefish chubs collected at fish smoking plants of the Milwaukee area. *Appl. Microbiol. 15*: 877.

Pastoni, F., Roggi, C., Caccialanza, G., and Gandini, C. (1986). Isolation of *Clostridium* species from honey. *Proceedings of World Congress on Foodborne Infections and Intoxications*, West Berlin, pp. 506–510.

Post, L. S., Lee, D. A., Solberg, M., Furgang, D., Specchio, J., and Graham, C. (1985). Development of botulinal toxin and sensory deterioration during storage of vacuum and modified atmosphere packaged fish fillets. *J. Food Sci. 50*: 990.

Prokhorovich, L. E., Saltykova, L. A., Malkina, Z. M., Gritsko, L. P., and Shenderouskaya, L. M. (1975). Botulism contamination of canned apricots and red peppers. *Konservnaya i Ovoshchesushil Naya Promyshlennost. 6*: 38.

Quarto, M., Armenise, E., and Attimonelli, D. (1983). Research on the presence of *Clostridium botulinum* in raw and home-prepared vegetables. *Igiene Mod. 80*: 384.

Riemann, H. (1963). Safe heat processing of canned cured meats with regard to bacterial spores. *Food Technol. 17*: 39.

Roberts, T. A., and Smart, J. L. (1976). The occurrence and growth of *Clostridium* spp. in vacuum-packed bacon with particular reference to *Cl. perfringens* (*welchii*) and *Cl. botulinum*. *J. Food Technol. 11*: 229.

Roberts, T. A., and Smart, J. L. (1977). The occurrence of clostridia, particularly *Clostridium botulinum*, in bacon and pork. In *Spore Research 1976* (A. N. Barker, J. Wolf, D. J. Ellar, G. J. Dring, and G. W. Gould, eds.) Academic Press, New York, p. 911.

Rouhbakhsh-Khaleghdoust, A. (1975). The incidence of *Clostridium botulinum* type E in fish and bottom deposits in the Caspian Sea coastal waters. *Pahlavi Med. J. 6*: 550.

Rozanova, L. I., Zemlyakov, V. L., and Mazokhina, N. N. (1972). Factors affecting *Clostridium botulinum* contamination of vegetables intended for preservation and of materials used. *Gigiena i Sanitariya 37*: 102.

Sakaguchi, G. (1989). University of Osaka Prefecture, Osaka, Japan. Personal communication.

Sebald, M., Jouglard, J., and Gilles, G. (1974). Botulisme humain de type B après ingestion de fromage. *Ann. Microbiol. (Inst. Pasteur) 125A*: 349.

Sugiyama, H., Mills, D. C., and Kuo, L-J. C. (1978). Number of *Clostridium botulinum* spores in honey. *J. Food Prot. 41*: 848.

Suhadi, F., Thayib, S. S., and Sumarsono, N. (1981). Distribution of *Clostridium botulinum* around fishing areas of the western part of Indonesian waters. *Appl. Environ. Microbiol. 41*: 1468.

Taclindo, C., Midura, T., Nygaard, G. S., and Bodily, H. L. (1967). Examination of prepared foods in plastic packages for *Clostridium botulinum*. *Appl. Microbiol. 15*: 426.

Thatcher, F. S., Erdman, I. E., and Pontefract, R. D. (1967). Some laboratory and regulatory aspects of the control of *Cl. botulinum* in processed foods. In *Botulism 1966 Proceedings of the Fifth International Symposium on Food Microbiology*, Moscow, July 1966 (M. Ingram and T. A. Roberts, ed.), Chapman and Hall Ltd., London, p. 511.

Vergieva, V., and Incze, K. (1979). The ecology of *Clostridium botulinum*. *Husipar 28*: 79.

4

Epidemiology of Human Foodborne Botulism

Andreas H. W. Hauschild

Health Protection Branch,
Health and Welfare Canada,
Ottawa, Ontario, Canada

I. INTRODUCTION

The aim of this chapter is to update and expand an earlier review, which included a discussion of human foodborne botulism worldwide (Hauschild, 1989). That review showed a close association between the frequency and type of botulism outbreaks and the occurrence of *Clostridium botulinum* in the environment. It also highlighted the epidemiology of botulism as a significant component in the ecology of the microorganism.

Some of the data presented in this chapter have been gleaned from publications. Others, particularly the more recent ones, were collected through personal communications, as indicated in Tables 1 and 4. These tables also list the countries from which surveillance data on botulism have been obtained. Countries with relatively frequent occurrences of botulism are shown in Table 1, along with data relating to the numbers of outbreaks and affected cases within given periods. Outbreaks are defined as incidents involving one or more botulism cases (CDC, 1979). The botulinum types and foods associated with the outbreaks are listed in Tables 2 and 3, respectively. Data from countries with infrequently recorded incidents of botulism are summarized in Table 4. Some of the information summarized in the tables is dated, either on account of communication problems or because epidemiological work on botulism was discontinued, e.g., in the Soviet Union (Prozorovsky, 1986). In many areas of the

Table 1 Recorded Foodborne Botulism

Country[a]	Period	Outbreaks	Cases Total	Cases Fatal[b]	Averages/year Outbreaks	Averages/year Cases	Ref.[c]
United States[d]	1971–89	272	597	63(11)	14	31	1,2
Alaska	1971–89	51	122	8(7)	3	6	1–3
Canada	1971–89	79	202	28(14)	4	11	4–6
Argentina	1980–89	16	36	13(36)	2	4	7–9
Poland	1984–87	1301	1791	46(3)	325	448	10,11
Czechoslovakia[e]	1979–84	17	20	0	3	3	12
Hungary	1985–89	31	57	1(2)	6	11	13
Yugoslavia[f]	1984–89	12	51		2	8	14
Germany							
East	1984–89	33	52	4(8)	6	9	15
West	1983–88	63	154	(4)[g]	10	26	16,17
Belgium	1982–89	11	25	1(4)	1	3	18,19
France	1978–89	175	304	7(2)	15	25	20
Italy[h]	1979–87		310			34	21
Spain	1969–88	63	198	11(6)	3	10	22,23
Portugal	1970–89	24	80	0	1	4	24,25
Denmark[i]	1984–89	11	16	2(12)	2	3	26,27
Norway	1961–90	19	42	3(7)	<1	1	28
U.S.S.R.	1958–64	95	328	95(29)	14	47	29
Iran	1972–74		314	(11)[j]		105	30
China	1958–83	986	4377	548(13)	38	168	31
Japan	1951–87	97	479	110(23)	3	13	32

[a]Or state.

[b]In brackets, as percentage of total cases.

[c]1) CDC, 1979; 2) Hatheway, 1990; 3) Wainwright et al., 1988; 4) Hauschild and Gauvreau, 1985; 5) HWC, 1987b, 1988–1990; 6) Smith, 1990; 7) Ciccarelli and Giménez, 1981; 8) Fernandez et al., 1986; 9) Giménez, 1990; 10) Anusz, 1986, 1987, 1988, 1989; 11) Szczawinsky, 1989; 12) Helcl, 1985; 13) Domján Kóvacz, 1990; 14) Kalember Radosavljevic, 1990; 15) Dittmann and Rasch, 1990; 16) Pöhn and Grossmann, 1987; 17) Teufel, 1990; 18) André, 1990; 19) Marchal et al., 1985; 20) Sebald, 1990; 21) Aureli, 1989; 22) MSC, 1984, 1986, 1987, 1988, 1989; 23) Tello, 1990; 24) Lecour, 1990; 25) Lecour et al., 1988, 1989; 26) Huss, 1990; 27) Jorgensen, 1990; 28) Skulberg, 1990; 29) Matveev et al., 1967; 30) Pourtaghva et al., 1975; 31) Ying and Shuyan, 1986; 32) Itoh, 1989.

[d]Alaska figures included.

[e]Slovakia not included.

[f]Slovenia and Macedonia not included.

[g]Based on 1986–88 data.

[h]Data from some southern provinces not included.

[i]Greenland included.

[j]Based on 18 fatalities among 170 investigated cases.

Table 2 *C. botulinum* Types Involved in Outbreaks[a]

Country	Outbreaks with type identified	Type (%)			
		A	B	E	Other[b]
United States	252	61	21	17	0.4(F)
Canada	76	4	8	88	0
Argentina	13	77	8	0	15(A + F; AF)
Poland	830[c]	3	94	3	0
Czechoslovakia	6	17	83	0	0
Hungary	31	0	100	0	0
Germany			>90[d]		
Belgium	11	0	55	9	36(B+C)
France	171	0	97	2	0.6(AB)
Italy[e]	15	20	60	7	13(A+B)
Spain	36	0	92	3	6(A+B)
Portugal	18	0	100	0	0
Denmark	11	0	0	100	0
Norway	19	0	47	47	5(F)[f]
U.S.S.R.	45	33	38	29	0
Iran	63	0	3	97	0
China	733	93	5	1	0.4(A+B)
Japan	97	2	2	96	0

[a]Addendum to Table 1.
[b]In parentheses, type(s) identified.
[c]No. of cases.
[d]Estimate (Lücke, 1983).
[e]Data for 1979–83 only.
[f]Type F suspected.

world, particularly where botulism occurs infrequently, epidemiological data on the disease are nonexistent.

There is no doubt that the outbreak summaries presented in this chapter are far from complete, but probably less so than summaries of most other foodborne illnesses. Estimates for the ratio of reported to actual foodborne cases are in the ranges of 1:100 (Edel et al., 1977; Hauschild and Bryan, 1980), 1:350 (Todd, 1989), and 1:>1000 (Archer and Kvenberg, 1985). Because botulism is generally so severe, it is more likely to be detected than some of the milder forms of food poisoning. Nevertheless, unrecognized and misdiagnosed cases of botulism are common. In a restaurant-related outbreak in Vancouver, Canada, the initial diagnoses for 28 patients included psychiatric illness, viral syndrome, laryngeal trauma, overexertion, and a variety of other maladies. Eleven of the

Table 3 Food Involved in Outbreaks[a]

Country	Outbreaks with food identified	Food (%) Meats	Fish	Fruits and vegetables	Other[b]	Food source (%) Home	Commerce
United States	222	16	17	59	9	92	8
Canada	75	72	20	8	0	96	4
Argentina	14	29	21	36	14	79	21
Poland	1500[c]	83	12	5	0	75	25
Czechoslovakia	14	72	7	14	7	100	0
Hungary	28	89	0	4	7[d]	100	0
Yugoslavia	8	100	0	0	0	100	0
Germany							
East	31	52	26	19	3	73	27
West	55	78	13	9	0	100	0
Belgium	8	75	12	12	0	62	38
France	123	89	3	6	2	88	12
Italy[e]	13	8	8	77	8		
Spain	48	38	2	60	0	90	10
Portugal	23	91	9	0	0	100	0
Denmark	10	100	0	0	0	100	0
Norway	19	16	84	0	0	100	0
U.S.S.R.	83	17	67	16	0	97	3
Iran	63	3	97	0	0		
China	958	10	0	86	4		
Japan	95	0	99	1	0	98	2

[a]Addendum to Table 1.
[b]Including mixed vehicles.
[c]No. of cases.
[d]Mixtures of meats and vegetables.
[e]Data for 1979–83 only.

cases remained misdiagnosed for over a month and were recognized only when a second group of cases could be linked to the same vehicle (St. Louis et al., 1988). Misdiagnoses until after death, most commonly as Guillain-Barré syndrome, chemical intoxication, stroke, or staphylococcal food poisoning, have often been reported (Badhey et al., 1986; CDC, 1979; Horwitz et al., 1975; Hughes et al., 1981). Also, investigating agencies often fail to transmit cases of botulism to reporting agencies. We must bear in mind, therefore, that the foregoing summaries of foodborne botulism likely present only a fraction of the actual outbreaks and cases.

Table 4 Countries with Few or No Recorded Outbreaks

Country	Period	Outbreaks	Cases		Types identified[a]	Reference[b]
			Total	Fatal		
Mexico	1975–89	2	4	3	A(1)	1
Guatemala	1961–67	6	18	2		2
Venezuela	1980–89	0				3
Peru	to 1988	1	16	2	B	4
Brazil	1958–90	4	19	9	A(4)	5,6
Chile	to 1989	1	7	4		7,8
Austria	1980–89	0				9
Switzerland	1973–89	5	57	0	A(1),B(2)	10–12
Bulgaria	1987–88		6			13
Greece	1975–89	0				14
Netherlands	1977–89	2	3	0	B(2)	15,16
Sweden	1969–89	6	6	0	A(1),E(3)	17,18
Finland	1975–87	1	2	0		13,19
Iceland	1981–89	2	7	0	B(2)	20
Ireland	1981–84	0				19
Britain	1955–89	3	33	3	A(1),[c]B(1),E(1)[c]	21–23
Israel	1948–89	1[c]	6	1	E	24,25
Taiwan	1986–90	5	17	2	A(2),B(1)	26,27
India	1970–90	0				28
Nigeria	to 1989	0				29
Kenya	to 1979	2	17	11	A(1)	30–32
Madagascar	to 1985	2	63	33	E(2)	33
Australia	1942–89	5	53	9	A(1)	34
New Zealand	to 1989	1	2	1	A	35

[a]In parentheses, no. of type identifications.
[b]1) Valdespino-Gómez, 1990; 2) Halbinger, 1973; 3) Garcia, 1989; 4) Quevedo, 1988; 5) Ferreira et al., 1987; 6) Serrano, 1990; 7) Chamorro et al., 1974; 8) Marambio, 1989; 9) Halbich, 1989; 10) Bernasconi et al., 1988; 11) Colebatch et al., 1989; 12) Schwab, 1990; 13) Teufel, 1990; 14) Tselikas, 1989; 15) Notermans, 1990; 16) Postma et al., 1989; 17) Hermanson, 1985; 18) Danielsson-Tham, 1990; 19) WHO, 1983, 1984, 1990; 20) Georgsson, 1990; 21) Ball et al., 1979; 22) Critchley et al., 1989; 23) O'Mahony et al., 1990; 24) Slater et al., 1989; 25) Telzak et al., 1989; 26) Chou, 1989; 27) DHT, 1990; 28) Narayan, 1990; 29) Rotimi, 1985; 30) Smith et al., 1979; 31) Nightingale and Ayim, 1980; 32) Ajode, 1990; 33) Vicens et al., 1985; 34) Murrell, 1990; 35) Till, 1990.
[c]Incriminated food manufactured in other countries.

Most of the reported botulism outbreaks have been linked to the responsible botulinum type (Table 2), but there are exceptions. The German records only list a few outbreaks with an identified type. The figure of >90% of outbreaks associated with type B (Table 2) merely presents an estimate (Lücke, 1983).

Most incidents could also be linked to the responsible food (Table 3). The vehicles involved have been divided into three groups: meats, fish, and fruits and vegetables. In the last category the commonly incriminated vehicle was a vegetable, whereas involvement of a fruit was the exception.

II. BOTULISM IN NORTH AMERICA

A. United States

The CDC records in Atlanta, GA (Hatheway, 1990), list 272 botulism outbreaks and 597 cases in the period of 1971–1989 (Table 1). Sixty-three (11%) of the cases were fatal. Botulism in Alaska accounted for approximately 20% of the U.S. total, with a death rate of 7%.

Table 2 shows that most of the outbreaks were caused by type A, followed by types B and E at similar frequencies. Type F was identified in a single incident in California. A different picture emerges when the figures for Alaska and the other 49 states are separated. Of a total of 43 type E outbreaks, 38 occurred in Alaska where type E is the predominant environmental type (Table 5). The remaining five type E outbreaks accounted for only 2.5% of the type-identified incidents in continental United States. Another pattern emerges when the epidemiological data for the eastern and western United States are separated, roughly along the 95th meridian (Table 5). Whereas causative type A and B

Table 5 Variation in *C. botulinum* Types as the Cause of Botulism Between Regions of the United States, 1971–89

Region	Outbreaks with type identified	Type (%)			
		A	B	E	F
Continental United States[a]	201	74	23	2	0.5
Alaska	49	10	12	78	0
Western United States[b]	133	88	11	1	1
Eastern United States[b]	68	46	49	6	0

[a]Two outbreaks in Hawaii and Puerto Rico not included.
[b]Divided roughly along 95° longitude, i.e., the western borders of Minnesota, Iowa, Missouri, Arkansas, and Louisiana.

strains were nearly balanced in the east, type A was responsible for eight times the number of outbreaks as type B in the west. This ratio is consistent with the predominance of type A spores in the soil of the western United States (Smith, 1978; Smith and Sugiyama, 1988).

Most of the implicated foods in the United States were fruits and vegetables (Table 3). Out of 136 outbreaks in this category, 132 were linked to vegetables and only 4 to fruits (blackberries, figs, apple sauce, and strawberries). The majority of the products had been home-preserved.

The recorded outbreaks from fruits and vegetables all occurred in continental United States and accounted for 75% of the outbreaks in this region (Table 6). In contrast, only fish, fish products (salmon eggs), and meats were responsible for outbreaks in Alaska, despite the common practice of vegetable home canning in that state (Wainwright et al., 1988). A single outbreak was linked to a combination of fish and whale meat.

Of 39 outbreaks from fish reported in the United States, 25 occurred in Alaska alone. Similarly, 22 out of 37 outbreaks from meats were reported from Alaska. Single outbreaks, both involving fish, were reported from Hawaii and Puerto Rico. The responsible types were A and B, respectively. The Hawaiian outbreak was the first recorded since statehood in 1959.

For 44 of the Alaskan outbreaks from fish and marine mammals, the causative type could be ascertained; in 36 of these (82%) the type involved was E (Table 7). On the other hand, the few outbreaks from land mammals were associated only with types A and B.

In their review of foodborne botulism in Alaska from 1947 to 1985, Wainwright et al. (1988) reported that 62% of the peccant foods had been fermented; 87% of the foods had been consumed raw, and 13% had been partially cooked. All the persons affected were native Alaskans, mostly Inuits. Foodborne botulism in continental United States shows no recognizable seasonality,

Table 6 Foods Involved in Outbreaks of Botulism in Mainland United States and Alaska, 1971–89

Region[a]	Outbreaks with food identified	Food (%)			
		Meats	Fish	Fruits and vegetables	Other[b]
Continental United States	182	8	7	75	10
Alaska	48	46	52	0	2

[a]Two outbreaks (both from fish) in Hawaii and Puerto Rico not included.
[b]Including mixed vehicles.

Table 7 Association of Incriminated Foods with *C. botulinum* Types in Alaska, 1971–89

Food	Type			
	A	B	E	ND[a]
Fish & fish eggs	3	3	18	1
Sea mammals[b]	1	0	18	0
Mix (fish & whale)	0	1	0	0
Land mammals[c]	1	2	0	0
ND[a]	0	0	2	1
Total	5	6	38	2

[a]ND, not determined.
[b]Seal, whale, and walrus.
[c]Beaver.

whereas outbreaks in Alaska occurred in seasonal clusters; 77% (33/43) of the outbreaks recorded from 1976 to 1989 happened in the 5-month period from June to October.

Fatality rates from 1951 to 1989 show a steady decline, from 34% in the 1951–60 period to the current 7% (Table 8). Improved medical services and a more general availability of antisera undoubtedly contributed to the decrease. Recognition of more nonfatal cases of botulism may be another factor in the declining death rate.

Deaths attributed to types A, B, and E are listed in Table 9. Type E showed the most dramatic decrease in death rate, particularly in the early 1970s. This decrease was attributed primarily to the general distribution of type E antiserum, which first became commercially available in 1964 (Hauschild and Gauvreau, 1985). Table 9 also confirms the relatively low numbers of fatal cases in botulism of type B (St. Louis et al., 1988).

Table 8 Fatality Rate of Botulism Cases in the United States, 1951–89

Period	Fatal cases/Total cases	Death rate (%)
1951–60	74/220	34
1961–70	42/181	23
1971–80	44/339	13
1981–89	19/258	7

Table 9 Fatality Rates of Botulism Cases of Types A, B, and E in the United States, 1961–89

	Fatal cases/Total cases			Death rate (%)		
Period	A	B	E	A	B	E
1961–70	11/38	4/23	12/32	29	17	38
1971–80	25/165	7/113	2/29	15	6	7
1981–89	15/137	0/33	4/68	11	0	6
Total	51/340	11/169	18/129	15	7	14

B. Canada

Of the 79 botulism outbreaks recorded in the 1971–89 period (Table 1), 69 (87%) occurred in northern native Canadian communities. As in Alaska, the prevalent type associated with the outbreaks was E.

The foods involved in the outbreaks are listed in Table 10. Half of these were raw or parboiled meats from marine mammals, i.e., seals, whales, and walrus, all affecting Inuits. Toxin production in these outbreaks was attributed to the widespread Inuit habit of keeping meats at ambient temperature for some time, often until it has decomposed to a desired flavor and texture (Hauschild and Gauvreau, 1985). Holding temperatures as low as 4–5°C may still allow *C. botulinum* type E to produce toxin.

The fermented Inuit products are known as *urraq* (seal flippers in oil), *muktuk* (meat of the white whale, *beluga*, with blubber and skin), *micerak* (mostly fat from seal, whale, or walrus), and *ogsuk* (mostly seal oil). These products all lack readily fermentable carbohydrates and therefore do not undergo a sufficiently rapid pH reduction to prevent toxigenesis. Lack of carbohydrates is also the reason for outbreaks from fermented salmon eggs affecting west coast Indians (Table 10)(Dolman, 1964). In a survey of freshly fermented salmon eggs (Hauschild and Gauvreau, 1985) the average pH was close to 6.0. Only the fermented eggs of the chum salmon (*Oncorynchus keta*) had a pH range of 4.2–5.1, which would inhibit or significantly delay development of *C. botulinum* type E. Only two outbreaks from salmon eggs have been reported during the 1981–89 period. This decrease may have resulted from repeated efforts to discourage the hazardous fermentation of salmon eggs (Bowmer and Wilkinson, 1976).

Non-native Canadians were involved in outbreaks from smoked fish, from preserved vegetables, and from pork. Only three incidents were linked to home-canned vegetables, which is in sharp contrast to the frequent involvement of these products in outbreaks in the United States. The incriminated commercial

Table 10 Foods, Botulinum Types, and Ethnic Groups Associated with Foodborne Botulism in Canada, 1971–89

Food	No. of outbreaks	No. of cases	No. of deaths	Ethnic group	Toxin type[a]
Meat from marine mammals					
Raw or parboiled	39	86	16	Inuit	E(37)
Fermented	10	16	3	Inuit	E(10)
Caribou meat, raw or parboiled	3	5	1	Inuit	E(2)
Ptarmigan,[b] raw or parboiled	1	1	0	Inuit	E(1)
Fish					
Raw, parboiled or dried	4	9	2	Inuit	B(1),E(3)
Smoked	3	5	1	Non-native	B(1)E(2)
Salmon eggs, fermented	8	17	4	Amerindians	E(8)
Preserved vegetables or pork					
Home-preserved	4	4	0	Non-native	A(2),B(2)
Commercial	3	44	0	Non-native	A(1),B(2)
Unknown	4	5	1	Inuit	E(4)

[a]In parentheses, no. of outbreaks associated with the type.
[b]Northern grouse.

products were bottled marinated mushrooms imported from the United States (Todd et al., 1974), bottled garlic in oil, also imported from the United States but temperature-abused locally (St. Louis et al., 1988), and in-house bottled mushrooms (HWC, 1987a).

Table 10 shows a distinct correlation between causative foods and ethnic groups: raw, parboiled, fermented, or dried meats, fish and salmon eggs only affected the native population, whereas non-natives were involved in outbreaks from smoked fish and otherwise preserved (canned, salted, or refrigerated) vegetables or pork.

Table 10 also shows an association between causative foods and botulinum types. Type E was almost consistently involved in outbreaks from marine mammals, caribou, fish, and fish eggs. The two exceptions, both associated with fish, were nonproteolytic (group II) strains of type B. Outbreaks from vegetables and pork were linked to types A and B only (Table 10).

Caribou generally do not carry the type E organism (HWC, 1985). It is likely, therefore, that the incriminated meat was contaminated through contact with soil, or with meat or oil from sea mammals (Hauschild and Gauvreau, 1985).

As in the United States, the death rate has decreased significantly during recent decades, i.e., from 58% during 1951–60 to 10% in the 1981–89 period (Table 11). The greatest reduction occurred after commercial type E antitoxin first became available in 1964.

Table 11 Fatality Rates of Botulism Cases in Canada, 1971–89

Period	Fatal cases/Total cases	Death rate (%)
1951–60	19/33	58
1961–70	30/66	45
1971–80	17/94	18
1981–89	11/108	10

The seasonality of Canadian outbreaks resembles that in Alaska: the number of incidents recorded in the 6-month period from May to October was about twice as high as during the rest of the year.

C. Restaurant-Associated Botulism in North America

The North American summaries of botulism outbreaks (Table 1) include seven incidents from foods served in restaurants between 1977 and 1987. At least 177 patrons were affected, and three died (Table 12). These outbreaks have recently been reviewed by Dodds (1990).

In two incidents (1 and 7) the incriminated foods had been heat-processed in-house by faulty techniques. As a result of incident No. 7, the *Sanitation Code* of the Canadian Restaurant and Food Service Association now states: "No in-house (non-commercial) canned or bottled foods may be served in restaurants." Some states have similar regulations (Todd, 1985).

Two outbreaks were attributed to temperature abuse of food ingredients, i.e., leftover baked potatoes wrapped in aluminum foil that had been held for up to 5 days at ambient temperature before being used in potato salad (incident No. 3) and commercial chopped garlic in soybean oil, held without refrigeration for some months until used to prepare garlic butter (incident No. 6). Sugiyama et al. (1981) showed that stab or surface inoculated baked potatoes in foil produced toxin within 3–4 days at 22°C. The chopped garlic contained no preservatives or acids, and its safety rested entirely on adequate refrigeration. Solomon and Kautter (1988) demonstrated that the inoculated product allowed toxigenesis at room temperature. A second botulism outbreak from chopped garlic in oil involving three cases was recorded in New York (FDA, 1989). As a result of these incidents, garlic in oil can now be sold in North America only if, in addition to refrigeration, the product is protected by a second barrier to botulinum growth, e.g., acidification.

In one outbreak (No. 4), toxigenesis was attributed to temperature abuse of the finished product (sauteed onions). The major contributor to an anaerobic environment of the onions was probably margarine (Solomon and Kautter, 1986).

Table 12 Restaurant-Associated Botulism in North America

Incident no.	Year	State or province	Cases Total	Cases Fatal	Type	Incriminated food	Conditions responsible for toxigenesis	Ref.[a]
1	1977	Michigan	59	0	B	Canned peppers	Underprocessing	Terranova et al., 1978
2	1978	New Mexico	34	2	A	Potato salad and/or three-bean salad	Not determined	CDC, 1978 Mann et al., 1983
3	1978	Colorado	8	0	A	Potato salad	Improper storage of ingredients (baked potatoes)	Seals et al., 1981 Sugiyama et al., 1981
4	1983	Illinois	28	1	A	Sauteed onions	Temperature abuse of finished product	MacDonald et al., 1985 Solomon and Kautter, 1986
5	1984	Louisiana	1	0	A	Baked potatoes	Not determined	MacDonald et al., 1986
6	1985	British Columbia	36	0	B	Garlic butter	Improper storage of ingredients (garlic in oil)	St. Louis et al., 1988
7	1987	British Columbia	11	0	A	Bottled mushrooms	Underprocessing and/or inadequate acidification	HWC, 1987

[a]Some data updated by more recent CDC records (Hatheway, 1989).

III. BOTULISM IN CENTRAL AND SOUTH AMERICA

A. Argentina

Argentina is the only Latin American country with a substantial number of recorded botulism outbreaks (Table 1). These occurred predominantly in the provinces of Mendoza and Buenos Aires, and almost exclusively between 30° and 40° latitude south (Ciccarelli and Giménez, 1981). In the 10-year period from 1980 to 1989 (Table 1), 16 outbreaks comprising 36 cases with a mortality rate of 36% have been reported (Giménez, 1990).

One significant factor in the relatively large number of deaths is undoubtedly the frequent involvement of type A (Table 2). The previous summary listed type A as the exclusive cause of 14 type-identified outbreaks in the 1970–80 period. In 1984, a strain of type B was implicated for the first time in a botulism outbreak (from home-canned meat and peas) in Argentina. A combination of types A and F, and a single strain of subtype AF, also designated Af (Fernandez et al., 1986), were associated with outbreaks from home-pickled trout in 1981 and 1982. Subtype AF produces toxins A and F in proportions of roughly 10:1 to 100:1. It has been isolated repeatedly from South American soil samples (Ciccarelli and Giménez, 1981; Giménez and Ciccarelli, 1978) but was never implicated in an outbreak until 1982. The remaining Argentine outbreaks were all caused by type A (Table 2).

The frequent association of type A with botulism in Argentina is consistent with extensive soil surveys (Ciccarelli and Giménez, 1981) which established type A also as the prevalent form of *C. botulinum* in the Argentine environment. The most commonly implicated foods were vegetables (Table 3), primarily home-canned peppers. The responsible meats were pickled *vizcacha* (a South American rodent) and goat meats. The fish comprised pickled trout and carp (Giménez, 1990).

B. Other Central and South American Countries

Reports of botulism from other Latin American countries have been sporadic. Two outbreaks were reported from Mexico (Table 4). The first incident, recorded in Mexico City in 1975, involved a single fatal case of type A botulism from canned tuna (CPZ, 1975). The second outbreak, in the state of Sonora in 1982, was associated with canned hot peppers and affected three patients, two of them fatally (Valdespino-Gómez, 1990). Halbich (1973) reported 18 cases with two deaths in Guatemala from 1961 to 1967 (Table 4) and four cases in Venezuela in 1964 and 1965. No outbreaks have been recorded in Venezuela from 1980 to 1989. The first reported outbreak in Peru occurred in 1988 (Table 4). It affected at least 16 people and was caused by locally merchandized hot dogs contaminated with *C. botulinum* type B (Quevedo, 1988).

The few confirmed outbreaks in Brazil all involved type A. Two relatively large outbreaks, one in the southern state of Rio Grande do Sul with nine cases and seven fatalities (Pereira-Filho, 1958), and one in Minas Gerais with eight cases and two fatalities (Ferreira et al., 1987), were caused by home-preserved fish and home-canned pork, respectively. Ferreira et al. (1987) and Serrano (1990) reported single type A cases involving an unidentified food and pickled vegetables, respectively. One suspected botulism incident from chicken paté in Rio de Janeiro affecting two persons with one fatality (PAHO, 1982; Serrano, 1982) remained bacteriologically unconfirmed.

A single outbreak from canned vegetables was reported from Chile (Chamorro et al., 1974). Seven patients were affected, four of them fatally (Table 4). Another incident in Santiago in 1981 was suspected as botulism, but remained unconfirmed (Marambio, 1989).

IV. BOTULISM IN EUROPE

A. Poland

The largest numbers of botulism outbreaks and cases worldwide have been reported from Poland (Table 1). However, the meticulous recording of botulism by Anusz (1986–1989) is undoubtedly a significant factor in these large numbers.

For nearly half the reported cases (830/1791), the responsible botulinum type could be identified; 94% of these were type B (Table 2). For 1500 of the 1791 cases, the associated foods were determined: 83% were meats, followed by fish (12%) and vegetables (5%) (Table 3).

Most of the implicated foods were canned (Table 13). Of these, meats and vegetables were canned mainly in the home, whereas 70% of the incriminated canned fish were of commercial origin (Table 13). The commercial products were preserved mainly in metal cans, while home-processors usually used glass jars (Anusz, 1987).

The common involvement of *C. botulinum* type B, particularly in outbreaks from meats, raises the question about the origin of the botulinum spores that

Table 13 Canned Foods Associated with Botulism in Poland, 1984–87

Incriminated canned products[a]	No. of cases	Origin or products (%)	
		Commercial	Home
Total	921	24	76
Meats	730	15	85
Fish	132	70	30
Vegetables	59	32	68

[a]Including food preserved in glass jars.

Table 14 Processed Meats Associated with Botulism in Poland, 1984–87

Incriminated meats	No. of cases	Origin of meats (%)	
		Industrial	Home
Total	1244	20	80
Canned meat	730	15	85
Cured ham	205	3	97
Fermented sausages	168	56	44
Other	141	23	77

developed and produced toxin in the canned products. It is commonly assumed that *C. botulinum* in Poland and neighboring countries is almost exclusively of the nonproteolytic (group II) kind (Berndt, 1978; Lücke, 1983; Schulze and Funke, 1981) with D_{100} values generally below 0.1 minute. Since any heat process, even in an open-water bath, should be sufficient to completely destroy contaminating type B spores of group II, the relevant contamination must occur after the heat process, likely through defective seams or rubber seals. Viable type E spores which are equally heat-sensitive can likewise enter the can or jar only after the heat process. Type E outbreaks from canned fish with demonstrated postprocess leakage have been documented (Stersky et al., 1980). It should be stressed that the assumed predominance of nonproteolytic *C. botulinum* type B in European countries had not been satisfactorily confirmed. Published soil surveys generally do not include testing of type B strains for proteolytic activity, and very few strains implicated in European outbreaks have been isolated and characterized. Complete identification of the causative organism in each botulism incident would be of great value to the food industry and to medical services.

Canned meats, ham, and fermented sausages accounted for nearly 90% of all the incriminated meats (Table 14). Except for the fermented sausages, the meats were mainly preserved in the home (Table 14).

As in Alaska and Canada, botulism in Poland shows a clear seasonality. The highest incidence was noted in the months of May to August, with peaks in June or July. A small annual peak may also be found in the winter (January/February).

B. Czechoslovakia

As in Poland, the prevalent type involved in Czech outbreaks was B (Table 2), and the most commonly incriminated foods were meats (Table 3), i.e., meat pies, sausages, smoked meats, and canned meats. All the foods had been preserved in the home, about half of them by canning (Helcl, 1985).

C. Hungary

An average of six outbreaks and 11 cases per annum have been recorded in Hungary from 1985 to 1989 (Table 1). This constitutes a significant increase from the previously reviewed period of 1961–85 (Hauschild, 1989). The increase started in the early 1980s and may be attributable to an improved surveillance system.

The Hungarian outbreaks have several aspects in common with those reported from Poland: in each of the 31 incidents listed in Table 1, type B was implicated as the cause (Table 2), and 89% of the incriminated vehicles were processed meats (Table 3). Another 7% were mixtures of meat and vegetables. None of the foods was of commercial origin (Domján-Kovács, 1990).

D. Yugoslavia

Botulism was recorded only sporadically in Yugoslavia prior to 1984. Since then, Kalember-Radosavljevic (1990) listed a total of 51 cases involved in a minimum of 12 incidents (Table 1). As in Poland and Hungary, the type most frequently implicated was B, followed by A (Kalember-Radosavljevic, 1990). Where the causative food could be identified, it was always a meat, usually ham (Table 3).

E. Germany

A total of 96 outbreaks and 206 cases were recorded during recent 6-year periods in East and West Germany (Table 1). On a population basis, the numbers of cases in the two regions were similar, but the relative number of outbreaks was somewhat higher in the east.

The German records only occasionally identify the responsible botulinum type; e.g., of the 33 incidents in East Germany, three were associated with type B, one with type E, and 29 remained untyped (Dittmann and Rasch, 1990). Lücke (1983) estimated the involvement of type B in German outbreaks as over 90%, comparable to the frequent implication of type B in France and Poland (Table 1). In contrast to the botulinum type, the responsible foods were identified in nearly all the outbreaks, primarily as meats (Table 3). About half the incriminated meats were smoked hams.

The main difference between the two regions lies in the origin of the incriminated foods: whereas 27% of the foods were commercially processed in East Germany, not a single commercial product was implicated in West Germany (Table 3).

F. Belgium

As in other European countries discussed above, most of the Belgian outbreaks were associated with type B (Table 2) and with meats (Table 3), generally smoked hams. The only type E outbreak, involving two cases with one death, was caused by imported Alaskan salmon. Contamination of the product was attributed to can perforation by a malfunctioning can-reforming machine, and postprocess entry of *C. botulinum*.

In four of the reported outbreaks, toxins of types B and C were both found in the serum of patients (Marchal et al., 1985; WHO, 1984; André, 1990). However, in view of its low toxicity to primates (Dolman and Murakami, 1961), any type C toxin in the presence of type B toxin in the patients' serum would seem insignificant in pathogenesis. In one of the outbreaks, the serum analysis was inconsistent with the analysis of the incriminated food (vegetable paté), which contained type B toxin only (Marchal et al., 1985).

G. France

Botulism in France has been recorded meticulously for years, largely through the efforts of Dr. M. Sebald of the Pasteur Institute (Carré et al., 1987; Sebald, 1970, 1990; Sebald et al., 1980). A comparison of the outbreak summary for France from 1978 to 1989 (Table 1) with the previous summary for 1978–84 (Hauschild, 1989) would suggest some decrease in annual cases during recent years. This is confirmed by Table 15, which shows a general decline in outbreaks, total cases, and fatal cases over the 12-year period.

Type B remained the predominant cause of outbreaks (Table 2), and meats accounted for 89% of the incriminated foods (Table 3). Among the meats, cured hams were the primary vehicle, accounting for 70% of all the foods and 78% of the meats.

Because of its relatively low water activity and its slow loss of sensory quality, ham is often held well above 5°C during curing, i.e., within the growth range of the common (group II) European form of *C. botulinum* type B. Growth is generally initiated in the inner ham, and since the salt has to diffuse into the meat from the surface, there is ample time for growth and toxigenesis before the salt concentration becomes inhibitory (Lücke, 1983, Lücke et al.,

Table 15 Decreasing Numbers of Recorded Botulism Outbreaks and Cases in France, 1978–89

Period	Outbreaks	Cases	
		Total	Fatal[a]
1978–81	73	138	5 (3.6)
1982–85	51	93	2 (2.2)
1986–89	51	73	0 (0)

[a]In parentheses, as percentage of total cases.

1982). To prevent botulism from surface-cured ham, Lücke (1983) recommended a curing temperature of 5°C and measures to keep the microbial contamination of the inner ham to a minimum. Nitrate as a curing salt ingredient does not delay or reduce the formation of toxin (Lücke et al., 1982).

In 1988, an outbreak from ham was attributed to *C. botulinum* type E. This type is a rare contaminant of meat from domestic animals, but, as a member of group II, its development in ham should resemble that of nonproteolytic (group II) type B strains. Only three other outbreaks, all associated with fish, were caused by type E.

In 1981, an unusual *C. botulinum* strain was isolated from a fatal case of botulism, and from the incriminated food, canned rabbit paté (Poumeyrol et al., 1983). So far, this is the only known strain that produces both type A and B toxins (Table 2).

Of the various implicated foods, 12% were of commercial origin, but the manufacturers were generally small, local establishments (*artisanats*). The average number of persons affected by the commercial products (2.5) was only marginally larger than the number affected in outbreaks from home-prepared foods (Sebald, 1990).

H. Italy

Only the total reported cases (Table 1) could be updated from the previous review (Hauschild, 1989). As in other European countries discussed so far, type B was the prevalent cause of botulism (Table 2), but, in contrast to these countries, meats were rarely implicated. Instead, 77% of the outbreaks were associated with vegetables (Table 3): green beans, mushrooms, eggplant, broccoli, and peppers. Most of these had been home-preserved in oil.

I. Spain

Botulism in Spain is recorded meticulously in annual summaries (MSC, 1984, 1986–1989). Compared to the previous review covering the period of 1969–83 (Hauschild, 1989), the current summary for 1969–1988 (Table 1) indicates increases in both annual outbreaks and cases. This is also demonstrated in Table 16. The increases have been credited to an improved surveillance system (MSC, 1989).

Nearly a third of the recorded incidents occurred in the southern region of Andalusia. Almost all the outbreaks (92%) were associated with type B (Table 2). As in Italy, most of them involved vegetables (Table 3), primarily beans, followed by asparagus and pepper. All the incriminated vegetables had been canned. Assuming that the implicated type B strains are of group II, the relevant contamination of the canned vegetables must have occurred after the heat process. One outbreak was attributed to fish, the only food associated with type E (MSC, 1986).

Table 16 Increasing Numbers of Recorded Botulism Outbreaks and Cases in Spain, 1969–88

Period	Outbreaks	Cases
1969–73	5	22
1974–78	6	16
1979–83	19	55
1984–88	33	105

J. Portugal

The data on Portuguese outbreaks (Tables 1–3) were reported from a single hospital in the northern region, the Hospital S. João in Oporto, where botulism appears to be more common than in central or southern Portugal (Lecour, 1990). As in Spain, the numbers of recorded outbreaks and cases have increased in recent years: in the 10-year period of 1970–79, five outbreaks with 19 cases were reported, compared to 19 outbreaks and 61 cases from 1980 to 1989. The responsible botulinum type was consistently B (Table 2). In contrast to the Spanish outbreaks, 91% of the Portuguese incidents were associated with meats (Table 3), none with fruit or vegetables. Of the 21 incriminated meats, 17 were smoked hams; the others were sausages (3) and bacon (1), all prepared in the home.

The following observations (Lecour et al., 1988) all suggest that the implicated type B strains belong to group II: (1) there were no fatalities among the 80 patients, and many of them required no treatment with antiserum; (2) onset of the symptoms was generally delayed, with incubation periods of up to 6 days; and (3) most responsible foods had a normal taste, indicating that no proteolysis had occurred.

In addition to the cases associated with Hospital S. João in Oporto, Malhado et al. (1982) reported four other incidents with a total of 15 cases from Lisbon between 1972 and 1980. As in Oporto, none of these cases were fatal. The responsible foods were ham (3) and mussels (1). Where the responsible type could be identified, it was again B (Malhado et al., 1982). Additional botulism cases have been treated in another hospital in Oporto, and in Coimbra, central Portugal (Lecour, 1990).

K. Scandinavia

Botulism in Scandinavia differs in several aspects from that in the rest of Europe, mainly in the relatively frequent implication of fish, and of *C. botulinum* type E.

Denmark

In the period of 1951–84, nearly 80% of the Danish outbreaks were caused by type E. About the same proportion of incidents was associated with fish (Hauschild, 1989). Most of the people affected were inhabitants of continental Denmark. In contrast, of the 11 outbreaks reported from 1984 to 1989 (Table 1), eight occurred in Greenland, three on the Faeroe Islands and none in continental Denmark. All the outbreaks were associated with type E. Where the responsible food was identified, it was always meat (Table 3): seal meat in seven Greenland incidents, and mutton in the three Faeroe outbreaks. It may be surprising that each of the three outbreaks involving land animals (sheep) should be associated with type E. However, the animals had been slaughtered by people employed in the fish industry, who may have contaminated the meat with spores of type E (Huss, 1990).

Norway

In the 30-year period from 1961 to 1990, 19 incidents of botulism involving 42 cases with three deaths were recorded in Norway (Table 1). With one exception where type F was the suspected cause (Table 2), the outbreaks (9 each) were all associated with either type B or type E.

Sixteen outbreaks (84%) were associated with one kind of fish product, raw fermented trout (*rakefisk*) (Skulberg, 1964). Although the pH decreases below 5.0 during fermentation, the decrease may be too slow to prevent toxigenesis. Safe processing depends on proper salt levels and rigid temperature control until the pH reaches 5.0 (Huss, 1981). In recent years, recorded incidents from *rakefisk* have declined: in the decades of 1961–70, 1971–80, and 1981–90, their numbers were 8, 5, and 3, respectively. Two of the three outbreaks from meats were caused by salted and dried hams, both contaminated with type B.

Sweden

Six incidents have been recorded in Sweden since 1969, five of them associated with fish (Table 4). Where the type could be identified, the fish-associated outbreaks involved type E. The remaining product was a blood sausage contaminated with type A (Hermanson, 1985). No further outbreaks have been reported in the last 9 years, i.e., since June of 1981 (Danielsson-Tham, 1990).

Finland

One single outbreak involving two nonfatal cases (tourists) was reported in 1981 (Table 4). The responsible food (sausages) was of foreign origin and had been brought to Finland by the visitors from their own country (WHO, 1983).

Iceland

Two outbreaks, both associated with nonproteolytic (group II) type B, have been reported (Georgsson, 1990). Seven cases were affected, none of them fatally (Table 4). In one incident, the responsible food was identified as blood sausage.

L. Other European Countries

No outbreaks were recorded in the last 10–15 years in Austria or Greece (Table 4). Six Bulgarian cases in 1987 and 1988 were listed, without further details, by the WHO surveillance program for control of foodborne illness in Europe (Teufel, 1990).

Five incidents with 57 cases were reported in Switzerland over a 17-year period (Table 4). Most of the cases (48) were involved in a single outbreak in 1973 (from Brie cheese contaminated with type B; see Chapter 10). Other incriminated foods were home-preserved beans and mushrooms and vacuum-packaged bratwurst (Bernasconi et al., 1988; Schwab, 1990). A rice salad, packaged by a Swiss company for international airlines, was the suspected cause of a single case of type A botulism in 1987 (Colebatch et al., 1989).

Only two botulism incidents (in 1981 and 1987) were recorded in the Netherlands from 1977 to 1989. Both were associated with sausages and *C. botulinum* type B.

A major type B outbreak in Lancashire and North Wales (U.K.) affected 27 cases with one death (Critchley et al., 1989). The responsible food was hazelnut yoghurt from a local dairy. The toxin was introduced with a yoghurt ingredient, i.e., canned hazelnut purée, a mixture of preroasted hazelnuts, starch, water, and some other ingredients. Toxin production was attributed to underprocessing of the canned purée (O'Mahony et al., 1990).

Two previous outbreaks in Britain (in 1955 and 1978) were both caused by imported products: pickled fish with type A toxin from Mauritius (Mackay-Scollay, 1958) and canned Alaskan salmon with type E toxin (Gilbert and Willis, 1980). The salmon incident affected four patients with two deaths and was attributed to postprocess entry of *C. botulinum* into a defective can (Ball et al., 1979).

V. BOTULISM IN THE SOVIET UNION

Recent publications in the U.S.S.R. on botulism deal with the subject in general (Papina, 1983) or with particular outbreaks (Shneider, 1988), but epidemiological work on botulism has been discontinued (Prozorovsky, 1986). The previous summary (Hauschild, 1989), based on a somewhat dated review by Matveev et al. (1967), therefore could not be updated.

In the period of 1958–64, types A, B, and E were implicated in approximately equal numbers of outbreaks (Table 2). Two thirds of the incidents were associated with fish (Table 3), and these occurred mainly in regions around the Black Sea and adjacent Sea of Azov, the Caspian Sea, and Lake Baykal. Surprisingly, type A was involved in as many outbreaks from fish as type E. Environmental surveys of Kravchenko and Shishulina (1967) identified only 8% of *C. botulinum* isolates as type A and 62% as type E.

Meat-associated outbreaks were all caused by home-cured ham, primarily in the Baltics and in the Ukraine. As in other adjacent European areas, e.g., Poland, nearly all these incidents involved type B (82%), and the mortality rate was low; 5% from ham compared to an overall mortality rate of 29% (Table 1).

Nearly half the vegetable-associated outbreaks were caused by home-preserved wild mushrooms, all of them linked to *C. botulinum* type B. The frequent involvement of mushrooms was confirmed more recently by Papina (1983). As in a number of other countries, e.g., Canada and Poland, botulism outbreaks in the U.S.S.R. showed a distinct seasonality, with a peak in the months of June to August (Matveev et al., 1967).

VI. BOTULISM IN ASIA

Foodborne botulism has been reported only from a few countries in Asia.

A. Israel

In 1987, Israel recorded its first incident of foodborne botulism since 1948. It affected six persons with one fatality, and was associated with *C. botulinum* type E. The incriminated food was *kapchunka*, a brined and air-dried freshwater fish, which had been processed commercially by a small establishment in New York City and brought home by Israeli tourists (Slater et al., 1989; Telzak et al., 1989). Prior to this incident, a single case of wound botulism was reported in 1984 (Slater et al., 1989).

B. Iran

Botulism presents a major health problem in Iran. Rouhbakhsh-Khaleghdoust and Pourtaghva (1977) described an outbreak from salt-cured carp eggs that affected 75 persons with 17 deaths. The latest published summary of foodborne botulism in Iran covers the period of 1972–74 (Table 1). Of the 314 recorded cases, 170 were thoroughly investigated. The latter comprised 63 outbreaks and included 18 deaths (Pourtaghva et al., 1975).

Of the 63 investigated outbreaks, 61 (97%) were associated with type E and with fish or fish products (Tables 2 and 3). Fleshy portions of the fish were responsible for only 10% of the outbreaks; the other 90% were caused by fish eggs (*ashbal*). The frequent involvement of *ashbal* has its cause in the method of preparation, which includes salt curing of the eggs for some months, often with the potential for toxin production. The eggs are eaten uncooked without further treatment, whereas the actual fish is usually cooked or steamed. Two of the 63 outbreaks involved type B and meats (ham and brined goat meat).

Pourtaghva et al (1975) reported a distinct outbreak peak in May/June. The peak is consistent with seasonal fish processing in late fall/early winter, and curing during the intervening months.

The involvement of types E and B in Iranian outbreaks is consistent with the incidence of the two types in the environment. The ratios at which types E and B were detected in aquatic sediments and in fish from the Caspian Sea were 92:8 and 96:4, respectively. No other types were found (Rouhbakhsh-Khaleghdoust, 1973; Hauschild, 1989).

C. China

Table 1 lists 986 outbreaks from 1958 to 1983. Nearly all of these were reported from northern China above 30° latitude, and 90% occurred in the northwestern province of Xinjiang (Sinkiang) alone (Ying and Shuyan, 1986).

Most of the type-identified outbreaks (93%) were associated with type A, followed by types B and E (Table 2). The prevalence of causative types varied within the country: in Xinjiang, 97% of the outbreaks were caused by type A; in the northern provinces of Ningxia, Shanxi, and Hebei, type B predominated, and the few incidents reported from the northeastern area were all associated with type E (Ying and Shuyan, 1986).

The responsible foods were mostly of plant origin (Table 3), i.e., fermented bean curd (74%) and a thick fermented bean sauce (12%). Toxin production occurs during fermentation (Ying and Shuyan, 1986), but the deficiencies in the fermentation process may be varied. According to Pedersen (1979), proper fermentation of the curd depends on an active mold culture, commonly a *Mucor* species, and on the salt level.

Meats were involved in 93 out of 958 food-identified outbreaks (Table 3). With the exception of two outbreaks associated with canned meats, all the meat-related incidents were caused by home-fermented beef.

The overall death rate was 13%, but with the availability of antisera starting in 1960, the ratio decreased, most significantly after 1974. Of 510 patients receiving symptomatic treatment but no antitoxin, 255 (50%) died, whereas among nearly 4000 patients treated with antitoxin, the fatality rate was only 8%.

In the treatment of botulism patients in Xinjiang, Ying and Shuyan (1986) observed that "the disease tended to be milder in cases in which the patient had consumed alcoholic beverages along with the contaminated food." Evidence of a beneficial effect of alcohol on the outcome of botulism remains anecdotal. In contrast to the observation of Ying and Shuyan, native people in Canadian Inuit communities often associated alcohol consumption with aggravated manifestations of botulism. In a Cape Dorset outbreak, two out of 18 Inuits sharing the incriminated food had consumed alcohol with the meal, and only these two became seriously ill with botulism (HWC, 1978).

As in Iran, botulism in China is seasonal, but the outbreaks reach a peak in the winter and spring when home-preserved foods replace fresh produce (Ying and Shuyan, 1986).

Gao et al. (1990) summarized 745 outbreaks with 2861 cases reported from 15 of the 30 provinces and autonomous regions in the period of 1958 to 1989. Basically, the outbreak patterns were similar to those described by Yin and Shuyan (1986), type A in vegetable products being the predominant cause. However, the relative numbers of outbreaks caused by types B (12%) and E (7%) and of outbreaks associated with meats and fish (totaling 28%) were higher than those reported by Ying and Shuyan (Tables 2 and 3).

Gao et al. (1990) confirmed that nearly all the reported outbreaks occurred north of 30° latitude. This pattern is consistent with environmental surveys, which showed 15% of soil and food samples from the northern part of China positive for *C. botulinum*, compared to 2.5% of the samples from southern China. Additionally, all but one of the 534 positive samples from the north were of types A, B, E, and F, whereas 45 of 58 positive samples from the south (78%) were types C and D (Gao et al., 1990).

D. Taiwan

The first botulism incident in Taiwan was reported in 1977. It affected five cases and was associated with type A (Tsai and Ting, 1977). The largest outbreak recorded in the 5-year period from 1986 to 1990 (Table 4) involved nine cases with two fatalities and was caused by commercially canned peanuts in glass jars containing type A toxin. Development of *C. botulinum* in the jars was attributed to serious underprocessing (Chou et al., 1988). Three incidents from home-preserved foods affected a total of seven nonfatal cases (DHT, 1987, 1990).

E. Japan

The majority of outbreaks listed in Table 1 occurred in the northern prefectures of Hokkaido, Aomori, and Akita. The prevalent cause of the outbreaks was type E (Table 2). With only one exception, the incidents were all associated with fish or fish products (Table 3). The most commonly incriminated products were *izushi* (88%) and *kirikomi* (7%) (Itoh, 1989).

In the preparation of *izushi*, fleshy pieces of fish are soaked in water for a few days with occasional water replacements for the removal of blood, packed tightly in a tub with cooked rice, diced vegetables, salt, vinegar, and spices, and left to ferment, often for 3 weeks or longer. The product is consumed without further cooking (Nakamura et al., 1956; Nakano and Kodama, 1968). Toxigenesis occurs only during the soaking and early fermentation stages; after that, the pH drops well below 5.0 (Sakaguchi, 1979; Sakaguchi et al., 1968). *Kirikomi* is prepared similarly but without cooked rice (Sakaguchi, 1989). Suggestions for preventing botulism from fermented fish products include (1) soaking the fish below 5°C, (2) use of high-quality raw materials, e.g., no fermentation of

Table 17 Decreasing Numbers of Recorded Botulism Outbreaks and Cases in Japan, 1952–87

Period	Outbreaks	Cases	
		Total	Fatal[a]
1952–60	33	179	57 (32)
1961–69	30	167	30 (18)
1970–78	22	71	10 (14)
1979–87	10	52	12 (23)

[a]In parentheses, as percentage of total cases.

ungutted fish, and (3) addition of lactic acid cultures to the fermentation mixtures to assure a rapid pH drop (Sakaguchi, 1979).

Four outbreaks were associated with types other than E: two with type A and two with type B. One outbreak each of types A and B could be linked to the responsible foods, both of commercial origin. The type A outbreak was caused by vacuum-packaged, stuffed lotus rhizome and affected 36 cases with 11 deaths in several prefectures (Hayashi et al., 1986; Sakaguchi, 1989). The type B outbreak involved imported bottled caviar, which affected 21 cases with 3 deaths (Fukuda et al., 1970).

Botulism outbreaks in Japan have become less frequent in recent years. A comparison of 9-year periods from 1952 to 1987 shows a general decline in outbreaks, cases and fatalities (Table 17). The slight rise in fatal cases during the 1979–87 period is due to the unusual incident from lotus rhizome, which alone caused 11 of the 12 fatalities. Reports of botulism outbreaks from other Asian countries are rare (Onul and Willke, 1989; Torpy, 1938).

VII. BOTULISM IN AFRICA

Reports of human botulism from a few African nations (Table 4) suggest that outbreaks on this continent are rare. However, they can be large and extremely severe (Vicens et al., 1985).

Smith et al. (1979) reported the first human botulism outbreak from Kenya. It was caused by sour milk contaminated with *C. botulinum* type A prepared in a gourd. Eleven people were affected, six of them fatally. The second outbreak in Kenya was caused by termites, a cultural food, and affected six cases with five deaths. Toxigenesis was attributed to keeping the termites in a sealed bag for 3 days (Nightingale and Ayim, 1980).

The first recorded outbreak in Madagascar occurred in 1982 and involved about 60 people with 30 deaths (Vicens et al., 1985). This was one of the rare type E outbreaks associated with a meat product, i.e., bologna manufactured in a local butchery. The event was followed in 1984 by an outbreak from fish, also associated with type E (Vicens et al., 1985). Three cases were affected, all of them fatally. Other incidents, both associated with meats, have been reported from Chad (Demarchi et al., 1958) and Rhodesia (Fleming, 1960).

VIII. BOTULISM IN AUSTRALIA AND NEW ZEALAND

From 1942 to 1983, five incidents of human foodborne botulism have been recorded in Australia (Table 4). Four of these were linked to vegetables, including imported canned mushrooms, and one to canned tuna.

Except for a suspected incident of botulism from corned beef contaminated with *C. botulinum* type A, which nevertheless remained inconclusive (Murrell, 1990), no outbreaks of human botulism have been reported from Australia since 1983. On the other hand, botulism in cattle remains an acute problem (Murrell, 1990).

Till (1990) reported one incident of type A botulism affecting two persons, one of them fatally, in New Zealand. The incriminated food was a batch of home-bottled fermented mussels and watercress, a traditional Maori dish.

IX. CONCLUSIONS

A. Location and Seasonality of Foodborne Botulism

The bulk of reported botulism outbreaks occurred in the northern hemisphere, particularly in countries north of the Tropic of Cancer (23.5°N). In the southern hemisphere, only Argentina reported a substantial number of outbreaks. These occurred south of the Tropic of Capricorn, predominantly between 30° and 40° south (Ciccarelli and Giménez, 1981).

Countries or states with frequent occurrences of botulism generally show an annual rise and fall in the incidence of outbreaks. In Alaska, Canada, Poland, the former U.S.S.R., and Iran, the outbreaks are at a peak in the period of May to October, whereas most of the Chinese outbreaks occur in the winter and early spring. No seasonality was apparent in the outbreak pattern of mainland United States.

B. Association between Location of Outbreaks and *C. botulinum* Types A, B, and E

Clostridium botulinum type E is the common cause of botulism in the colder regions of the northern hemisphere: Alaska, Canada, Greenland, Scandinavia, parts of the former Soviet Union, Iran, and northern Japan. The foods associated with these outbreaks are mostly fish and marine mammals.

Types A and B are generally the cause of outbreaks in the more temperate zones, with one of the two often predominating over the other. Type A is linked to the majority of outbreaks in the western United States, Argentina, Brazil, and China, whereas type B is prevalent in outbreaks of European countries, namely, Poland, Czechoslovakia, Hungary, Yugoslavia, Germany, Belgium, France, Italy, Spain, and Portugal. Occasionally, both types have been implicated in the same incident.

Two exceptional strains, isolated in outbreaks in France and Argentina, each produced two different toxins. The French strain, subtype AB, formed toxins A and B (Poymeyrol et al., 1983), and the Argentine strain, subtype AF, produced toxins A and F (Fernandez et al., 1986).

In general, the global pattern of outbreaks involving types A,B, and E is consistent with the distribution of the three types in the environment and their prevalence in geographic regions (Dodds and Hauschild, 1989).

C. Involvement in Types C, D, F, and G in Foodborne Botulism

Involvement of type F in botulism outbreaks is rare (Table 2). The latest reported incident occurred in California in 1986 (Hatheway, 1990). Type F was detected in combination with type A in an Argentine outbreak in 1981 (Giménez, 1990) and was the suspected cause of an outbreak in Norway in 1970 (Skulberg, 1990). Moller and Scheibel (1960) and Midura et al. (1972) reported earlier outbreaks involving type F of group I and group II, respectively.

Infrequent type F outbreaks are consistent with the rare occurrence of this type in the environment (Dodds and Hauschild, 1989) and with a relatively low lethality of its toxin in primates. The oral toxicity of toxin F to monkeys is in the order of 5×10^4 mouse LD_{50}/kg, which compares to a range of about 5×10^2 to 2×10^3 LD_{50}/kg for toxins A, B, and E (Dolman and Murakami, 1961; Morton, 1961; Ono et al., 1970).

Type C was identified, in combination with type B, in four reported outbreaks in Belgium (Table 2). However, because of its very low toxicity to primates, the presence of toxin C is probably not a significant factor in pathogenesis. Like toxin D, its lethality in monkeys is in the 10^5–10^6 LD_{50}/kg range (Dolman and Murakami, 1961). Types C and D have both been implicated in earlier outbreaks (Demarchi et al., 1958; Matveev et al., 1967; Meyer et al., 1953; Prévot et al., 1955), but either type as the cause of human botulism is in some doubt (Hauschild, 1989; Roberts and Gibson, 1979). The toxicity of toxin G is somewhat higher (comparable to F; Ciccarelli et al., 1977), but type G has never been implicated in foodborne botulism.

D. Association Between Location of Outbreaks and Causative Foods

Meats were the primary cause of botulism in Canada, the northern Danish territories (Greenland and Faeroe Islands), and in the various European countries where type B was commonly associated with outbreaks, except in Italy and Spain. In Canada and Greenland, the meats were predominantly of marine origin and affected primarily the native population. In continental Europe, the major single causes of botulism were home-cured smoked ham, particularly in France, Belgium, Germany, and Portugal, and canned meats, namely in Poland.

In the United States, Argentina, Spain, Italy, and China, the most frequently implicated foods were vegetables. These were commonly home-canned, except in China where they were fermented.

In Alaska, Norway, Sweden, the Soviet Union, Iran, and Japan, the majority of outbreaks were associated with fish. In general, the responsible botulinum type was E, except in Norway and the Soviet Union, where type E outbreaks were balanced by approximately the same number of outbreaks involving type B and type A, respectively.

The association between regions and causative foods is largely attributable to the contamination of common food sources with botulinum spores, e.g., marine mammals and fish in the colder regions with *C. botulinum* type E, and vegetables in the western United States with type A spores from the soil. Another, probably equally important factor is the local preference for particular foods and traditional, albeit unsafe, methods of preserving them. Such methods include the fermentation of seal meat, trout and other fish, or salmon eggs without adequate fermentable carbohydrates, and the salt curing of ham with insufficient cooling. Improvements to traditional food preservation methods should be one of the most effective means of preventing botulism. Already, efforts towards consumer education in British Columbia, Norway, and Japan seem to have reduced significantly the incidence of outbreaks from salmon eggs, *rakefisk*, *izushi*, and other fermented products.

E. Fatality Rates Associated with Type B Botulism

Outbreaks in Europe involving predominantly type B of group II show fatality rates of 0–8% (Table 1), with an overall rate of 2–3%. Relatively low death rates are also associated with type B of group I. In the United States where type B strains are predominantly of group I, the overall death rate from 1961 to 1988 in type B outbreaks was 7%, compared to 15% and 13% in type A and type E outbreaks, respectively (Table 9). The low fatality rates are consistent with comparatively less severe symptoms resulting from type B intoxication, and a relatively low affinity to the receptor site on the presynaptic membrane (Gangarosa, 1972; Hughes et al., 1981; Sebald and Saimot, 1973).

REFERENCES

Ajode, K. A. (1990). Ministry of Health, Nairobi, Kenya. Personal communication.

André, P. (1990). Institut d'Hygiène et d'Epidémiologie, Ministère de la Santé Publique et de la Famille. Brussels, Belgium. Personal communication.

Anusz, Z. (1986). Zatrucia jadem kielbasianym. *Przegp. Epidemiol. 40*: 89.

Anusz, Z. (1987). Zatrucia jadem kielbasianym—1985 rok. *Przegp. Epidemiol. 41*: 78.

Anusz, Z. (1988). Zatrucia jadem kielbasianym—1986 rok. *Przegp. Epidemiol. 42*: 65.

Anusz, Z. (1989). Zatrucia jadem kielbasianym—1987 rok. *Przegp. Epidemiol. 43*: 64.

Archer, D. L., and Kvenberg, J. E. (1985). Incidence and cost of foodborne diarrheal disease in the United States. *J. Food Prot. 48*: 887.

Aureli, P. (1989). Istituto Superiore di Sanita. Rome, Italy. Personal communication.

Badhey, H., Cleri, D. J., D'Amato, R. F., Vernaleo, J. R., Veinni, V., Tessler, J., Wallman, A. A., Mastellone, A. J., Giuliani, M., and Hochstein, L. (1986). Two fatal cases of type E adult food-borne botulism with early symptoms and terminal neurological signs. *J. Clin. Microbiol. 23*: 616.

Ball, A. P., Hopkinson, R. B., Farrell, I. D., Hutchison, J. G. P., Paul, R., Watson, R. D. S., Page, A. J. F., Parker, R. G. F., Edwards, C. W., Snow, M., Scott, D. K., Leone-Ganado, A., Hastings, A., Ghosh, A. C., and Gilbert, R. J. (1979). Human botulism caused by *Clostridium botulinum* type E: The Birmingham outbreak. *Quart. J. Med. 191*: 473.

Bernasconi, C., Nadal, D., Wüst, J., Lips, U., and Boltshauser, E. (1988). Food-borne botulism: An uncommon disorder. *Helv. Paediat. Acta. 43*: 515.

Berndt, S. F. (1978). Botulismus. *Med. Klin. 73*: 879.

Bowmer, E. J., and Wilkinson, D. A. (1976). Botulism in Canada, (1971–74). *Can. Med. Assoc. J. 115*: 1084.

Carré, H., Gledel, J., Poumeyrol, M., Sebald, M., Thomas, G., and Veit, P. (1987). Enquete sur un foyer de botulisme. Necessité du respect des bonnes pratiques professionnelles. *Med. Nutr. 23*: 391.

CDC (Centers for Disease Control) (1978). Botulism—New Mexico. *MMWR 27*: 138.

CDC (Centers for Disease Control) (1979). Botulism in the United States (1899–1977). *Handbook for Epidemiologists, Clinicians, and Laboratory Workers*. U.S. Department of Health, Education and Welfare, CDC, Atlanta, GA.

Chamorro, H., Alvarez, M., Mariaca, N., Lisboa, C., and Muñoz, V. (1974). Botulismo: I. Aspectos clinicos de una intoxicacion colectiva. *Rev. Méd. Chile 102*: 120.

Chou, J. H. (1989). National Quarantine Service, Department of Health. Taipei, Taiwan. Personal communication.

Chou, J. H., Hwang, P. H., and Malison M. D. (1988). An outbreak of type A foodborne botulism in Taiwan due to commercially preserved peanuts. *Int. J. Epidemiol. 17*: 899.

Ciccarelli, A. S., and Giménez, D. F. (1981). Clinical and epidemiological aspects of botulism in Argentina. In *Biomedical Aspects of Botulism* (G. E. Lewis, ed.), Academic Press, New York, p. 291.

Ciccarelli, A. S., Whaley, D. N., McCroskey, L. M., Giménez, D. F., Dowell, V. R., and Hatheway, C. L. (1977). Cultural and physiological characteristics of *Clostridium botulinum* type G and the susceptibility of certain animals to its toxin. *Appl. Environ. Microbiol. 34*: 843.

Colebatch, J. G., Wolff, A. H., Gilbert, R. J., Mathias, C. J., Smith, S. E., Hirsch, N., and Wiles, C. M. (1989). Slow recovery from severe foodborne botulism. *Lancet* *2*(8673): 1216.

CPZ (Centro Panamericano de Zoonosis) (1975). Food-borne diseases. *Boletin Informativo II*. Engl. ed., November.

Critchley, E. M. R., Hayes, P. J., and Isaacs, P. E. T. (1989). Outbreak of botulism in North West England and Wales, June 1989. *Lancet* *2*(8667): 849.

Danielsson-Tham, M. -L. (1990). Swedish University of Agricultural Sciences, Uppsala, Sweden. Personal communication.

Demarchi, J., Mourgues, C., Orio, J., and Prévot, A. -R. (1958). Existence du botulisme humain de type D. *Acad. Nat. Méd. Bull. 142*: 580.

DHT (Department of Health, Taiwan) (1987). Two outbreaks of botulism in aborigines associated with home-preserved meats—Nantou and Ilan counties. *Epidemiol. Bull.* *3*: 29.

DHT (Department of Health, Taiwan) (1990). Case report of A botulism intoxication—Taipei City. *Epidemiol. Bull. 6*: 93.

Dittmann, D., and Rasch, G. (1990). Zentralinstitut für Hygiene, Mikrobiologie und Epidemiologie der DDR. Berlin, Germany. Personal communication.

Dodds, K. L. (1990). Restaurant-associated botulism outbreaks in North America. *Food Control 1*: 139.

Dodds, K. L., and Hauschild, A. H. (1989). Distribution of *Clostridium botulinum* in the environment and its significance in relation to botulism. In *Recent Advances in Microbial Ecology. Proceedings, 5th International Symposium on Microbial Ecology* (T. Hattori, Y. Ishida, Y. Maruyama, R. Y. Morita, and A. Uchida, eds.), Japan Scientific Societies Press, Tokyo, Japan, p. 472.

Dolman, C. E. (1964). Botulism as a world health problem. In *Botulism*. Proceedings of Symposium, Cincinnati, Ohio (K. H. Lewis and K. Cassel, eds.), Public Health Service Publ. No. 999-FP-1, p. 5.

Dolman, C. E., and Murakami, L. (1961). *Clostridium botulinum* type F with recent observations on other types. *J. Infect. Dis. 109*: 107.

Domján-Kovács, H. (1990). Veterinary and Food Investigating Service. Budapest, Hungary. Personal communication.

Edel, W., van Schothorst, M., and Kampelmacher, E. H. (1977). *Salmonella* and salmonellosis—the present situation. *Proceedings, International Symposium on Salmonella and Prospects for Control* (D. A. Barnum, ed.), University of Guelph, Guelph, Ontario, Canada, p. 1.

FDA (Food and Drug Administration) (1989). *HHS News* P89–20. FDA, Washington, DC.

Fernandez, R. A., Ciccarelli, A. S., Arenas, G. N., and Giménez, D. F. (1986). Primer brote de botulismo por *Clostridium botulinum* subtipo Af. *Rev. Argentina Microbiol.* *18*: 29.

Ferreira, M. S., Nishioka, S. A., de Almeida, A. B., Silveira, P. V. P., de Souza, M. C., Storti, P. C., Zenebon, O., Gelli, D. S., and de Souza, A. (1987). Botulismo: Consideraçóes acerca de oito casos ocorridos no triángulo mineiro, Minas Gerias, Brasil. *Rev. Inst. Med. Trop. São Paulo 29*: 137.

Fleming, R. H. C. (1960). Report on four cases of botulism in Gatooma, S. Rhodesia. *Centr. Afr. J. Med. 6*: 91.

Fukuda, T., Kitao, T., Tanikawa, H., and Sakaguchi, G. (1970). An outbreak of type B botulism occurring in Miyazaki Prefecture. *Jpn. J. Med. Sci. Biol. 23*: 243.

Gangarosa, E. J. (1972). Botulism. In *Infectious Diseases* (P. Hoeprich, ed.), Harper & Row, New York, p. 1031.

Gao, Q. Y., Huang, Y. F., Wu, J. G., Liu, H. D., and Xia, H. Q. (1990). A review of botulism in China. *Biomed. Environ. Sci. 3*: 326.

Garcia, G. (1989). Universidad Central de Venezuela, Maracay, Venezuela. Personal communication.

Georgsson, F. (1990). Hollustuvernd Rikisins, Reykjavik, Iceland. Personal communication.

Gilbert, R. J., and Willis, A. T. (1980). Botulism. *Commun. Med. 2*: 25.

Giménez, D. F. (1990). 9 de Julio, Mendoza, Argentina. Personal communication.

Giménez, D. F., and Ciccarelli, A. S. (1978). New strains of *Clostridium botulinum* subtype Af. *Zentralbl. Bakteriol. Parasitenk. Infektionskr. Hyg. I. Abt. Orig. A 240*: 215.

Halbich, H. (1990). Bundesministerium für Gesundheit. Vienna, Austria. Personal communication.

Halbinger, R. E. (1973). *Clostridium botulinum* in fishery products. In *Microbial Safety of Fishery Products* (C. O. Chichester and H. D. Graham, eds.), Academic Press, New York, p. 191.

Hatheway, C. L. (1990). Centers for Disease Control. Atlanta, GA. Personal communication.

Hauschild, A. H. W. (1989). *Clostridium botulinum*. In *Foodborne Bacterial Pathogens* (M. P. Doyle, ed.), Marcel Dekker, New York, p. 111.

Hauschild, A. H. W., and Bryan, F. L. (1980). Estimate of cases of food- and water-borne illness in Canada and the United States. *J. Food Prot. 43*: 435.

Hauschild, A. H. W., and Gauvreau, L. (1985). Food-borne botulism in Canada, 1971–84. *Can. Med. Assoc. J. 133*: 1141.

Hayashi, K., Sakaguchi, S., and Sakaguchi, G. (1986). Primary multiplication of *Clostridium botulinum* type A in mustard—miso stuffing of "karashi-renkon" (deep-fried mustard-stuffed lotus root). *Int. J. Food Microbiol. 3*: 311.

Helcl, J. (1985). Institut Hygieny a Epidemiologie. Prague, Czechoslovakia. Personal communication.

Hermanson, I. (1985). Karolinska Sjukhuset. Stockholm, Sweden. Personal communication.

Horwitz, M. A., Marr, J. S., Merson, M. H., Dowell, V. R., and Ellis, J. M. (1975). A continuing common-source outbreak of botulism in a family. *Lancet 2*(7940): 861.

Hughes, J. M., Blumenthal, J. R., Merson, M. H., Lombard, G. L., Dowell, V. R., and Gangarosa, E. J. (1981). Clinical features of types A and B food-borne botulism. *Ann. Intern. Med. 95*: 442.

Huss, H. H. (1981). *Clostridium botulinum type E and botulism*. Ministry of Fisheries, Technical University, Lyngby, Denmark.

Huss, H. H. (1990). Ministry of Fisheries, Technical University. Lyngby, Denmark. Personal communication.

HWC (Health and Welfare Canada) (1978). Botulism at Cape Dorset, N.W.T. *Can. Dis. Weekly Rep. 4*: 77.

HWC (Health and Welfare Canada) (1985). Botulism in Canada—caribou meat as a source of intoxication? *Can. Dis. Weekly Rep. 11*: 93.

HWC (Health and Welfare Canada) (1987a). Restaurant-associated botulism from in-house bottled mushrooms—British Columbia. *Can. Dis. Weekly Rep. 13*: 35.

HWC (Health and Welfare Canada) (1987b). Botulism in Canada—summary for 1986. *Can. Dis. Weekly Rep. 13*: 47.

HWC (Health and Welfare Canada) (1988). Botulism in Canada—summary for 1987. *Can. Dis. Weekly Rep. 14*: 41.

HWC (Health and Welfare Canada) (1989). Botulism in Canada—summary for 1988. *Can. Dis. Weekly Rep. 15*: 78.

HWC (Health and Welfare Canada) (1990). Botulism in Canada—summary for 1989. *Can. Dis. Weekly Rep. 16*: 89.

Itoh, T. (1989). Tokyo Metropolitan Research Laboratory of Public Health. Tokyo, Japan. Personal communication.

Jorgensen, B. (1990). Landslaegeembedet i Gronland, Nuuk/Godthåb, Greenland. Personal communication.

Kalember-Radosavljevic, M. (1990). Military Medical Academy, Belgrade, Yugoslavia. Personal communication.

Kravchenko, A. T., and Shishulina, L. M. (1967). Distribution of Cl. botulinum in soil and water in the U.S.S.R. In *Botulism 1966* (M. Ingram and T. A. Roberts, eds.), Chapman & Hall Ltd., London, p. 13.

Lecour, H. (1990). School of Medicine. Hospital S. João. Porto, Portugal. Personal communication.

Lecour, H., Ramos, M. H., Almeida, B., and Barbosa, R. (1988). Food-borne botulism. A review of 13 outbreaks. *Arch. Intern. Med. 148*: 578.

Lecour, H., Ramos, M. H., Almeida, B., Barbosa, R., and Almeida, J. (1989). Human botulism, an oft-forgotten disease. *Third European Congress on Anaerobic Bacteria and Infections*, Munich, Germany, p. 45.

Lücke, F. -K. (1983). Botulismus in Europa—epidemiologische und experimentelle Befunde. *Neue Erkenntnisse über die Erreger mikrobieller Lebensmittel-Vergiftungen. Schweiz. Ges. Lebensmittelhyg. 13*: 30.

Lücke, F. -K., Hechelmann, H., and Leistner, L. (1982). Botulismus nach Verzehr von Rohschinken. *Fleischwirtschaft 62*: 203.

MacDonald, K. L., Spengler, R. F., Hatheway, C. L., Hargrett, N. T., and Cohen, M. L. (1985). Type A botulism from sauteed onions. Clinical and epidemiologic observations. *JAMA 253*: 1275.

MacDonald, K. L., Cohen, M. L., and Blake, P. A. (1986). The changing epidemiology of adult botulism in the United States. *Am. J. Epidemiol. 124*: 794.

Mackay-Scollay, E. M. (1958). Two cases of botulism. *J. Pathol. Bacteriol. 75*: 482.

Malhado, J. A., de Mello Vieira, J., de Sousa, M., and Proença, R. (1982). A propósito de un estudio de cuatro brotes de botulismo en Portugal. *Rev. Clin. Espan. 164*: 173.

Mann, J. M., Lathrop, G. D., and Bannerman, J. A. (1983). Economic impact of a botulism outbreak. *J. Am. Med. Assoc. 249*: 1299.

Marambio, E. (1989). Instituto de Salud Publica. Santiago, Chile. Personal communication.

Marchal, A., de Maeyer, S., and Yde, M. (1985). Le botulisme en Belgique à propos de quelques foyers récents. *Rev. Méd. Brux. 6*: 690.

Matveev, K. I., Nefedjeva, N. P., Bulatova, T. I., and Sokolov, I. S. (1967). Epidemiology of botulism in the U.S.S.R. In *Botulism 1966* (M. Ingram and T. A. Roberts, eds.), Chapman & Hall, London, p. 1.

Meyer, K. F., Eddie, B., York, G. K., Collier, C. P., and Townsend, C. T. (1953). *Clostridium botulinum* type C and human botulism. *Atti del 6 Congresos Internazionale di Microbiologia, Rome 2*: 123.

Midura, T. F., Nygaard, G. S., Wood, R. M., and Bodily, H. L. (1972). *Clostridium botulinum* type F: Isolation from venison jerky. *Appl. Microbiol. 24*: 165.

Moller, V., and Scheibel, I. (1960). Preliminary report on the isolation of an apparently new type of *Cl. botulinum. Acta Pathol. Microbial. Scand. 48*: 80.

Morton, H. E. (1961). *The toxicity of Clostridium botulinum type A toxin for various species of animals, including man.* Institute for Cooperative Research, University of Pennsylvania, Philadelphia.

MSC (Ministerio de Sanidad y Consumo) (1984). Botulismo en España (1969–1983). *Bol. Epidemiol. Semanal. Sem. 15*: 113.

MSC (Ministerio de Sanidad y Consumo) (1986). Botulismo en España. Años 1984–1985. *Bol. Epidemiol. Semanal. Sem. 9*: 65.

MSC (Ministerio de Sanidad y Consumo) (1987). Botulismo en España. Año 1986. *Bol. Epidemiol. Semanal. Sem. 17–18*: 121.

MSC (Ministerio de Sanidad y Consumo) (1988). Botulismo en España. Año 1987. *Bol. Epidemiol. Semanal. Sem.* 25–26.

MSC (Ministerio de Sanidad y Consumo (1989). Botulismo en España. Año 1988. *Bol. Epidemiol. Semanal. Sem.* 19–20.

Murrell, W. G. (1990). University of Sydney, Sydney, N.S.W., Australia. Personal communication.

Nakamura, Y., Iida, H., Saeki, K., Kanzawa, K., and Karashimada, T. (1956). Type E botulism in Hokkaido, Japan. *Jpn. J. Med. Sci. Biol. 9*: 45.

Nakano, W., and Kodama, E. (1968). On the reality of "izushi", the causal food of botulism, and on its folkloric meaning. In *Toxic Microorganisms* (M. Herzberg, ed.), Proceedings, 1st U.S.—Japan Conference on Toxic Microorganisms, Honolulu, HI, p. 388.

Narayan, K. G. (1990). Birsa Agricultural University, Ranchi, India. Personal communication.

Nightingale, K. W., and Ayim, E. N. (1980). Outbreak of botulism in Kenya after ingestion of white ants. *Br. Med. J. 281*: 1682.

Notermans, S. (1990). National Institute of Public Health and Environmental Protection. Bilthoven, The Netherlands. Personal communication.

O'Mahony, M., Mitchell, E., Gilbert, R. J., Hutchinson, D. N., Begg, N. T., Rodhouse, J. C., and Morris, J. E. (1990). An outbreak of foodborne botulism associated with contaminated hazelnut yoghurt. *Epidemiol. Infect. 104*: 389.

Ono, T., Karashimada, T., and Iida, H. (1970). Studies on the serum therapy of type E botulism. III. *Jpn. J. Med. Sci. Biol. 23*: 177.

Onul, M., and Willke, A. (1989). Foodborne botulism and its epidemiological features seen in the last few years in our country (Turkey). *Mikrobiyol. Bült. 23*: 284.

PAHO (Pan American Health Organization) (1982). Botulism in the Americas. *Epidemiol. Bull. 3*: 1.

Papina, G. V. (1983). Botulism [Russ.]. *Medits. Sestra. 42*: 4.

Pedersen, C. S. (1979). *Microbiology of Food Fermentations*, 2nd ed. AVI Publ., Westport, CT.

Pereira-Filho, M. J. (1958). Diagnóstico biológico do surto de botulismo humano do pronto socorro de pôrto alegre. *Medicina Cirugia 19*: 52.

Pöhn, H. P., and Grossmann, R. (1987). Ausbrüche von Lebensmittelinfektionen und mikrobiell bedingten Intoxikationen in der Bundesrepublik Deutschland einschl. Berlin (West) 1983–1986. *Öff. Gesundh. -Wes. 49*: 577.

Postma, T. J., Hazenberg, G. J., Driessen, J. J. M., and Faes, T. J. C. (1989). Een patient met botulisme veroorzaakt door *Clostridium botulinum* type B. *Ned. Tijdschr. Geneeskd. 133*: 2137.

Poumeyrol, M., Billon, J., Delille, F., Haas, C., Marmonier, A., and Sebald, M. (1983). Intoxication botulique mortelle due à une souche de *Clostridium botulinum* de type AB. *Méd. Mal. Infect. 13*: 750.

Pourtaghva, M., Machoun, A., Fatollah-Zadeh, Khodadoust, A., Taeb, H., Farzam, H. and Farhangui. (1975). Le botulisme en Iran. *Méd. Mal. Infect. 5*: 536.

Prévot, A. -R., Terrasse, J., Daumail, J., Cavaroc, M., Riol, J., and Sillioc, R. (1955). Existence en France du botulisme humain de type C. *Bull. Acad. Nat. Méd. 139*: 355.

Prozorovsky, S. (1986). Gamaleya Institute, Moscow, U.S.S.R. Personal communication.

Quevedo, F. (1988). First outbreak of human botulism reported in Peru. Unpublished report.

Roberts, T. A., and Gibson, A. M. (1979). The relevance of *Clostridium botulinum* type C in public health and food processing. *J. Food Technol. 14*: 211.

Rotimi, V. O. (1985). University of Lagos, Lagos, Nigeria. Personal communication.

Rouhbakhsh-Khaleghdoust, A. (1973). The incidence of *Clostridium botulinum* type E in fish and bottom deposits in the Caspian Sea coastal waters. *Pahlavi Med. J. 6*: 550.

Rouhbakhsh-Khaleghdoust, A., and Pourtaghva, M. (1977). A large outbreak of type E botulism in Iran. *Trans. R. Soc. Trop. Med. Hyg. 71*: 444.

St. Louis, M. E., Peck, S. H. S., Bowering, D., Morgan, G. B., Blatherwick, J., Banerjee, S., Kettyls, G. D. M., Black, W. A., Milling, M. E., Hauschild, A. H. W., Tauxe, R. V., and Blake, P. A. (1988). Botulism from chopped garlic: delayed recognition of a major outbreak. *Ann. Int. Med. 108*: 363.

Sakaguchi, G. (1979). Botulism. In *Food-borne Infections and Intoxications*, 2nd ed. (H. Riemann and F. L. Bryan, eds.), Academic Press, New York, p. 389.

Sakaguchi, G. (1989). Overview of botulism in humans and lower animals. In *Recent Advances in Microbial Ecology. Proceedings, 5th International Symposium on Microbial Ecology*. (T. Hattori, Y. Ishida, Y. Maruyama, R. Y. Morita, and A. Uchida, eds.), Japan Scientific Societies Press, Tokyo, Japan, p. 467.

Sakaguchi, G., Sakaguchi, S., and Karashimada, T. (1968). Characterization of *Clostridium botulinum* type E toxin in "izushi." In *Toxic Microorganisms* (M. Herzberg, ed.), Proceedings, 1st U.S.—Japan Conference on Toxic Microorganisms, Honolulu, HI, p. 393.

Schulze, K., and Funke, R. (1981). Ein Fall von Botulismus unter dem besonderen Aspekt der Diagnose am Lebensmittel. *Arch. Lebensmittelhyg. 32*: 83.

Schwab, H. (1990). Bundesamt für Gesundheitswesen. Bern, Switzerland. Personal communication.

Seals, J. E., Snyder, J. D., Edell, T. A. Hatheway, C. L., Johnson, C. J., Swanson, R. C., and Hughes, J. M. (1981). Restaurant-associated type A botulism: Transmission by potato salad. *Am. J. Epidemiol. 113*: 436.

Sebald, M. (1970). Sur le botulisme en France de 1956 à 1970. *Bull. Acad. Nat. Méd. 154*: 703.

Sebald, M. (1990). Institut Pasteur, Paris, France. Personal communication.

Sebald, M. and Saimot, G. (1973). Toxémie botulique. Intérêt de sa mise en évidence dans le diagnostic du botulisme humain de type B. *Ann. Microbiol.* (Inst. Pasteur) *124A*: 61.

Sebald, M., Billon, J., Cassaigne, R., Rosset, R., and Poumeyrol, G. (1980). Le botulisme en France—incidence, mortalité, aliments responsables avec étude des foyers dus à un aliment qui n'est pas de préparation familiale. *Méd. Nutr. 16*: 262.

Serrano, A. M. (1982). Um provável surto de botulismo humano no Brasil. *Hig. Aliment. 1*: 59.

Serrano, A. M. (1990). Universidade Estadual de Campinas, Campinas, S. P., Brazil. Personal communication.

Shneider, V. P. (1988). Repeated botulism. *Klin. Medit.* (U.S.S.R.) *66*: 126.

Skulberg, A. (1964). *Studies on the Formation of Toxin by Clostridium botulinum*. Veterinary College of Norway, Oslo, Norway.

Skulberg, A. (1990). Norwegian Food Research Institute. Ås, Norway. Personal communication.

Slater, P. E., Addiss, D. G., Cohen, A., Leventhal, A., Chassis, G., Zehavi, H., Bashari, A., and Costin, C., (1989). Foodborne botulism: An international outbreak. *Int. J. Epidemiol. 18*: 693.

Smith, B. (1990). McGill University, Montreal, Quebec, Canada. Personal communication.

Smith, L. DS. (1978). The occurrence of *Clostridium botulinum* and *Clostridium tetani* in the soil of the United States. *Health Lab. Sci. 15*: 74.

Smith, L. DS., and Sugiyama, H. (1988). *Botulism: The Organism, Its Toxins, the Disease* 2nd ed. Charles C. Thomas, Springfield, IL.

Smith, D. H., Timms, G. L., and Refai, M. (1979). Outbreak of botulism in Kenyan nomads. *Ann. Trop. Med. Parasitol. 73*: 145.

Solomon, H. M., and Kautter, D. A. (1986). Growth and toxin production by *Clostridium botulinum* in sauteed onions. *J. Food Prot. 49*: 618.

Solomon, H. M., and Kautter, D. A. (1988). Outgrowth and toxin production by *Clostridium botulinum* in bottled chopped garlic. *J. Food Prot. 51*: 862.

Stersky, A., Todd, E., and Pivnick, H. (1980). Food poisoning associated with post-process leakage (PPL) in canned foods. *J. Food Prot. 43*: 465.

Sugiyama, H., Woodburn, M., Yang, K. H., and Movroydis, C. (1981). Production of botulinum toxin in inoculated pack studies of foil-wrapped baked potatoes. *J. Food Prot. 44*: 896.

Szczawinski, J. (1989). Warsaw Agricultural University, Warsaw, Poland. Personal communication.

Tello, O. (1990). Centro Nacional de Epidemiologia, Ministerio de Sanidad y Consumo, Madrid, Spain. Personal communication.

Telzak, E. E., Bell, E. P., Kautter, D. A., Crowell, L., Budnick, L. D., Morse, L. D., and Schultz, S. (1990). An international outbreak of type E botulism due to uneviscerated fish. *J. Infec. Dis. 161*: 340.

Terranova, W., Breman, J. G., Locey, R. P., and Speck, S. (1978). Botulism type B: Epidemiologic aspects of an extensive outbreak. *Am. J. Epidemiol 108*: 150.

Teufel, P. (1990). Institut für Veterinärmedizin, Bundesgesundheitsamt, Berlin, Germany. Personal communication.

Till, D. (1990). New Zealand Communicable Disease Centre, Porirua, New Zealand. Personal communication.

Todd, E. C. D. (1985). Economic loss from foodborne disease outbreaks associated with foodservice establishments. *J. Food Prot. 48*: 169.

Todd, E. C. D. (1989). Preliminary estimates of costs of foodborne disease in Canada and costs to reduce salmonellosis. *J. Food Prot. 52*: 586.

Todd, E., Chang, P. C., Hauschild, A., Sharpe, A., Park, C., and Pivnick, H. (1974). Botulism from marinated mushrooms. *Proceedings, IV International Congress on Food Science and Technology*. Valencia, Spain Vol. 3, p. 182.

Torpy, C. D. (1938). Suspected botulism: A report on two cases. *Ind. Med. Gaz. 73*: 600.

Tsai, C. -C., and Ting, S. -C. (1977). Ecological studies of *Clostridium botulinum* in soils of Taiwan. *J. Formosan Med. Assoc. 76*: 563.

Tselikas, E. (1989). Public Health Division, Ministry of Health, Welfare and Social Security, Athens, Greece, Personal communication.

Valdespino-Gómez, J. L. (1990). Instituto Nacional de Diagnostico y Referencia Epidemiologicos. Mexico City, Mexico. Personal communication.

Vicens, R., Rasolofonirina, N., and Coulanges, P. (1985). Premiers cas humains de botulisme alimentaire à Madagascar. *Arch. Inst. Pasteur Madagascar 52*: 11.

Wainwright, R. B., Heyward, W. L., Middaugh, J. P., Hatheway, C. L., Harpster, A. P., and Bender, T. R. (1988). Food-borne botulism in Alaska, 1947–1985: Epidemiology and clinical findings. *J. Infect. Dis. 157*: 1158.

WHO (World Health Organization) (1983). *WHO Surveillance Programme for Control of Foodborne Infections and Intoxications in Europe*. Second report. Robert von Ostertag Institute, Berlin, Germany.

WHO (World Health Organization) (1984). *WHO Surveillance Programme for Control of Foodborne Infections and Intoxications in Europe*. Third report, 1982. Robert von Ostertag Institute, Berlin, Germany.

WHO (World Health Organization) (1990). *WHO Surveillance Programme for Control of Foodborne Infections and Intoxications in Europe*. Fourth report, 1983/1984. Robert von Ostertag Institute, Berlin, Germany.

Ying, S., and Shuyan, C. (1986). Botulism in China. *Rev. Infect. Dis. 8*: 984.

5

Worldwide Incidence and Ecology of Infant Botulism

Karen L. Dodds

Health Protection Branch,
Health and Welfare Canada,
Ottawa, Ontario, Canada

I. INTRODUCTION

Since the first two cases of infant botulism were diagnosed in the United States (Pickett et al., 1976), approximately 1000 cases of infant botulism have been reported from around the world, predominantly (>90%) from the United States. The disease, which affects otherwise healthy children usually less than 1 year old, is characterized by constipation, generalized weakness, and various neurological disorders (Arnon, 1989). While most patients require hospitalization, fatal cases are rare. After its recognition as a disease, it quickly became evident that infant botulism presented a new pathogenic role for *Clostridium botulinum*.

Classical, or foodborne botulism, recognized since the 1890s (van Ermengem, 1897), is an intoxication caused by the ingestion of food with preformed toxin. The nature of the foods most often associated with foodborne botulism usually preclude the involvement of infants. The etiology of infant botulism involves ingestion of clostridial spores and germination, multiplication, and toxigenesis in the infant's intestines. The most prevalent source of spores appears to be the general environment. The only food implicated thus far is honey, which occasionally carries *C. botulinum* spores in large numbers (see Chapter 3). Originally limited to *C. botulinum* types A and B, the toxin types and organisms associated with the disease now include types E and F and toxigenic *Clostridium butyricum* and *Clostridium baratii*.

II. PATHOGENESIS

Infant botulism results from the production of botulinum toxin in the infant's intestines after germination and outgrowth in situ of ingested *C. botulinum* spores. This differs from classical, foodborne botulism, which results from the ingestion of preformed toxin.

One of the most remarkable features of infant botulism is the age at onset. Over 90% of hospitalized cases in the United States have been between 2 weeks and 6 months of age, and 50% had onset between 1 and 3 months of age (Arnon, 1989). As this disease continues to be recorded, the age of the oldest patients increases (Hatheway, 1990): it was 5½ months in 1978, 8½ months in 1981, 12 months in both 1986 and 1987, and, in 1990, a 2-year-old was identified as having infant botulism. Sudden infant death syndrome (SIDS) cases and cases of salmonellosis in infants show a similar age distribution.

The restricted age range of infant botulism appears to be directly related to the vulnerability of the infant's intestines to colonization by *C. botulinum* (Arnon, 1986). Studies using mice have established that the normal intestinal microflora of adults prevents *C. botulinum* colonization of the gut (Moberg and Sugiyama, 1979). Adult germ-free mice or adult mice treated with antibiotics can be colonized by botulinum spores, but this susceptibility is lost when the mice acquire intestinal organisms indigenous to conventional adult mice (Burr and Sugiyama, 1982). Also, healthy infant mice are susceptible to colonization with *C. botulinum* spores but only for a very limited age range (Sugiyama and Mills, 1978). The period of susceptibility is also related to the infant's diet and indigenous intestinal flora. In general, infants fed human milk have acidic feces (pH 5.1–5.4) with large numbers of *Bifidobacterium* spp. and low concentrations of facultative or strictly anaerobic bacteria (Arnon, 1986). Formula-fed infants have less acidic feces (pH 5.9–8.0) with lower numbers of *Bifidobacterium* spp. but larger numbers of other anaerobes (*Bacteroides* spp., clostridia, anaerobic streptococci), and of facultative anaerobic bacteria. Long et al. (1985) observed that all infants hospitalized for infant botulism in Pennsylvania had been solely or primarily breast-fed, and therefore concluded that breast-feeding is a "risk factor." However, the evidence is not definitive. It must be noted that human milk contains immune factors, which probably help protect the infant from botulism. Arnon et al. (1981) also noted that a significantly higher number of hospitalized infant botulism patients were primarily breast-fed than were healthy control infants. In contrast, all California infants whose sudden deaths were attributed to *C. botulinum* infection were primarily formula-fed. These findings led them to suggest associations between breast-feeding and the more gradual onset form of infant botulism, and between formula-feeding and sudden infant death attributable to *C. botulinum* infection. In California, the mean age of onset of infant botulism for formula-fed babies was significantly less than for breast-fed infants (Arnon et at., 1982). It is quite likely that the perturbation in the gut ecosystem caused by the introduction of foods other than milk, or changing from human milk to formula, plays a role in the susceptibility of infants to *C. botulinum* (Arnon, 1986).

CDC Atlanta has suggested a new category for botulism, "classification unknown," to include cases occurring in persons older than 1 year of age in which no specific food has been implicated (Morris and Hatheway, 1980). It was postulated that at least some adult patients had *C. botulinum* intestinal infections with secondary intoxication. This possibility was confirmed by Chia et al. (1986), who found a causal connection between an adult patient with infant-type botulism and a food source with viable *C. botulinum* but no preformed toxin. *C. botulinum* toxin and organisms of the same type were demonstrated in the patient's stool almost 3 weeks after exposure to the food source. The patient had had recent gastrointestinal surgery coupled with perioperative antibiotic therapy, which probably compromised her resistance to *C. botulinum*. Similar cases have since been described (McCroskey and Hatheway, 1988).

III. TOXIN TYPES INVOLVED

Most cases of infant botulism, and all of the early cases, were due to type A or B organisms of the proteolytic group I (Midura, 1979). A type F case was reported from New Mexico in 1979 (Davis et al., 1979). However, the organism that produced the type F toxin was not *C. botulinum*, but was culturally and biochemically identical to *C. baratii*, with the exceptions that it produced the neurotoxin and fermented trehalose (Hall et al., 1985). The first two cases of type E infant botulism occurred in Rome, Italy (Aureli et al., 1986). The toxigenic organism was difficult to isolate because its characteristics differed from those of *C. botulinum* type E. Except for its toxigenicity, the isolated organism corresponds to *C. butyricum* (McCroskey et al., 1986). The discovery of these two novel strains producing botulinum toxin increases the confusion regarding the taxonomy of *C. botulinum*, bringing to six the number of physiologically distinct groups of clostridia that can produce botulinum neurotoxin (see Chapter 1). With the exception of the cases caused by these novel strains and a recent case in Japan caused by type C *C. botulinum* (Oguma et al., 1990), all cases of infant botulism have been caused by *C. botulinum* strains of the proteolytic group I (Hatheway and McCroskey, 1987). This may be important regarding the pathogenesis of the disease. The strains of group I grow and produce toxin optimally at temperatures approaching body temperature (37°C). Strains of *C. baratii* and *C. butyricum* also grow optimally at, or near, body temperature. Strains of the nonproteolytic group II have lower optimal temperatures (30–35°C) and are therefore likely to be less competitive than group I strains in establishing themselves in the intestinal environment (Hatheway, 1988).

In a study of SIDS cases in Switzerland, type G was found in three cases and incriminated as the cause of death in two of them (Sonnabend et al., 1985). The same study also found three cases of type F and one case of mixed types F and A. The pattern of enteric colonization differed between types: while type G and its toxin were found only in the small intestine, type F organisms and toxin were present only in the large intestine.

IV. INCIDENCE

Infant botulism has been reported from many developed countries including Argentina (Lentini et al., 1984), Australia (Shield, et al., 1981; Ryan, 1987), Canada (Hauschild et al., 1987), Chile (Erazo et al., 1987), Czechoslovakia (Neubauer and Milacek, 1981), England (Turner et al., 1978), France (Paty et al., 1987), Italy (Aureli et al., 1986), Japan (Noda et al., 1988), Spain (Tortosa et al., 1986), Sweden (Jansson et al., 1985), Taiwan (Wang et al., 1988), and the United States (Arnon, 1989) (Table 1). The highest incidence by far has been reported in the United States, with the annual number of infant botulism cases now exceeding the number of foodborne and wound botulism cases (Anonymous, 1987). California reports approximately half of the cases (Arnon, 1989) (Table 1). This distinct geographic disparity probably reflects both regional variation in physician awareness and the worldwide distribution of *C. botulinum* spore types.

V. ENVIRONMENTAL ASPECTS

There are several lines of evidence suggesting that the greatest exposure of infants to spores is from the environment. Since *C. botulinum* is widely distributed in soil and dust, people are always at some risk of ingesting spores. However, it appears that healthy children and adults are completely resistant to this form of exposure to *C. botulinum*, whereas many infants are not (Chin et al., 1979).

There is a strong association between the worldwide distribution of spore types and the incidence of infant botulism and, especially, the causative type. This is best demonstrated in the United States, where the distribution of cases by toxin type reflects the known distribution in the soils (Arnon, 1989; Smith, 1978). Type A cases predominate from the Rocky Mountains westward, while type B cases predominate in the east. Also, in Argentina all cases of infant botulism have been type A (Giménez, 1990), which is the most prevalent type in Argentine soils (Ciccarelli and Giménez, 1981). The situation in Japan may appear contradictory; all cases of infant botulism have been type A, although type E predominates in the environment. However, group II strains have not been associated with infant botulism, and the cases in Japan have all been associated

Table 1 Worldwide Reports of Infant Botulism[a]

Country	Number of cases of infant botulism	Number of fatalities	Toxin type	Honey implicated in cases: Possible	Honey implicated in cases: Confirmed	Ref.[b]
Argentina	23[c]	6	A	?[d]	?	1
Australia	11	0	A, B	1	No	2–4
Canada	4	0	A, B	1	1	5
Chile	1	0	?	No		6
Czechoslovakia	1	0	B	No		7
England	2	0	A, BF	No		8,9
France	1	0	B	No		10
Italy	4	1	A, B, E	4	No	11
Japan	10[e]	0	A, C	9	6	12,13
Spain	1	0	B	1	?	14
Switzerland	1	0	A	No		15
Sweden	1	0	A	?	?	16
United States	932	?	A, B, F	Yes	~20%	17,18
Taiwan	1	0	B	No		19

[a]To June 30, 1990. Does not include SIDS cases.
[b]1) Giménez, 1990; 2) Shield et al., 1981; 3) Murrell and Stewart, 1983; 4) Ryan, 1987; 5) Hauschild et al., 1987; 6) Erazo et al., 1987; 7) Neubauer and Milacek, 1981; 8) Turner et al., 1978; 9) Smith et al., 1989; 10) Paty et al., 1987; 11) Aureli et al., 1989; 12) Sakaguchi, 1989; 13) Oguma et al., 1990; 14) Tortosa et al., 1986; 15) Gautier et al., 1988; 16) Jansson et al., 1985; 17) Hatheway, 1990; 18) Arnon, 1986; 19) Wang et al., 1988.
[c]All cases except one reported from the Province of Mendoza.
[d]Association and/or result not published in report.
[e]Three cases were not laboratory confirmed.

with the feeding of honey. Japan imports most of its honey from China (Sakaguchi, 1989), where type A predominates.

More direct studies have also implicated the environment as the source of spores causing infant botulism. *C. botulinum* has been isolated from environmental samples such as soil and vacuum-cleaner dust in comparable frequency from both case-associated and control homes in California, but in every instance in which it was isolated from a case home, the environmental isolate had the same toxin type as the isolate from the patient (Arnon et al., 1981; Chin et al., 1979). Similarly, *C. botulinum* was isolated from environmental samples associated with three of four cases of infant botulism in Australia; in each case, the toxin type was the same as that isolated from the patient (Murrell and Stewart,

1983). A study in Utah (Thompson et al., 1980) found that 11 of 12 affected infants resided within one or two blocks of a site of soil disruption due to agriculture or construction, and most homes and their neighborhoods were dusty. An environmental source of *C. botulinum* spores was also demonstrated by a study of three cases of infant botulism which occurred within 800 meters of each other between June 1981 and September 1984 in a trailer park in Colorado (Istre et al., 1986). A great majority of soil samples (91%) from the trailer park and vacuum-cleaner dust (77%) from homes yielded *C. botulinum* type A, the same type associated with all three patients.

VI. IMPLICATED FOODS

One of the earliest studies of infant botulism (Midura and Arnon, 1976) examined numerous food items consumed by three patients. Honey ingested by one infant contained *C. botulinum* spores, but no preformed toxin. Older family members who had also consumed the honey remained healthy. No *C. botulinum* organisms or toxin were found in any other infant foods available for testing, including breast milk and vitamin supplements. Honey continues to be the only food implicated in infant botulism. Spores of *C. botulinum* have been isolated repeatedly from leftover honey fed to afflicted infants, and their toxin type consistently matched that of the isolates from the infants' stools (Hauschild et al., 1988). By 1984, 20 such cases had been reported in the United States. The California Department of Health Services now distributes a pamphlet advising against giving babies honey during their first year. This is the position of Health and Welfare Canada, the Centers for Disease Control, the American Academy of Pediatrics, and the Sioux Honey Association, the world's largest honey producer. Because of the association between honey and infant botulism, several surveys for spores in honey have been published. These are detailed in Chapter 3 and are briefly summarized here (Table 2). In general, surveys of honey in the United States, Europe, and Asia report spore levels in the order of <1–10 per kg, while Canadian surveys indicate a lower spore load. In contrast to the general surveys, some honey samples associated with infant botulism have been heavily contaminated with botulinal spores. A Canadian sample associated with illness, and other samples from the same production lot, contained 10^3–10^4 botulinal spores per kg (Hauschild et al., 1988). Midura (1979) reported concentrations of 5×10^3–8×10^4 per kg in samples associated with infant botulism. Infant foods other than honey have been examined, and both corn syrup (Kautter et al., 1982) and dry cereal (Hauschild et al., 1988) have been positive for botulinal spores. However, in contrast to honey, it is very unlikely that they would cause illness because the numbers of spores found have been very low. In addition, no food other than honey has been implicated in infant botulism.

Table 2 Summary of Surveys for *C. botulinum* Spores in Honey

Location	Approx. MPN per kg[a]	Types identified	Association with illness	Ref.[b]
United States	2	A, B	no	1–5
	8×10^4	A, B	yes	2
Canada	<0.2		no	6
	3	A	no	7
	8×10^3	A	yes	6
Europe	6	A	no	8–14
Asia	8	A, B, C	no	14
Central and South America	21	A, C, D	no	11,14

[a]Average of MPN counts listed for location in Table 5, Chapter 3.
[b]1) Sugiyama et al., 1978; 2) Midura et al., 1979; 3) Huhtanen et al., 1981; 4) Kautter et al., 1982; 5) Guilfoyle and Yager, 1983; 6) Hauschild et al., 1988; 7) Hauschild personal communication, 1991; 8) Hartgen, 1980; 8) Flemmig and Stojanowic, 1980; 9) Aureli et al., 1983; 10) Pastoni et al., 1986; 11) Hetland, 1986; 12) Berry et al., 1987; 13) Sakaguchi, 1989.

VII. CLINICAL ASPECTS

While most patients require hospitalization, the clinical spectrum of infant botulism has expanded to include some outpatient cases and a small portion of SIDS cases. The initial symptoms are less clear-cut than those of foodborne botulism (Arnon et al., 1977; Johnson et al., 1979; Neubauer and Milacek, 1981; Pickett et al., 1976; Thompson et al., 1980; Turner et al., 1978). The most common, and usually the earliest, symptom is constipation. After several days to a week of constipation, medical attention is generally requested. Often the infants also show a generalized weakness and a weak cry. Other symptoms may include feeding difficulty and poor sucking, lethargy, lack of facial expression, irritability, and progressive "floppiness." Ocular dysfunctions, including ptosis and dilated and sluggish pupils, usually become evident during the course of the disease but are not always noted on initial examination. Respiratory arrests occur frequently but are seldom fatal.

Diagnosis of infant botulism is often difficult due to the lack of specific symptoms and variation in severity at first examination (Johnson et al., 1979) and requires the identification of *C. botulinum* toxin in the serum or toxin and/or organisms in stools from a patient exhibiting the clinical symptoms. Stools from affected individuals usually contain moderate to high toxin levels ($10–10^5$

mouse lethal doses per gram), and up to 10^8 viable organisms per gram (Paton et al., 1983; Takahashi et al., 1988; Wilcke et al., 1980). Toxin is seldom found in the serum of patients (Hatheway, 1979; Johnson et al., 1979; Midura, 1979). Excretion of the organism and toxin in stools continues long after the onset of symptoms, often even after the infant's apparent recovery. Toxin has been detected in a patient for as long as 138 days and organisms for as long as 158 days after the onset of illness (Turner et al., 1978).

Various studies of healthy infants have indicated that *C. botulinum* is not part of their normal resident intestinal microflora (Arnon, 1985; Midura, 1979; Hatheway and McCroskey, 1987; Stark and Lee, 1982; Wilcke et al., 1980). This is also indicated by the fact that patients recovering from infant botulism do eventually clear *C. botulinum* from their intestines (Midura and Arnon, 1976; Shield et al., 1981). Therefore, finding *C. botulinum* and/or botulinum toxin in the feces of an infant showing clinical signs of botulism reliably confirms the diagnosis.

Optimal treatment for infant botulism consists primarily of high quality supportive care with close attention to nutrition and pulmonary aid (Arnon et al., 1977; Johnson et al., 1979). Approximately one quarter of all infants affected require mechanical ventilation. Many infants require gavage feeding. Experience with early cases showed that antitoxin is not required for complete recovery (Arnon et al., 1979; Pickett et al., 1976; Shield et al., 1981; Turner et al., 1978). Similarly, the use of antibiotics does not appear to affect the course or outcome of infant botulism. When antibiotics such as ampicillin and penicillin were administered, either orally or parenterally, patients continued to excrete *C. botulinum* and toxin in their feces (Arnon et al., 1977; Pickett et al., 1976; Turner et al., 1978).

With time, recovery appears to be complete. There have been only two reports of incomplete recovery. Gautier et al. (1988) described a patient with a dysrhythmic EEG associated with abnormal auditory responses with implied brain stem dysfunction. Jones et al. (1990) described an infant who showed persistent neurological changes after otherwise recovering from infant botulism.

VIII. SIDS

A possible link between infant botulism and SIDS was suggested as early as 1976 by Midura and Arnon, who noted the similarity between the sudden respiratory arrest of an infant botulism patient and SIDS. The recognition that botulinum toxin is produced by *C. botulinum* in the intestines of patients, and often in sufficient quantities to cause respiratory arrest, led Arnon et al. (1978) to examine necropsy specimens from SIDS cases for *C. botulinum* and its toxin. They found *C. botulinum* organisms in 10 of 211 SIDS cases (4.9%). Toxin was detected in 2 of the 10 culture-positive cases. They further found that the age distribution of 62 hospitalized infant botulism patients was similar to that of their 211 SIDS cases and that of other series of SIDS cases.

C. botulinum of type A was isolated from one case of SIDS that occurred in Manitoba (Hauschild et al., 1983). In another study of necropsy specimens in Switzerland, botulinum toxin was found in 9 of 59 SIDS cases (15%) (Sonnabend et al., 1985). In this study, the toxin was detected in four serum samples. Since toxin is not normally detected in serum, this might suggest that these patients had very high toxin levels, precipitating acute episodes of infant botulism. However, in seven of the specimens, the toxin types were very unusual: three were type F, three were type G, and one was type C. Preliminary results of a study involving SIDS cases in Montreal, Canada, have found three of 62 colon samples positive for *C. botulinum* type A (Dodds and Delage, unpublished results).

The isolation of *C. botulinum* type G from a SIDS case (Sonnabend et al., 1981) was the first association of type G with possible illness. If type G plays a role in SIDS, these cases may go undetected for two reasons: unlike all other types of *C. botulinum*, type G does not produce a positive lipase reaction on egg yolk agar, the agar commonly used in screening tests for presumptive *C. botulinum*; and, because type G only produces small amounts of toxin in culture, toxin is often not demonstrable until 7 days, longer than the routine 4-day incubation (Hatheway and McCroskey, 1987).

While there are probably many different etiologies leading to SIDS, it does appear that infant botulism may explain a small portion of these cases.

IX. PREVENTION

Since honey is the only food that has been implicated in infant botulism and is not essential for the nutrition of infants, it is recommended that honey not be fed to infants under one year of age. It would also seem prudent to exclude other nonessential sweeteners that occasionally carry small numbers of *C. botulinum* spores, e.g., corn syrup, from infants' diets. However, the most prevalent source of spores is the infant's general environment, which should be clean and relatively dust-free. Additional preventive measures will depend on a better understanding of the host and environmental determinants of this disease.

X. SUMMARY

While much has been learned about infant botulism since it was first recognized in 1976, some important aspects have yet to be investigated. It is quite likely that physicians are not universally aware of the disease and may not always recognize the symptoms. To date, the disease has only been reported from 14 developed countries, but the distribution of infant botulism is probably more global in nature, since *C. botulinum* spores are ubiquitous worldwide (Dodds and Hauschild, 1989). Further study on the disease incidence may reveal important factors regarding host susceptibility. Although the disease is not as severe as foodborne botulism and its mortality rate is lower, long and expensive hospital stays and the connection between infant botulism and SIDS are two reasons for continued study of the disease.

REFERENCES

Anonymous (1987). Summary of notifiable diseases United States 1986. *MMWR, 35(55)*: 3,51.

Arnon, S. S. (1989). Infant botulism. In *Anaerobic Infections in Humans* (S. M. Finegold and W. L. George, eds.), Academic Press, Inc., Toronto, p. 601.

Arnon, S. S. (1986). Infant botulism: Anticipating the second decade. *J. Infect. Dis. 154*: 201.

Arnon, S. S. (1985). Infant botulism. In *Clostridia in Gastrointestinal Disease* (S. P. Borriello, ed.), CRC Press, Inc., Boca Raton, FL, p. 40.

Arnon, S. S., Damus, K., and Chin, J. (1981). Infant botulism: Epidemiology and relation to sudden infant death syndrome. *Epidemiologic Rev. 3*: 45.

Arnon, S. S., Damus, K., Thompson, B., Midura, T. F., and Chin, J. (1982). Protective role of human milk against sudden death from infant botulism. *J. Pediatr. 100*: 568.

Arnon, S. S., Midura, T. F., Clay, S. A., Wood, R. M., and Chin, J. (1977). Infant botulism: Epidemiological, clinical, and laboratory aspects. *JAMA 237*: 1946.

Arnon, S. S., Midura, T. F., Damus, K., Wood, R. M., and Chin, J. (1978). Intestinal infection and toxin production by *Clostridium botulinum* as one cause of sudden infant death syndrome. *Lancet 1*: 1273.

Arnon, S. S., Werner, S. B., Faber, H. K., and Farr, W. H. (1979). Infant botulism in 1931: Discovery of a misclassified case. *Am. J. Dis. Child. 133*: 580.

Aureli, P., Fenicia, L., Creti, R., Bertini, E., Vigevano, F., di Capua, M., and Pirozzi, N. (1989). Botulismo infantile in Italia. Aspetti clinici e microbiologici di 5 casi. *Riv. Ital. Peditr. 15*: 442.

Aureli, P., Fenicia, L., Pasolini, B., Gianfranceschi, M., McCroskey, L. M., and Hatheway, C. L. (1986). Two cases of type E infant botulism caused by neurotoxigenic *Clostridium butyricum* in Italy. *J. Infect. Dis. 154*: 207.

Aureli, P., Ferrini, A. M., and Negri, S. (1983). Ricerca delle spore di *Cl. botulinum* nel miele. *Riv. Soc. Ital. Sci. Alim. 12*: 457.

Berry, P. R., Gilbert, R. J., Oliver, R. W. A. (1987). Some preliminary studies on low incidence of infant botulism in the United Kingdom. *J. Clin. Pathol. 40*: 121.

Burr, D. H., and Sugiyama, H. (1982). Susceptibility to enteric botulinum colonization of antibiotic-treated adult mice. *Infect. Immun. 36*: 103.

Chia, J. K., Clark, J. B., Ryan, C. A., and Pollack, M. (1986). Botulism in an adult associated with food-borne intestinal infection with *Clostridium botulinum. N. Engl. J. Med. 315*: 239.

Chin, J., Arnon, S. S., and Midura, T. F. (1979). Food and environmental aspects of infant botulism in California. *Rev. Inf. Dis. 1*: 693.

Ciccarelli, A. S., and Giménez, D. F. (1981). Clinical and epidemiological aspects of botulism in Argentina. In *Biomedical Aspects of Botulism* (G. E. Lewis, ed.), Academic Press, New York, p. 291.

Davis, D., DeLaTorre, L., Kazemier, A., Snyder, R., Chavez, T., Pincomb, B., Hall, J., Skeels, M., Woolfolk, C., Madson, R., Hoffman, R. E., and Burkhart, M. (1979). Type F infant botulism—New Mexico. *MMWR, 29*: 85.

Dodds, K. L., and Hauschild, A. H. (1989). Distribution of *Clostridium botulinum* in the environment and its significance in relation to botulism. In *Recent Advances in Microbial Ecology* (T. Hattori, Y. Ishida, Y. Maruyama, R. Y. Morita, and A. Uchida, eds.), Japan Scientific Societies Press, Tokyo, p. 472.

Erazo, R., Marmabio, E., Cordero, J., Fielbaum, O., and Trivino, X. (1987). Botulismo infantil: Comunicacion de un caso. *Rev. Med. Chi., 115*: 344.

Flemmig, R., and Stojanowic, V. (1980). Untersuchungen von Bienenhonig auf *Clostridium botulinum* sporen. *Arch. Lebensmittelhyg. 31*: 179.

Gautier, E., Gallusser, A., and Despland, P. A. (1988). Botulisme infantile. *Helv. Paediat. Acta. 43*: 521.

Giménez, D. F. (1990). Personal communication. Miramar, Argentina.

Guilfoyle, D. E., and Yager, J. F. (1983). Survey of infant foods for *Clostridium botulinum* spores. *J. Assoc. Off. Anal. Chem. 66*: 1302.

Hall, J. D., McCroskey, L. M., Pincomb, B. J., and Hatheway, C. L. (1985). Isolation of an organism resembling *Clostridium baratii* which produces type F botulinal toxin from an infant with botulism. *J. Clin. Microbiol. 21*: 654.

Hartgen, V. H. (1980). Untersuchungen von Honigproben auf Botulinustoxin. *Arch. Lebensmittelhyg. 31*: 178.

Hatheway, C. L. (1990). Personal communication. Centers for Disease Control, Atlanta, Georgia.

Hatheway, C. L. (1988). Botulism. In *Laboratory Diagnosis of Infectious Diseases—Principles and Practice* (A. Balows, W. J. Hausler, M. Ohashi, et al., eds.), Vol. 1, Springer-Verlag, New York, p. 111.

Hatheway, C. L. (1979). Laboratory procedures for cases of suspected infant botulism. *Rev. Infect. Dis. 1*: 647.

Hatheway, C. L., and McCroskey, L. M. (1987). Examination of feces and serum for diagnosis of infant botulism in 336 patients. *J. Clin. Microbiol. 25*: 2334.

Hauschild, A. H. W., Gauvreau, L., and Black, W. A. (1983). Botulism in Canada—summary for 1982. *Can. Dis. Wkly. Rpt. 9–14*: 53.

Hauschild, A. H. W., Gauvreau, L., and Black, W. A. (1987). Botulism in Canada—Summary for 1986. *Can. Dis. Wkly. Rpt. 13–11*: 47.

Hauschild, A. H. W., Hilsheimer, R., Weiss, K. F., and Burke, R. B. (1988). *Clostridium botulinum* in honey, syrups and dry infant cereals. *J. Food Prot. 51*: 892.

Hetland, A. (1986). *Clostridium botulinum* sporer i norskprodusert honning? *Norsk. Vet. 98*: 725.

Hoffman, R. E., Pincomb, B. J., Skeels, M. R., and Burkhart, M. J. (1982). Type F infant botulism. *Am. J. Dis. Child. 136*: 270.

Huhtanen, C. N., Knox, D., and Shimanuka, H. (1981). Incidence and origin of *Clostridium botulinum* spores in honey. *J. Food Prot. 44*: 812.

Istre, G., Compton, R., Novotny, T., Young, J. E., Hatheway, C. L., and Hopkins, R. S. (1986). Infant botulism: Three cases in a small town. *Am. J. Dis. Child. 140*: 1013.

Jansson, L., Bjerre, I., Schalen, C., and Mardh, P. A. (1985). Forsta fallet av infantil botulism i Sverige. *Lakartidningen 82*: 955.

Johnson, R. O., Clay, S. A., and Arnon, S. S. (1979). Diagnosis and management of infant botulism. *Am. J. Dis. Child. 133*: 586.

Jones, S., Huma, Z., Huagh, C., Young, Y., Starer, F., and Sinclair, L. (1990). Central nervous system involvement in infantile botulism. *Lancet 335*: 228.

Kautter, D. A., Lilly, T., Solomon, H. M., and Lynt, R. K. (1982). *Clostridium botulinum* spores in infant foods: A survey. *J. Food Prot. 45*: 1028.

Lentini, E., Fernandez, R., Ciccarelli, A. S., and Giménez, D. F. (1984). Botulismo en al lactante: Una nueva enfermedad? *Arch. Arg. Pediatr. 82*: 197.

Long, S. S., Gajewski, J. L., Brown, L. W., and Gilligan, P. H. (1985). Clinical, laboratory, and environmental features of infant botulism in southeastern Pennsylvania. *Pediatr. 75*: 935.

McCroskey, L. M., and Hatheway, C. L. (1988). Laboratory findings in four cases of adult botulism suggest colonization of the intestinal tract. *J. Clin. Microbiol. 26*: 1052.

McCroskey, L. M., Hatheway, C. L., and Fenicia, L. (1986). Characterization of an organism that produces type E botulinal toxin but which resembles *Clostridium butyricum* from the feces of an infant with type E botulism. *J. Clin. Microbiol. 23*: 201.

Midura, T. F. (1979). Laboratory aspects of infant botulism in California. *Rev. Infect. Dis. 1*: 652.

Midura, T. F., and Arnon, S. S. (1976). Infant botulism: Identification of *Clostridium botulinum* and its toxins in faeces. *Lancet 2*: 934.

Midura, T. F., Snowden, S., Wood, R. M., and Arnon, S. S. (1979). Isolation of *Clostridium botulinum* from honey. *J. Clin. Microbiol. 9*: 282.

Moberg, L. J., and Sugiyama, H. (1979). Microbial ecological basis of infant botulism as studied with germfree mice. *Infect. Immun. 25*: 653.

Morris, G. M., and Hatheway, C. L. (1980). Botulism in the United States 1979. *J. Infect. Dis. 142*: 302.

Murrell, W. G., and Stewart, B. J. (1983). Botulism in New South Wales, 1980–1981. *Med. J. Aust. 1*: 13.

Neubauer, M., and Milacek, V. (1981). Infant botulism type B in central Europe. *Zentralbl. Bakteriol. Mikrobiol. Hyg. (A) 250*: 540.

Noda, H., Sugita, K., and Koike, A. (1988). Infant botulism in Asia. *Am. J. Dis. Child. 142*: 125.

Oguma, K., Yokota, K., Hayashi, S., Takeshi, K., Kumagai, M., Itoh, N., Tachi, N., and Chiba, S. (1990). Infant botulism due to *Clostridium botulinum* type C toxin. *Lancet 336*: 1449.

Pastoni, F., Roggi, C., and Caccialanza, G. (1986). Isolation of *Clostridium* species from honey. *Proceedings World Congress of Foodborne Infections and Intoxications*, West Berlin, p. 506.

Paton, J. C., Lawrence, A. J., and Steven, I. M. (1983). Quantities of *Clostridium botulinum* organisms and toxin in feces and presence of *Clostridium botulinum* toxin in the serum of an infant with botulism. *J. Clin. Microbiol. 17*: 13.

Paty, E., Valdes, L., Harpey, J. P., Doré, F., Hubert, P., Roy, C., and Caille, B. (1987). Un cas de botulisme chez un nourrisson de 11 mois. *Arch. Fr. Pediatr. 44*: 129.

Pickett, J., Berg, B., Chaplin, E., and Brunstetter-Shafer, M. (1976). Syndrome of botulism in infancy: clinical and electrophysiologic study. *N. Engl. J. Med. 295*: 770.

Ryan, P. J. (1987). Infant botulism—the first reported case from Queensland. *Med. J. Aust. 146*: 105.

Sakaguchi, G. (1989). Personal communication. University of Osaka Prefecture, Osaka, Japan.

Shield, L. K., Wilkinson, R. G., and Ritchie, M. (1981). Infant botulism in Australia—a case report. *Aust. Paediatr. J. 17*: 59.

Smith, G. E., Hinde, F., Westmoreland, D., Berry, P. R., and Gilbert, R. J. (1989). Infantile botulism. *Arch. Dis. Child. 64*: 871.

Smith, L. D. (1978). The occurrence of *Clostridium botulinum* and *Clostridium tetani* in the soil of the United States. *Health Lab. Sci. 15*: 74.

Sonnabend, O. A. R., Sonnabend, W., and Heinzle, R. (1981). Isolation of *Clostridium botulinum* type G and identification of type G botulinal toxin in humans: Report of five sudden unexpected deaths. *J. Infect. Dis. 143*: 22.

Sonnabend, O. A. R., Sonnabend, W. F. F., and Krech, U. (1985). Continuous microbiological and pathological study of 70 sudden and unexpected infant deaths: Toxigenic intestinal *Clostridium botulinum* infection in 9 cases of sudden infant death syndrome. *Lancet 1*: 237.

Stark, P. L., and Lee, A. (1982). Clostridia isolated from the feces of infants during the first year of life. *J. Pediatr. 100*: 362.

Sugiyama, H., and Mills, D. C. (1978). Intraintestinal toxin in infant mice challenged intragastrically with *Clostridium botulinum* spores. *Infect. Immun. 21*: 59.

Sugiyama, H., Mills, D. C., and Kuo, L. J. (1978). Number of *Clostridium botulinum* spores in honey. *J. Food Prot. 41*: 848.

Takahashi, M., Shimizu, T., Ooi, K., Noda, H., Nasu, T., and Sakaguchi, G. (1988). Quantification of *Clostridium botulinum* type A toxin and organisms in the feces of a case of infant botulism and examination of other related specimens. *Japan J. Med. Sci. Biol. 41*: 21.

Thompson, J. A., Glasgow, L. A., Warpinski, J. R., and Olson, C. (1980). Infant botulism: clinical spectrum and epidemiology. *Pediatr. 66*: 936.

Tortosa, P. T., Villalta, E. M., Caamano, J. R., Cano, C. L., Mira, A. P., and Borrajo, E. (1986). Botulismo del lactante. Presentacion de un caso. *An. Esp. Pediatr. 24*: 193.

Turner, H. D., Brett, E. M., Gilbert, R. J., Ghosh, A. C., and Liebeschuetz, H. J. (1978). Infant botulism in England. *Lancet 1*: 1277.

van Ermengem, E. (1897). Ueber einen neuen anaeroben Bacillus und seine Beziehungen zum Botulismus. *Z. Hyg. 26*: 1.

Wang, C. C., Chu, M. L., Liou, W. Y., Twu, P. H., Lin, C. H., and Lee, C. L. (1988). Infant botulism: report of a case. *J. Formosan Med. Assoc. 87*: 919.

Wilcke, B. W., Midura, T. F., and Arnon, S. S. (1980). Quantitative evidence of intestinal colonization by *Clostridium botulinum* in four cases of infant botulism. *J. Infect. Dis. 141*: 419.

II

CONTROL OF *CLOSTRIDIUM BOTULINUM* IN FOODS

6

Principles of Control

Jungho Kim

Sunchon National University, Sunchon, Republic of Korea

Peggy M. Foegeding

North Carolina State University, Raleigh, North Carolina

I. INTRODUCTION

Botulism is a neuroparalytic disease caused by toxigenic strains of *Clostridium botulinum*. Although the number of botulism cases and the mortality rate have decreased in recent years, botulism still remains a significant public health hazard. Four clinical forms of botulism are currently recognized, i.e., foodborne, infant, wound, and unclassified botulism (see Chapter 1). In this chapter, the discussion will concentrate on the control of foodborne botulism.

The following are the primary events leading to foodborne botulism. First, the food must be contaminated with the spores or cells of toxigenic *C. botulinum*. Generally, contamination is due to the presence of *C. botulinum* in the environment where the food is produced, harvested, processed, or stored. Second, the cells or spores must survive the food-processing treatment. Alternatively, postprocessing contamination must occur. Next, the organism must multiply and produce toxin in the food. For this, the food must have an environment or microenvironment favorable for germination and outgrowth of the spores, and for growth and toxin production of the vegetative cells. Finally, the food must be consumed without sufficient cooking to destroy the heat-labile toxin.

Foodborne botulism can be effectively controlled if one or more of the following methods are employed: keeping the food, food-processing unit, and container from contamination and/or removal of contaminating *C. botulinum* cells and spores; destruction of existing *C. botulinum* cells and spores; prevention of spore germination and outgrowth and/or vegetative cell growth and toxin production; and destruction of produced toxin. It would be ideal if we could completely prevent the food and food-contact environment from being contaminated with *C. botulinum* during harvesting, processing, and storage. However, because of its wide distribution in nature, it is often impossible to keep the organism from contaminating foods. Therefore, a more realistic goal is to keep the contamination level as low as possible and to remove or inactivate contaminating organisms as early in processing as possible. This goal is achieved by proper application of the hazard analysis critical control point (HACCP) concept (Bauman, 1990) and Good Manufacturing Practices (GMP) (CFR, 1990a). Assuming that the contamination level is kept as low as possible, *C. botulinum* spores should be destroyed or kept from germination, growth, and toxin production by properly applied food-processing and storage processes as detailed in this and subsequent chapters. Destruction of botulinum toxin is addressed in Chapter 3 of this volume. For consumers, the last resort for avoiding botulism is not to eat food with any indication of spoilage and to practice safe food handling in the home. Thus, it remains an important responsibility for public health agencies to educate the consumers not to eat suspicious foods, to handle foods safely, and to properly cook foods.

II. SPORE STRUCTURE

Most theories of how spores resist inactivation treatments are closely linked to the spore ultrastructure and composition (Gombas, 1983; Gould, 1983). The structure of a spore is briefly described here. (More interested readers may consult Murrell, 1969, 1988, and Russell, 1982.)

At the spore center is the *core* or *protoplast*, containing the DNA, RNA, enzymes, and other components that the spore will require upon germination. The core is also the location of components that may be involved in spore resistance and/or dormancy. Surrounding the core is the *inner spore membrane*, which contains the enzymes required for the synthesis of the next layer, the *germ cell wall*. The germ cell wall has lysozyme sensitivity similar to the vegetative cell wall and is believed to contain vegetative peptidoglycan. Upon germination, the core, inner spore membrane, and germ cell wall become the cytoplasm, membrane, and wall, respectively, of the newly emerged vegetative cell.

Exterior to the germ cell wall is the *cortex*. The peptidoglycan of the cortex has a unique sensitivity to lysozyme and is chemically different from the germ cell wall peptidoglycan. The *outer spore membrane*, which contains the enzymes required for cortex synthesis, encircles the cortex and has opposite polarity from the inner spore membrane. The *spore coats* are generally rigid, polypeptide-rich layers, which may entrap sporangial cytoplasmic components.

The spore coats provide a permeability barrier to some compounds. The *exosporium* is an outer membranous layer, which may be loosely or tightly fitted around the spore. Upon germination, the cortex is hydrolyzed, and the outer membrane, spore coats, and exosporium are discarded by the emerging cell.

Dipicolinic acid (DPA) is a unique constituent of bacterial spores and has been isolated from spores of all *Clostridium* and *Bacillus* species analyzed. DPA is located primarily in the core and associated in a chelated form with calcium. The DPA and calcium contents of spores of different species range from about 5 to 15% and 1 to 3% of the dry weight, respectively (Murrell, 1969). The role of DPA in heat resistance is somewhat controversial. Many researchers implicated DPA as an agent in establishing heat resistance of spores (Church and Halvorson, 1959; Hashimoto et al., 1960). Day and Costilow (1964) indicated that heat resistance developed coincidentally with synthesis of DPA. However, Levinson et al. (1961) found no direct relationship between heat resistance and DPA or calcium content, but a good agreement was found between the calcium:DPA ratio and heat resistance of spores. Grecz et al. (1972) indicated a correlation between the thermodynamic stabilities of the Ca-DPA chelate in model systems and the *in vivo* heat resistance of Ca-DPA–containing spores of *C. botulinum*. Hanson et al. (1972) indicated that DPA is a nonessential determinant of heat resistance but may function in the maintenance of heat resistance.

Another unique constituent of bacterial spores is a family of small, acid-soluble spore proteins (SASPs) located in the spore protoplast (for a review, see Setlow, 1988). SASPs are synthesized late in sporulation and degraded early in spore germination (Setlow, 1978, 1985). One function of SASPs is to supply, by their degradation, amino acids for protein synthesis during spore germination. The α/β-type SASPs are associated with spore DNA (Francesconi et al., 1988) and also play a key role in the resistance of spores to UV light (Mason and Setlow, 1986, 1987). The amount of SASPs varies from 8 to 20% of total spore protein, depending on the species examined.

III. CONTROL OF *C. BOTULINUM* BY INACTIVATION

C. botulinum vegetative cells and spores, which may contaminate food processing environments, packaging materials, and foods, can be inactivated by treatments involving heat, gas, ionizing radiation, and chemicals. Because of the substantial resistance of *C. botulinum* spores to those treatments and because *C. botulinum* likely enters foods in the spore and not the vegetative state, discussion will be concentrated on the inactivation of spores rather than vegetative cells. Priority will be given to the discussion of the mechanism of action, mechanism of the spore resistance, and factors affecting the resistance of the spores to such treatments.

A. Chlorine Compounds

Among the chemical sanitizers commonly used in the food industry, chlorine compounds, which are widely used, are most effective against bacterial spores. Chlorine compounds used in the food industry include sodium hypochlorite (NaOCl), calcium hypochlorite ($Ca(OCl)_2$), gaseous chlorine (Cl_2), chlorine dioxide (ClO_2), chlorinated trisodium phosphate, and chloramine compounds (chloramine T, dichloramine T, etc.). Chlorine compounds consist of a mixture of hypochlorite ions (OCl^-), molecular chlorine (Cl_2), hypochlorous acid (HOCl), and other active chlorine compounds in aqueous solution. The antimicrobial effectiveness of chlorine compounds results from the strong oxidizing properties of chlorine If chlorine is reduced to chloride ($Cl + e^- \rightarrow Cl^-$) by inorganic reducing substances or organic material (other than amine), it loses its antimicrobial activity. Among the chlorine compounds, sodium and calcium hypochlorites give the most active chlorine in sanitizing solutions and are most widely used in the food industry.

Mechanism of Sporicidal Activity

Chlorine may act as a sporicidal agent by disrupting the spore coat as well as underlying layers, thereby altering spore permeability. Wyatt and Waites (1975) found that chlorine disrupts coats of *Clostridium* and *Bacillus* spores by combining with and removing protein. Electron microscopy studies with chlorine-treated *B. anthracoides* and *B. cereus* indicated that chlorine initially caused separation of the spore coat from the cortex, followed by sequential dissolution of the spore layers (Galanina et al., 1979; Kulikovskii, 1976; Kulikovsky et al., 1975). Protein was removed from hypochlorite-treated *C. botulinum*, *C. bifermantans*, *B. cereus*, and *B. subtilis* (Foegeding and Busta, 1983c; Waites et al., 1972; Wyatt and Waites, 1971). Chlorine-treated *Clostridium* and *Bacillus* spores lost DPA, calcium, RNA, and DNA to the surrounding medium (Dye and Mead, 1972; Kulikovsky et al., 1975). Hypochlorite treatment permits germination of spores with lysozyme, which is not possible unless procedures are used to alter the coats, making the spores permeable to lysozyme (Foegeding and Busta, 1983c; Wyatt and Waites, 1975). This finding further supports disruption and removal of spore coat protein and exposure of spore cortex as a result of hypochlorite treatment.

Pretreatment of spores with sublethal concentrations of chlorine may increase the sensitivity of spores to heat or chemical treatment (Dye and Mead, 1972; Foegeding and Busta, 1983c). The increased sensitivity may be due to altered coat and cortex structures (Foegeding, 1983). Treatment of *C. bifermantans*, *B. subtilis*, and *B. cereus* spores with low concentrations of chlorine has been shown to prevent or reduce the rate of germination, but spores that germinated were able to form vegetative cells (Wyatt and Waites, 1975). At higher concentrations of chlorine, where the germination mechanism was inactivated, lysozyme initiated germination; but the spores were damaged to the extent that no outgrowth occurred (Wyatt and Waites, 1975). A reduced rate of germination after hypochlorite treatment was also observed in *C. botulinum* (Foegeding and Busta, 1983a). Sublethal doses of hypochlorite changed the requirements for germination and subsequent multiplication of *C. botulinum* (Foegeding and Busta, 1983b, d).

Resistance of Spores to Chlorine Compounds and Factors Affecting Sporicidal Activity

C. botulinum spores in general are more resistant than non–spore-forming bacteria, which are readily inactivated after a few minutes at a chlorine concentration of 1 ppm, but they are less resistant than *Bacillus* spores, which can survive for longer than 10 minutes at a chlorine concentration of 10 ppm (Ito et al., 1967). *C. botulinum* strains vary in their resistance to free available chlorine (Table 1). Factors affecting the sporicidal activity of chlorine compounds include the type of organism, type and concentration of chlorine compound used, exposure time, temperature, pH, and presence of organic material. The more heat-resistant types also appear to be more chlorine resistant (Ito et al., 1967). Dye and Mead (1972) compared the sporicidal activity of hypochlorite and

Table 1 Resistance of *C. botulinum* Spores to Chlorine Compounds

Strain	Free available chlorine concentration (ppm)	pH	Time[a] (min)	Ref.
A	4.5	6.5	10.5	Ito and Seeger, 1980
B[b]	4.5	6.5	12.0	
B[c]	4.5	6.5	5.5	
C	4.5	6.5	3	
E	4.5	6.5	6.0	
F	4.5	6.5	8	
12885A	4.5	3.5	2.1	Ito et al., 1968
	4.5	5.0	4.3	
	4.5	7.0	8.5	
	4.5	8.0	24.0	
Saratoga E	4.5	3.5	1.1	Ito et al., 1967
	4.5	5.0	2.8	
	4.5	6.5	4.0	
	4.5	8.0	17.0	
	4.5	10.0	>30.0	

[a] Time required to inactivate 99.99% spore population.
[b] Proteolytic strain.
[c] Nonproteolytic strain.

chloramine T against *Clostridium* spores. The data indicated that combined available chlorine as chloramine T was less sporicidal than free available chlorine from sodium or calcium hypochlorite. The sporicidal activity of free available chlorine increases as pH decreases (see Table 1). The sporicidal activity of chlorine compounds increases with exposure time, concentration of chlorine, and temperature. The presence of organic material decreases the sporicidal activity of free available chlorine. Unlike free available chlorine, ClO_2 is not highly reactive with organic compounds or ammonia; hence, its use is favored where organic loads are expected to be high (Foegeding, 1983).

Use of Chlorine Compounds as Sanitizers

Chlorine compounds are used effectively to sanitize food-processing equipment and food containers, for washing and conveying raw food products, and for cooling heat-sterilized cans. The use of chlorinated water sprays at selected locations in a processing plant reduces or prevents accumulation of microorganisms and off-odors. For sanitizing equipment, solutions of 100–200 μg of hypochlorite/ml are usually held in or circulated through equipment for at least 2 minutes. Fogging or high-pressure spraying of tanks or enclosed vats with solutions of about 200 μg of hypochlorite/ml are also used commonly. For cooling heat-sterilized cans, 1–2 μg of chlorine/ml of cooling water is recommended (Luh and Kean, 1975). It is important to maintain chlorine sanitizer concentrations at levels that are lethal to the spores present. It should also be kept in mind that chlorination cannot substitute for good sanitary practices and good plant operations.

B. Hydrogen Peroxide (H_2O_2)

Hydrogen peroxide (H_2O_2) has been recognized as a bactericidal agent for many years and is used as a bactericide in foods such as milk and eggs. It is also used as a bleaching and sterilizing agent in fish paste products and boiled noodles in Japan (Ishiwata et al., 1971). There is renewed interest in the utilization of H_2O_2 for sterilization of food processing equipment and food packaging materials, especially in aseptic packaging processes (Stevenson and Shafer, 1983).

Mechanism of Sporicidal Activity

Many reports suggest that the mode of action of H_2O_2 is not due to the H_2O_2 molecule itself, but to the production of the hydroxyl free radical, a powerful oxidant (Turner, 1983). Activation of peroxide to the hydroxyl radical is necessary for sporicidal activity (King and Gould, 1969). The hydroxyl radical, being highly reactive, can attack membrane lipids, DNA, and other essential cell components (Turner, 1983). Like chlorine compounds, H_2O_2 removes protein from spores (Russell, 1982). Spores of high resistance are known to have a thicker cortex and a smaller protoplast (Waites et al., 1980).

Resistance of Spores to H_2O_2 and Factors Affecting Sporicidal Activity of H_2O_2

Considering the importance of *C. botulinum* in public health, not much information about the resistance of the spores of *C. botulinum* to H_2O_2 is available. Ito et al. (1973) compared the resistance of *C. botulinum* spores to those of *C. sporogenes* and *Bacillus* strains. When spores were exposed to 30% H_2O_2 at 30°C, *C. botulinum* type B spores were most resistant, followed by spores of *C. botulinum* type A, *C. sporogenes*, and *Bacillus* strains. At 35% H_2O_2 and 87.8°C, the most resistant *C. botulinum* strain (169B) was less resistant than *Bacillus* strains. Thus, organisms that were resistant to H_2O_2 at 30°C were not necessarily the most resistant at the higher temperature. Leaper (1988) compared the resistance to 17.7% H_2O_2 of 15 spore crops of *C. botulinum* type A and type B prepared using several sporulation media and conditions. Spores of *C. botulinum* type B (strain 115) exhibited a range of resistance at 20°C (D-values of 2.6–13.7 min), dependent on the sporulation medium. An increase in the concentration of H_2O_2 and/or exposure temperature increased the sporicidal activity. At low concentrations, H_2O_2 solutions are bactericidal but not highly sporicidal. To obtain rapid sporicidal activity, relatively high concentrations (e.g., 35%) are needed. At room temperature, H_2O_2 is not rapidly sporicidal. The Q_{10} value for the sporicidal rate of H_2O_2 has been reported to range from 1.6 (Swartling and Lindgren, 1968) to 2.5 (Cerf and Metro, 1977). Other factors affecting the sporicidal activity of H_2O_2 include pH, ions, ultrasonic waves, ultraviolet irradiation, and the state of spores (wet or dry) (Stevenson and Shafer, 1983). The effect of pH on sporicidal activity has been related to the stability of H_2O_2 at various pH levels (Curran et al., 1940). H_2O_2 has maximal stability at pH 3.5–4.0 but becomes increasingly unstable at very low or very high pH values (Schumb et al., 1955). Ions, such as Cu^{2+} and Fe^{2+}, that accelerate the decomposition of peroxide are known to increase the sporicidal activity of H_2O_2 (Gould and Hitchins, 1963; King and Gould, 1969). Ultrasonic waves and ultraviolet radiation produce synergistic effects on the sporicidal activity of H_2O_2 (Ahmed and Russell, 1975; Bayliss and Waites, 1979, 1982). A U.S. patent by Peel and Waites (1981) utilizes a sterilization method employing UV-irradiated solutions of H_2O_2.

Use of H_2O_2 in Aseptic Packaging

Swartling and Lindgren (1968) used H_2O_2 to destroy spores on packaging material in an aseptic filling and packaging system. While such systems have been operating in Europe for many years, a U.S. regulation permitting use of H_2O_2 for sterilizing food contact surfaces was not enacted until 1981 (Davis and Dignan, 1983).

Several aseptic filling and packaging systems utilize H_2O_2 in conjunction with heat and/or other sterilants. Since regulations require low residuals (generally less than 0.1 ppm), hot air is commonly used to dissipate the H_2O_2. The temperatures attained during such operations should result in rapid destruction of spores while some H_2O_2 is still present. It is known that H_2O_2 treatment prior to heat treatment reduces the resistance of spores to heat so that a mild heat treatment will inactivate injured spores (Toledo et al., 1973). In one commercial system the packaging material passes through a 23–25% H_2O_2 bath at 60–80°C; this requires approximately 0.13–0.15 minutes (von Bockelmann, 1973). Another commercial process operates at room temperature with a 15–20% H_2O_2 bath and a passage time of 0.05–0.07 minutes (von Bockelmann, 1973). The sterilization efficiency of these systems in terms of the survival of *C. botulinum* spores was not specified. The danger from anaerobic organisms such as *C. botulinum* in aseptically packaged foods using plastic or paper-based packaging material can be reduced by using materials with a relatively high permeability to oxygen, a low vacuum in the container (Toledo et al., 1973), acidity of the food, and other factors. Worldwide, aseptically packaged foods have a history of safety.

C. Thermal Inactivation

Thermal inactivation is the most common method of sterilizing food products. Compared to other means of sterilization, heat generally is the most appropriate, practical, and convenient. Moist and dry heat are used for sterilization. These processes appear to kill bacteria and their spores by different mechanisms (Russell, 1982). Moist heat is more effective than dry heat. Considering the microorganisms of public health significance, *C. botulinum* is among the most difficult to destroy. Thus, in low-acid, hermetically sealed foods, the primary objective of the sterilization processes is the elimination of this organism.Joslyn (1983) has reviewed the kinetics and molecular events of heat sterilization.

Mechanism of Sporicidal Activity

The mechanism of heat inactivation of bacterial spores is still not completely understood. When spores are lethally heated, intracellular constituents are released, and there is a progressive loss of DPA and calcium, the rate of release being temperature dependent (Brown and Melling, 1973; Hunnell and Ordal, 1961; Rode and Foster, 1960). However, the death of spores proceeds faster than the release of DPA. It was suggested that the leakage occurs as a result of damage to a permeability barrier (Russell, 1982). Hashimoto et al. (1972) suggested that the primary cause of thermal inactivation of spores is due to physical and chemical alterations, which interfere with the absorption of water into the core during germination. Flowers and Adams (1976) suggested that a spore structure destined to become cell membrane or cell wall is the site of thermal injury. The apparent death of spores may be due to the inability of spores to germinate or the inability after germination to outgrow. Dry heat is believed to cause death by the destructive oxidation of cell components (Ernest, 1977).

Mechanism of Heat Resistance

Reviews on spore heat resistance mechanisms include those of Gerhardt (1988), Gerhardt and Marquis (1989), Gombas (1983), Gould (1983, 1986), Lindsay et al. (1985), Murrell (1981, 1988), and Warth (1985). Thermoresistance of spores is attributable to at least three physiochemical determinants that affect the protoplast: dehydration, mineralization, and thermal adaptation (Gerhardt and Marquis, 1989).

It is generally thought that the mechanism of heat resistance involves restricting the mobility of heat-labile components of the spore core, e.g., proteins and genetic material, thereby preventing denaturation (Gombas, 1983). Many researchers agree that dehydration of the protoplast is responsible for this and renders heat-labile components less sensitive to irreversible thermal denaturation. Dehydration of the protoplast is thought to be the only determinant necessary for the heat resistance of spores (Gerhardt and Marquis, 1989). However, there is no general agreement on how the dehydration is achieved, and no single proposed mechanism alone is consistent with all the observed properties of the spore. It is possible that all of the proposed mechanisms are involved to some extent (Gerhardt and Murrell, 1978). Once protoplast dehydration is obtained, its maintenance requires an intact cortex, but not a coat or exosporium (Gerhardt and Marquis, 1989).

Mineralization of the protoplast, usually by calcium, affects heat resistance. Heat resistance of *Bacillus* spores was positively related to calcium content and inversely to magnesium content (Murrell and Warth, 1965). *C. botulinum* spores containing an increased iron content were more susceptible to thermal inactivation than were spores with normal iron levels or spores containing increased zinc or manganese. Spores formed with added Fe or Cu also appeared less able to repair heat-induced injuries than spores with added manganese or zinc (Kihm et al., 1990). Mineralization of spores is accompanied by a reduction in the water content of the protoplast, and at least part of the increased heat resistance associated with mineralization is due to dehydration (Beaman and Gerhardt, 1986). At the upper and lower limits of protoplast water content in fully hydrated spores, however, mineralization acts independently. Mineralization is apparently attained by active transport through the sporangia membrane and passive flow into the developing spore. Although often associated with minerals, DPA is not thought to be necessary for attaining heat resistance, but may have a role in retaining the resistance (Gerhardt and Marquis, 1989).

The spores of thermophilic species are inherently more resistant than those of mesophilic or psychrophilic species (Warth, 1978). However, increasing the sporulation temperature can produce additional heat resistance (Heinen and Lauwers, 1981; Khoury et al., 1987; Williams and Robertson, 1954). In spores, both the inherent and imposed components of thermal adaptation are reflected in the protoplast water content over much, but not all, of the temperature range (Beaman and Gerhardt, 1986).

Heat Activation vs. Inactivation

Generally, bacterial spores exhibit a logarithmic survival curve when exposed to heat. However, strictly exponential thermal death curves are the exception, and deviations from the logarithmic order of death are rather common. Various explanations for this phenomenon have been suggested (Cerf, 1977; Han et al., 1976; Stumbo, 1973). Initial shoulders or initial rises in viable counts are common deviations that may be attributable to heat activation of the spores.

Curran and Evans (1945) noticed that sublethal heat could induce dormant spores to germinate—a phenomenon referred to as heat activation. The temperature and time needed for heat activation vary from one type, strain, or spore preparation of *C. botulinum* to another. *C. botulinum* type A spores undergo maximum activation by heat treatment at 80°C for 10–20 minutes (Smith, 1977). Treatments with calcium dipicolinate, ethanol, and low pH are also known to activate spores (Keynan and Evenchik, 1969). Activation of bacterial spores is a reversible process that causes changes in germination rate, requirements for germination, spore composition, morphology, permeability, and metabolic activity (Keynan and Evenchik, 1969). Activated spores retain heat resistance and refractivity. In food processing, mild, sublethal heat treatments may constitute activation treatments for *C. botulinum* spores, thereby enhancing the ability of the spores to form vegetative cells.

Heat Resistance of Spores and Factors Affecting Resistance

Because of the importance of heat processing in the preparation of preserved foods and the ability of *C. botulinum* to cause fatal foodborne illness, the resistance of *C. botulinum* spores to heat has been studied more thoroughly than that of most other species. Heat resistance is commonly expressed as the decimal reduction time, or D-value. The D-value is the time required to inactivate 90% of the population at a specified temperature. D-values can be calculated by the formula:

$$D = \frac{t}{\log a - \log b}$$

where a is the initial number of spores and b is the number of remaining spores after time t. D-values vary considerably among *C. botulinum* strains, even within the same group or type. They also depend on the method (temperature and medium) of spore production, pretreatment of spores, e.g., with irradiation or hypochlorite, and the heating menstruum. Inadequate recovery systems may result in artificially low D-values. Because the heat resistance of *C. botulinum* spores is dependent on a multitude of factors, the survival of the spores should be checked whenever changes are made in the formulation of the product or in the procedure. The spores of *C. botulinum* types A and B are the most heat resistant, their resistance being characterized by $D_{121°C}$ values in the range of

0.10–0.20 minutes (Stumbo, 1973). They are of particular concern in the sterilization of canned low-acid foods. The canning industry adopted a D-value of 0.2 minutes at 121°C as a standard in calculating the required thermal process (Hauschild, 1989). The heat resistance of type E spores is comparatively low, with $D_{82.2°C}$ values in the range of 0.1–3.0 min. The D-values of some strains of *C. botulinum* are given in Table 2.

The z_D-value is the temperature change necessary to bring about a 10-fold change in the D-value:

$$z_D = \frac{T_2 - T_1}{\log D_1 - \log D_2}$$

where D_1 and D_2 are the D-values at temperatures T_1 and T_2, respectively. The z-values for botulinal spores of most resistant strains are in the order of 10°C (18°F), which has been adopted as a standard (Hauschild, 1989). However, actual values of strains may differ by several degrees (Lynt et al., 1981; Perkins et al., 1975). The minimum thermal process applied to commercial low-acid canned foods is 12 D, based on the most heat-resistant *C. botulinum* strains. Despite the deviation of botulinal strains from the standard D- and z_D-values, adherence of the canning industry to the 12-D concept has ensured the production of safe products in the past (Hauschild, 1989).

Strains As described in Chapter 1, spores of proteolytic *C. botulinum* strains are more resistant to heat than spores of nonproteolytic strains. Nonproteolytic type B strains have slightly higher heat resistance than type E stains, but none of the nonproteolytic type B strains demonstrate heat resistance comparable to that of proteolytic type B strains (Scott and Bernard, 1982). Proteolytic strains of types F, B, and A have similar heat resistances (Lynt et al., 1981). *C. botulinum* type G strains vary considerably in their heat resistance; some are comparable to proteolytic strains of type A, B, and F (D-values of 0.14–0.19 minutes at 121°C), but others are similar to nonproteolytic strains of types B and E (Lynt et al., 1984). Unlike other *C. botulinum* types, strains of type G have z_D-values of 38–49°C (Lynt et al., 1984).

Although the heat resistance of nonproteolytic *C. botulinum* spores is much lower than that of proteolytic strains, survival of nonproteolytic strains is of concern in products preserved by pasteurization and refrigeration (e.g., various cured meats or crabmeat) because of their ability to grow at refrigerated temperature (Hauschild, 1989). In the pasteurization of fish and fish products such as smoked fish, the process should accomplish about a 10 log cycle reduction in the initial population of the more resistant type E strains (Stumbo, 1973).

Sporulation Temperature and Medium Composition Although most clostridia produce spores of maximal heat resistance when grown at temperatures somewhat lower than the optimal growth temperature (Smith, 1977), Sugiyama

Table 2 Heat Resistance of *Clostridium botulinum* Spores

Strains[a]	Medium	Temp. (°C)	D (min)	Ref.
Type A				
62 A	tomato juice	110	0.95	Odlang and Pflug, 1977a
62 A	buffer (pH 7)	110	2.8	Odlang and Pflug, 1977a
62 A	buffer (pH 7)	112.8	1.23	Ito et al., 1967
62 A	buffer (pH 7)	115.5	0.74	Schmidt, 1964
62 A	buffer (pH 7)	121	0.20	Schmidt, 1964
A 16037	tomato juice	104.4	6.0	Odlang and Pflug, 1977a
A 16037	tomato juice	110	1.55	Odlang and Pflug, 1977a
A 16037	tomato juice	115.6	0.39	Odlang and Pflug, 1977a
Type B				
P	buffer (pH 7)	82.2	483, 868	Scott and Bernard, 1982
NP	buffer (pH 7)	82.2	1.49–32.3	Scott and Bernard, 1982
B 15580	tomato juice	110	0.74	Oldlang and Pflug, 1977a
B 15580	buffer (pH 7)	110	1.36	Odlang and Pflug, 1977a
213 B	buffer (pH 7)	112.8	1.32	Ito et al., 1967
32 B	buffer (pH 7)	112.8	0.15	Ito et al., 1967
213 B	buffer (pH 7)	122	0.13	Gaze and Brown, 1988
Type C				
C strains	buffer (pH 7)	104	0.02–0.90	Segner and Schmidt, 1971
Type E				
E strains	buffer (pH 7)	77	0.77–1.95	Ito et al., 1967
E strains	water	80	<0.33–1.25	Roberts and Ingram, 1965a
E strains	crabmeat	77	2.38–4.07	Lynt et al., 1977
E strains	crabmeat	82	0.49–0.74	Lynt et al., 1977
Type F				
NP	buffer (pH 7)	71	31.08–42.41	Lynt et al., 1979
NP	buffer (pH 7)	77	1.66–6.64	Lynt et al., 1979
NP	buffer (pH 7)	82	0.25–0.84	Lynt et al., 1979
P	buffer (pH 7)	98.9	12.19–23.22	Lynt et al., 1981
P	buffer (pH 7)	104.4	3.55–6.33	Lynt et al., 1981
P	buffer (pH 7)	110	1.45–1.82	Lynt et al., 1981
Type G				
HL	buffer (pH 7)	82.2	1.8–5.9	Lynt et al., 1984
HR	buffer (pH 7)	98.9	1.19–1.51	Lynt et al., 1984
HR	buffer (pH 7)	115.5	0.25–0.29	Lynt et al., 1984
HR	buffer (pH 7)	121	0.14–0.19	Lynt et al., 1984

[a]P, proteolytic strains; NP, nonproteolytic strains; HL, heat labile spores; HR, heat stable spores

(1951) found that spores of types A and B formed at 37°C were of higher heat resistance than those formed at 24, 29, or 41°C. With an increase in sporulation temperature, there may be an increase in the cation/DPA ratio of the spore, which could affect heat resistance (Lechowich and Ordal, 1962).

The composition of the sporulation medium is also important. Reducing the concentration of divalent cations of Fe, Ca, and Mn below certain levels lowers the heat resistance (Amaha and Ordal, 1957; Sugiyama, 1951). Spores formed in the presence of high concentrations of Fe or Cu are more sensitive to thermal inactivation than are native spores or spores formed in the presence of high concentrations of Mn or Zn (Kihm et al., 1990). Sugiyama (1951) reported that fatty acids in the growth medium also influence the heat resistance of spores; the longer the fatty acid carbon chain, the higher the resistance.

Effect of Pretreatment Prior exposure of spores to irradiation, hypochlorite, or H_2O_2 influences their resistance to heat. Gombas and Gomez (1978) found that *C. botulinum* type A spores were more sensitive to heat following exposure to sublethal doses of ionizing radiation, perhaps due to radiation-induced chain breaks in the peptidoglycan polymer (Stegeman et al., 1977).

Heating Menstruum The heat resistance of bacterial spores is markedly affected by the composition and pH of the suspending medium and by the presence of any antibacterial agents during the heating process.

Spores are more sensitive to heating at pH extremes than near neutrality (Roberts and Hitchins, 1969). Xezones and Hutchings (1965) determined the heat resistance of *C. botulinum* 62A suspended in spaghetti sauce at pH 4, 5, 6, and 7 and found D-values at 115.5°C of 0.128, 2.6, 4.91, and 5.15 minutes, respectively. Using *C. botulinum* type A and B strains, Odlaug and Pflug (1977a) reported D-values in tomato juice (pH 4.2) three times lower than in phosphate buffer (pH 7), whereas the z-values were hardly affected.

The heat resistance of the spores increases with decreasing water activity. Spores of *C. botulinum* type E had a $D_{110°C}$ value of less than 0.1 second at a high a_w, but this value rose more than 100,000 times as the a_w was lowered to 0.2–0.3 (Murrell and Scott, 1966). Spores of type B responded similarly, with the D-value markedly peaking in the a_w = 0.2–0.3 region.

Fats and proteins are known to be protective (Lücke, 1985). Molin and Snygg (1967) found that *C. botulinum* spores suspended in lipids were more resistant to heat than spores suspended in phosphate buffer. This protective effect could not be explained solely in terms of a low heat conductivity (Russell, 1982; Smith, 1977). Antibacterial agents such as subtilin or nisin lower the D-values of spores of *C. botulinum* type A and B strains by 35–85% (O'Brien et al., 1956), although it is not entirely clear whether the effect was on the thermal inactivation process or the prevention of outgrowth of the heated spores (Smith, 1977).

Recovery Conditions Injured spores can pose a problem in estimating the efficacy of any sterilizing procedure. Since heat-injured spores are more exacting in their requirements than normal spores (Roberts and Hitchins, 1969), measured D-values are also affected by the choice of recovery condition. Factors that affect the recovery of heat-injured spores include the composition and pH of the recovery medium, the presence of inhibitors, and the temperature and time of incubation (Foegeding and Busta, 1981; Russell, 1982).

Lysozyme (or egg yolk as a source of lysozyme) may increase the recovery of heat-injured spores because it lyses the spore cortex, enhancing germination. The recovery of nonproteolytic *C. botulinum* spores was improved by up to 4 log cycles by the addition of lysozyme to the recovery medium (Alderton et al., 1974; Scott and Bernard, 1985; Sebald and Ionesco, 1972). The effect of lysozyme on the recovery of proteolytic strains was found to be small; increases up to 2 log cycles in severely damaged spores have been observed (Alderton et al., 1974; Hauschild and Hilsheimer, 1977). The role of lysozyme in recovery has been attributed to the replacement of thermally inactivated germination enzymes (Scott and Bernard, 1985). The apparent low heat resistance of spores of type E may be somewhat misleading, because heat inactivation of these spores cannot be taken for granted, unless stimulation of germination by lysozyme has been attempted (Sebald and Ionesco, 1972).

Olsen and Scott (1946, 1950) reported that heated *C. botulinum* spores were sensitive to inhibitors in the recovery medium, probably unsaturated fatty acids. Added starch increased the recovery of heated spores, probably due to absorption of inhibitors. Odlaug and Pflug (1977c) showed that sodium bicarbonate and sodium thioglycollate in the recovery medium gave higher colony counts. Sodium bicarbonate is known to stimulate germination of *C. botulinum* (Treadwell et al., 1958). NaCl in the recovery medium reduces the colony counts of heated spores. The sensitivity of *C. botulinum* type B spores to NaCl increased with the severity of the heat treatment (Jarvis et al., 1976).

When heat-injured spores of *C. botulinum* are transferred to an appropriate medium, greater recovery occurs at incubation temperatures below the optimal temperature for unheated spores (Williams and Reed, 1942). Following heat treatment, spores may take longer than usual to germinate and, consequently, growth of vegetative cells may be delayed.

D. Ionizing Radiation

Because of public concern about its safety, irradiation is one of the most thoroughly studied technologies in the history of food processing. The major concerns have been the perceived possibility of induced radioactivity and the formation of carcinogenic or other toxic substances in the food. After extensive research and testing, health and safety authorities in 36 countries have approved irradiation of more than 49 different foods; 21 of these countries are currently applying the process (IAEA, 1989). Food irradiation is approved in the United States for certain food applications (FDA, 1986), most recently for treatment of raw poultry and pork. The goal of irradiation is not always sterility, nor is the target necessarily *C. botulinum* specifically. When radappertization (radiation sterilization) is required, the goal is to reduce the number of viable spores of the more irradiation-resistant *C. botulinum* by 12 log cycles.

Types and Units of Radiation

Among the types of ionizing radiation, gamma rays from radioactive isotopes such as cobalt-60, X-rays, and electrons from linear accelerators can be used for food preservation. Gamma- and X-irradiation are similar forms of electromagnetic radiation. Cobalt-60 has been the most acceptable radiation source because of its availability, price, and properties. It emits both gamma and beta rays. However, the energy level of the beta ray is relatively low for food sterilization, and gamma rays are usually the only effective ones.

The energy of radiation is expressed in rad (radiation absorbed dose) or Gy (gray) units. A dose of one rad is obtained when 0.01 joule (J) of radiation energy is absorbed per kilogram of material. One Gy is equal to 100 rad.

Mechanism of Sporicidal Activity and Resistance to Irradiation

Ionizing radiation can cause a wide variety of physical and biochemical effects in microorganisms. It is most likely, however, that the primary cellular target that governs the loss of viability is DNA (Ginoza, 1967). Irradiation-induced injury of *C. botulinum* spores causes single-strand breaks in DNA (Durban et al., 1974; Grecz and Grice, 1978; Grecz et al., 1978). Unless they are repaired, the structural defects in DNA are lethal to the bacterial cell (Ginoza, 1967). Besides the direct effect of irradiation on DNA, free radicals produced by radiolysis of water are also detrimental to spores. The reducing radicals, ·H and $\cdot H_2$, seem to have little effect, but the oxidizing radical, ·OH, is highly lethal to spores of *C. botulinum* (Grecz and Upadhyay, 1970).

Single-strand breaks in DNA can be repaired in spores during postirradiation germination (Terano et al., 1969). Repair of single-strand breaks is not affected by chloramphenicol (Terano et al., 1971) and therefore is not dependent upon protein synthesis. Direct repair of single-strand breaks during or after irradiation has been observed in highly resistant *C. botulinum* 33A spores at 0°C and in the absence of germination. In a sensitive strain (31B), no significant repair was observed (Durban et al., 1974). During germination, the irradiation resistance of spores is lost, but it may be restored by quickly suspending the spores in a sucrose solution. This finding suggests that the water content of the spore core is a major factor in the resistance of spores to radiation (Gould, 1978). However, some findings contradict this hypothesis (Iwasaki et al., 1974; Tallentire and

Powers, 1963; Tallentire et al., 1974). In spores, the UV photoproducts differ from the thymine-containing cyclobutane dimers produced in vegetative cells (Donnellan and Setlow, 1965), as do the repair processes (Donnellan and Stafford, 1968).

Resistance of Spores to Radiation and Factors Affecting Resistance

Irradiation survival curves of *C. botulinum* spores show a shoulder followed by an exponential order of death. It is likely that the shoulder of the curve corresponds to the period during which irradiation damage is mild enough to be repaired (Roberts and Hitchins, 1969). Tailing-off phenomena on irradiation survivor curves, due to a small number of highly resistant spores, have also been noticed (Anellis et al., 1965a,b). The production of irradiation-resistant mutants is one possible reason for the tailing-off phenomenon (Wheaton and Pratt, 1962).

The irradiation D-value (irradiation dose required to inactivate 90% of the population at a specified temperature) can be used to show the relative resistances of microorganisms. Anellis et al. (1969) found considerable variation with different levels of irradiation. Irradiation D-values for *C. botulinum* spores are given in Table 3. Factors that influence the radiation resistance of bacterial spores include type and strain of organism, preirradiation treatment of spores, O_2 availability, irradiation temperature, composition of the suspending medium, and recovery condition.

Type and Strain of Organism Among the clostridia, *C. botulinum* types A and B spores are the most radiation resistant (Anellis and Koch, 1962; Thornley, 1963). D-values of proteolytic *C. botulinum* strains at −50 to −10°C, the preferred temperature range for radappertization, are in the range of 0.2–0.45 Mrad (2.0–4.5 kGy) in both neutral buffers and foods (Dempster, 1985; Grecz et al., 1971; Kreiger et al., 1983). In contrast to its heat resistance, type E is only marginally more sensitive to gamma irradiation than proteolytic strains; its D-values are in the 0.1–0.2 Mrad range (Erdman et al., 1961).

Preirradiation Treatment of Spores Morgan and Reed (1954) and Kempe (1955) found that a preirradiation heat-shocking treatment did not affect the lethality of subsequent irradiation of *C. botulinum* spores, whereas Anderson et al. (1967) showed that heated spores of *C. botulinum* type A were more resistant than unheated spores.

Oxygen The presence of O_2 during or after an irradiation process can markedly influence the response of bacterial spores, with the highest sensitivity being shown in the presence of O_2 or air and the lowest resistance under anaerobic conditions (Russell, 1982).

Irradiation Temperature In the broad range of −200 to +50°C, D-values change approximately 1 krad/°C (Grecz et al., 1965, 1971). D-values for *C. bot-*

Table 3 Radiation Resistance of *C. botulinum* Spores

Spore type	Suspending medium	Temperature (°C)	D-value (Mrad)	Ref.
A	buffer (pH 7)	[a]	0.224–0.344	Anellis and Koch, 1962
A	bacon	−17.8	0.188–0.221	Anellis et al., 1965b
A	beef	95	0.121–0.237	Grecz, 1966
A	beef	0	0.34–0.41	Grecz, 1966
A	beef	−196	0.590–0.712	Grecz, 1966
A	water	18–23	0.12–0.14	Roberts and Ingram, 1965b
B	buffer (pH 7)	[a]	0.120–0.349	Anellis and Koch, 1962
B	water	18–23	0.11	Roberts and Ingram, 1965b
B	beef stew	22	0.238	Schmidt et al., 1960
B	beef	5	0.34–0.40	Kempe and Graikoski, 1962
A & B	buffer (pH 7)	−20	0.18–0.38	Denny et al., 1959
A & B	beef	−30 ± 10	0.25–0.36	Anellis et al., 1975
A & B	beef steak	27	0.30–0.34	Schmidt and Nank, 1960
A & B	beef	−29	0.40–0.68	Wheaton et al., 1961
A & B	codfish cake	−30 ± 10	0.070–0.331	Anellis et al., 1972
A & B	chicken, ham, pork	−30 ± 10	0.34–0.38	Kreiger et al., 1983
C	water	18–23	0.14	Roberts and Ingram, 1965b
D	water	18–23	0.22	Roberts and Ingram, 1965b
E	buffer (pH 7)	18–23	0.129–0.134	Roberts and Ingram, 1965b
E	water	18–23	0.08–0.16	Roberts and Ingram, 1965b
E	beef stew	22	0.122–0.144	Schmidt et al., 1960
F	water	18–23	0.25	Roberts and Ingram, 1965b

[a]Treatment started at 2–4°C, temperature rose 9–12°C during treatment.

ulinum 62A spores increased from 0.268 Mrad to 0.350 Mrad to 0.400 Mrad when the temperature was decreased from 20°C to −80°C to −196°C (Rowley et al., 1968). However, depending upon the irradiation environment, D-values may increase with an increase in temperature, starting at about 20°C (Grecz et al., 1965, 1971). At low temperatures, the production and action of ionizing radicals, which are primarily responsible for the deterioration of flavor and odor of irradiated food, are much reduced. Radiation has little direct deleterious effect on food quality and retains its sporicidal effect even at low temperatures. Therefore, irradiation of food at low temperatures may yield the desired destruction of spores by direct hits without the undesired secondary effect of radical formation (Smith, 1977). However, the influence of pH and temperature on the survival of irradiated *C. botulinum* spores is complicated. Very definite peaks and troughs of survival are found at various combinations of temperature and hydrogen ion concentration (Grecz and Upadhyay, 1970). Also, ionizing radicals are apparently produced even at temperatures as low as −196°C (Smith, 1977).

Composition and pH of the Suspending Medium Generally, *C. botulinum* spores are more sensitive when irradiated in phosphate buffer than in various foods (Anellis et al., 1965a,b; El-Bisi et al., 1967). The effect of pH on the radiation resistance of spores (Edwards et al., 1954) is slight, except possibly at pH values below 5 (Farkas et al., 1967). Food additives can act as sensitizers or protectors. The radiation dose required to inactivate *C. botulinum* spores in beef was reduced by the presence of NaCl, $CaCl_2$, alkyl isothiocyanate, nutmeg, sodium citrate, sodium nitrate, sodium nitrite, and EDTA (Anderson et al., 1967; Krabbenhoft et al., 1964).

Recovery Conditions General requirements for the recovery of heat- and radiation-injured *C. botulinum* spores are similar (Hauschild, 1989). The presence of NaCl during irradiation was without effect on the radiation resistance of *C. botulinum* spores, but increasing doses of gamma irradiation rendered spores increasingly sensitive to postirradiation inhibition by NaCl in the plating medium (Chowdhury et al., 1976; Rowley et al., 1974). Irradiation-resistant strains also are more resistant to NaCl than irradiation-sensitive strains (Kiss et al., 1978). The resuscitation of irradiated spores is improved significantly by magnesium in the recovery medium, which may be attributed to the magnesium dependence of a DNA ligase involved in DNA repair (Foegeding and Busta, 1981).

E. Ethylene Oxide (ETO)

Ethylene oxide (also called epoxyethane or dimethylene oxide) is the most widely used gaseous sterilizing agent. ETO in the vapor phase is used to treat certain low-moisture foods and to sterilize aseptic packaging materials (Toledo, 1975). The use of ETO is limited to dry foods, because the reaction with water to form ethylene glycol rapidly depletes the concentration of ETO in high-moisture foods (Lindsay, 1985). The ability of ETO to penetrate certain packaging materials allows foods to be sterilized in sealed containers.

Mechanism of Sporicidal Activity

ETO destroys all forms of microorganisms, including spores and viruses, but the mechanism of action is poorly understood. Phillips (1977) suggested that the activity of ETO in both spores and vegetative cells was due to alkylation of sulfhydryl, amino, carboxyl, phenolic, and hydroxyl groups. Experimental evidence indicates that the reaction of ETO with nucleic acids is the primary cause of sporicidal and bactericidal activity. Winarno and Stumbo (1971) showed that the lethal effect of ETO on *C. botulinum* type A spores was due to alkylation to guanine and adenine components of DNA, and suggested the observed impairment of RNA and protein synthesis to be an indirect effect of this.

Resistance of Spores to ETO and Factors Affecting Resistance

Generally, spores are much more resistant to inactivation treatments than are vegetative cells. However, the sensitivity of spores to ETO is comparable to that of some vegetative cells. Even highly resistant spores are only about five times more resistant than the most sensitive vegetative cells. The ability of ETO to penetrate may play some role in the effective action against spores. *C. botulinum* appears to be one of the most resistant organisms to ETO, with a D-value of 11.8 minutes (Winarno and Stumbo, 1971) compared to 14.5–15.3 minutes for the most resistant *B. subtilis* under the same condition (700 mg/l ETO, 40°C, 30–35% relative humidity).

The major factors affecting the sporicidal activity of ETO and other gases are gas concentration, temperature, relative humidity (RH), exposure time, and carrier material (Caputo and Odlaug, 1983). Kuzminski et al. (1969) studied the effect of temperature and RH on inactivation of *C. botulinum* 62A spores by 700 mg/l of ETO (Table 4). The sporicidal activity increased as temperature increased and RH decreased. At the lowest RH studied (3%), the Q_{10} value was 2.21, compared to a Q_{10} value of 1.67 at 73% RH. Winarno and Stumbo (1971) reported D-values (700 mg/l ETO, 40°C) for *C. botulinum* 62A of 6.0, 7.4 and 11.8 min at 11, 23, and 33% RH respectively. Savage and Stumbo (1971) reported a D-value (700 mg/l ETO, 40°C, 47% RH) of 11.0 to 12.0 min for *C. botulinum*. The sporicidal activity of ETO increases with increasing concentration of ETO up to a certain range (500 mg/l for *Bacillus globigii*), but does not increase beyond that (Caputo and Odlaug, 1983). The ETO resistance of spores is affected by the carrier material for the spores. Caputo and Odlaug (1983)

Table 4 Ethylene Oxide Resistance of *C. botulinum* 62A Spores[a]

% RH	40°C	50°C	60°C	70°C
3	4.30	1.90	ND[b]	ND[b]
23	8.18	4.65	2.60	1.60
33	13.75	5.75	2.65	1.74
53	10.20	5.60	2.80	1.80
73	8.80	5.10	3.00	1.95

[a]The resistance is expressed as D-values in min. Ethylene oxide concentration was 700 ± 20 mg/1.
[b]Not done.

suggested that the ratio of reactants (i.e., ETO, water) delivered to the spores during exposure is affected by the absorbing characteristics of the carrier material.

Use of ETO

ETO is used to control microorganisms and insects in ground spices or other natural seasoning material. In the United States, residual ETO cannot exceed 50 μg/g of spice. Other food applications include dried coconut, black walnuts, starch, and packaging materials. The vapors of ETO are flammable and explosive, and ETO is diluted with CO_2 or fluorocarbon to reduce the hazard (Caputo and Odlaug, 1983). The concentration of ETO is controlled by controlling the pressure of the gas mixture in the equipment.

ETO reacts with water to form toxic ethylene glycol. ETO is also known to react with naturally occurring chlorides in food to form toxic chlorohydrins (Wesley et al., 1965), the level of which is regulated in some countries. Thus, it is not used to sterilize spices or foods that contain NaCl. Another consideration in the use of ETO is the possible adverse effect on vitamins, including riboflavin, niacin, and pyridoxine (Bakerman et al., 1956). Sometimes propylene oxide (PO) is used instead of ETO because the hydration of PO results in nontoxic propylene glycol. However, P0 is not as effective as ETO and reacts with chlorides to form toxic chlorohydrins.

VI. CONTROL OF *C. BOTULINUM* BY PREVENTION OF GROWTH AND TOXIN PRODUCTION

It is not always practical or desirable to inactivate *C. botulinum* spores to prevent botulism since such harsh treatments may reduce the organoleptic and/or nutritional quality of the food and increase the processing cost. Also, the presence of spores in the food does not lead to the production of botulinum toxin unless the food provides a proper environment for germination and growth. Therefore, *C. botulinum* is controlled in many foods by inhibition rather than destruction. Inhibition of germination, growth, and toxin production is usually achieved by controlling more than one environmental factor such as temperature, pH, oxidation-reduction potential, O_2 availability, water activity, and food preservatives.

Since most spores can germinate under conditions in which vegetative growth is impossible, the important consideration for safety of a food is whether *C. botulinum* can grow and produce toxin (Sperber, 1982). In this section, the germination process and the factors affecting germination of *C. botulinum* spores will be briefly discussed. Emphasis will be given to the discussion of the factors affecting growth and toxin production.

A. Prevention of Germination

Spore Germination and Outgrowth

Three sequential processes are involved in the transformation of a dormant bacterial spore into a vegetative cell: activation, germination, and outgrowth. Activation of bacterial spores was discussed earlier in this chapter (see Sec. III.C).

Germination is an irreversible process in which activated spores are converted into vegetative cells. Within minutes, the initiation of germination leads to the loss of most of the unique spore properties and constituents. In general, the accepted sequence of germination events is loss of heat resistance, loss of chemical resistance, loss of calcium and DPA, excretion of peptidoglycan, increase in stainability, loss of refractivity, and decrease in optical density (Gould, 1970). Most of the physical changes observed in germinating spores (decrease in optical density, phase darkening, etc.) are the results of depolymerization and excretion of spore constituents and uptake of water (Gould, 1970; Hsieh and Vary, 1975). The initiation of germination (or trigger reaction) is thought to be a physiochemical event involving a protein receptor molecule on the spore cytoplasmic membrane (Cheng et al., 1978; Vary, 1973). The receptor responds to the presence of germinants, including certain amino acids (e.g., L-cysteine, L-alanine, L-proline), ions (e.g., Mn^{+2}), surfactants (e.g., *n*-dodecyclamine), glucose, and calcium DPA chelate. Lactate and bicarbonate have also been reported as co-germinants for *C. botulinum* spores (Ando, 1973a; Foegeding and Busta, 1983a,b).

Germinated spores become swollen and shed their coats to allow the young vegetative cells to emerge, elongate, and divide (Gould, 1962). Outgrowth is the process in which a vegetative cell develops from a germinated spore, and it occurs when germination takes place in a medium that can support vegetative growth (Hansen et al., 1970). A number of substances, including sorbic acid and nitrite, inhibit germination and/or outgrowth of *C. botulinum*.

Factors Affecting Germination of Spores

The germination process has been found to respond differently to changes in the environment when compared to the outgrowth and vegetative growth stages (Sperber, 1982).

Germinants Alanine or cysteine, a carbon source, and bicarbonate allow complete and rapid germination of most activated *C. botulinum* spores. The most effective carbon sources are lactate, glucose, galactose, inosine, and fructose; mannose and xylose are less effective (Ando, 1971). Other alpha-hydroxy acids can substitute for lactate (Ando, 1974), which is useful in recovery of injured spores (Foegeding and Busta, 1983b). Treadwell et al. (1958) demonstrated the necessity of bicarbonate or CO_2 to stimulate *C. botulinum* germination. The rate and extent of germination in chemically defined media increased with increasing concentrations of bicarbonate for several strains of *C. botulinum* (Ando, 1973a; Rowley and Feeherry, 1970). King and Gould (1971) suggested that the active stimulant was CO_2 and not bicarbonate.

Oxidation-Reduction Potential (Eh) The oxidation-reduction potential of the medium is not a critical factor. Ando and Iida (1970) demonstrated that type E spores could germinate at an Eh of +414 mV, while growth could occur only at an Eh below +200 mV. Compounds such as sodium thioglycollate and cysteine, which function as reducing agents, have been reported to either stimulate or inhibit germination (Barker and Wolf, 1971; Holland et al., 1969; Rowley and Feeherry, 1970).

pH The optimum pH for *C. botulinum* spore germination is around 7.0 (Ando, 1973a; Rowley and Feeherry, 1970). The lower limiting pH value for germination is 4.6 for types A and B and 4.8 for type E (Gould, 1969). The upper limiting pH value is around 9.0 (Smith, 1977). However, under practical conditions in foods, the limiting pH is influenced by other factors inherent to the food system (Smoot and Pierson, 1982).

Temperature Clostridial spores can germinate over a wide range of temperatures. The optimum temperature for *C. botulinum* germination is 37°C (Ando, 1973a; Ando and Iida, 1970), but type A spores are known to germinate at 4°C (Ando, 1973a; Rowley and Feeherry, 1970) and type E spores below 4°C (Grecz and Arvay, 1982; Schmidt et al., 1961). Germination of *C. botulinum* 62A in a chemically defined medium occurs at temperatures up to 70°C, with the optimum being 37°C (Rowley and Feeherry, 1970). Type E spores have been shown to germinate rapidly at 50°C (Grecz and Arvay, 1982).

Water Activity Under otherwise optimum conditions for growth, spores of *C. botulinum* germinate at a_w levels as low as 0.89 (Baird-Parker and Freame, 1967). Under suboptimal conditions, germination is more sensitive to low a_w levels (Baird-Parker and Freame, 1967; Ohye et al., 1967).

Preservatives There has been a great interest in chemical food preservatives capable of inhibiting germination of *C. botulinum* spores. Most research in this area has dealt with preservatives that can be used in meat products such as cured pork and beef products, smoked fish, etc., which do not necessarily receive sterilizing heat treatments (Smoot and Pierson, 1982).

Nitrite was one of the earliest food preservatives tested for its ability to inhibit *C. botulinum* spore germination. Nitrite is capable of inhibiting spore germination in certain specific laboratory media systems, but at the levels allowed in foods, it does not seem to inhibit spore germination (Smoot and Pierson, 1982). At low levels, sodium nitrite is known to stimulate germination of *C. botulinum* type A and E spores (Ando, 1973b, 1974). Sofos et al. (1979a) and Benedict (1980) have reviewed the effect of nitrite on germination of *C. botulinum* spores.

NaCl inhibits germination of various species of *Clostridium*, but only at concentrations that would be organoleptically unacceptable in food products (Smoot and Pierson, 1982). Increasing the level of NaCl up to 8% in a chemically defined medium progressively decreased the rate and extent of *C. botulinum* type A spore germination over a 3-hour period but could not prevent germination over longer periods (Ando, 1973b). Type E spores were inhibited by 5% NaCl in the same chemically defined medium (Ando, 1974). Germination of type B spores was inhibited at sucrose concentrations above 38% (Beers, 1957). Water activities of these media were not determined.

Sorbates have been used in foods mainly to inhibit the growth of yeasts, molds, and certain bacteria. Sorbates can also replace a high proportion of nitrite added to cured meat products as an inhibitor of *C. botulinum* (Sofos et al., 1979a; Widdus and Busta, 1982). Sofos et al. (1979b) reported inhibition of *C. botulinum* spore germination (loss of heat resistance) by 0.2% sorbic acid in a chicken frankfurter–like emulsion. However, Ando (1973b, 1974) observed germination of type A and E spores to be only slightly affected by 1% sodium or potassium sorbate. The pH of the germination medium was 6.7, which is above the effective pH limit of about 6.5 for sorbic acid and its salts.

EDTA inhibited germination of *C. botulinum* 62A spores in a fish homogenate at a concentration of 2.5 mM over a pH range of 6.5–8.1 (Winarno et al., 1971). In contrast, Ando (1973b) reported that 5.0 mM EDTA was ineffective in preventing type A spore germination in a laboratory medium.

Some spice oils (Ismaiel and Pierson, 1990) inhibit *C. botulinum* spore germination, and several other food preservatives have antibotulinal activities targeted at germination or subsequent developmental stages (Busta and Foegeding, 1983). These specific antibotulinal agents often enhance control of *C. botulinum*.

B. Prevention of Growth and Toxin Production

Since the control of *C. botulinum* in most foods is achieved by the inhibition of growth and toxin production, the requirements of these processes have been studied extensively. Botulinum toxins are produced under conditions that permit vegetative growth (Baird-Parker, 1971). In most foods, *C. botulinum* is inhibited by a combination of factors rather than a single factor (Hauschild, 1989).

Factors Affecting Growth and Toxin Production

Nutritional requirements for *C. botulinum* are complex and include several amino acids, growth factors, and mineral salts. The organism grows best in rich organic media, such as meat, but without competition it can grow in relatively poor media such as vegetables and fruits (Sofos et al., 1979a). Therefore, it must be assumed that *C. botulinum* may grow and produce toxin in any food unless inhibitory factors are present. The main factor limiting growth and toxin production of *C. botulinum* in foods are temperature, pH, water activity, redox potential (O_2 availability), added food preservatives, and competing microflora. Proteolytic and nonproteolytic strains show different responses to some of these factors. The effects of these factors are interrelated, i.e., changing one factor alters the sensitivity or tolerance to the others. Combinations of inhibitory factors are more restrictive to growth and toxin production than each factor alone (Ohye and Christian, 1967).

Temperature Because of the importance of low-temperature storage, growth studies have generally focused on defining minimum growth temperatures (Hauschild, 1989). The lower temperature limit for the growth of proteolytic strains is approximately 10°C (50°F). The upper temperature limit is 45–50°C, and the optimum is about 40°C (Hauschild, 1989). Tanner and Oglesby (1941) studied 17 strains of *C. botulinum* types A and B and found that all grew at 15°C, few grew at 10°C, and none at 5°C. Smelt and Haas (1978) showed that all nine strains studied could grow at 12°C, but none grew at 10°C. Ohye and Scott (1953) studied the influence of temperature on the growth rates of 20 type A and B strains; the optimum was 37°C, with very slow growth at 12.52°C and fairly rapid growth at 45°C. *C. botulinum* type A grew very slowly at 48°C, at which temperature the toxin was slowly inactivated (Bonventure and Kempe, 1959). Weakly proteolytic *C. botulinum* type G grows at but not below 12°C in both laboratory broth and crabmeat (Solomon et al., 1982).

Nonproteolytic strains grow and produce toxin under certain conditions at refrigeration temperatures. Therefore, adequate refrigeration alone is not a complete safeguard against botulism (Sperber, 1982). The lower temperature limit for growth of these strains is 3.3°C, the upper limit 40–45°C, and the optimum temperature 30–35°C (Hauschild, 1989). Four strains of type E grew and produced toxin in beef stew within 36 days at 3.3°C (38°F), but toxin was not produced in 104 days at 2.2°C (36°F) (Schmidt et al., 1961). Similarly, type E grew in fresh oysters and clams stored at 4°C, but not at 2°C (Patel et al., 1978). Three nonproteolytic strains (two type F, one type B), produced toxin at 3.3°C (Eklund et al., 1967a,b).

The time required for growth and toxin production at refrigeration temperatures is dependent on the substrate (Ohye and Scott, 1957; Solomon et al., 1977, 1982), size of spore inoculum (Cockey and Tatro, 1974; Eklund et al., 1967a,b), previous heat treatments (Solomon et al., 1982), and temperature of incubation (Eklund et al., 1967a,b). In general, several weeks of incubation are needed for toxigenesis at these temperatures (Eklund et al., 1967a; Simunovic et al., 1985; Solomon et al., 1977). Nonproteolytic strains may have sufficient time to grow in refrigerated foods if the shelf life is extensive, whereas proteolytic strains can grow only if temperature abuse occurs. Mild heat treatments and vacuum or modified atmosphere packaging may select for *C. botulinum* and allow toxicity to occur in organoleptically acceptable products. Thus, *C. botulinum* may pose a hazard in refrigerated foods with extended shelf life (Conner et al., 1989; Notermans et al., 1990).

pH. It is generally accepted that *C. botulinum* will not grow and produce toxin in foods with a pH of 4.6 or lower. Current U.S. government regulations state that canned food at pH 4.6 or lower is safe without the rigorous conventional commercial sterilization (CFR, 1990b).

The minimum pH requirement for growth of proteolytic *C. botulinum* is in the range of 4.6–4.8, although for many strains it may be well over 5.0 (Hauschild, 1989). For nonproteolytic strains the limit is about pH 5.0 (Ito and Chen, 1978; Post et al., 1985; Segner et al., 1966). However, Raatjes and Smelt (1979), Smelt et al. (1982), Tanaka (1982), Tsang et al. (1985), Wong et al. (1988), and Young-Perkins and Merson (1987) have reported growth and toxin production by *C. botulinum* at pH 4.36–4.0. Young-Perkins and Merson (1987) concluded that pH alone was not a suitable index for predicting toxicity endpoints in media with a high protein content. They proposed that a minimum titratable acidity (for a specific acidulant) might better define the acidification conditions that inhibit *C. botulinum* spore outgrowth and toxin production. Factors that influence the acid tolerance of *C. botulinum* include strain, medium composition, growth temperature, inoculum size, the acidulant used, presence of preservatives, and prior heat and/or other treatments (Blocher and Busta, 1983). Other environmental factors such as water activity and redox potential are also known to influence the acid tolerance of *C. botulinum*. The growth of other organisms such as yeasts and molds may affect the growth of *C. botulinum* at low pH by changing the environment. Several of the acid or acidified foods incriminated in botulism outbreaks were found to contain other acid-tolerant microorganisms (Odlaug and Pflug, 1978).

Early work by Dozier (1924) with 19 strains of *C. botulinum* types A and B indicated that cell inocula grew at pH 4.87, while the minimum pH for growth from spore inoculum was 5.01. Later work by Baird-Parker and Freame (1967) indicated about equal minimum pH values for growth from cell to spore inocula. Odlaug and Pflug (1977b) reported that types A and B spores did not grow in tomato juice stored at 90°F (32.2°C) for 6 months at pH 4.2. Ito et al. (1978) reported that canned figs became toxic at pH 5.45 but not at pH 5.35. Segner et al. (1966) found the minimum pH for growth of type E in a variety of culture media to be 5.2–5.3. Lerke (1973) demonstrated growth at 10^4 type E spores/g in crabmeat cocktail at pH 4.95 but not at pH 4.86.

Townsend et al. (1954) investigated the minimum pH for growth of *C. botulinum* 62A and 213B in 12 different food products. The minimum pH for toxin production ranged from 4.84 for a vegetable juice to 5.44 for a prune pudding. Tanaka (1982) reported toxin production by types A and B in media containing high concentrations of proteins at pH values well below 4.6. From the observation that toxin production occurred only when large amounts of precipitated protein were present, he postulated that the presence of small, localized pockets of higher pH values within the matrices of precipitated protein might be responsible for the growth of the organism at low pH. These preferable microenvironments may become enlarged with progressing growth (Hauschild, 1989). Young-Perkins and Merson (1987) showed that type A produced toxin at a pH as low as 4.07 and the minimum pH for toxin production decreased with increasing concentrations of soy protein. Although spores and germinated cells were occasionally observed within protein aggregates, toxin also developed in HCl-acidified samples agitated to disrupt clumps and maintain the soy in suspension. The authors suggested that the protein may function (1) as a reducing agent, allowing the cell to overcome the apparent limiting redox potential, (2) as a source of essential metabolites, and (3) as a buffer, retarding acidification of the cell interior. Wong et al. (1988) found that beef protein at pH values below 4.6 permitted germination, outgrowth, and toxin production to a greater extent than soy protein. They suggested that compositional differences between beef and soy protein might be the reason for the disparate rates of growth and toxin production. At present, only canned, pasteurized, acidic, or acidified fruits and vegetables are protected by acidity alone. Because of their relatively low protein content, these products do not allow *C. botulinum* to grow at or below pH 4.6, and adherence of the food industry to this pH level has ensured that canned acidic fruits and vegetables are safe.

Hauschild et al. (1975) demonstrated the influence of marination in acetic acid on the growth of type B spores. Uninoculated mushrooms contained a large enough natural inoculum to become toxic at pH 5.47. The mushrooms became toxic at pH 5.04 when 10^5 spores/g were added, with one of 12 samples becoming toxic at pH 4.77. Wong et al. (1988) found that reducing the inoculum size from 10^7 to 10^4 spores/ml into sterile beef media acidified with hydrochloric acid or citric acid delayed but did not prevent spore outgrowth and toxin production at pH levels below 4.6.

Various acidulants used in foods show different antimicrobial activities. The antimicrobial effects of a particular acid are a function of the physical properties of the acid (pKa, molecular weight, number of carboxyl groups), pH, acid concentration, and the organism being tested. The acid may exert its primary effects by altering the pH of the system or have specific effects due to the undissociated acid form. The antimicrobial activity of fatty acids with 1–14 carbon atoms is known to correlate with the undissociated fraction but not with the acid anion (Sofos and Busta, 1981). The amount and type of acid constituents naturally present in the system prior to acidification may also be critical in determining the degree of inhibition in the final system (Blocher and Busta, 1983). Smelt et al. (1982) found that citric acid was less inhibitory than lactic or acetic acids at pH 4.4. In soy suspensions co-inoculated with *Bacillus* and *C. botulinum*, the time required for detectable botulinum toxin in the presence of acetic, lactic, and citric/hydrochloric acids was 14, 12 and 4 weeks, respectively. Tsang et al. (1985) reported that type E spores were capable of growth and toxin production at 26°C in laboratory medium acidified with citric acid at pH 4.20. Growth and toxin production were not detected at pH values below 5.0 when acetic acid was used. Other environmental factors such as water activity and redox potential affect the limiting pH. Dodds (1989) reported that botulinum toxin was produced in vacuum-packed potatoes at pH levels ≥ 4.75, ≥ 5.25, and ≥ 5.75 when the a_w was ≥ 0.970, 0.965, and 0.960, respectively. Toxin was not detected when the a_w was 0.955.

Upper pH limits for growth and toxin production of *C. botulinum* lie in the range of pH 8–9 (Hobbs, 1976) but are of little or no practical consequence in food safety.

Salt and Water Activity Water activity is the ratio of vapor pressure of the food to the vapor pressure of pure water ($a_w = P/P_O$) at the same temperature. Simply viewed, water activity expresses the amount of water available to the microorganism. Denny et al. (1969) found that growth of *C. botulinum* types A and B in canned bread was dependent on the water activity of the bread, but not on its moisture content. When various kinds of breads were inoculated with 2.0×10^4 spores/g of *C. botulinum* types A and B and incubated at 85°F for one year, none of 73 cans with a_w values < 0.95 became toxic, while 28 of 56 cans with $a_w > 0.955$ became toxic.

Salt (NaCl) remains one of the most important factors in the control of foodborne *C. botulinum*. The inhibitory effect of NaCl is largely based on the depression of the water activity. Therefore, the effectiveness of salt depends on its concentration in the aqueous phase, often referred to as *brine concentration* [% brine = (% NaCl $\times$ 100)/(% H_2O + % NaCl)] (Hauschild, 1989). Reported minimum water activities for growth of *C. botulinum* in NaCl-containing foods or media are 0.94–0.96 for types A and B and about 0.97 for type E (Baird-Parker and Freame, 1967; Denny et al., 1969; Emodi and Lechowich, 1969; Marshall et al., 1971; Ohye and Christian, 1967). The minimum a_w of 0.94 and 0.97 corresponds to a salt (brine) concentration of approximately 10% and 5%, respectively (Hauschild, 1989; Sperber, 1982).

The type of solute used to control water activity influences growth and toxin production. Generally, NaCl, KCl, glucose, and sucrose show similar patterns, while glycerol permits growth at lower a_w values (Sperber, 1983). Baird-Parker and Freame (1967) reported that replacement of NaCl with glycerol as a_w depressant reduced growth-limiting a_w levels by up to 0.03 units. The limiting a_w level with glycerol was 0.93 for types A and B and 0.94 for type E. Slight reductions in limiting water activity levels were also noted when NaCl was partially replaced with a mixture of KCl and Na_2SO_4 (Ohye et al., 1967). In an effort to replace NaCl with KCl to reduce the sodium intake, Pelroy et al. (1985) studied the effects of NaCl, KCl, or an equimolar mixture of NaCl plus KCl on toxin production by *C. botulinum* type E in smoked white fish. Used alone, NaCl and KCl had equivalent inhibitory effects, but the equimolar mixture of the salts was slightly less inhibitory. They speculated that a substrate containing a relative balance of Na^+ and K^+ may be more favorable for growth and toxin production by *C. botulinum* than one containing an excess of either ion.

Emodi and Lechowich (1969) found that 38.5% sucrose or 22.5% glucose inhibited growth from spores of four strains of type E. Water activities for these media were not determined.

Water activity limits in foods depend on a variety of other control factors such as pH, redox potential, and added food preservatives. The limiting a_w may be raised significantly with added preservatives or by increasing acidity (Baird-Parker and Freame, 1967; Hauschild and Hilsheimer, 1979).

Redox Potential and Gaseous Atmosphere The redox potential (Eh) of a food is determined by the balance of oxidizing and reducing agents. In most cases, a high Eh in foods is due to the presence of O_2 (Hungate, 1969). In addition to the redox potential, the poising capacity of the medium is likely to affect the growth of anaerobes (Lund and Wyatt, 1984). The optimum Eh for the growth of *C. botulinum* is −350 mV (Smoot and Pierson, 1979), but reported maximum Eh values for initiation of growth range from +30 to +250 mV (Ando and Iida, 1970; Huss et al., 1979; Morris, 1976; Smoot and Pierson, 1979). When high numbers of spores are present, their metabolic activity during germination may reduce the Eh to a level permitting growth of vegetative cells (Lund and Wyatt, 1984). Spore germination can occur at much higher Eh values than growth (Ando and Iida, 1970) and is known to decrease Eh (Smith, 1975). Other control factors such as NaCl and acidity will lower the maximum Eh level for growth (Smoot and Pierson, 1979).

Since *C. botulinum* is an anaerobic organism, it is a common assumption that *C. botulinum* cannot grow in foods that are exposed to O_2 or in foods with a high redox potential. Actually, the Eh of most foods exposed to O_2 is usually low enough to permit growth of *C. botulinum* (Sperber, 1982). Christiansen and Foster (1965) found the same rate of toxin production by *C. botulinum* type A in sliced bologna, whether or not it was vacuum packaged. Ajmal (1968) reported that *C. botulinum* type E inoculated into meat and fish produced toxin under aerobic and anaerobic conditions of incubation. Sugiyama and Yang (1975) inoculated spores of *C. botulinum* into fresh mushrooms, which were then packaged and stored at 20°C. The respiration of mushrooms removed free oxygen and allowed the spores to germinate and produce toxin in 3–4 days.

Vacuum-packaging of perishable foods has been a concern because it might facilitate *C. botulinum* growth. Pace and Krumbiegel (1973) reported the rate of toxin production to be higher in vacuum-packaged fish than in fish packaged without vacuum. The slowest rate was in unpackaged products. However, Johannsen (1965) stated that the formation of toxin in vacuum-packaged products is often lower than in those not packaged in vacuum. He proposed that the microaerophilic environment in vacuum-packed products encourages the growth of lactobacilli, which antagonize growth and toxin formation of *C. botulinum*.

Modified atmosphere packaging is used to extend the shelf life and improve the quality of foods. Depending upon the particular atmosphere, it may promote, prevent, or have no effect on *C. botulinum* spores. CO_2 may inhibit or stimulate *C. botulinum* (Foegeding and Busta, 1983e), depending upon the concentration. Atmospheres containing altered levels of CO_2 must be considered with caution (Eklund, 1982; Hintlian and Hotchkiss, 1986; Post et al., 1985b; Sperber, 1982; Solomon et al., 1990).

Inhibition of Growth and Toxin Production by Preservatives

Nitrite Sodium and potassium salts of nitrite and nitrate are used to cure meats and fish. Although color development and fixing, flavor enhancing, and antioxidant properties are important organoleptic considerations in curing muscle foods, the most important role of nitrite in cured food products is the inhibition of *C. botulinum*. Nitrite is the functional constituent. While nitrate has no antibotulinal activity, it is partially converted to nitrite by endogenous microorganisms (Christiansen et al., 1973; Hustad et al., 1973). Inhibition of *C. botulinum* in cured meats by nitrite has been extensively studied, and many reviews of this subject have been published (Benedict, 1980; Christiansen, 1980; Holley, 1981; Sofos et al., 1979a).

The mechanism(s) by which nitrite inhibits botulinum toxin production is still unknown. Johnston et al. (1969) proposed various possibilities to account for the inhibitory effect of nitrite, including enhanced destruction of spores by heat, stimulation of spore germination during heating followed by heat inactivation of germinated spores, inhibition of germination after heating, and production of more inhibitory substances from nitrite. Ingram (1974, 1976) evaluated these possibilities and found that most of them were not effective in real food systems. Sofos et al. (1979a) proposed that the effectiveness of nitrite is due to its inhibition of spore outgrowth and subsequent multiplication and not to inhibition of germination. Tompkin et al. (1978b) proposed that nitric oxide may react with an essential iron-containing compound (such as ferredoxin) in germinated spores, thus interfering with energy metabolism. This theory is supported by findings that iron added to meats reduces the antibotulinal effect of nitrite (Tompkin et al., 1978b, 1979a). The effect of iron is reversed by the addition of EDTA to the meat. Benedict (1980) proposed that nitrite may inhibit outgrowth and subsequent toxin production by three interacting mechanisms: (1) oxidation of cellular biochemicals within vegetative cells and spores, (2) restricted use of iron (or other essential metal ions) through inhibition of solubilization, transport, or assimilation, and (3) cell surface membrane activity limiting substrate transport by the growing cell. Recent studies indicate that nitrite and nitrosyl compounds inhibit the phosphoroclastic system in *C. sporogenes* (Payne et al., 1990a,b). This system, which provides ATP to the cell, also functions in *C. botulinum*.

The effectiveness of nitrite is dependent upon complex interactions among pH, NaCl levels, heat treatment, time and temperature of storage, spore inoculum level, and the composition of the food (Cook and Pierson, 1983). Nitrite has been shown to be more effective with decreasing pH, indicating that undissociated nitrous acid may be an active component (Roberts and Ingram, 1973; Shank et al., 1962; Tarr, 1941). The maximum effectiveness of nitrite does not correspond to a pH value equal to or lower than the pKa value of 3.4. Presumably, this is due to the accelerated decomposition of nitrous acid at pH values close to the pKa of nitrite (Sinskey, 1979). Roberts et al. (1976) reported that the concentration of nitrite required to inhibit a mixture of *C. botulinum* type A and B spores in a pasteurized pork slurry decreased from 300 to 40 μg/g as the concentration of NaCl increased from 1.8% to 3.5%. Increases in the spore inoculum level can override the inhibitory effect of nitrite and allow toxin production (Christiansen et al., 1973; Hustad et al., 1973; Ingram, 1974).

Nitrite delays, but does not necessarily prevent, botulinal outgrowth and subsequent toxin production. The effectiveness of nitrite and the duration of its effect may depend upon a balance between the rate of nitrite depletion and the death rate of spores that have germinated (Christiansen, 1980). The rate of depletion depends on product formulation, pH, and time and temperature relationships during processing and subsequent storage (Fox and Nicholas, 1974). Nitrite is depleted more rapidly at low pH or high temperature. Extended storage is known to deplete nitrite and decrease the degree of botulinal inhibition (Tompkin et al., 1978c). Chelating agents such as ascorbate, EDTA, and cysteine are known to enhance the efficacy of nitrate by chelating iron (Christiansen, 1980; Tompkin et al., 1979a). High concentrations of ascorbate may reduce nitrite inhibition by depleting nitrite (Tompkin et al., 1979b).

One of the concerns surrounding the use of nitrite in curing is the formation of nitrosamines via reaction of nitrous acid with secondary amines (Lijinsky and Epstein, 1970). Nitrosamines can also be formed from nitrite by microorganisms, including clostridia (Ayanaba and Alexander, 1973; Mills and Alexander, 1976). The carcinogenic and mutagenic properties of nitrosamines have been demonstrated and extensively reviewed (Gomez et al., 1974; Lijinsky, 1976), and the occurrence of nitrosamines in various foods has been documented (Crosby and Sawyer, 1976; Gray and Randell, 1979; Sebranek and Cassens, 1973). Furthermore, nitrous acid is a known mutagen (Zimmermann, 1977), and nitrite has been implicated as an agent that causes cancer in experimental animals (Newberne, 1979). Sofos and Busta (1980) and Widdus and Busta (1982) reviewed alternatives to the use of nitrite as an antibolutinal agent. Some of the proposed alternatives are irradiation, sorbate, sodium hypophosphite, fumarate esters, acidulation by lactic acid–producing bacteria, alpha-tocopherol, and ascorbates.

Sorbic Acid Sorbic acid and its salts, which are widely used to inhibit yeasts and molds, have been extensively examined as a possible partial replacement for nitrite in cured meats (Liewen and Marth, 1985; Sofos and Busta, 1980, 1981, 1983; Sofos et al., 1979a; Widdus and Busta, 1982). Studies have shown that potassium sorbates at a concentration of 2600 mg/kg could be used to replace a high proportion of nitrite (Huhtanen and Feinberg, 1980; Huhtanen et al., 1983; Roberts et al., 1982).

Although species of clostridia were reported to be relatively resistant to sorbic acid (Emard and Vaughn, 1952), and sorbic acid has been known to be utilized as a carbon source by proteolytic strains of *C. botulinum* (York and Vaughn, 1954, 1955), sorbic acid at the appropriate pH will delay botulinal growth and toxin production (Sofos et al., 1979b). Potassium sorbate inhibits germination and outgrowth of *C. botulinum* spores, as well as growth of vegetative bacteria (Cook and Pierson, 1983; Sofos et al., 1986), but the concentrations necessary to inhibit germination are higher (Blocher and Busta, 1985; Smoot and Pierson, 1981). The use of sorbic acid to reduce the risk of botulinal growth in foods therefore depends on its ability to inhibit the growth of vegetative cells (Lund et al., 1987). Sorbic acid is thought to interfere with a variety of enzyme systems essential for outgrowth and cell metabolism (York and Vaughn, 1964).

The inhibitory effect of sorbate on the growth of *C. botulinum* increases with a decrease in pH between 7.0 and 4.9, and the effect is dependent on the concentration of undissociated sorbic acid (Lund et al., 1987). Sorbic acid has a pKa value of 4.80 and is most effective at pH $<$ 6.0–6.5. The synergistic inhibitory effect of sorbic acid and nitrite is also pH dependent and was not observed above pH 6.2 (Sofos et al., 1980). Sorbic acid currently is used in foods such as bakery products and cheese primarily for its antimycotic activity.

Parabens Parabens are alkyl esters of parahydroxybenzoic acid and are also called parasepts of PHB esters. Parabens are more effective than benzoic acid over a wide pH range and are considered safer for human consumption. The methyl and propyl parabens are generally recognized as safe (GRAS). Current FDA regulations permit up to 0.1% in certain foods (Cook and Pierson, 1983). Dymicky and Huhtanen (1979) found that the effectiveness of *C. botulinum* inhibition increased with increasing ester chain length. The undecyl ester has the same aqueous solubility as the methyl ester but is about 3000 times more inhibitory. The effect of medium appears to have a significant role in the effectiveness of parabens. Although Robach and Pierson (1978) and Reddy et al. (1982) demonstrated inhibition of growth and toxin production from *C. botulinum* spore inocula in laboratory media, Pierson and Reddy (1982) found that parabens had little effect against *C. botulinum* spores in comminuted pork and wieners. Gonsalves (1968) found that parabens in an aqueous system (Pork Infusion Agar) had increasing inhibitory properties as the ester chain length increased. However, parabens in an oil-containing food (fish homogenate) were less inhibitory as the chain length increased. The increasing chain length gives the parabens a higher partition coefficient, and they become more soluble in oil or fat than in water. Therefore, the effective concentration in the aqueous phase of fat-containing foods may decrease with increasing chain length. Parabens are used as microbial preservatives in baked goods, soft drinks, beer, jams, jellies, olives, pickles, and syrups.

Nisin and Other Bacteriocins Nisin, a polypeptide antimicrobial agent produced by certain strains of *Streptococcus lactis*, is known to be effective against *C. botulinum* (Hurst, 1981; Rayman et al., 1981; Scott and Taylor, 1981a,b), although spores of *C. botulinum* are more resistant than spores of many other bacteria (Sperber, 1982). Since the first use of nisin in Swiss cheese to prevent spoilage by *Clostridium butyricum*, nisin at levels as high as 500 μg/ml has been used in Europe and elsewhere as a general anticlostridial agent in a number of foods, mostly dairy products (Hurst, 1981).

In culture media, nisin is more effective in the range of pH 5.5–6.0 than in the range of pH 7.0–8.0 (Scott and Taylor, 1981b). Somers and Taylor (1987) found that nisin is effective in pasteurized process cheese spreads. Addition of nisin allowed formulation of cheese spread with reduced sodium levels (addition of 1.4% disodium phosphate and no added NaCl) or slightly higher moisture levels. Its use in cheese spreads recently has been approved in the United States (Hauschild, 1989). However, the usefulness of nisin as an antibotulinal agent in meat products is less promising. Nisin at levels up to 550 ppm (550 μg/ml) in combination with 60 ppm of nitrite failed to prevent outgrowth of *C. botulinum* spores in pork slurries adjusted to pH 5.8 (Rayman et al., 1983). Reducing the pH enhanced nisin activity.

Addition of nisin at levels up to 500 ppm extended the shelf life of chicken frankfurter emulsions challenged with *C. botulinum* by only a week at 27°C (Taylor et al., 1985). In bacon, nisin alone had no antibotulinal effect (Taylor and Somers, 1985). The limited effectiveness of nisin in meat emulsions may reflect lack of stability in the emulsion, which is probably the result of nisin binding to meat particles (Scott and Taylor, 1981a; Taylor et al., 1985). An early investigation implicated germination as the process inhibited by nisin (Treadwell et al., 1958), but it is now well established that outgrowth is inhibited (Holley, 1981).

Subtilin is produced by certain strains of *Bacillus subtilis* and is similar to nisin in structure and composition. Subtilin is stable in acid environments and resistant to heat (30–60 min at 121 °C) (Jay, 1983). It is active against *C. botulinum* (Anderson and Michener, 1950; Le Blanc et al., 1953). Tylosin inhibits outgrowth of germinating bacterial spores, is effective against *C. sporogenes*

PA3679 (Wheaton and Hays, 1964), and is stable during heating and storage (Jay, 1983). Other bacteriocins are produced by microorganisms common in foods, but their antibotulinal activity in foods has not been reported. Except for nisin, bacteriocins are not currently permitted for direct addition to foods.

Phenolic Antioxidants Butylated hydroxyanisole (BHA), butylated hyroxytoluene (BHT), tertiary butylhydroquinone (TBQH), and propyl gallate (PG) are synthetic phenolic compounds currently approved for use in foods as antioxidants. These phenolic antioxidants are known to inhibit germination and growth from *C. botulinum* inocula in laboratory media (Pierson and Reddy, 1982; Reddy et al., 1982; Tompkin et al., 1979a), but they are less effective in meat systems. Although Reddy et al. (1982) found that BHA, BHT, and TBHQ at concentrations of 200–400 μg/ml inhibited outgrowth of *C. botulinum* spores in laboratory media; in a comminuted pork product, growth occurred within 7 days (Pierson and Reddy, 1982). It is possible that phenolic antioxidants are less inhibitory to spores in food systems because of their lipid solubility (Cook and Pierson, 1983). 8-Hydroxyquinoline at a concentration of 200 ppm or in combination with a low concentration of sodium nitrite (40 ppm) inhibited the growth and toxin production of *C. botulinum* in comminuted pork for 60 days at 27°C (Pierson and Reddy, 1982).

Polyphosphates Polyphosphates are used in food, especially processed meat, poultry, and seafood, to achieve certain functional objectives rather than microbiological control. Current U.S. regulations limit addition of phosphates to amounts needed to achieve functionality (Tompkin, 1984; Wagner, 1986). Polyphosphates, especially sodium acid pyrophosphate (SAPP), have been shown to enhance inhibition of *C. botulinum* by other antibotulinal agents such as nitrite and sorbate (Barbut et al., 1987; Ivey and Robach, 1978; Nelson et al., 1983; Wagner and Busta, 1983, 1984). However, Roberts et al. (1981a) observed that 0.3% Curaphos 700 (a commercial polyphosphate blend) increased the probability of botulinum toxin production in pork slurries at pH 5.5–6.3. The same concentration of Curaphos 700 inhibited toxin production in high pH (6.3–6.8) pork slurries (Roberts et al., 1981b). Madril and Sofos (1986) postulated that this observation is related to metal ion sequestration and inhibition of clostridia by nitrite. The antimicrobial activities of nitrite and SAPP seem to be interdependent, and since the rate of nitrite depletion is increased with reduced pH (Christiansen, 1980), then the lower pH treatment may not be very effective in preventing growth of clostridia. Diphosphates are known to be more effective than triphosphates or longer chained polyphosphates (Nelson et al., 1983). Another form of phosphate, sodium hypophosphite, also inhibits *C. botulinum* (Pierson et al., 1981). Either 3000 mg/kg alone or 1000 mg/kg of sodium hypophosphite with a low concentration of sodium nitrite (40 mg/kg) was as effective as 120 mg/kg of sodium nitrite for inhibiting botulinum toxin production in bacon.

The mechanism of antibotulinal activity of polyphosphates is not known. Nelson et al. (1983) postulated that polyphosphates may act as sequestrants and enhance the antibotulinal activity of nitrite in a manner similar to isoascorbate. Post et al. (1963) and Elliot et al. (1964) suggested that phosphates may inhibit microbial growth by sequestering or interfering with metal ions. Wagner and Busta (1985, 1986) proposed that SAPP may be inhibiting actual production or function of the protease-activating botulinum toxin.

Ascorbates Ascorbate (vitamin C) or isoascrobate (erythorbate) are used in cured meat products to accelerate the curing process. They also stabilize color, act as synergists to antioxidants, and suppress nitrosamine formation (Hauschild, 1989; Sofos and Busta, 1980). Although ascorbates have no independent antibotulinal activity, they have been shown to enhance the inhibitory activity of nitrite against *C. botulinum*. Tompkin et al. (1978a, 1979b) reported that a combination of 200 ppm isoascorbate and 50 ppm nitrite was as effective as 156 ppm nitrite alone. Similar results were reported by Ashworth and Spencer (1972) and Johnston and Loynes (1971). At higher concentrations, ascorbate may reduce the inhibitory activity of nitrite (Bowen and Deibel, 1974). Tompkin et al. (1979b) proposed that ascorbates may enhance the activity of nitrite by sequestering metal ions and decrease the activity of nitrite by accelerating its depletion. Ascorbic acid is naturally present in many foods such as fruits, in addition to being added to processed meats.

EDTA, other chelating agents (Tompkin et al., 1978a, 1979a), and cysteine (Johnston and Loynes, 1971) also enhance the effectiveness of nitrite against *C. botulinum*. EDTA is approved for use in a variety of foods including processed meats, poultry, seafood, and vegetables.

Smoke The preservative effect of smoking is based on the combination of drying and deposition of chemicals from the thermal decomposition of wood. Several hundred chemicals, including acids, alcohols, aldehydes, ketones, phenols, carbonyl compounds, waxes, and tars, have been identified in the smoke condensate (Boyle et al., 1988; Sink and Hsu, 1977). While smoking of meat appears to have little effect on *C. botulinum* spores (Christiansen et al., 1968; Hauschild, 1989), both generated smoke and liquid smoke, in combination with NaCl, were effective inhibitors of outgrowth and toxin production by *C. botulinum* types A and E in hot-processed fish (Eklund, 1982; Eklund et al., 1982). The minimum NaCl concentrations (in the water phase) required to inhibit toxin production by surface-inoculated type A or type E spores were decreased in the presence of smoke products from 4.6 to 2.8%, and from 3.7 to less than 2%, respectively. However, smoking should not be considered as a substitute for NaCl or refrigerated storage.

Other Inhibitors Sulfur dioxide (added as sodium metabisulfite) has been shown to delay *C. botulinum* outgrowth in perishable canned comminuted pork

(Tompkin et al., 1980), but rather high concentrations (>100 μg/g) were necessary to achieve significant inhibition. Up to 450 μg/g sulfur dioxide is permitted in the United Kingdom. Sodium nitrite partially reduced the efficacy of sulfur dioxide. Sulfur dioxide or sulfites are permitted in a variety of foods including flour, many fruits, fruit juices, dried fruits and vegetables, syrup, pickles, wine, and beer. It is not permitted in most countries for use in foods considered to be a good source of vitamin B_1 and is only permitted in meat and fish products in some countries.

Hall and Maurer (1986) reported that mace, bay leaf, and nutmeg extracts inhibited botulinum toxin production. DeWit et al. (1979) found that garlic oil and onion oil at a concentration of 1000 μg/g of meat slurry inhibited toxin production by *C. botulinum* type A. The inhibition was not complete, and toxin production by types B and E was not inhibited. Ismaiel (1988) reported that garlic oil inhibited germination of *C. botulinum* types A, B, and E in laboratory media, but had no significant effect on outgrowth and toxin production. The oils of black pepper, clove, cinnamon, and origanum inhibited vegetative growth in laboratory media. Origanum oil at 400 ppm in combination with 50–100 ppm sodium nitrite significantly delayed growth and toxin production in vacuum-packaged comminuted pork (Ismaiel, 1988).

The *n*-monoalkyl maleates and fumarates, especially those esterified with high (C_{13}–C_{18}) alcohols, were reported to be effective against *C. botulinum* 62A (Dymicky et al., 1987). Lactate salts (sodium, calcium, potassium, or ammonium) are effective in preventing growth of *C. botulinum* in cooked meat and comminuted raw turkey (Ande and Underhill, 1987; Maas et al., 1989). Incorporation of 3-(4-tosylsulfonyl) acrylonitrile at 20–50 ppm in combination with 40 ppm of sodium nitrite inhibited growth of *C. botulinum* in meat products (Dahlstrom, 1988).

Effect of Other Microorganisms

C. botulinum growth and toxin production may be directly or indirectly affected by endogenous or added microorganisms. The growth of some microorganisms may change the environment and allow *C. botulinum* to grow and produce toxin under otherwise unfavorable conditions. Some microorganisms may inhibit *C. botulinum*, either by changing the environment unfavorably or by producing specific inhibitory substances.

The growth of acid-tolerant molds, such as *Penicillin, Cladosporium, Mycoderma, Trichosporon*, and *Aspergillus*, in acidic foods has been shown to increase the pH, at least in their vicinity, to a level that would allow *C. botulinum* to grow and produce toxin (Huhtanen et al., 1976; Odlaug and Pflug, 1978, 1979). Acidity gradients from near neutrality under the mycelial mat on the surface to pH < 4.6 at the bottom of the containers have been observed in tomato juice inoculated with molds (Huhtanen et al., 1976; Odlaug and Pflug, 1979).

The availability of O_2 was a limiting factor for the development of mycelial mats and pH gradients. Several of the acid or acidified foods incriminated in botulism outbreaks contained acid-tolerant microorganisms such as yeasts, molds, and others (Odlaug and Pflug, 1978).

Lactic acid bacteria, such as *Lactobacillus*, *Pediococcus*, and *Streptococcus* (now *Lactococcus*), have been shown to produce acid and inhibit *C. botulinum* growth and toxin production in meat products (Tanaka et al., 1980, 1985). Addition of a fermentable carbohydrate may be required to ensure enough acid production. The inhibitory efficacy of this system depends on the initial pH, type and level of inoculum, type and level of carbohydrate, buffering capacity of the product, and the presence of inhibitory compounds (Conner et al., 1989). Bacteriocin production by lactic acid bacteria may be responsible for some of the inhibition. Substances inhibitory to *C. botulinum* are also known to be produced by *Bacillus licheniformis* (Wentz et al., 1967), *C. perfringens* (Smith, 1975), enterococci (Kafel and Ayres, 1969), and nontoxigenic organisms closely resembling *C. botulinum* (Anastasio et al., 1971) but have no practical application at present.

Interaction of Control Factors

The preceding discussion details the many environmental control factors that influence the ability of *C. botulinum* to develop in a food. It is not possible for control factors to function independently because temperature, water activity, pH, redox potential, and other factors such as preservatives all affect *C. botulinum* in foods. When these controls are applied at the limits allowing *C. botulinum* development and toxin production, they often function synergistically but may also have additive or antagonistic effects (see Scott, 1989; Mossell, 1983).

CONCLUSIONS

The history of the food-processing industry demonstrates effective control of *C. botulinum*. The causes of botulism incidents have generally been identified, and successful corrective actions have been taken to avoid recurrences. However, the food-processing industry is dynamic and continuously responds to the needs and desires of consumers by developing new products. Thus, challenges for controlling *C. botulinum* in food products will continue to present themselves. An understanding of the basic genetics and physiology of the organism and the principles and mechanisms of action of existing controls are essential in meeting these challenges.

REFERENCES

Ahmed, F. I. K., and Russell, C. (1975). Synergism between ultrasonic waves and hydrogen peroxide in the killing of micro-organisms. *J. Appl. Bacteriol.* *39*:31.

Ajmal, M. (1968). Growth and toxin production of *Clostridium botulinum* type E. *J. Appl. Bacteriol. 31*:120.

Alderton, G., Chen, J. K., and Ito, K. A. (1974). Effect of lysozyme on the recovery of heated *Clostridium botulinum* spores. *Appl. Microbiol. 27*:613.

Amaha, M., and Ordal, Z. J. (1957). Effect of divalent cations in the sporulation medium on the thermal death of *Bacillus coagulans* var. *thermoacidurans. J. Bacteriol. 74*:596.

Anastasio, K. L., Soucheck, J. A., and Sugiyama, H. (1971). Boticinogeny and action of the bacteriocin. *J. Bacteriol. 107*:143.

Ande, C. F., and Underhill, F. A. (1987). A food stuff having a surface coating of lactate salt. European Patent Application EP 0 250 074 A2.

Andersen, A. A., and Michener, H. D. (1950). Preservation of foods with antibiotics. I. The complementary action of subtilin and mild heat. *Food Technol. 4*:188.

Anderson, A. W., Corlett, D. A., and Krabbenhoft, K. L. (1967). The effects of additives on radiation resistance of *Cl. botulinum* in meat. In *Botulism 1966* (M. Ingram and T. A. Roberts, eds.), Chapman and Hall, London, p. 76.

Ando, Y. (1971). The germination requirement of spores of *Clostridium botulinum* type E. *Jpn. J. Microbiol. 15*:515.

Ando, Y. (1973a). Studies on germination of spores of clostridial species capable of causing food poisoning. I. Factors affecting the germination of spores of *Clostridium botulinum* type A on a chemically defined medium. *J. Food Hyg. Soc. Jpn. 14*:457.

Ando, Y. (1973b). Studies on germination of spores of clostridial species capable of causing food poisoning. II. Effects of some chemical food additives on the growth from spores of *Clostridium botulinum* type E. *J. Food Hyg. Soc. Jpn. 14*:462.

Ando, Y. (1974). Alpha-hydroxy acids as co-germinants for some clostridial spores. *Jpn. J. Microbiol. 18*:100.

Ando, Y., and Iida, H. (1970). Factors affecting the germination of spores of *Clostridium botulinum* type E. *Jpn. J. Microbiol. 14*:361.

Anellis, A., and Koch, R. B. (1962). Comparative resistance of strains of *Clostridium botulinum* to gamma rays. *Appl. Microbiol. 10*:326.

Anellis, A., Berkowitz, D., Jarboe, C., and El-Bisi, H. M. (1969). Radiation sterilization of prototype military foods. III. Pork loin. *Appl. Microbiol. 18*:604.

Anellis, A., Berkowitz, D., Kemper, D., and Rowley, D. B. (1972). Production of types A and B spores of *Clostridium botulinum* by the biphasic method: Effect on spore population, radiation resistance, and toxigenicity. *Appl. Microbiol. 23*:734.

Anellis, A., Grecz, N., and Berkowitz, D. (1965a). Survival of *Clostridium botulinum* spores. *Appl. Microbiol. 13*:397.

Anellis, A., Grecz, N., Huber, D. A., Berkowitz, D., Schneider, M. D., and Simon, M. (1965b). Radiation sterilization of bacon for military feeding. *Appl. Microbiol. 13*:37.

Anellis, A., Shattuck, E., Rowley, D. B., Ross, E. W., Jr., Whaley, D. N., and Dowell, V. R., Jr. (1975). Low-temperature irradiation of beef and methods for evaluation of radappertization process. *Appl. Microbiol. 30*:811.

Ashworth, J., and Spencer, R. (1972). The Perigo effect in pork. *J. Food Technol. 7*:111.

Ayanaba, A., and Alexander, M. (1973). Microbial formation of nitrosamines in vitro. *Appl. Environ. Microbiol. 25*:862.

Baird-Parker, A. C. (1971). Factors affecting the production of bacterial food poisoning toxins. *J. Appl. Bacteriol. 34*:181.

Baird-Parker, A. C., and Freame, B. (1967). Combined effect of water activity, pH and temperature on the growth of *Clostridium botulinum* from spore inocula. *J. Appl. Bacteriol. 30*:420.

Bakerman, H., Romine, M., Schricker, J. A., Takahashi, S. M., and Mickelson, O. (1956). Stability of certain B-vitamins exposed to ethylene oxide in the presence of chlorine chloride. *J. Agr. Food Chem. 4*:956.

Barbut, S., Maurer, A. J., and Thayer, D. W. (1987). Gamma-irradiation of *Clostridium botulinum* inoculated turkey frankfurters formulated with different chloride salts and polyphosphates. *J. Food Sci. 52*:1137.

Barker, A. N., and Wolf, J. (1971). Effects of thioglycollate on the germination and growth of some clostridia. In *Spore Research 1971* (A. N. Barker, G. W. Gould, and J. Wolf, eds.), Academic Press, New York, p. 95.

Bauman, H. (1990). HACCP: Concept, development, and application. *Food Technol. 44*(5):156.

Bayliss, C. E., and Waites, W. M. (1979). The combined effect of hydrogen peroxide and ultraviolet irradiation on bacterial spores. *J. Appl. Bacteriol. 47*:263.

Bayliss, C. E., and Waites, W. M. (1982). Effect of simultaneous high intensity ultraviolet irradiation and hydrogen peroxide on bacterial spores. *J. Food Technol. 17*:467.

Beaman, T. C., and Gerhardt, P. (1986). Heat resistance of bacterial spores correlated with protoplast dehydration, mineralization, and thermal adaptation. *Appl. Environ. Microbiol. 52*:1242.

Beers, R. J. (1957). Effect of moisture activity on germination. In *Spores* (H. O. Halvorson, ed.) American Institute of Biological Sciences, Washington, DC, p. 45.

Benedict, R. C. (1980). Biochemical basis for nitrite-inhibition of *Clostridium botulinum* in cured meat. *J. Food Prot. 43*:877.

Blocher, J. C., and Busta, F. F. (1983). Bacterial spore resistance to acid. *Food Technol. 37*(12):87.

Blocher, J. C., and Busta, F. F. (1985). Multiple modes of inhibition of spore germination and outgrowth by reduced pH and sorbate. *J. Appl. Bacteriol. 59*:469.

Bonventre, P. F., and Kempe, L. L. (1959). Physiology of toxin production by *Clostridium botulinum* types A and B. III. Effect of pH and temperature during incubation on growth, autolysis, and toxin production. *Appl. Microbiol. 7*:374.

Bowen, V. G., and Deibel, R. H. (1974). Effects of nitrite and sodium ascorbate on botulinal toxin formation in wieners and bacon. In *Proceeding of Meat Industry Research Conference*, Am. Meat Inst. Found., Chicago, IL, pp. 63–68.

Boyle, D. L., Sofos, J. N., and Maga, J. A. (1988). Inhibition of spoilage and pathogenic microorganisms by liquid smoke from various woods. *Lebensm.-Wiss. Technol. 21*:54.

Brown, M. R. W., and Melling, J. (1973). Release of dipicolinic acid and calcium and activation of *Bacillus stearothermophilus* as a function of time, temperature and pH. *J. Pharm. Pharmacol. 25*:478.

Busta, F. F., and Foegeding, P. M. (1983). Chemical food preservatives. In *Disinfection, Sterilization and Preservation* (S. S. Block, ed.), Lea and Febiger, Philadelphia. pp. 656-694.

Caputo, R. A., and Odlaug, T. E. (1983). Sterilization with ethylene oxide and other gases. In *Disinfection, Sterilization, and Preservation*, 3rd ed. (S. S. Block, ed.), Lea & Febiger, Philadelphia, p. 47.

Cerf, O. (1977). Tailing of survival curves of bacterial spores. *J. Appl. Bacteriol. 42*:1.

Cerf, O., and Metro, F. (1977). The tailing survival curves of *Bacillus licheniformis* spores treated with hydrogen peroxide. *J. Appl. Bacteriol. 42*:405.

CFR (1990a). Current food manufacturing practice in manufacturing, packaging, or holding human food. Code of Federal Regulations, Title 21, Part 110, U.S. Government Printing Office, Washington, DC.

CFR (1990b). Acidified foods. Code of Federal Regulations, Title 21, Part 114, U.S. Government Printing Office, Washington, DC.

Cheng, Y. E., Fitz-James, P., and Aronson, A. I. (1978). Characterization of a *Bacillus cereus* protease mutant defective in an early stage of spore germination. *J. Bacteriol. 133*:393.

Chowdhury, M. S. U., Rowley, D. B., Anellis, A., and Levinson, H. S. (1976). Influence of postirradiation incubation temperature on recovery of radiation-injured *Clostridium botulinum* 62A spores. *Appl. Environ. Microbiol. 32*:172.

Christiansen, L. N., Deffner, J., Foster, E. M., and Sugiyama, H. (1968). Survival and outgrowth of *Clostridium botulinum* type E spores in smoked fish. *Appl. Microbiol. 16*:133.

Christiansen, L. N. (1980). Factors influencing botulinal inhibition by nitrite. *Food Technol. 34*(5):237.

Christiansen, L. N., and Foster, E. M. (1965). Effect of vacuum packaging on growth of *Clostridium botulinum* and *Staphylococcus aureus* in cured meats. *Appl. Microbiol. 13*:1023.

Christiansen, L. N., Johnston, R. W., Kautter, D. A., Howard, J. W., and Aunan, W. J. (1973). Effect of nitrite and nitrate on toxin production by *Clostridium botulinum* and on nitrosamine formation in perishable canned comminuted cured meat. *Appl. Microbiol. 25*:357(Erratum *26*:653).

Church, B. D., and Halvorson, H. (1959). Dependence of heat resistance of bacterial endospores on the dipicolinic acid content. *Nature 183*:124.

Cockey, R. R., and Tatro, M. C. (1974). Survival studies with spores of *Clostridium botulinum* type E in pasteurized meat of the blue crab, *Callinectes sapidus*. *Appl. Microbiol.* 27:629.

Conner, D. E., Scott, V. N., Bernard, D. T., and Kautter, D. A. (1989). Potential *Clostridium botulinum* hazards associated with extended shelf-life refrigerated foods: A review. *J. Food Safety 10*:131.

Cook, F. K., and Pierson, M. D. (1983). Inhibition of bacterial spores by antimicrobials. *Food Technol. 37*(11):115.

Crosby, N. T., and Sawyer, R. (1976). N-Nitrosoamines: A review of chemical and biological properties and their estimation in foodstuffs. *Adv. Food Res. 22*:1.

Curran, H. R., and Evans, F. R. (1945). Heat activation inducing germination in the spores of thermotolerant and thermophile aerobic bacteria. *J. Bacteriol. 49*:335.

Curran, H. R., Evans, F. R., and Leviton, A. (1940). The sporicidal action of hydrogen peroxide and the use of crystalline catalase to dissipate residual peroxide. *J. Bacteriol. 40*:423.

Dahlstrom, R. V. (1988). Method for preservation of meat products. European Patent EP 0 122 062 B1.

Davis, R. B., and Dignan, D. M. (1983). Use of hydrogen peroxide sterilization in packaging foods. *Assoc. Food Drug Off. Quart. Bull. 47*:12.

Day, L. E., and Costilow, R. N. (1964). Physiology of the sporulation process in *Clostridium botulinum*. I. Correlation of morphological changes with catabolic activities, synthesis of dipicolinic acid, and development of heat resistance. *J. Bacteriol. 88*:690.

Dempster, J. F. (1985). Radiation preservation of meat and meat products: A review. *Meat Sci. 12*:61.

Denny, C. B., Bohrer, C. W., Perkins, W. E., and Townsend, C. E. (1959). Destruction of *Clostridium botulinum* by ionizing radiation. I. Neutral phosphate at room and freezing temperatures. *Food Res. 24*:44.

Denny, C. B., Goeke, D. J., Jr., and Sternberg, R. (1969). Inoculation tests of *Clostridium botulinum* in canned breads with specific reference to water activity. Res. Rep. No. 4–69, National Canners Association, Washington, DC.

DeWit, J. C., Notermans, S., Gorin, N., and Kampelmacher, E. H. (1979). Effect of garlic oil or onion oil on toxin production by *Clostridium botulinum* in meat slurry. *J. Food Prot. 42*:222.

Dodds, K. L. (1989). Combined effect of water activity and pH on inhibition of toxin production by *Clostridium botulinum* in cooked, vacuum-packed potatoes. *Appl. Environ. Microbiol. 55*:656.

Donnellan, J. E., Jr., and Setlow, R. B. (1965). Thymine photoproducts but not thymine dimers found in ultraviolet-irradiated bacterial spores. *Science 149*:308.

Donnellan, J. E., Jr., and Stafford, R. S. (1968). The ultraviolet photochemistry and photobiology of vegetative cells and spores of *Bacillus megaterium*. *Biophys. J. 8*:17.

Dozier, C. C. (1924). Optimum and limiting hydrogen-ion concentration for *B. botulinus* and quantitative estimation of its growth. XVI. *J. Infect. Dis. 35*:105.

Durban, E., Grecz, N., and Farkas, J. (1974). Direct enzymatic repair of deoxyribonucleic acid single-strand breaks in dormant spores. *J. Bacteriol. 118*:129.

Dye, M., and Mead, G. C. (1972). The effect of chlorine on the viability of clostridial spores. *J. Food Technol. 7*:173.

Dymicky, M., Bencivengo, M., Buchanan, R. L., and Smith, J. L. (1987). Inhibition of *Clostridium botulinum* 62A by fumarates and maleates and relationship of activity to some physicochemical constants. *Appl. Environ. Microbiol. 53*:110.

Dymicky, M., and Huhtanen, C. N. (1979). Inhibition of *Clostridium botulinum* by p-hydroxybenzoic acid n-alkyl esters. *Antimicrob. Agents Chemother. 15*:798.

Edwards, R. B., Peterson, L. J., and Cunnings, D. G. (1954). The effect of cathode rays on bacteria. *Food Technol. 8*:284.

Eklund, M. W. (1982). Significance of *Clostridium botulinum* in fishery products preserved short of sterilization. *Food Technol. 36*(12):107.

Eklund, M. W., Pelroy, G. A., Paranjpye, R., Peterson, M. E., and Teeny, F. M. (1982). Inhibition of *Clostridium botulinum* types A and E toxin production by liquid smoke and NaCl in hot-process smoke-flavored fish. *J. Food Prot. 45*:935.

Eklund, M. W., Poysky, F. T., and Wieler, D. I. (1967a). Characteristics of *Clostridium botulinum* type F isolated from the Pacific coast of the United States. *Appl. Microbiol. 15*:1316.

Eklund, M. W., Wieler, D. I., and Poysky, F. T. (1967b). Growth and toxin production of nonproteolytic type B *Clostridium botulinum* at 3.3 to 5.6°C. *J. Bacteriol. 93*:1461.

El-Bisi, H. M., Snyder, O. P., and Levin, R. E. (1967). Radiation death kinetics of *Cl. botulinum* spores at cryogenic temperatures. In *Botulism 1966* (M. Ingram, and T. A. Roberts, eds.), Chapman and Hall, London, p. 89.

Elliot, R. P., Straka, R. P., and Garibaldi, H. A. (1964). Polyphosphate inhibition of growth of Pseudomonads from poultry meat. *Appl. Microbiol. 12*:517.

Emard, L. O., and Vaughn, R. H. (1952). Selectivity of sorbic acid media for the catalase-negative lactic acid bacteria and clostridia. *J. Bacteriol. 63*:487.

Emodi, A. S., and Lechowich, R. V. (1969). Low temperature growth of type E *Clostridium botulinum* spores. I. Effects of sodium chloride, sodium nitrite and pH. *J. Food Sci. 34*:78.

Erdman, I. E., Thatcher, F. S., and MacQueen, K. F. (1961). Studies on the irradiation of microorganisms in relation to food preservation. I. The comparative sensitivities of specific bacteria of public health significance. *Can. J. Microbiol. 7*:199.

Ernst, R. R. (1977). Sterilization by heat. In *Disinfection, Sterilization, and Preservation* (S. S. Block, ed.), Lea & Febiger, Philadelphia, p. 481.

Farkas, J., Kiss, I., and Andrassy, E. (1967). The survival and recovery of irradiated bacterial spores as affected by population density and some external factors. In *Radiosterilization of Medical Products*, International Atomic Energy Agency, Vienna, p. 343.

FDA (1986). Irradiation in the production, processing, and handling of foods; final rule. *Fed. Reg. 51*(75):13376.

Flowers, R. S., and Adams, D. M. (1976). Spore membrane(s) as the site of damage within heated *Clostridium perfringens* spores. *J. Bacteriol. 125*:429.

Foegeding, P. M. (1983). Bacterial spore resistance to chlorine compounds. *Food Technol. 37*(11):100.

Foegeding, P. M., and Busta, F. F. (1981). Bacterial spore injury—An update. *J. Food Prot. 45*:776.

Foegeding, P. M., and Busta, F. F. (1983a). Hypochlorite injury of *Clostridium botulinum* spores alters germination responses. *Appl. Environ. Microbiol. 45*:1360.

Foegeding, P. M., and Busta, F. F. (1983b). Proposed role of lactate in germination of hypochlorite-treated *Clostridium botulinum* spores. *Appl. Environ. Microbiol. 45*:1369.

Foegeding, P. M., and Busta, F. F. (1983c). Proposed mechanism for sensitization by hypochlorite treatment of *Clostridium botulinum* spores. *Appl. Environ. Microbiol. 45*:1374.

Foegeding, P. M., and Busta, F. F. (1983d). Differing L-alanine germination requirements of hypochlorite-treated *Clostridium botulinum* spores from two crops. *Appl. Environ. Microbiol. 45*:1415.

Foegeding, P. M. and Busta, F. F. (1983e). Effect of carbon dioxide, nitrogen and hydrogen gases on germination of *Clostridium botulinum* spores. *J. Food Prot. 46*:987.

Fox, J. B., Jr., and Nicholas, R. A. (1974). Nitrite in meat: Effect of various compounds of loss of nitrite. *J. Agr. Food Chem.* *22*:302.

Francesconi, S. C., MacAlister, T. J., Setlow, B., and Setlow, P. (1988). Immunoelectron microscopic localization of small, acid-soluble spore proteins in sporulating cells of *Bacillus subtilis. J. Bacteriol.* *170*:5963.

Galanina, L. A., Mityushina, L. L., Duda, V. I., and Bekhtereva, M. N. (1979). Electron microscopic study of *Bacillus anthracoides* spores under normal conditions and upon exposure to chloroactive disinfectant. *Microbiol.* *48*:368.

Gaze, J. E., and Brown, K. L. (1988). The heat resistance of spores of *Clostridium botulinum* 213B over the temperature range 120 to 140°C. *Int. J. Food Sci. Technol.* *23*:373.

Gerhardt, P. (1988). The refractory homeostasis of bacterial spores. In *Homeostatic Mechanisms in Microorganisms* (R. Whittenbury, J. G. Banks, G. W. Gould, and R. B. Board, eds.), Bath University Press, Bath, United Kingdom, p. 41.

Gerhardt, P., and Marquis, R. E. (1989). Spore thermoresistance mechanisms. In *Regulation of Procaryotic Development* (I. Smith, R. A. Slepecky, and P. Setlow, eds.), American Society Microbiol., Washington, DC, p. 43.

Gerhardt, P., and Murrell, W. G. (1978). Basis and mechanisms of spore resistance: A brief preview. In *Spores VII* (G. H. Chambliss, and J. C. Vary, eds.), American Society Microbiol., Washington, DC, p. 18.

Ginoza, W. (1967). The effects of ionizing radiation on nucleic acids of bacteriophages and bacterial cells. *Annu. Rev. Microbiol.* *21*:325.

Gombas, D. E. (1983). Bacterial spore resistance to heat. *Food Technol.* *37*(11):105.

Gombas, D. E., and Gomez, R. F. (1978). Sensitization of *Clostridium perfringens* spores to heat by gamma radiation. *Appl. Environ. Microbiol.* *36*:403.

Gomez, R. F., Johnston, M., and Sinskey, A. J. (1974). Activation of nitrosomorpholine and nitrosopyrrolidine to bacterial mutagens. *Mutat. Res.* *24*:5.

Gonsalves, A. A. (1968). The efficacy of the methyl, propyl, butyl and heptyl esters of parahydroxybenzoic acid as inhibitors of *Clostridium botulinum* types A, B and E. Ph.D. thesis, University of Massachusetts, Amherst.

Gould, G. W. (1962). Microscopical observations on the emergence of cells of *Bacillus* spp. from spores under different cultural conditions. *J. Appl. Bacteriol.* *25*:35.

Gould, G. W. (1969). Germination. In *The Bacterial Spore* (G. W. Gould and A. Hurst, eds.), Academic Press, London, p. 397.

Gould, G. W. (1970). Germination and the problems of dormancy. *J. Appl. Bacteriol.* *33*:34.

Gould, G. W. (1978). Practical implications of compartmentalization and osmotic control of water distribution in spores. In *Spores VII* (G. Chambliss, and J. C. Vary, eds.), American Society Microbiol., Washington, DC, p. 21.

Gould, G. W. (1983). Mechanism of resistance and dormancy. In *The Bacterial Spore*, Vol. 2 (A. Hurst, and G. W. Gould, eds.), Academic Press, London, p. 173.

Gould, G. W. (1986). Water and survival of bacterial spores. In *Membranes, Metabolism and Dry Organisms* (A. C. Leopold, ed.), Cornell University Press, Ithaca, NY, p. 143.

Gould, G. W., and Hitchins, A. D. (1963). Sensitization of bacterial spores to lysozyme and hydrogen peroxide with agents which rupture disulfide bonds. *J. Gen. Microbiol.* *33*:413.

Gray, J. I., and Randall, C. J. (1979). The nitrite/N-nitrosamine problem in meats: An update. *J. Food Prot.* *42*:168.

Grecz, N. (1966). Theoretical and applied aspects of radiation D-values for spores of *Clostridium botulinum*. In *Food Irradiation*, International Atomic Energy Agency, Vienna, p. 307.

Grecz, N., and Arvay, L. H. (1982). Effect of temperature on spore germination and vegetative cell growth of *Clostridium botulinum*. *Appl. Environ. Microbiol.* *43*:331.

Grecz, N., and Grice, J. (1978). The use of alkaline sucrose gradient sedimentation of DNA to study injury and repair of bacterial spores. *Spore Newsletter* *6*:61.

Grecz, N., and Upadhyay, J. (1970). Radiation survival of bacterial spores in neutral and acid ice. *Can. J. Microbiol.* *16*:1045.

Grecz, N., Snyder, O. P., Walker, A. A., and Anellis, A. (1965). Effect of temperature of liquid nitrogen on radiation resistance of spores of *Clostridium botulinum*. *Appl. Microbiol.* *13*:527.

Grecz, N., Tang, T., and Rajan, K. S. (1972). Relation of metal chelate stability to heat resistance of bacterial spores. In *Spores V* (H. O. Halvorson, R. Hanson, and L. L. Campbell, eds.), American Society Microbiol., Washington, DC, p. 53.

Grecz, N., Walker, A. A., Anellis, A., and Berkowitz, D. (1971). Effect of irradiation temperature in range −196 to 95C on the resistance of spores of *Clostridium botulinum* 33A in cooked beef. *Can. J. Microbiol.* *17*:135.

Grecz, N., Wiatr, C., Durban, E., Kang, T., and Farkas, J. (1978). Bacterial spores: Biophysical aspects of recovery from radiation injury. *J. Food Process. Preserv.* *2*:315.

Hall, M. A., and Maurer, A. J. (1986). Spice extracts, lauricidin, and propylene glycol as inhibitors of *Clostridium botulinum* in turkey frankfurter slurries. *Poultry Sci.* *65*:1167.

Han, Y. W., Zhang, H. I., and Krochta, J. M. (1976). Death rates of bacterial spores: Mathematical models. *Can. J. Microbiol.* *22*:295.

Hansen, J. N., Spiegelman, G., and Halvorson, H. O. (1970). Bacterial spore outgrowth: Its regulation. *Science* *168*:1291.

Hanson, R. S., Curry, M. V., Garner, J. V., and Halvorson, H. O. (1972). Mutants of *Bacillus cereus* T that produce thermoresistant spores lacking dipicolinate and have low levels of calcium. *Can. J. Microbiol.* *18*:1139.

Hashimoto, T., Black, S. H., and Gerhardt, P. (1960). Development of fine structure, thermostability, and dipicolinate during sporogenesis in a bacillus. *Can. J. Microbiol.* *6*:203.

Hashimoto, T., Frieben, W. R., and Conti, S. F. (1972). Kinetics of germination of heat-injured *Bacillus cereus* spores. In *Spores V* (H. O. Halvorson, R. Hanson, and L. L. Campbell, eds.), American Society Microbiol., Washington, DC, p. 409.

Hauschild, A. H. W. (1989). *Clostridium botulinum*. In *Foodborne Bacterial Pathogens* (M. P. Doyle, ed.), Marcel Dekker, New York, p. 111.

Hauschild, A. H. W., Aris, B. J., and Hilsheimer, R. (1975). *Clostridium botilinum* in marinated products. *Can. Inst. Food Sci. Technol.* *8*:84.

Hauschild, A. H. W., and Hilsheimer, R. (1977). Enumeration of *Clostridium botulinum* in meats by a pour-plate procedure. *Can. J. Microbiol.* *23*:829.

Hauschild, A. H. W., and Hilsheimer, R. (1979). Effect of salt content and pH on toxigenesis by *Clostridium botulinum* in caviar. *J. Food Prot.* *42*:245.

Heinen, W., and Lauwers, A. M. (1981). Growth of bacteria at 100°C and beyond. *Arch. Microbiol. 129*:127.

Hintlian, C. B., and Hotchkiss, J. H. (1986). The safety of modified atmosphere packaging: A review. *Food Technol. 40*(12):70.

Hobbs, G. (1976). *Clostridium botulinum* and its importance in fishery products. *Adv. Food Res. 22*:135.

Holland, D., Barker, A. N., and Wolf, J. (1969). Factors affecting germination of clostridia. In *Spores IV* (L. L. Campbell, ed.), American Society Microbiol., Bethesda, MD, p. 317.

Holley, R. A. (1981). Review of the potential hazard from botulism in cured meats. *Can. Inst. Food Sci. Technol. J. 14*:183.

Hsieh, L. K., and Vary, J. C. (1975). Peptidoglycan hydrolysis during initiation of spore germination in *Bacillus megaterium*. In *Spores VI* (P. Gerhardt, R. N. Costilow, and H. L. Sadoff, eds.), American Society Microbiol., East Lansing, MI, p. 465.

Huhtanen, C. N., and Feinberg, J. I. (1980). Sorbic acid inhibition of *Clostridium botulinum* in nitrite-free poultry frankfurters. *J. Food Sci. 45*:453.

Huhtanen, C. N., Feinberg, J. I., Trenchard, H., and Phillips, J. G. (1983). Acid enhancement of *Clostridium botulinum* inhibition in ham and bacon prepared with potassium sorbate and sorbic acid. *J. Food Prot. 46*:807.

Huhtanen, C. N., Naghski, J., Custer, C. S., and Russell, R. W. (1976). Growth and toxin production by *Clostridium botulinum* in moldy tomato juice. *Appl. Environ. Microbiol. 32*:711.

Hungate, R. E. (1969). A roll tube method of cultivation of strict anaerobes. In *Methods in Microbiology*, Vol. 3B (J. R. Norris, and D. W. Ribbons, eds.), Academic Press, London, p. 117.

Hunnell, J. W., and Ordal, Z. J. (1961). Cytological and chemical changes in heat killed and germinated bacterial spores. In *Spores II* (H. O. Halvorson, ed.), Burgess Pub. Co., Minneapolis, MN, p. 101.

Hurst, A. (1981). Nisin. *Adv. Appl. Microbiol. 27*:85.

Huss, H. H., Schaeffer, L., Pederson, A., and Jepsen, A. (1979). Toxin production by *Clostridium botulinum* type E in relation to the measured oxidation-reduction potential (Eh), packaging method and associated microflora. In *Advances in Fish Science and Technology* (J. J. Connell, ed.), Fishing News (Books Ltd.), Farnham, Surrey, United Kingdom, p. 476.

Hustad, G. O., Cerveny, J. G., Trenk, H., Deibel, R. H., Kautter, D. A., Fazio, T., Johnston, R. W., and Kolari, O. E. (1973). Effect of sodium nitrite and sodium nitrate on botulinal toxin production and nitrosamine formation in wieners. *Appl. Microbiol. 26*:22.

IAEA (1989). IAEA news features food processing by irradiation: World facts and trends. International Atomic Energy Agency, Vienna.

Iida, H. (1968) Activation of *Clostridium botulinum* toxin by trypsin. In *Proceedings of the 1st U.S.-Japan Conference on Toxic Microorganisms*, Honolulu, HI, pp. 336–340.

Ingram, M. (1974). The microbiological effects of nitrite. In *Proceedings of International Symposium on Nitrite on Meat Products*, Pudoc, Wageningen, The Netherlands, pp. 63–75.

Ingram, M. (1976). The microbial role of nitrite in meat products. In *Microbiology in Agriculture, Fisheries, and Food* (F. A. Skinner, and J. G. Can, eds.), Academic Press, London, p. 1.

Ishiwata, H., Watanabe, Y., and Tanimura, A. (1971). Studies on the formation of abnormal substances in foods by treatments with hydrogen peroxide (1): Effect of oxidation products of amino acids on the growth of *Escherichia coli*. *J. Food Hyg. Soc. Jpn.* *12*:512.

Ismaiel, A. A. (1988). The inhibition of *Clostridium botulinum* growth and toxin production by essential oils of spices [Abstract of thesis, *FSTA 21*(4):4T28]. Ph.D. thesis, Virginia Polytechnic Institute and State University, Blacksburg, VA.

Ismaiel, A. A., and Pierson, M. D. (1990). Inhibition of germination, outgrowth, and vegetative growth of *Clostridium botulinum* 67B by spice oils. *J. Food Prot.* *53*:755.

Ito, K. A., and Chen, J. K. (1978). Effect of pH on growth of *Clostridium botulinum* in foods. *Food Technol.* *32*(6):71.

Ito, K. A., Chen, J. K., Seeger, M. L., Unverferth, J. A., and Kimball, R. N. (1978). Effect of pH on the growth of *Clostridium botulinum* in canned figs. *J. Food Sci.* *43*:1634.

Ito, K. A., Denny, C. B., Brown, C. K., Yao, M., and Seeger, M. L. (1973). Resistance of bacterial spores to hydrogen peroxide. *Food Technol.* *27*(11):58.

Ito, K. A., and Seeger, M. L. (1980). Effects of germicides on microorganisms in can cooling waters. *J. Food Prot.* *43*:484.

Ito, K. A., Seeger, M. L., Bhorer, C. W., Denny, C. B., and Bruch, M. K. (1968). The thermal and germicidal resistance of *Clostridium botulinum* types A, B, and E spores. In *Proceedings of the 1st U.S.-Japan Conference on Toxic Microorganisms*, Honolulu, HI, p. 410.

Ito, K. A., Seslar, D. J., Mercer, W. A., and Meyer, K. F. (1967). The thermal and chlorine resistance of *Clostridium botulinum* types A, B, and E spores. In *Botulism 1966* (M. Ingram, and T. A. Roberts, eds.), Chapman and Hall, London, p. 108.

Ivey, F. J., and Robach, M. C. (1978). Effect of ascorbic acid and sodium nitrite on *Clostridium botulinum* outgrowth and toxin production in canned comminuted pork. *J. Food Sci.* *43*:1782.

Iwasaki, A., Tallentire, A., Kimber, B. F., and Powers, E. L. (1974). The influence of added H_2O and D_2O on anoxic radiation sensitivity in bacterial spores. *Radiation Res.* *57*:306.

Jarvis, B., Rhodes, A. C., King, S. E., and Patel, M. (1976). Sensitization of heat-damaged spores of *Clostridium botulinum* type B to sodium chloride and sodium nitrite. *J. Food Technol.* *11*:41.

Jay, J. M. (1983). Antibiotics as food preservatives. In *Food Microbiology* (A. H. Rose, ed.), Academic Press, London, p. 117.

Johannsen, A. (1965). *Clostridium botulinum* type E in foods and the environment generally. *J. Appl. Bacteriol.* *28*:90.

Johnston, M. A., and Loynes, R. (1971). Inhibition of *Clostridium botulinum* by sodium nitrite as affected by bacteriological media and meat suspensions. *Can. Inst. Food Sci. Technol. J.* *4*:179.

Johnston, M. A., Pivnick, H., and Samson, J. M. (1969). Inhibition of *Clostridium botulinum* by sodium nitrite in a bacteriological medium and in meat. *Can. Inst. Food Sci. Technol. J.* 2:52.

Joslyn, L. (1983). Sterilization by heat. In *Disinfection, Sterilization and Preservation* (S. S. Block, ed.), Lea and Febiger, Philadelphia. pp. 3–46.

Kafel, S., and Ayres, J. C. (1969). The antagonism of enterococci on other bacteria in canned hams. *J. Appl. Bacteriol. 32*:217.

Kempe, L. L. (1955). Combined effect of heat and radiation in food sterilization. *Appl. Microbiol. 3*:346.

Kempe, L. L., and Graikoski, J. T. (1962). Gamma ray sterilization and residual toxicity studies on ground beef inoculated with spores of *Clostridium botulinum. Appl. Microbiol. 10*:31.

Keynan, A., and Evenchik, Z. (1969). Activation. In *The Bacterial Spore* (G. W. Gould, and A. Hurst, eds.), Academic Press, London, p. 359.

Khoury, P. H., Lombardi, S. J., and Slepecky, R. A. (1987). Perturbation of the heat resistance of bacterial spores by sporulation temperature and ethanol. *Curr. Microbiol. 15*:15.

Kihm, D. J., Hutton, M. T., Hanlin, J. H., and Johnson, E. A. (1990). Influence of transition metals added during sporulation on heat resistance of *Clostridium botulinum* 113B spores. *Appl. Environ. Microbiol. 56*:681.

King, W. L., and Gould, G. W. (1969). Lysis of bacterial spores with hydrogen peroxide. *J. Appl. Bacteriol. 32*:481.

King, W. L., and Gould, G. W. (1971). Mechanisms of stimulation of germination of *Clostridium sporogenes* spores by bicarbonate. In *Spore Research 1971* (A. N. Barker, G. W. Gould, and J. Wolf, eds.), Academic Press, New York, p. 71.

Kiss, I., Rhee, C. O., Grecz, N., Roberts, T. A., and Farkas, J. (1978). Relation between radiation resistance and salt sensitivity of spores of five strains of *Clostridium botulinum* types A, B, and E. *Appl. Environ. Microbiol. 35*:533.

Krabbenhoft, K. L., Corlett, D. A., Jr., Anderson, A. W., and Elliker, P. R. (1964). Chemical sensitization of *Clostridium botulinum* spores to radiation in meat. *Appl. Microbiol. 12*:424.

Kreiger, R. A., Snyder, O. P., and Pflug, I. J. (1983). *Clostridium botulinum* ionizing radiation D-value determination using a micro food sample system. *J. Food Sci. 48*:141.

Kulikovskii, A. V. (1976). Structural and biochemical changes in spores of *Bacillus cereus* produced by caustic soda and hypochlorite. *Microbiology 45*:113.

Kulikovsky, A., Pankratz, H. S., and Sadoff, H. L. (1975). Ultrastructural and chemical changes in spores of *Bacillus cereus* after action of disinfectants. *J. Appl. Bacteriol. 38*:39.

Kuzminski, L. N., Howard, G. L., and Stumbo, C. R. (1969). Thermochemical factors influencing the death kinetics of spores of *Clostridium botulinum* 62A. *J. Food Sci. 34*:561.

Leaper, S. (1988). Resistance of spores to hydrogen peroxide and a preliminary investigation on microbial levels on packaging materials for aseptic systems. Technical Memorandum No. 504, Campden Food & Drink Res. Association, Chipping, Campden, United Kingdom.

LeBlanc, F. R., Devlin, K. A., and Stumbo, C. R. (1953). Antibiotics in food preservation. I. The influence of subtilin on the thermal resistance of spores of *Clostridium botulinum* and putrefactive anaerobe 3679. *Food Technol. 7*:181.

Lechowich, R. V., and Ordal, Z. J. (1962). The influence of the sporulation temperature on the heat resistance and chemical composition of bacterial spores. *Can. J. Microbiol. 8*:287.

Lerke, P. (1973). Evaluation of potential risk of botulism from seafood cocktails. *Appl. Microbiol. 25*:807.

Levinson, H. S., Hyatt, M. T., and Moore, F. E. (1961). Dependence of the heat resistance of bacterial spores on the calcium:dipicolinic acid ratio. *Biochem. Biophys. Res. Commun. 5*:417.

Liewen, M. B., and Marth, E. H. (1985). Growth and inhibition of microorganisms in the presence of sorbic acid: A review. *J. Food Prot. 48*:364.

Lijinsky, W. (1976). Reaction of drugs with nitrous acid as a source of carcinogenic nitrosamines. *Cancer Res. 34*:255.

Lijinsky, W., and Epstein, S. S. (1970). Nitrosamines as environmental carcinogens. *Nature 225*:21.

Lindsay, J. A., Murrell, W. G., and Warth, A. D. (1985). Spore resistance and the basic mechanism of heat resistance. In *Sterilization of Medical Products* (L. E. Harris, and A. J. Skopek, eds.), Johnson & Johnson, Botany, New South Wales, Australia, p. 162.

Lindsay, R. C. (1985). Food additives. In *Food Chemistry*, 2nd ed. (O. R. Fennema, ed.), Marcel Dekker, New York, p. 629.

Lücke, F–K. (1985). Heat inactivation and injury of *Clostridium botulinum* spores in sausage mixtures. In *Fundamental and Applied Aspects of Bacterial Spores* (G. J. Dring, D. J. Ellar, and G. W. Gould, eds.), Academic Press, London, p. 409.

Luh, B. S., and Kean, C. E. (1975). Canning of vegetables. In *Commercial Vegetable Processing* (B. S. Luh, and J. G. Woodroof, eds.), AVI, Westport, CT, p. 195.

Lund, B. M., George, S. M., and Franklin, J. G. (1987). Inhibition of type A and type B (Proteolytic) *Clostridium botulinum* by sorbic acid. *Appl. Environ. Microbiol. 53*:935.

Lund, B. M., and Wyatt, G. M. (1984). The effect of redox potential, and its interaction with sodium chloride concentration, on the probability of growth of *Clostridium botulinum* type E from spore inocula. *Food Microbiol. 1*:49.

Lynt, R. K., Kautter, D. A., and Solomon, H. M. (1979). Heat resistance of nonproteolytic *Clostridium botulinum* type F in phosphate buffer and crabmeat. *J. Food Sci. 44*:108.

Lynt, R. K., Kautter, D. A., and Solomon, H. M. (1981). Heat resistance of proteolytic *Clostridium botulinum* type F in phosphate buffer and crabmeat. *J. Food Sci. 47*:204.

Lynt, R. K., Kautter, D. A., and Solomon, H. M. (1982). Differences and similarities among proteolytic and nonproteolytic strains of *Clostridium botulinum* types A, B, E and F: A review. *J. Food Prot. 45*:466.

Lynt, R. K., Solomon, H. M., and Kautter, D. A. (1984). Heat resistance of *Clostridium botulinum* type G in phosphate buffer. *J. Food Prot. 47*:463.

Lynt, R. K., Solomon, H. M., Lilly, T., Jr., and Kautter, D. A. (1977). Thermal death time of *Clostridium botulinum* in meat of blue crab. *J. Food Sci. 42*:1022.

Maas, M. R., Glass, K. A., and Doyle, M. P. (1989). Sodium lactate delays toxin production by *Clostridium botulinum* in cook-in-bag turkey products. *Appl. Environ. Microbiol. 55*:2226.

Madril, M. T., and Sofos, J. N. (1986). Interaction of reduced NaCl, sodium acid pyrophosphate and pH on the antimicrobial activity of comminuted meat products. *J. Food Sci. 51*:1141.

Marshall, B. J., Ohye, D. F., and Christian, J. H. B. (1971). Tolerance of bacteria to high concentrations of NaCl and glycerol in the growth medium. *Appl. Microbiol. 21*:363.

Mason, J. M., and Setlow, P. (1986). Evidence for an essential role for small, acid-soluble, spore proteins in the resistance of *Bacillus subtilis* spores to ultraviolet light. *J. Bacteriol. 167*:174.

Mason, J. M., and Setlow, P. (1987). Different small, acid-soluble proteins of the a/b type have interchangeable roles in the heat and ultraviolet radiation resistance of *Bacillus subtilis* spores. *J. Bacteriol. 169*:3633.

Mills, A. L., and Alexander, M. (1976). N-Nitrosamine formation by cultures of several microorganisms. *Appl. Environ. Microbiol. 31*:892.

Molin, N., and Snygg, G. B. (1967). Effect of lipid materials on heat resistance of bacterial spores. *Appl. Microbiol. 15*:1422.

Morgan, B. H., and Reed, J. M. (1954). Resistance of bacterial spores to gamma irradiation. *Food Res. 19*:357.

Morris, J. G. (1976). Oxygen and obligate anaerobe. *J. Appl. Bacteriol. 40*:229.

Mossel, D. A. A. (1983). Essentials and perspectives of the microbial ecology of foods. In *Food Microbiology: Advances and Prospects* (T. A. Roberts and F. A. Skinner, eds.), Academic Press, London, p. 1.

Murrell, W. G. (1969). Chemical composition of spores and spore structures. In *The Bacterial Spore* (G. W. Gould, and A. Hurst, eds.), Academic Press, London, p. 215.

Murrell, W. G. (1981). Biophysical studies on the molecular mechanisms of spore heat resistance and dormancy. In *Sporulation and Germination* (H. S. Levinson, A. L. Sonenshein, and D. J. Tipper, eds.), American Society Microbiol., Washington, DC, p. 64.

Murrell, W. G. (1988). Bacterial spores—nature's ultimate survival package. In *Microbiology in Action* (W. G. Murrell, and I. R. Kennedy, eds.), John Wiley, New York, p. 311.

Murrell, W. G., and Scott, W. J. (1966). The heat resistance of bacterial spores at various water activities. *J. Gen. Microbiol. 43*:411.

Murrell, W. G., and Warth, A. D. (1965). Composition and heat resistance of bacterial spores. In *Spores III* (L. L. Campbell, and H. O. Halvorson, eds.), American Society Microbiol., Ann Arbor, MI, p. 1.

Nelson, K. A., Busta, F. F., Sofos, J. N., and Wagner, M. K. (1983). Effect of polyphosphates in combination with nitrite-sorbate or sorbate on *Clostridium botulinum* growth and toxin production in chicken frankfurter emulsions. *J. Food Prot. 46*:846.

Newberne, P. M. (1979). Nitrite promotes lymphoma incidence in rats. *Science 204*:1079.

Notermans, S., Dufrenne, J., and Lund, B. M. (1990). Botulism risk of refrigerated, processed foods of extended durability. *J. Food Prot. 53*:1020.

O'Brien, R. T., Titus, D. S., Devlin, K. A., Stumbo, C. R., and Lewis, J. C. (1956). Antibiotics in food preservation. II. Studies on the influence of subtilin and nisin on the thermal resistance of food spoilage bacteria. *Food Technol. 10*:352.

Odlaug, T. E., and Pflug, I. J. (1977a). Thermal destruction of *Clostridium botulinum* spores suspended in tomato juice in aluminum thermal death time tubes. *Appl. Environ. Microbiol. 34*:23.

Odlaug, T. E., and Pflug, I. J. (1977b). Effect of storage time and temperature on the survival of *Clostridium botulinum* in acid media. *Appl. Environ. Microbiol. 34*:30.

Odlaug, T. E., and Pflug, I. J. (1977c). Recovery of spores of *Clostridium botulinum* in yeast extract agar and pork infusion agar after heat treatment. *Appl. Environ. Microbiol. 34*:377.

Odlaug, T. E., and Pflug, I. J. (1978). *Clostridium botulinum* and acid foods. *J. Food Prot. 41*:566.

Odlaug, T. E., and Pflug, I. J. (1979). *Clostridium botulinum* growth and toxin production in tomato juice containing *Aspergillus gracilis. Appl. Environ. Microbiol. 37*:496.

Ohye, D. F., and Christian, J. H. B. (1967). Combined effects of temperature, pH, and water activity on growth and toxin production by *Clostridium botulinum* types A, B, and E. In *Botulism 1966* (M. Ingram and T. A. Roberts, eds.), Chapman and Hall, London, p. 217.

Ohye, D. F., Christian, J. H. B., and Scott, W. J. (1967). Influence of temperature on the water relations of growth of *Cl. botulinum* type E. In *Botulism 1966* (M. Ingram, and T. A. Roberts, eds.), Chapman & Hall, London, p. 136.

Ohye, D. F., and Scott, W. J. (1953). The temperature relations of *Clostridium botulinum*, types A and B. *Aust. J. Biol. Sci. 6*:178.

Ohye, D. F., and Scott, W. J. (1957). Studies in the physiology of *Clostridium botulinum* type E. *Aust. J. Biol. Sci. 10*:85.

Olsen, A. M. and Scott, W. J. (1946). Influence of starch in media used for the detection of heated bacterial spores. *Nature 157*:337.

Olsen, A. M., and Scott, W. J. (1950). The enumeration of heated bacterial spores. I. Experiments with *Clostridium botulinum* and other species of *Clostridium. Austral. J. Sci. Res. Series B 3*:219.

Pace, P. J., and Krumbiegel, E. R. (1973). *Clostridium botulinum* and smoked fish production: 1963–1972. *J. Milk Food Technol. 36*:42.

Patel, H. R., Patel, P. C., and York, G. K. (1978). Growth and toxicogenesis of *Clostridium botulinum* type E on marine mollusks at low temperatures. *J. Food Sci. Technol. India 15*:231.

Payne, M. J., Glidewell, C., and Cammack, R. (1990b). Interactions of iron-thiol-nitrosyl compounds with the phosphoroclastic system of *Clostridium sporogenes* by nitrite. *J. Gen. Microbiol. 136*:2077.

Payne, M. J., Woods, L. F. J., Gibbs, P., and Cammack, R. (1990a). Electron paramagnetic resonance spectroscopic investigation of the inhibition of the phosphoroclastic system of *Clostridium sporogenes* by nitrite. *J. Gen. Microbiol. 136*:2067.

Peel, J. L., and Waites, W. M. (1981). Improvements in methods of sterilization. U.S. Patent 4,289,728.

Pelroy, G. A., Scherer, A., Peterson, M. E., Paranjpye, R., and Eklund, M. W. (1985). Inhibition of *Clostridium botulinum* type E toxin formation by potassium chloride and sodium chloride in hot-process (smoked) whitefish (*Coregonus clupeaformis*). *J. Food Protect. 48*:971.

Perkins, W. E., Ashton, D. H., and Evancho, G. M. (1975). Influence of the z value of *Clostridium botulinum* on the accuracy of process calculations. *J. Food Sci. 40*:1189.

Phillips, C. R. (1977). Gaseous sterilization. In *Disinfection, Sterilization, and Preservation*, 2nd ed. (S. S. Block, ed.), Lea & Febiger, Philadelphia, p. 592.

Pierson, M. D., and Reddy, N. R. (1982). Inhibition of *Clostridium botulinum* by antioxidants and related phenolic compounds in comminuted pork. *J. Food Sci. 47*:1926.

Pierson, M. D., Rice, K. M., and Jadlocki, J. F. (1981). Sodium hypophosphite inhibition of *Clostridium botulinum*. In *Proceedings of the 27th European Meat Research Workers Congress*, Vienna, Vol. 2, p. 651.

Post, L. S., Amoroso, T. L., and Solberg, M. (1985a). Inhibition of *Clostridium botulinum* type E in model acidified food systems. *J. Food Sci. 50*:966.

Post, F. J., Krishnamurity, G. B., and Flanagan, M. D. (1963). Influence of sodium hexamataphosphate on selected bacteria. *Appl. Microbiol. 11*:430.

Post, L. S., Lee, D. A., Solberg, M., Furgang, D., Specchio, J. and Graham, C. (1985b). Development of botulinal toxin and sensory deterioration during storage of vacuum and modified atmosphere packaged fish fillets. *J. Food Sci. 50*:990.

Raatjes, G. J. M., and Smelt, J. P. P. M. (1979). *Clostridium botulinum* can grow and form toxin at pH values lower than 4.6. *Nature 281*:398.

Rayman, M. K., Aris, B., and Hurst, A. (1981). Nisin: A possible alternative or adjunct to nitrite in the preservation of meats. *Appl. Environ. Microbiol. 41*:375.

Rayman, K., Malik, N., and Hurst, A. (1983). Failure of nisin to inhibit outgrowth of *Clostridium botulinum* in a model cured meat system. *Appl. Environ. Microbiol. 46*:1450.

Reddy, N. R., Pierson, M. D., and Lechowich, R. V. (1982). Inhibition of *Clostridium botulinum* by antioxidants, phenols, and related compounds. *Appl. Environ. Microbiol. 43*:835.

Robach, M. C., and Pierson, M. D. (1978). Influence of para-hydroxybenzoic acid esters on the growth and toxin production of *Clostridium botulinum* 10755A. *J. Food Sci. 43*:787.

Roberts, T. A., Gibson, A. M., and Robinson, A. (1981a). Factors controlling the growth of *Clostridium botulinum* types A and B in pasteurized, cured meats. I. Growth in pork slurries prepared from "low" pH meat (pH 5.5–6.3). *J. Food Technol. 15*:239.

Roberts, T. A., Gibson, A. M., and Robinson, A. (1981b). Factors controlling the growth of *Clostridium botulinum* types A and B in pasteurized, cured meats. II. Growth in pork slurries prepared from "high" pH meat (range 6.3–6.8). *J. Food Technol. 16*:337.

Roberts, T. A., Gibson, A. M., and Robinson, A. (1982). Factors controlling the growth of *Clostridium botulinum* types A and B in pasteurized, cured meats. III. The effect of potassium sorbate. *J. Food Technol. 17*:307.

Roberts, T. A., and Hitchins, A. D. (1969). Resistance of spores. In *The Bacterial Spore* (G. W. Gould, and A. Hurst, eds.), Academic Press, London, p. 611.

Roberts, T. A., and Ingram, M. (1965a). The resistance of spores of *Clostridium botulinum* type E to heat and radiation. *J. Appl. Bacteriol. 28*:125.

Roberts, T. A., and Ingram, M. (1965b). Radiation resistance of spores of *Clostridium* species in aqueous suspension. *J. Food Sci. 30*:879.

Roberts, T. A., and Ingram, M. (1973). Inhibition of growth of *Cl. botulinum* at different pH values by sodium chloride and sodium nitrite. *J. Food Technol* *8*:467.

Roberts, T. A., Jarvis, B. J., and Rhodes, A. C. (1976). Inhibition of *Clostridium botulinum* by curing salts in pasteurized pork slurry. *J. Food Technol.* *11*:25.

Rode, L. J., and Foster, J. W. (1960). Induced release of dipicolinic acid from spores of *Bacillus megaterium. J. Bacteriol.* *79*:650.

Rowley, D. B., Anellis, A., Wierbicki, E., and Baker, A. W. (1974). Status of the radappertization of meats. *J. Milk Food Technol.* *37*:86.

Rowley, D. B., El-Bisi, H. M., Anellis, A., and Snyder, O. P. (1968). Resistance of *Clostridium botulinum* spores to ionizing radiation as related to radappertization of food. In *Proceedings of the 1st U.S.-Japan Conference on Toxic Microorganisms*, Honolulu, HI, pp. 459–467.

Rowley, D. B., and Feeherry, F. (1970). Conditions affecting germination of *Clostridium botulinum* 62A spores in a chemically defined medium. *J. Bacteriol.* *104*:1151.

Russell, A. D. (1982). *The Destruction of Bacterial Spores.* Academic Press, London.

Savage, R. A., and Stumbo, C. R. (1971). Characteristics of progeny of ethylene oxide treated *Clostridium botulinum* type 62A spores. *J. Food Sci.* *36*:182.

Schmidt, C. F. (1964). Spores of *C. botulinum*: Formation, resistance, germination. In *Botulism* (K. H. Lewis and K. Cassel, Jr., eds.), U.S. Public Health Service, Cincinnati, OH, p. 69.

Schmidt, C. F., and Nank, W. K. (1960). Radiation sterilization of food. I. Procedures for the valuation of the radiation resistance of spores of *Clostridium botulinum* in food products. *Food Res.* *25*:321.

Schmidt, C. F., Lechowich, R. V., and Folinazzo, J. F. (1961). Growth and toxin production by type E *Clostridium botulinum* below 40°F. *J. Food Sci.* *26*:626.

Schmidt, C. F., Nank, W. K., and Lechowich, R. V. (1960). Radiation sterilization of food. 2. Aspects of growth, sporulation and radiation resistance of spores of *Clostridium botulinum* type E. Paper presented at the 20th Annual Meeting of the Institute of Food Technologists.

Schumb, W. C., Satterfield, C. N., and Wentworth, R. L. (1955). Stabilization. In *Hydrogen Peroxide*, Reinhold Pub. Corp., New York, p. 515.

Scott, V. N. (1989). Interaction of factors to control microbial spoilage of refrigerated foods. *J. Food Prot.* *52*:431.

Scott, V. N., and Bernard, D. T. (1982). Heat resistance of spores of non-proteolytic type B *Clostridium botulinum. J. Food Prot.* *45*:909.

Scott, V. N., and Bernard, D. T. (1985). The effect of lysozyme on the apparent heat resistance of nonproteolytic type B *Clostridium botulinum. J. Food Safety* *7*:145.

Scott, V. N., and Taylor, S. L. (1981a). Effect of nisin on the outgrowth of *Clostridium botulinum* spores. *J. Food Sci.* *46*:117.

Scott, V. N., and Taylor, S. L. (1981b). Temperature, pH, and spore load effects on the ability of nisin to prevent the outgrowth of *Clostridium botulinum* spores. *J. Food Sci.* *46*:121.

Sebald, M., Ionesco, H., and Prevot, A. R. (1972). Germination 1zP-dependante des spores de *Clostridium botulinum* type E. *C. Rend Acad. Sci. Paris* *275*:2175.

Sebranek, J. G., and Cassens, R. G. (1973). Nitrosamines: A review. *J. Milk Food Technol.* *36*:76.

Segner, W. P., and Schmidt, C. F. (1971). Resistance of spores of marine and terrestrial strains of *Clostridium botulinum* type C. *Appl. Microbiol. 22*:1030.

Segner, W. P., Schmidt, C. F., and Boltz, J. K. (1966). Effect of sodium chloride and pH on the outgrowth of spores of type E *Clostridium botulinum* at optimal and sub-optimal temperatures. *Appl. Microbiol. 14*:49.

Setlow, P. (1978). Degradation of dormant spore protein during germination of *Bacillus megaterium* spores. In *Limited Proteolysis in Microorganisms* (G. N. Cohen, and H. Holzer, eds.), U.S. Department of Health Education and Welfare, Washington, DC, p. 109.

Setlow, P. (1985). Protein degradation during bacterial spore germination. In *Fundamental and Applied Aspects of Bacterial Spores* (D. J. Ellar, ed.), Academic Press, London, p. 285.

Setlow, P. (1988). Small, acid-soluble spore proteins of *Bacillus* species: structure, synthesis, genetics, function, and degradation. *Ann. Rev. Microbiol. 42*:319.

Shank, J. L., Silliker, J. H., and Harper, R. H. (1962). The effect of nitric oxide on bacteria. *Appl. Microbiol. 10*:185.

Simunovic, J., Oblinger, J. L., and Adams, J. P. (1985). Potential for growth of nonproteolytic types of *Clostridium botulinum* in pasteurized restructured meat products: A review. *J. Food Prot. 48*:265.

Sink, J. D., and Hsu, L. A. (1977). Chemical effects of smoke processing on frankfurter manufacture and storage characteristics. *J. Food Sci. 42*:1489.

Sinskey, A. J. (1979). Preservatives added to foods. In *Nutritional and Safety Aspects of Food Processing* (S. R. Tannenbaum, ed.), Marcel Dekker, New York, p. 369.

Smelt, J. P. P. M., and Haas, H. (1978). Behavior of proteolytic *Clostridium botulinum* types A and B near the lower temperature limits of growth. *Europ. J. Appl. Microbiol. Biotechnol. 5*:143.

Smelt, J. P. P. M., Raatjes, G. J. M., Growther, J. S., and Verrips, C. T. (1982). Growth and toxin formation by *Clostridium botulinum* at low pH values. *J. Appl. Bacteriol. 52*:75.

Smith, L. D. S. (1975). Inhibition of *Clostridium botulinum* by strains of *Clostridium perfringens* isolated from soil. *Appl. Microbiol. 30*:319.

Smith, L. D. S. (1977). The spores. In *Botulism, The Organism, Its Toxins, The Disease*, Charles C. Thomas, Springfield, IL, p. 34.

Smith, M. V. (1975). The effect of oxidation-reduction potential on the outgrowth and chemical inhibition of *Clostridium botulinum* type E spores. Ph.D. thesis, Virginia Polytechnic Institute, Blacksburg, VA.

Smoot, L. A., and Pierson, M. D. (1979). Effect of oxidation-reduction potential on the outgrowth and chemical inhibition of *Clostridium botulinum* 10755A spores. *J. Food Sci. 44*:700.

Smoot, L. A., and Pierson, M. D. (1981). Mechanisms of sorbate inhibition of *Bacillus cereus* T and *Clostridium botulinum* 62A spore germination. *Appl. Environ. Microbiol. 42*:477.

Smoot, L. A., and Pierson, M. D. (1982). Inhibition and control of bacterial spore germination. *J. Food Prot. 45*:84.

Sofos, J. N., and Busta, F. F. (1980). Alternatives to the use of nitrite as an antibotulinal agent. *Food Technol. 34*(5):244.

Sofos, J. N., and Busta, F. F. (1981). Antimicrobial activity of sorbate. *J. Food Prot. 44*:614.

Sofos, J. N., and Busta, F. F. (1983). Sorbates. In *Antimicrobials in Foods* (A. L. Branen, and P. M. Davidson, eds.), Marcel Dekker, New York, p. 141.

Sofos, J. N., Busta, F. F., and Allen, C. E. (1979a). Botulism control by nitrite and sorbate in cured meats: A review. *J. Food Prot. 42*:739.

Sofos, J. N., Busta, F. F., and Allen, C. E. (1979b). Sodium nitrite and sorbic acid effects on *Clostridium botulinum* spore germination and total microbial growth in chicken frankfurter emulsions during temperature abuse. *Appl. Environ. Microbiol. 37*:1103.

Sofos, J. N., Busta, F. F., and Allen, C. E. (1980). Influence of pH on *Clostridium botulinum* control by sodium nitrite and sorbic acid in chicken emulsions. *J. Food Sci. 45*:7.

Sofos, J. N., Pierson, M. D., Blocher, J. C., and Busta, F. F. (1986). Mode of action of sorbic acid on bacterial cells and spores. *Int. J. Food Microbiol. 3*:1.

Solomon, H. M., Kautter, D. A., Lilly, T., and Rhodehamel, J. (1990). Outgrowth of *Clostridium botulinum* in shredded cabbage at room temperature under a modified atmosphere. *J. Food Prot. 53*:831.

Solomon, H. M., Kautter, D. A., and Lynt, R. K. (1982). Effect of low temperature on growth of nonproteolytic *Clostridium botulinum* types B and F and proteolytic type G in crabmeat and broth. *J. Food Prot. 45*:516.

Solomon, H. M., Lynt, R. K., Lilly, T., Jr., and Kautter, D. A. (1977). Effect of low temperature on growth of *Clostridium botulinum* spores B in meat of the blue crab. *J. Food Prot. 40*:5.

Somers, E. B., and Taylor, S. L. (1987). Antibotulinal effectiveness of nisin in pasteurized process cheese spreads. *J. Food Prot. 50*:842.

Sperber, W. H. (1982). Requirements of *Clostridium botulinum* for growth and toxin production. *Food Technol. 36*(12):89.

Sperber, W. H. (1983). Influence of water activity of foodborne bacteria—A review. *J. Food Prot. 46*:142.

Stegeman, H., Mossel, D. A. A., and Pilnick, W. (1977). Studies on the sensitizing mechanism of pre-irradiation to a subsequent heat treatment on bacterial spores. In *Spore Research 1976* (A. Barker, J. Wolf, D. J. Ellar, G. J. Dring, and G. W. Gould, eds.), Academic Press, London, p. 565.

Stevenson, K. E., and Shafer, B. D. (1983). Bacterial spore resistance to hydrogen peroxide. *Food Technol. 37*(11):111.

Stumbo, C. R. (1973). *Thermobacteriology in Food Processing*, 2nd ed., Academic Press, New York.

Sugiyama, H. (1951). Studies on factors affecting the heat resistance of spores of *Clostridium botulinum. J. Bacteriol. 62*:81.

Sugiyama, H., and Yang, K. H. (1975). Growth potential of *Clostridium botulinum* in fresh mushrooms packaged in semipermeable plastic film. *Appl. Microbiol. 30*:964.

Swartling, P., and Lindgren, B. (1968). The sterilizing effect against *Bacillus subtilis* spores of hydrogen peroxide at different temperatures and concentrations. *J. Dairy Res. 35*:423.

Tallentire, A., and Powers, E. L. (1963). Modification of sensitivity to X-irradiation by water in *Bacillus megaterium. Radiat. Res. 20*:270.

Tallentire, A., Hayes, J. R., Kimber, B. F., and Powers, E. L. (1974). H_2O and D_2O sorption on spores of *Bacillus megaterium*. *Radiat. Res. 57*:300.

Tanaka, N. (1982). Toxin production by *Clostridium botulinum* in media at pH lower than 4.6. *J. Food Prot. 45*:234.

Tanaka, N., Meske, L., Doyle, M. P., Traisman, E., Thayer, D. W., and Johnston, R. W. (1985). Plant trials of bacon made with lactic acid bacteria, sucrose and lowered sodium nitrite. *J. Food Prot. 48*:679.

Tanaka, N., Traisman, E., Lee, M. H., Cassens, R. G., and Foster, E. M. (1980). Inhibition of botulinum toxin formation in bacon by acid development. *J. Food Prot. 43*:450.

Tanner, F. W., and Oglesby, E. W. (1941). Influence of temperature on growth and toxin production by *Clostridium botulinum*. *Food Res. 6*:481.

Tarr, H. L. A. (1941). Bacteriostatic action of nitrites. *Nature 147*:417.

Taylor, S. L., and Somers, E. B. (1985). Evaluation of the antibotulinal effectiveness of nisin in bacon. *J. Food Prot. 48*:949.

Taylor, S. L., Somers, E. B., and Krueger, L. A. (1985). Antibotulinal effectiveness of nisin-nitrite combinations in culture medium and chicken frankfurter emulsions. *J. Food Prot. 48*:234.

Terano, H., Tanooka, H., and Kadota, H. (1969). Germination-induced repair of single-strand breaks of DNA in irradiated *Bacillus subtilis* spores. *Biochem. Biophys. Res. Comm. 37*:66.

Terano, H., Tanooka, H., and Kadota, H. (1971). Repair of radiation damage to deoxyribonucleic acid in germinating spores of *Bacillus subtilis*. *J. Bacteriol. 106*:925.

Thornley, H. J. (1963). Radiation resistance among bacteria. *J. Appl. Bacteriol. 26*:334.

Toledo, R. T. (1975). Chemical sterilants for aseptic packaging. *Food Technol. 29*(5):102.

Toledo, R. T., Escher, F. E., and Ayres, J. C. (1973). Sporicidal properties of hydrogen peroxide against food spoilage organisms. *Appl. Microbiol. 26*:592.

Tompkin, R. B. (1984). Indirect antimicrobial effects in foods: Phosphates. *J. Food Safety 6*:13.

Tompkin, R. B., Christiansen, L. N., and Shaparis, A. B. (1978a). Antibotulinal role of isoascorbate in cured meat. *J. Food Sci. 43*:1368.

Tompkin, R. B., Christiansen, L. N., and Shaparis, A. B. (1978b). Enhancing nitrite inhibition of *Clostridium botulinum* with isoascorbate in perishable canned cured meat. *Appl. Environ. Microbiol. 35*:59.

Tompkin, R. B., Christiansen, L. N., and Shaparis, A. B. (1978c). Effect of prior refrigeration on botulinal outgrowth of perishable canned cured meat. *Appl. Environ. Microbiol. 35*:863.

Tompkin, R. B., Christiansen, L. N., and Shaparis, A. B. (1979a). Iron and the antibotulinal efficacy of nitrite. *Appl. Environ. Microbiol. 37*:351.

Tompkin, R. B., Christiansen, L. N., and Shaparis, A. B. (1979b). Isoascorbate level and botulinal inhibition in perishable canned cured meat. *J. Food Sci. 44*:1147.

Tompkin, R. B., Christiansen, L. N., and Shaparis, A. B. (1980). Antibotulinal efficacy of sulfur dioxide in meat. *Appl. Environ. Microbiol. 39*:1096.

Townsend, C. T., Yee, L., and Mercer, W. A. (1954). Inhibition of the growth of *Clostridium botulinum* by acidification. *Food Res. 19*:536.

Treadwell, P. E., Jann, G. J., and Salle, A. (1958). Studies on factors affecting the rapid germination of spores of *Clostridium botulinum. J. Bacteriol. 76*:549.

Tsang, N., Post, L. S., and Solberg, M. (1985). Growth and toxin production by *Clostridium botulinum* in model acidified systems. *J. Food Sci. 50*:961.

Turner, F. J. (1983). Hydrogen peroxide and other oxidant disinfectants. In *Disinfection, Sterilization and Preservation*, 3rd ed. (S. S. Block, ed.), Lea & Febiger, Philadelphia, p. 240.

Vary, J. C. (1973). Germination of *Bacillus megaterium* spores after various extraction procedures. *J. Bacteriol. 116*:797.

von Bockelmann, B. (1973). Aseptic packaging of milk. *Alimenta 12*:119.

Wagner, M. K. (1986). Phosphates as antibotulinal agents in cured meats: A review. *J. Food Prot. 49*:482.

Wagner, M. K., and Busta, F. F. (1983). Effect of sodium acid pyrophosphate in combination with sodium nitrite or sodium nitrite/potassium sorbate in *Clostridium botulinum* growth and toxin production in beef/pork frankfurter emulsions. *J. Food Sci. 48*:990.

Wagner, M. K., and Busta, F. F. (1984). Inhibition of *Clostridium botulinum* growth from spore inocula in media containing sodium acid pyrophosphate and potassium sorbate with or without added sodium chloride. *J. Food Sci. 49*:1588.

Wagner, M. K., and Busta, F. F. (1985). Inhibition of *Clostridium botulinum* 52A toxicity and protease activity by sodium acid pyrophosphate in media systems. *Appl. Environ. Microbiol. 50*:16.

Wagner, M. K., and Busta, F. F. (1986). Association of [^{32}P] with *Clostridium botulinum* 52A vegetative cells following growth in a medium containing sodium dihydrogen [^{32}P]-pyrophosphate. *J. Food Prot. 49*:352.

Waites, W. M., Bayliss, C. E., and King, N. R. (1980). The effect of sporulation medium on spores of *Clostridium bifermentans. J. Gen. Microbiol. 116*:271.

Waites, W. M., King, N. R., and Bayliss, C. E. (1972). Effect of alkali treatment on the germination and morphology of *Clostridium bifermentans*. In *Spores V* (H. O. Halvorson, R. Hanson, and L. L. Campbell, eds.), American Society Microbiol., Washington, DC, p. 430.

Warth, A. D. (1978). Relationship between the heat resistance of spores and the optimum and maximum growth temperatures of *Bacillus* species. *J. Bacteriol. 134*:699.

Warth, A. D. (1985). Mechanisms of heat resistance. In *Fundamental and Applied Aspects of Bacterial Spores* (G. J. Dring, D. J. Ellar, and G. W. Gould, eds.), Academic Press, London, p. 209.

Wentz, M. W., Scott, R. A., and Vennes, J. W. (1967). *Clostridium botulinum* type F: Seasonal inhibition by *B. lichenformis. Science 155*:89.

Wesley, F., Rourke, B., and Darbishire, O. (1965). The formation of persistent toxic chlorohydrins in foodstuffs by fumigation with ethylene oxide and propylene oxide. *J. Food Sci. 30*:1307.

Wheaton, E., and Hays, G. L. (1964). Antibiotics and the control of spoilage in canned foods. *Food Technol. 18*:549.

Wheaton, E., and Pratt, G. B. (1962). Radiation survival curves of *Clostridium botulinum* spores. *J. Food Sci. 27*:327.

Wheaton, E., Pratt, G. B., and Jackson, J. M. (1961). Radioresistance of five strains of *Clostridium botulinum* in selected food products. *J. Food Sci. 24*:345.

Widdus, R., and Busta, F. F. (1982). Antibotulinal alternatives to the current use of nitrite in foods. *Food Technol. 36*(11):105.

Williams, O. B., and Reed, J. M. (1942). The significance of the incubation temperature of recovery cultures in determining spore resistance to heat. *J. Infect. Dis. 71*:225.

Williams, O. B., and Robertson, W. J. (1954). Studies on heat resistance. VI. Effect of temperature of incubation at which formed on heat resistance of aerobic thermophilic spores. *J. Bacteriol. 67*:377.

Winarno, F. G., and Stumbo, C. R. (1971). Mode of action of ethylene oxide on spores of *Clostridium botulinum* 62A. *J. Food Sci. 36*:892.

Winarno, F. G., Stumbo, C. R., and Hays, K. M. (1971). Effect of EDTA on the germination of and outgrowth from spores of *Clostridium botulinum* 62A. *J. Food Sci. 36*:781.

Wong, D. M., Young-Perkins, K. E., and Merson, R. L. (1988). Factors influencing *Clostridium botulinum* spore germination, outgrowth, and toxin formation in acidified media. *Appl. Environ. Microbiol. 54*:1446.

Wyatt, L. M., and Waites, W. M. (1971). Studies with spores of *Clostridium bifermentans*: Comparison of germination mutants. In *Spore Research 1971* (A. N. Baker, G. W. Gould, and J. Wolf, eds.), Academic Press, London, p. 123.

Wyatt, L. R., and Waites, W. M. (1975). The effect of chlorine on spores of *Clostridium bifermentans*, *Bacillus subtilis* and *Bacillus cereus. J. Gen. Microbiol. 89*:337.

Xezones, H., and Hutchings, I. J. (1965). Thermal resistance of *Clostridium botulinum* (62A) spores as affected by fundamental food constituents. *Food Technol. 19*:1003.

York, G. K., II, and Vaughn, R. H. (1954). Use of sorbic acid enrichment media for species of *Clostridium. J. Bacteriol. 68*:739.

York, G. K., II, and Vaughn, R. H. (1955). Resistance of *Clostridium parabotulinum* to sorbic acid. *Food Res. 20*:60.

York, G. K., II, and Vaughn, R. H. (1964). Mechanisms in the inhibition of microorganisms by sorbic acid. *J. Bacteriol. 88*:441.

Young-Perkins, K. E., and Merson, R. L. (1987). *Clostridium botulinum* spore germination, outgrowth and toxin production below 4.6; interactions between pH, total acidity and buffering capacity. *J. Food Sci. 52*:1084.

Zimmerman, F. K. (1977). Genetic effects of nitrous acid. *Mutat. Res. 39*:127.

7

Control in Meat and Meat Products

Friedrich-Karl Lücke

Fachhochschule, Fulda, Germany

Terry A. Roberts

Agricultural and Food Research Council Institute of Food Research, Reading, Berkshire, United Kingdom

I. TYPES OF MEAT AND MEAT PRODUCTS

A. Fresh (Unprocessed) Meats

Fresh meat provides an excellent growth substrate to most foodborne microorganisms. In addition to some fermentable carbohydrates, it contains sufficient amounts of free amino acids, lactic acid, and vitamins to support growth of even fastidious bacteria (Dainty, 1982, 1985; Dainty et al., 1983; Lawrie, 1985). The types of microorganisms found on fresh meat are largely determined by the pH value, by the availability of O_2 in the atmosphere immediately above the meat surface and in the surface tissues, and by temperature.

The pH value attained after postmortem glycolysis depends on the glycogen content of the tissue and the specific activity and stability of the enzymes involved in the conversion of glycogen to lactate. Chicken skin, connective tissue, and certain muscles of pigs have a low lactic acid content and a high ultimate pH value. Consequently, they spoil more rapidly than red muscles with pH values between 5.5 and 5.8. However, in the pH range above 5.5, growth of clostridia in tissues is little affected by pH.

The redox potential of red meat has long been believed to be important in determining the nature of microbial spoilage, but its exact role is not clear. After slaughter and bleeding, O_2 supply to the muscle is stopped, and residual O_2 is rapidly consumed by respiratory enzymes in the muscle. Hence, deep muscle tissue becomes anaerobic, and clostridial spores may germinate and grow, particularly in connective tissues, which contain less lactic acid than red muscles. This results in a condition known as *bone taint*. However, microbial growth in deep musculature is rare except when refrigeration is inadequate (Gill, 1979; Gill et al., 1978; Roberts and Mead, 1985). Clostridia are not regularly present in deep tissues, and today's chilling methods usually lower the internal temperature of large cuts to a level not conducive to clostridial growth.

A commonly held view that the deep tissues of carcasses from healthy animals are likely to contain viable bacteria has led to regulations governing, for example, the length of time between slaughter and evisceration and the cooling of carcasses after slaughter. Although there are several reports of relatively large numbers of bacteria in deep tissues (Narayan, 1966; Nowicki, 1976; Pusztai, 1970), there is doubt about the sampling techniques that could have allowed accidental contamination during removal of the sample. Where very careful aseptic techniques have been used, it has been possible to obtain large quantities of sterile tissue from carcasses (Buckley et al., 1976; Ockerman et al., 1969). Traditional thinking has greatly exaggerated the importance of deep tissue contamination as a threat to public health or a limiting factor in shelf life.

Clostridial spores are more resistant than vegetative bacteria to the antibacterial action of blood and lymph, and spores can often be isolated from the liver, spleen, and lymph nodes of healthy animals. Spores introduced into the bloodstream are rapidly eliminated from the general circulation, but they accumulate in the liver and may survive for many weeks (Gill et al., 1981). The few spores occasionally found in livers probably originate from cuts and abrasions in the live animal. They pose no health hazard, provided livers are cooled after slaughter. The most commonly occurring spores, those of *Clostridium perfringens*, are controlled by cooling to 15°C within 8 hours of slaughter, because the spores seem to germinate slowly and *C. perfringens* grows poorly below 15°C (Gill et al., 1981). Such a cooling regime would also serve to control the germination and growth of proteolytic *Clostridium botulinum*.

Clostridia may also grow on the surface of meat but only during prolonged storage at temperatures above 20°C where mesophilic clostridia can compete with other bacteria (Gill and Newton, 1980). Even on vacuum-packed beef, where more favorable growth conditions might be expected, proteolytic strains of *C. botulinum* grew at 25°C only after severe spoilage was evident (Hauschild et al., 1985). Psychrotrophic, nonproteolytic *C. botulinum* strains have a higher growth potential on fresh meats, even in unpackaged foods, and have been responsible for outbreaks of meatborne botulism among the Inuit population in northern Canada and Alaska (Hauschild, 1989). The Inuits, however, appear to be the only people regularly eating raw, putrid meat. Accordingly, no case of botulism due to the consumption of fresh meat has ever been reported from any other part of the world, even from countries where consumption of raw unprocessed meats is common (e.g., Germany, Belgium, and the Netherlands).

B. Processed Meats

Intrinsic factors affecting the survival and growth of microorganisms in processed meats include the water activity (a_w), the pH value, and the presence of antimicrobial compounds such as nitrite. These factors are determined by the properties of the meat and the formulation of the products. Microorganisms present may then be partially or completely destroyed by a heat treatment, and survivors may be inhibited by extrinsic factors such as low temperature. Generally, several factors including pH, available water, temperature, and added preservatives act in combination in the preservation of meats. Despite enormous research efforts, our understanding of the relative importance of factors previously identified as being important in meat preservation remains relatively poor. Our understanding of the control of *C. botulinum* is certainly inadequate in the current climate of rapidly changing production and storage technologies, the increased length of the food distribution chain, and consumer demands for less heavily preserved products with a longer shelf life.

Table 1 lists types of processed meats which, if properly manufactured, are not spoiled by microorganisms. It also summarizes the contribution of various intrinsic and extrinsic factors to their microbiological stability at ambient temperature. Table 2 lists types of perishable meats and the factors contributing to their microbiological safety.

II. CONTAMINATION OF MEAT WITH *C. BOTULINUM* SPORES AND ITS PREVENTION

C. botulinum spores are naturally present in meats, but the frequency of occurrence and the factors that determine their presence and distribution are poorly documented (see Chapter 3). This is mainly due to technical difficulties in the detection and enumeration of *C. botulinum* spores, which still rely upon testing for botulinum toxin in mice, unless monoclonal antibodies are available. In most countries, monoclonal antibody against botulinum toxin is not available commercially. Most of the clostridia that occur in raw meats are harmless putrefactive mesophiles. However, from time to time spores of *C. botulinum* are present, and this must be taken into account by all meat processors. There have been relatively few surveys of *C. botulinum* (summarized by Hauschild, 1989; Tompkin, 1980). Estimates of levels of contamination range from less than 0.1 spore/kg to 7 spores/kg. These estimates are influenced by the size of sample examined and the bacteriological method used. In some cases, the samples taken were very small, and the methods used are not universally considered reliable for the detection of *C. botulinum*. The higher estimates used large samples and reliable methods.

Table 1 Shelf-stable Meat Products and the Contribution of Various Factors to Their Microbiological Stability

Product type	Inhibition and inactivation of microorganisms by					
	heat	low a_w	low pH	nitrite	smoke	chilling
Raw dried meats, not fermented						
Westphalian-type ham	–	+++	+	–	+	–[a]
Parma-type ham	–	+++	+	–	–	–[a]
Other (biltong, jerky, Chinese-type dry sausage)	–	+++	+	–	–	–[a]
Dry fermented sausage						
German- or Danish-type	–	++	++	–[a]	+	–
French- or Italian-type	–	+++	+[a]	–[a]	–	–
Canned unsalted meat	+++	–	–	–	–	–
Shelf-stable meats heated in hermetically sealed containers (SSCCMs)						
Ham and shoulder	+	+	+	++	–	–
Luncheon meat, frankfurter-type sausages	++	++	+	++	–	–
Liver- and blood-containing sausage mixtures	+++	++	±	–	–	–
Pasteurized dried meats						
German-type Brühdauerwurst, genuine Italian mortadella	+	++	+	+	+	–[a]
Pasteurized acidified meats	+	+	+++	–	–	–

[a]Factor important during early stages of ripening/drying process.

The more frequent occurrence of botulism from pork products than from other meats implies more frequent contamination of pork than of beef or lamb. Poultry and poultry products have not been associated with human botulism. However, the range of poultry products marketed has increased enormously, often drawing on processing experience in the red meat industry, and using similar formulations, and the likelihood of the occasional presence of *C. botulinum* spores should not be ignored.

Surveys for the presence of *C. botulinum* must be regarded as inadequate to establish a reasonable estimate of low incidence. In view of the quantity of meat in processed products, more reliable estimates of the occurrence of *C. botulinum*

Table 2 Perishable Meat Products and the Contribution of Various Factors to Their Microbiological Safety

Product type	Inhibition and inactivation of microorganisms by					
	heat	low a_w	low pH	nitrite	smoke	chilling
Raw unground meats						
Bacon	−	+	+	+	+	+
Fermented sausage						
Undried, spreadable	−	+	+	+	±	+
Perishable meats heated in hermetically sealed containers						
Ham and shoulder	+	+	+	+	−	+
Luncheon meat, frankfurter-type sausages	++	+	+	+	−	+
Liver- and blood-containing sausage mixtures	++	+	±	−	−	+
Cooked meats, handled and/or sliced after cooking						
Ham and shoulder	+	+	+	+	+	+
Frankfurter- and bologna-type sausages	+	+	+	+	+	+
Liver-containing sausages	+	+	±	±	−	+
Blood-containing sausages	+	+	−	−	−	+
Pasteurized acidified meats	+	+	++	−	−	+

spores are needed. This will not be possible until specific methods for *C. botulinum* and its toxin become generally available, obviating the need for toxin tests in animals.

No means are currently available to guarantee the absence of *C. botulinum* from raw meat and poultry. The low incidence of spores in beef and lamb suggest that the eating habits of the animal are contributing factors. Pigs are more likely to ingest soil than cattle and sheep, and therefore more likely to ingest spores that occur naturally in soil. When animals are raised intensively, the situation differs, and the likelihood of ingestion of spores will be dependent on the feed and local hygiene. In addition, pig carcasses are scalded, a process that may result in contamination of their interior by spores from the scalding water (Jones et al., 1984). If scalding and dehairing are combined in a single process, the aorta may become contaminated with spore-containing scalding water as deep as the pelvic region (Troeger and Woltersdorf, 1987). Proper scalding procedures may eliminate this problem. Careful removal of soil from the animals may minimize the occurrence of *C. botulinum* spores in meat.

III. EPIDEMIOLOGY OF MEATBORNE BOTULISM

Botulism is relatively frequent in areas where preservation of foods in a home or farm setting is common and where such preserved products are regularly eaten without prior heating (see Chapter 4). Home processing of meat is much more common in continental Europe than in the United Kingdom or in the United States. Furthermore, meat inspection in countries like Germany effectively eliminates the risk of porkborne trichinellosis, and the population in general does not hesitate to eat raw pork products. Tompkin (1980) tabulated outbreaks of botulism from meat and poultry products that occurred until 1978. The highest incidence was in Poland, Germany, and France (Table 3). With few exceptions, all meat-mediated outbreaks in which the toxin type could be identified were due to type B botulinum toxin, and the fatality rate was 2–5%.

In Poland, between 200 and 500 cases of botulism are reported each year (about one case per 100,000 people). According to Anusz (1975, 1980), about 80% of the incriminated products were home processed, and in 90%, meat products were involved. This high incidence also reflects the agricultural and economic situation in Poland in the 1970s, which necessitated home preservation of meats to cover the needs in times of shortage. The most hazardous practice (reported for 60–70% of cases) is cooking pork in glass jars (*wecks*) and storing the

Table 3 Countries in Which Ingestion of Toxic Meat Products Accounts for More than Half of Reported Outbreaks of Botulism

Country	Period	Outbreaks with food identified	% outbreaks due to meats
Poland	1979–83	[a]	87
France	1978–84	83	86
Portugal[b]	1970–84	12	83
West Germany	1971–82	[c]	>75
Czechoslovakia	1979–84	14	72
Canada	1971–85	63	70[d]
Hungary	1961–85	58	67
Belgium	1982–84	5	60

[a]2434 cases.
[b]Lecour et al., 1988.
[c]499 cases.
[d]Mostly from marine mammals.
Source: Hauschild, 1989.

jars under inadequate refrigeration. Skoczek and Mierzejewski (1979) detected *C. botulinum* (proteolytic type B strains) in 9 of 1023 swollen cans containing various meats, and Mierzejewski and Palec (1981) reported that the contents of 38% of experimentally inoculated cans became toxic before evident spoilage.

In West Germany, the number of cases of botulism slowly declined during the last 30 years from about 60 to about 25 per year. In parallel, many households have given up canning and curing of foods. A higher percentage of foods are now preserved at home by freezing. The data available (Berndt, 1978; Buhl and Hertel, 1971; Meyer, 1928) suggest that the ingestion of toxic meat products accounted for about 70% of these cases of botulism. Most of the cases have been linked to the ingestion of toxic raw hams, while a minor proportion was due to canned sausage mixtures, particularly those containing liver or blood.

In France, 64% of botulism outbreaks could be traced to the consumption of home- or farm-salted raw hams (Sebald et al., 1980). Like the German-type products, such hams were dry salted, and no brine was injected into the muscles. The incidence of meatborne botulism was highest during the German occupation in World War II. It is not surprising that in shortage situations many mistakes are made during slaughtering and curing, and spoiled products are more likely to be eaten than in times of abundant food supply. The French experience (Tompkin, 1980) indicates that formation of botulinum toxin in raw hams was favored by (1) improper treatment of pigs before slaughter (increased number of spores in the bloodstream, depletion of glycogen, and high final pH of the meat), (2) improper slaughtering hygiene, (3) slaughtering during the warm season and absence of cooling facilities, and (4) the shortage of salt during the war.

Salted pork appears to be the most important vehicle for meatborne botulism in Belgium, Norway, and Spain. It should be remembered that the first strain of *C. botulinum* ever isolated and described in detail was from a raw ham that had caused an outbreak of botulism in the village of Ellezelles, Belgium (van Ermengem, 1897).

Meat from marine mammals has been frequently implicated in outbreaks of botulism in the native Inuit population of northern Canada and Alaska. As stated earlier, the Inuits appear to be the only people regularly eating raw, putrid meat (Hauschild, 1989).

IV. CONTROL OF *C. BOTULINUM* IN MEAT PRODUCTS

A. Control in Meat-Containing Model Systems and in Cooked Perishable Meat Products

Most of the experiments performed to evaluate the risk of formation of botulinum toxin in meats have been performed in meat-containing model systems such as meat slurries (Rhodes and Jarvis, 1976) or comminuted cured lean pork

(Christiansen et al., 1973) pasteurized in hermetically sealed containers. These products are comparable to pasteurized ham. To extrapolate from these results to other meat products, one must be aware that most perishable meats become recontaminated by psychrotrophic spoilage bacteria upon subsequent handling, slicing, and packaging. Psychrotrophic lactobacilli are most competitive on perishable meat products. They tend to lower the risk of botulinum toxin formation by acid production if the products are stored under insufficient refrigeration. This has been observed with bacon (Hauschild, 1982) and liver sausage (Hauschild et al., 1982). Tanaka et al. (1985) added sugar and lactic acid bacteria to prevent development of *C. botulinum* during temperature abuse of bacon.

This approach to the control of *C. botulinum* is attractive because it does not rely upon the use of chemical preservatives, which might have other undesirable properties. However, in terms of assurance and control of food safety over a long period, it is unproven. Some researchers believe that it should not yet be relied upon because its effectiveness must be a function of the natural level of contamination by microorganisms, which is not under control, and of other components in the formulation, which may have an effect on the rate of growth of the lactic acid bacteria, as well as the temperature and duration of storage.

Control of Proteolytic Strains

Botulism from cured meat products is rare, but there is concern because we cannot guarantee the absence of *C. botulinum* from the raw meat or from ingredients that might be contaminated by soil. It must therefore be assumed that *C. botulinum* might be present, and appropriate controls must be applied.

The proteolytic strains of *C. botulinum* closely resemble the common group of *putrefactive anaerobes*, from which they can only be differentiated by the production of toxin (see Chapter 1). Because these putrefactive anaerobes and their spores possess very similar physiological properties to *C. botulinum*, numerous publications detail their tolerance of growth-inhibiting systems to avoid having to work with the more hazardous *C. botulinum* (Pivnick et al., 1969; Riemann, 1963).

In principle, *C. botulinum* can be controlled in a variety of ways. If it were present as vegetative cells, heating to pasteurization temperatures would be adequate to kill it along with other vegetative pathogens. However, is is normally present as spores, presumably as contamination from soil carried by raw agricultural products, or introduced as contamination from the food-processing environment.

In the early days of food preservation, outbreaks of botulism were common. Because the high resistance of spores of *C. botulinum* to heating was not appreciated, foods were frequently heated insufficiently. Canning was gradually put onto a scientific basis after the investigations of Esty and Meyer (1922), who heated very high concentrations of spores of proteolytic strains of *C. botulinum* at different temperatures. From their results it was calculated that 10^{12} spores of

C. botulinum would be killed by heating at 121°C for 3 minutes. This process, and equivalent processes, are known as a *botulinum cook*, or a 12 D process (D = decimal reduction time) (see Chapter 6).

The heat resistance of spores of *C. botulinum* is greatly affected by the medium in which the spores are heated and by the recovery medium. Heat resistance is reduced at low pH values, hence foods that are naturally acid and those that have been acidified can be rendered safe with much lower heat processes than the botulinum cook (Odlaug and Pflug, 1977, 1978; Roberts and Hitchins, 1969). The heat resistance of *C. botulinum* spores is increased at low water activities and in the presence of protein and fat (Lücke, 1985; Simunovic et al., 1985).

If a product cannot be heated sufficiently to kill *C. botulinum* with a high degree of certainty, an alternative method of control must be found, such as storage at refrigeration temperatures. However, good temperature control at all stages of food production, distribution, and retailing is rare. Additionally, foods occasionally suffer temperature abuse during distribution, and the control of temperature under domestic conditions is poorly documented.

The complexity of the control of *C. botulinum* where a full botulinum cook cannot be used, and where temperature control cannot be relied upon, is illustrated by pasteurized cured meats. These meats, most commonly pork products, contain viable bacterial spores that are unable to grow in the product, but which grow readily if subcultured into laboratory medium. Additionally, if bacterial spores from media are inoculated into the product, they grow readily. Since the product has been lightly heated, it is inevitable that spores of *Clostridium* spp. and *Bacillus* spp. survive, together with enterococci (fecal streptococci), which are relatively heat resistance for vegetative bacteria.

Pig-breeding programs have successfully selected pigs that convert feed more efficiently and grow rapidly to a leaner pig. In some breeds and particular crossbreeds, this has also resulted in some muscles, particularly those in the shoulder (collar), being of exceptionally high pH. It is not unusual for the pH of some small muscles to be close to 7.0, and much cured meat product is of higher pH than similar product even 10 years ago. The effect of pH alone on the growth of *C. botulinum* is small, but pH is important in determining the effectiveness of sodium chloride and sodium nitrite, and a change from 5.8 to 7.0 is very significant, considering the levels of nitrite commonly used.

In addition, levels of salt in cured meat products have fallen steadily during the past 10–15 years. The traditional salty product is no longer preferred by consumers, especially younger consumers, and the lobby to eat less salt has encouraged producers to market cured meat products that are very lightly preserved.

Nitrite is antimicrobial, conferring considerable bacteriological stability to many cured products, its activity increasing with reducing pH. It also contributes to the cured meat flavor, which differs from the taste of salted pork. Redox potential (Eh) also affects the activity of nitrite against some bacteria; anaerobic conditions (reduced Eh) increase its inhibitory effect. Bacterial growth in cured meats is also affected by the chemical composition of the cure (added carbohydrates, phosphates, etc.), the nature of any smoking and cooking processes, and the temperature and gas atmosphere during storage. Whether these factors interact to control microbial growth, or merely act in combination, is difficult to interpret. Microbial growth is largely determined by the pH and a_w of the product, the concentration of nitrite present, and the temperature of storage. In fact, if a_w and pH are known, is is broadly possible to predict whether a meat product will be stable at a given temperature.

The proceedings of several symposia and a number of comprehensive reviews dealing with the control of *C. botulinum* in cured meat products have been published in recent years (Benedict, 1980; Krol and Tinbergen, 1974; NAS/NRC, 1981, 1982; Overview, 1980, 1982, 1983; Pierson and Smoot, 1982; Roberts et al., 1981a, 1982; Sofos and Busta, 1983; Sofos et al., 1979; Tinbergen and Krol, 1977; Tompkin, 1983). The history of curing was reviewed by Binkerd and Kolari (1975), and the changes in production and marketing, which had consequences for the bacteriological stability and safety of the products, were addressed by Cerveny (1980). Trends in the use of nitrite in the United Kingdom were reviewed by Hannan (1981). The increased use of refrigeration, the direct application of nitrite rather than nitrate, and the pumping of curing salts all led to improvements in product stability. Canned cured products and cured products packaged in films of low O_2 permeability are now common and enjoy an excellent record of stability and safety, although at the time of their introduction some observers had severe reservations as to whether they would be bacteriologically safe. Phosphates are incorporated to minimize shrinkage, reduce cook-out losses, and improve texture with respect to slicing, but whether the consumer appreciates these industry oriented advantages is a matter of conjecture. Ascorbates (or isoascorbates) were included primarily to accelerate the formation of the cured meat pigment and improve color stability. Ascorbates/isoascorbates and certain phosphates may also improve bacteriological stability.

In the late 1960s, with the increasing availability of more sensitive analytical methods, nitrosamines were detected at very low levels in a variety of foods, including some cured meat products. Their known carcinogenicity in laboratory animals, even though used at levels 6–9 orders of magnitude greater, prompted demands to reduce to a minimum the quantities of nitrite used in cured meat products and to exclude it entirely wherever it was not essential for bacterial stability and safety, i.e., the control of *C. botulinum*. These demands threatened the future of the cured meat industries of many countries and led to a relatively huge investment in research into the chemical fate of nitrite in cured meats (Cassens et al., 1979; Sebranek and Cassens, 1973) and its activity against *C. botulinum*.

Progress in identifying the role of nitrite has been thoroughly, and critically, reviewed by Tompkin (1983), the suggested mechanisms summarized by Roberts and Gibson (1986), and the mechanisms of action reviewed by Roberts and Dainty (1991) and Roberts et al. (1991). The antimicrobial effect of nitrite has been related to nitrous acid formation (Castellani and Niven, 1955) and is therefore greatly affected by the pH value. Spore outgrowth is more sensitive to nitrite than germination.

There is disagreement as to whether the nitrite added or the residual nitrite is the more important factor governing the control of *C. botulinum* in cured meats. There are numerous reports that the likelihood of toxin production by *C. botulinum* is better predicted from the initial nitrite level than the residual (Christiansen et al., 1973). It has been suggested that prevention of growth from spores of *C. botulinum* is the consequence of germinated spores losing viability while nitrite remains, implying that the safety of canned meat is dependent upon sufficient residual nitrite remaining until the number of viable cells decreases below that needed to initiate growth (Christiansen et al., 1978).

Nitrite has no effect on the heat resistance of spores, and concentrations used in meat products do not induce germination. The presence of NaCl and nitrite in the recovery medium reduces the numbers of spores able to grow. Efforts have been made to identify whether the nitrite remaining or the nitrite that became undetectable chemically is responsible for preventing spore outgrowth. The role of isoascorbate in enhancing the antibotulinal effect of nitrite has been confirmed (Tompkin et al., 1978a, b, 1979).

It was proposed that the stability of meat systems inoculated with *C. botulinum* was dependent on the relative rates of germination of spores and of loss of nitrite. If germination occurred while nitrite remained, spores would be inhibited. If nitrite levels were reduced, e.g., by refrigerated storage, and the product subsequently temperature-abused, *C. botulinum* was able to grow and produce toxin (Christiansen et al., 1978). However, this explanation does not explain why *C. botulinum* spores grew and produced toxin in the presence of residual nitrite when no isoascorbate was present, but did not do so when isoascorbate was present initially and no residual nitrite could be detected even after a few days. The rate of germination of the spores would not explain the stability of the nitrite-free product for 6 months at temperatures able to support growth and toxin production.

The importance of the iron contents of meats was recognized. In meats with naturally high levels of iron, e.g., liver and beef hearts, the antibotulinal effect of nitrite was reduced (Tompkin et al., 1978c). If the iron was removed, for example, by sequestering with ascorbate, isoascorbate, or EDTA, the antibotulinal effect of nitrite was restored to the expected level.

It is now universally recognized that controlling growth of *C. botulinum* in cured meat products is the consequence of several factors acting in combination. The term *synergistic* is still frequently used, but evidence of true synergy between factors is lacking, and *in combination* seems more appropriate. Those factors known to combine to prevent microbial growth include brine concentration, pH of the product, nitrite (input or residual), severity of the heat treatment, the number of spores in the inoculum (or the unprocessed product in commercial conditions), temperature of storage, the nature of the competing microorganisms, available iron in the product (sometimes reflected by the type of meat in the product), and the presence of other additives, e.g., ascorbate/isoascorbate, phosphates, and nitrate (Roberts et al., 1981a, b, 1982; Tompkin, 1983; Tompkin et al., 1978d).

The statistical significance of factors interacting to prevent clostridial growth was first demonstrated by Riemann (1963) using spores of PA 3679 in luncheon meat. Subsequently, several examples of interacting factors appearing to explain the control of growth of clostridia were published, e.g., in shelf-stable canned luncheon meat (Pivnick et al., 1969) and by review of published literature (Riemann et al., 1972). Baird-Parker and Freame (1967) demonstrated that reducing the pH from 7.0 to 6.0 or 5.5 lowered the salt tolerance of vegetative cells of *C. botulinum* types A, B, and F. Roberts and Ingram (1973) showed the triple interaction between salt, nitrite, and pH at 35°C against *C. botulinum* types A, B, E, and F in laboratory medium, illustrating that boundaries could be drawn between growth-promoting and growth-inhibiting conditions. Hence, in principle, a similar degree of control of *C. botulinum* growth may be obtained by many different combinations of pH, salt, and nitrite, coupled with different processes and storage conditions.

In comprehensive studies that led to the development of a mathematical model to predict the probability of toxin production, factors significantly reducing toxin production included increasing salt, increasing nitrite, adding isoascorbate, adding nitrate, increasing the heat treatment, and decreasing the storage temperature (see Roberts et al., 1981a). While most of these might have been anticipated, the significance of isoascorbate was confirmation of earlier claims (Tompkin et al., 1978a, b, 1979).

The significant effect of nitrate contradicted earlier reports that nitrate per se had no antimicrobial action, although it had proved beneficial in limiting clostridial growth in vacuum-packaged bacon of high pH (Roberts and Smart, 1976). The possible antimicrobial role of phosphates has been critically reviewed by Tompkin (1984).

With the common availability of powerful desktop computers, considerable advances have been made in modeling microbial responses (see Chapter 14). Multiple linear regression had been used by Christiansen et al. (1973) to predict the probability of toxin production under specified conditions. They concluded that toxin production could most accurately be predicted from the level of initial (i.e., input) nitrite rather that the residual nitrite. Logistic regression has also been applied to data on toxin production to develop a model that would predict the probability of toxin production (Roberts et al., 1981c, 1982; Robinson et al., 1982).

Despite research carried out over the past 20 years, we still have no unifying concept that is universally accepted to explain the control of *C. botulinum* in meat products. One obvious difference between commerce and the laboratory is the number of spores of *C. botulinum* in the inoculum of experimental studies compared with their relatively rare occurrence in the commercial product. In most experiments there is consistently less growth and less toxin production with a low inoculum than with a high inoculum. With the low natural occurrence of *C. botulinum* in the meat supply, how relevant are laboratory studies with levels 10- to 1000-fold higher?

Control of Nonproteolytic Strains

Possible hazards from nonproteolytic strains of *C. botulinum* in pasteurized meat products have been discussed by Lücke et al. (1981) and by Simunovic et al. (1985). Nonproteolytic spores are much more sensitive to heat than those of proteolytic strains. Consequently, they are not a problem for shelf-stable canned meats. However, decimal reduction times (calculated from thermal death times) can be as high as 32 minutes at 80°C for some nonproteolytic type B strains (Scott and Bernard, 1982). This means that these spores may occasionally survive the pasteurization process given to many perishable meat products.

Lücke (unpublished data) compared type B strains isolated from toxic raw hams with other nonproteolytic *C. botulinum* type B strains and found similar biochemical properties: none degraded meat particles, all fermented mannose and sucrose, and the toxicity of all strains could be markedly increased by trypsin treatment. At 80°C, the initial inactivation rate was rapid (decimal reduction times at 80°C were 2.6–4.6 minutes for all strains). As observed by Sebald et al. (1972) and by Scott and Bernard (1985), survivor curves showed considerable tailing, particularly when lysozyme or egg yolk was added to the recovery medium.

The nature and causes of this tailing phenomenon are the subject of intense research. If they are a general property of all nonproteolytic strains of *C. botulinum*, then many heat processes in common use would not destroy the expected number of these spores. It is also well established that the spores are more commonly present in raw foods than expected from our, admittedly poor, understanding of the occurrence of proteolytic strains of *C. botulinum*. In modern food processing, many foods are heated gently to maintain eating and nutritional quality, using processes that apparently do not significantly reduce numbers of spores of nonproteolytic strains of *C. botulinum*. Those lightly heated foods are then stored under refrigeration, probably under temperature-time regimes that raise the possibility of growth of nonproteolytic *C. botulinum*, especially if those foods are temperature abused during distribution, storage, or in the home. Consideration of the degree of hazard of such processes is under debate. It might be argued that a wide range of such products have already been produced and marketed under commercial conditions where temperature abuse must have occasionally occurred, and there is no history of botulism. Others wish to be more certain that the formulation of the food, the process, and the storage conditions are able to control *C. botulinum* should spores be present.

All nonproteolytic strains examined by Lücke (unpublished data) were inhibited by about 4% brine ($a_w < 0.97$), but still had growth potential in products of about 3% brine ($a_w = 0.977$) at 8°C. Under these conditions, the lag phase was very long, and growth was only evident after a minimum of 2 weeks. This means that there should be little hazard from products recontaminated with psychrotrophic spoilage microorganisms after processing.

In the experiments of Pivnick and Bird (1965), *C. botulinum* type E occasionally grew in sliced vacuum-packed meats with a brine concentration below 4% and a low nitrite content, particularly in certain jellied meats containing too little sugar to allow for a pH reduction by the spoilage microorganisms. Only the latter products allowed growth at 10°C, and they sometimes became toxic before gross spoilage was evident.

There is still debate as to whether toxin production by *C. botulinum* can occur before spoilage or only after spoilage is evident. If the latter were the case, there would be some safeguard because the consumer would be warned by spoilage that the product might not be safe. Many detailed investigations have provided contradictory results, but there are sufficient numbers of examples of toxin having been produced before spoilage is evident to state firmly that the occurrence of spoilage before toxin production should not be relied upon.

Lücke et al. (1981) prepared liver sausage mixtures and bologna-type mixtures of a_w 0.977–0.985 and varying amounts of nitrite, inoculated them with spores of nonproteolytic *C. botulinum*, and cooked the products in cans to about

Table 4 Protection of Canned Brühwurst[a] (Bologna-Type Sausage) Mixtures from Toxigenesis by Nonproteolytic *C. botulinum* (3% brine; pH 6.0; Heated to Give Approx. 1 Log Reduction of Spore Numbers) Stored for 6 Months

Storage temp. (°C)	Protection[b] from toxin formation by nonproteolytic *C. botulinum* with indicated amount of $NaNO_2$ added			
	0 mg/kg	40 mg/kg	62 mg/kg	83 mg/kg
15	<4.5	<4.5	<4.5	<4.5
10	<4.4	4.6	5.2	6.2
8	4.6	4.9	6.4	6.0
5	6.8	7.3	>7.3	>8.0

[a]Sausage mixtures made from raw meat, pork back fat, and ice.
[b]Log (1/P), where P is the probability for a given spore to both survive and develop in the product.
Source: Lücke, unpublished data.

76°C core temperature. This process inactivated all vegetative microorganisms and destroyed about 90% of the inoculated spores. Irrespective of nitrite addition, the first cans of liver sausage swelled after about 2 weeks at 15°C, 4 weeks at 10°C, and 12 weeks at 8°C. One of 20 cans was toxic after 6 months at 5°C. In Brühwurst (bologna-type sausage) mixtures with no added nitrite, growth was observed after 1, 2 and 3 weeks at 15, 10, and 8°C, respectively, and some cans became toxic after about 16 weeks at 5°C. Nitrite addition gave significantly better protection from toxin formation (Table 4). However, the actual protection of the consumer is much better than suggested by these data because pasteurized perishable meats are rarely stored longer than 2–4 weeks at 7°C or below. Furthermore, they are virtually never heated in hermetically sealed containers but are recontaminated with spoilage organisms during handling and slicing.

B. Control in Raw Dried, Unfermented Meat Products

As outlined above, most of the outbreaks of botulism in Germany and France have been traced to raw hams that had been salted or cured on a domestic scale without brine injection. The observations by van Ermengem (1897), Meyer and Gunnison (1929), Zeller (1959), and Ralovich and Barna (1966) indicated that nonproteolytic type B strains were the etiological agents in most cases. Schulze and Funke (1981) isolated such a strain from a toxic ham. This strain and subsequent *C. botulinum* type B isolates from toxic raw hams had biochemical and technologically relevant properties similar to those of *C. botulinum* type B strains from aquatic habitats.

Lücke et al. (1982) inoculated large bone-in hams with spores of *C. botulinum* and subjected them to a curing process similar to commercial practice. They did not observe growth of proteolytic strains, whereas a mixture of spores of nonproteolytic *C. botulinum* strains (types B and E) produced toxin when curing and salt equilibration was at 8°C, but not at 5°C. During the subsequent aging period, toxin was formed if the a_w in the interior was above 0.97. No effect of nitrite and nitrate on toxin formation was observed.

Lechovich et al. (1977) cured country-style hams inoculated with proteolytic strains of *C. botulinum*. They found no growth and toxin formation, even in the absence of nitrite and nitrate. Likewise, Delényi et al. (1981) did not observe any influence of nitrate and nitrite on growth of *C. botulinum* type B when they cured 1-kg pieces of raw pork.

In agreement with epidemiological data, Lücke et al. (1982) concluded that most if not all cases of botulism with raw ham as the incriminated food were due to insufficient cooling during salting or to removing the hams from refrigeration before sufficient salt had penetrated to the interior of the meat. They recommended that (1) only hams with a pH below 5.8 should be used for salting (hams of higher pH need extended salting times and provide better conditions for bacterial growth), (2) salting, curing, and salt equilibration should all be carried out at 5°C or below, and (3) the hams should be left at this temperature until the internal a_w is below 0.96 (corresponding to approximately 4.5% salt).

For the manufacture of dried meats such as South African biltong, South American charqui, pastirma, and Chinese-type dried sausage, it is also important to prevent growth of pathogens while the a_w is still high. To be dried at ambient or elevated temperature, these products must be cut into thin strips to facilitate rapid moisture removal (see Leistner, 1987, for review).

C. Control of *C. botulinum* in Fermented Sausages

Not a single outbreak of botulism worldwide could be traced to the ingestion of fermented sausage; even the Norwegian outbreak quoted by Tompkin (1980) was due to a local specialty, *surpølse*, prepared from cooked ingredients (M. Yndestad, Veterinary College of Norway, personal communication). Experiments with inoculated summer sausage (Christiansen et al., 1975; Kueper and Trelease, 1974) demonstrated that a risk of formation of botulinum toxin only exists if the product is kept at elevated temperatures (27°C) for extended periods without a significant drop in pH. This situation is extremely unlikely in commercial practice. Incze and Delényi (1979) reported some transient toxin formation in experimentally inoculated salami; however, they added the spores by means of a syringe and may thus have created niches with conditions more favorable for growth. *C. botulinum* did not grow during fermentation of experimentally inoculated sausages, even in the absence of nitrite or nitrate (Lücke et al., 1983; Nordal and Gudding, 1975). Lücke et al. (1983) could not even provoke toxin formation if they delayed acid formation, raised the temperature to 25°C, or omitted the drying step.

D. Control of *C. botulinum* in Shelf-Stable Meat Products Heated in Hermetically Sealed Containers

As pointed out previously, meat products heated in sealed containers and subsequently stored under insufficient refrigeration have been involved in various outbreaks of botulism, particularly in Poland. This is due to several reasons:

1. An autoclave with enough capacity and suitable control devices is rarely available to small-scale meat processors (households, farms, small butcher shops).
2. The consumer tends to store products in cans or sealed glass jars under insufficient or no refrigeration.
3. Nowadays, few households have cellars or food storage rooms in which the temperature seldom rises above 15°C.
4. A botulinum cook ($F_0 > 2.5$) of sausages or other salted and/or cured meat causes undesirable changes in appearance and flavor.

In most of these outbreaks the product had not received an adequate heat process. If an adequate process had been given, even lack of refrigeration would not have created a hazard.

In commercial practice, only uncured, lightly salted products receive a botulinum cook ($F_0 > 4$). These include meat pieces in gravy and other products in which separation of jelly and fat is accepted. To avoid deterioration of product appearance and flavor, even large-scale, fully equipped meat processors generally cook their shelf-stable salted and/or cured products (SSCCMs) to F_0 values between 0.1 and 1.5 only. Such processes reduce the number of viable spores of proteolytic *C. botulinum* strains by about one to seven log cycles. Consequently, the lower heat intensity must be compensated for by inhibition of surviving spores by intrinsic factors, including NaCl (low a_w), low pH, and nitrite.

The more clostridial spores present, the more anticlostridial agents (NaCl, acid, nitrite) are required to inhibit them (Riemann, 1963). Hence, Hauschild (1982) expressed the degree of protection of a SSCCM as the logarithm of the number of spores required in the raw material to enable one single spore to survive, outgrow, and produce toxin during storage of the product. This number is the reciprocal of the probability P of a given spore to survive and give rise to toxin formation. P may be calculated by dividing the experimentally determined most probable number of spores capable of surviving, outgrowth, and toxin formation in the product by the inoculum level. Protection, be it by destruction or by inhibition of spores, is therefore expressed as $Pr = \log(1/P)$.

Experimental and empirical data show that, in comparison to pasteurized meats (heated to 70–80°C), less NaCl, less nitrite, and/or less acid is required in SSCCMs to prevent the development of surviving spores at ambient temperature: there are fewer spores to inhibit than in pasteurized products, and a sublethal heat treatment sensitizes a major proportion of the spores to the effects of inhibitory agents (Duncan and Foster, 1968; Ingram and Roberts, 1971; Jarvis et al., 1976; Pivnick and Thacker, 1970; Roberts and Ingram, 1966).

SSCCMs in international trade include luncheon meats, ham, and shoulder, as well as frankfurter or Vienna-type sausages in brine. In addition, various sausage mixtures cooked in hermetically sealed containers (cans, jars, autoclavable casings) are on the market, particularly in Germany.

A typical shelf-stable luncheon meat is currently formulated with 4% brine (equivalent to an a_w of 0.970) and 100–150 mg sodium nitrite/kg and heated to F_0 values around 0.5 (Hauschild and Simonsen, 1985). Inoculated pack studies gave protection values of 7–8 (Pivnick et al., 1969; data recalculated by Hauschild, 1982) with only marginal differences between batches formulated with 75 and 150 mg sodium nitrite/kg. Such products have so far never caused botulism. In an attempt to quantify the safety margin of different meat products, Hauschild and Simonsen (1985) considered the production volume and concluded that the actual protection of the consumer (*safety units*, defined as the logarithm of cans marketed per can causing illness) is above 7. They also provided guidelines for the input of NaCl and nitrite and the heat process (see below).

Shelf-stable canned cured ham and shoulder are formulated with less NaCl than luncheon meat and, for organoleptic reasons, subjected to a milder heat treatment (center temperatures around 100°C). Nevertheless, such products have so far never caused botulism, and the actual protection of the consumer is of the same order of magnitude as that for luncheon meat (Hauschild and Simonsen, 1985). Due to the lack of inoculated pack studies, it cannot be decided whether hams and shoulders really need less NaCl and heat to prevent formation of botulinum toxin. However, it seems more likely that the addition of more NaCl and a more severe heat treatment gives products with unacceptable sensory characteristics, and the more lenient recommendations for these products given by Hauschild and Simonsen (F_o value 10-fold lower than for luncheon meat with the same brine concentration) merely reflect the technological feasibility.

Shelf-stable frankfurter or Vienna-type sausages in brine are formulated with even less NaCl (2.4–3% brine, corresponding to a_w values of 0.982–0.977) but receive a more severe heat process (F_o = 0.5–2). On the basis of the good safety record and the estimated production volume, Hauschild and Simonsen (1985) recommended a brine concentration of 2.5% in conjunction with 150 mg ingoing sodium nitrite/kg and an F_o value of 1.5.

In central Europe, a large variety of meat products are heated in hermetically sealed containers. Many of these products, in particular those produced in households, farmyards, or small butcher shops, are only marginally stable at room temperature and have caused botulism, particularly in Poland (see above). Other salted or cured meat products heated in impervious containers (cans, jars, autoclavable casings) are produced on an industrial scale, especially in Germany, and have a good safety record. To assess the safety of German-type meat products heated in hermetically sealed containers, Lücke and Hechelmann (1986) studied the microbiological stability of various commercial and experimentally inoculated products during prolonged storage without refrigeration. Some protection values (log 1/P, experimentally determined as described by Hauschild, 1982) are given in Tables 5 and 6. Similar values were obtained by determining the time at 100°C necessary to prevent toxin formation in 50% of the containers and converting the thermal death times to apparent decimal reduction times and probabilities of spore survival and development. This indicated that the assumption of logarithmic spore destruction and a z-value of 10°C was realistic. From measurements of core temperatures and z = 10, it could be estimated that the cooking protocol (2 hours in a boiling waterbath) commonly used by small processors for 200-g cans (73 mm diameter) indeed gives F_o values of 0.3–0.4.

Table 5 shows that common formulations and processes for German-type cured meat products cooked in hermetically sealed containers gave protection

Table 5 Protection from *C. botulinum* of German-Type Canned Sausage Mixtures Heated to $F_o = 0.34$ and Stored for 90 Days

Product	pH	% Brine[a]	a_w	Storage temp. (°C)	Protection[b] from *C. botulinum* toxin formation with indicated amount of $NaNO_2$ added: 0 mg/kg	83 mg/kg
Brühwurst[c]	6.0	2.4	0.982	21	4.0	6.9
Brühwurst	6.0	2.8	0.979	21	<3.9	7.1
Brühwurst	6.1	3.0	0.977	21	n.d.	5.8
Brühwurst	5.9[d]	3.0	0.977	21	n.d.	6.3
Brühwurst	5.6[d]	3.0	0.977	21	n.d.	7.0
Brühwurst	6.15	3.8	0.972	21	6.1	>6.8
Brühwurst	6.15	3.8	0.972	30	5.1	6.6
Leberwurst[e]	6.3	2.6	0.980	21	2.3	2.5
Leberwurst	6.15	3.9	0.971	21	4.2	n.d.
Leberwurst	6.2	4.0	0.970	21	3.6	3.5
Leberwurst	6.15	4.1	0.969	21	6.3	>6.3
Leberwurst	6.1	4.5	0.966	21	>5.7	>5.7
Leberwurst	6.1	4.5	0.966	30	3.8	5.0

[a]Calculated from a_w according to Krispien et al., 1978.
[b]Log (1/P), where P is the probability for a given spore to both survive and develop in the product.
[c]Sausage mixtures made from raw meat, pork back fat, and ice.
[d]pH adjusted with gluconic acid (added as glucono-delta-lactone).
[e]Liver sausage mixtures, made from precooked pork and 25% liver.
Source: Lücke and Hechelmann, 1986.

values between 2 and 7. At a constant F_o value and a_w, liver sausage was up to 4 log units less protected than Brühwurst. This difference was due partly to a slower rate of heat inactivation and injury in the liver sausage and partly to the reduced effectiveness of nitrite when severely heated in the presence of liver (Lücke, 1985). Addition of about 80 mg sodium nitrite/kg, however, was effective in Brühwurst, particularly in products of a high a_w, where protection was improved by about 3 units.

Table 5 also shows that even a moderate reduction of a_w improved protection, particularly in liver sausage and nitrite-free Brühwurst. To protect the product from 10^8 spores, the a_w of nitrite-containing Brühwurst mixtures had to be adjusted to below 0.97, that of liver sausage to below 0.96, irrespective of nitrite addition. From the thermal resistance of the spores in the products, it can be estimated that to be as well protected, products with higher a_w values must be

Table 6 Protection of Canned Luncheon Meat and Bologna-Type Sausage Mixtures (*Brühwurst*) from *C. botulinum*

			Protection from *C. botulinum* during extended unrefrigerated storage[b] with indicated amount of $NaNO_2$ added			
F_o	% brine	a_w	0 mg/kg	75–85 mg/kg	150 mg/kg	Ref.
0.34	2.4	0.982	4.0	6.9	n.d.	Lücke and Hechelmann, 1986[c]
0.34	3.8	0.972	6.1	>6.8	n.d.	Lücke and Hechelmann, 1986[c]
0.64	3.6	0.973	n.d.	7.6	7.6	Pivnick et al., 1969[d]
0.64	4.1	0.969	n.d.	7.5	7.8	Pivnick et al., 1969[d]
0.64	4.6	0.966	n.d.	7.6	7.8	Pivnick et al., 1969[d]

[a]Measured or calculated from percent brine according to Krispien et al., 1978.
[b]Log (1/P), where P is the probability for a given spore to both survive and develop in the product.
[c]Data obtained with experimentally inoculated canned bologna-type sausage mixtures heated in boiling water.
[d]Recalculated by Hauschild (1982); data obtained with experimentally inoculated canned luncheon meat.

processed to higher F_o values, which may exceed 1.5 for liver sausage mixtures with a_w values of 0.98.

Adjusting the pH to below 6.0 with gluconic acid (Table 5) improved the protection of Brühwurst mixtures only slightly, while increasing jelly separation considerably. However, the pH of the mixtures should not exceed 6.5, because the antimicrobial effect of nitrite becomes very small at elevated pH values. Blood-containing products with pH above 6.5 and a high content of nitrite-inactivating haemoglobin iron provide almost ideal growth conditions for clostridia and are safe only if they received a botulinum cook or if their a_w was adjusted to the lower growth limit of proteolytic *C. botulinum* strains ($a_w < 0.95$).

Addition of up to 0.3% sodium hypophosphite had no effect on the safety of liver sausage mixtures cooked to $F_o = 0.34$ (Lücke, 1983).

The risk of toxin formation was greater at 30°C than at 21°C. If corrections are made for differences in the heat process and the incubation temperature, protection of Brühwurst was similar to that of luncheon meat of comparable formulation (Table 6) (Hauschild, 1982; Pivnick et al., 1969).

Lücke and Hechelmann (1986) also carried out a survey of the safety of canned sausage mixtures and autoclaved sausages on the West German market.

Only 62% of 310 containers showed neither microbial nor sensoric changes during 28 days at 25°C; 6% of them were spoiled by clostridia. This was hardly surprising because manufacturers recommended chilled storage of most canned sausage mixtures and indicated a shelf life of autoclaved sausages of only 6–8 weeks without refrigeration. The majority of the products investigated were lightly salted and cooked only to F_o 0.2–0.5. To remain stable, all Brühwurst samples had to be prepared with 70–90 mg of sodium nitrite, adjusted to an a_w below 0.97, and cooked to $F_o > 0.4$, while nitrite-free Brühwurst, as well as liver sausage and blood sausage mixtures, had to have a_w values below 0.96 to be safe. However, the majority of the products investigated did not meet these requirements and are often stored with little or no refrigeration. Indeed, canned liver sausage mixtures, pâté, and blood sausage have been incriminated in some outbreaks of botulism in various countries (see Tompkin, 1980), while products of the luncheon meat/Brühwurst type have a much better record of safety.

Table 7 compares the recommendations given by Hauschild and Simonsen (1985) and by Lücke and Hechelmann (1986) for the manufacture of SSCCMs (protected from at least 10^8 spores of *C. botulinum*). The fact that botulism due to SSCCMs is very rare, in spite of the poor protection of many products, supports Riemann's (1963) and Hauschild and Simonsen's (1985) view that SSCCMs can be considered safe if they are protected from 10^8 or more spores of *C. botulinum*. It also indicates that the recommendations of Lücke and Hechelmann (1986) are not too lenient. In particular, there is no evidence from the literature that the protection of SSCCMs is significantly reduced by lowering the nitrite addition to 75 mg/kg.

Table 7 Recommendations for the Manufacture of Shelf-Stable Meat Products Heated in Hermetically Sealed Containers

	For luncheon meats and chopped meats[a]		For mixtures of[b]	
Safety factor	Alternative 1	Alternative 2	Bologna-type sausage	Liver sausage, blood sausage, low-nitrite sausages
F_o	0.5	1.0	>0.4	<0.4
a_w	≤0.963	≤0.97	<0.97	<0.96
% brine	≥5	≥4	>4	>5.4
pH	not specified		<6.5	<6.5
$NaNO_2$ added (mg/kg)	150	150	80–100	0–100

[a]From Hauschild and Simonsen, 1985.
[b]From Lücke and Hechelmann, 1986.

E. Control of *C. botulinum* in Shelf-Stable Pasteurized Dried or Acidified Meat Products

If meat products are cooked to center temperatures in the 70–90°C range only, spores of proteolytic *C. botulinum* strains are not destroyed or injured to any extent, and the safety of the products at ambient temperature relies entirely on inhibition of the spores. In genuine Italian mortadella and in German *Brühdauerwurst*, this is empirically achieved by lowering the a_w by formulation or by a suitable drying process. As expected from experiments in broth, the a_w must be below 0.95 to provide protection from *C. botulinum* (Table 8).

Some meat products (brawns) are protected during storage at ambient temperature if the pH is adjusted to 5.0 or below by addition of vinegar jelly (Leistner, 1985). Results of Blanche-Koelensmid and van Rhee (1968) indicated that a mildly cooked meat product (*Gelderse Rookworst*) may be made shelf stable by lowering the pH to 5.4 in conjunction with an a_w below 0.97. However, this necessitates the addition of plenty of water-retaining additives and a second pasteurization in the vacuum pack to avoid spoilage by vegetative contaminants.

F. Control in Bacon

Pork is converted to bacon in many countries, but the processes and products differ considerably. In the United Kingdom, bacon is made from pork from the back, the middle, or the shoulder. In North America, it tends to be made from the belly, and therefore contains a higher proportion of fat. The process varies from the traditional *dry cure* in which the curing salts are rubbed on the surface of the pork, to the Wiltshire process in which the back is pumped with brine, then immersed in a cover brine in a cool cellar for up to several weeks, to mod-

Table 8 Growth of *C. botulinum* (Proteolytic Strains) in Mortadella-Type Sausage[a] at 25°C

Inoculum (spores/g)	a_w	Days of storage until evidence of clostridial growth
100	0.977	15
100	0.956	42
100	0.950	>150[b]
5000	0.977	11
5000	0.956	36
5000	0.950	>150[b]

[a]50 mg $NaNO_2$ added/kg; heated to 75°C in the center.
[b]No toxic sausages found.
Source: Leistner et al., 1980.

ern rapid processes where sliced pork is cured, sometimes in vacuum packs. Hence, although the product is called bacon, it may vary considerably in its properties.

Historically, botulism has not been associated with the consumption of bacon (Tompkin, 1980), even when vacuum packed and inevitably, from time to time, subjected to temperature abuse. This is attributable mainly to the low water activity of bacon and to the frying process, which would likely destroy any botulinum toxin. As well, the proteolytic strains, the most commonly occurring types of *C. botulinum*, are unable to grow below 10°C, and the proliferation of lactic acid bacteria in the predominantly vacuum-packed product lowers the pH sufficiently to inhibit clostridia. If bacon is made from pork of high pH, as would occur if shoulder pork is used, growth of clostridia can occur if the product is temperature abused (Roberts and Smart, 1976).

Attempts to find acceptable alternatives to nitrite or substances that would enable levels of nitrite to be reduced include the use of potassium sorbate, thoroughly reviewed by Sofos and Busta (1980), irradiation (Rowley et al., 1983), and a wide variety of esters of *p*-hydroxybenzoic acid, phenolic antioxidants, esters of fumaric acid, fatty acids and esters, and a variety of other compounds summarized in Roberts and Gibson (1986). Tanaka et al. (1985) were able to reduce the level of nitrite in bacon and yet maintain the same degree of protection against *C. botulinum* by adding lactic acid bacteria (*Lactobacillus plantarum* or *Pediococcus acidilactici*) and a fermentable carbohydrate. The safety of this process, the Wisconsin process, is due to an early decrease in pH.

V. THE ROLE OF NITRITE IN PREVENTION OF MEATBORNE BOTULISM

After the carcinogenic effects of nitrosamines became known in the late 1960s, more and more consumers and regulatory bodies questioned the necessity of treating meats with nitrite. This greatly stimulated research into the beneficial effect of nitrite in meats. A complete ban of nitrite could have serious consequences to the manufacturers of meat products because their products would look different (no curing color), taste different (no curing aroma), and age much more rapidly (no antioxidative effect). The meat industry and meat researchers realized that the strongest argument in favor of using nitrite is the inhibition of foodborne pathogens. This greatly stimulated research on the behavior of *C. botulinum* in meats. As summarized by Hauschild (1982, 1989) and in the present article, nitrite contributes to the protection of many meat products from growth and toxin formation by *C. botulinum*. The effect, however, is greatly influenced by pH, type, and concentration of reducing agents and iron. The availability of iron forms capable of binding and inactivating nitrite is in turn influenced by the type of animal tissue processed, the heat process, and the presence of iron-sequestrating agents. Hence, it can be stated that nitrite is not a very reliable hurdle against *C. botulinum*, and one should not assume that the use of nitrite can compensate for poor processing conditions. For example, it is doubtful that the residual nitrite in SSCCMs would be able to sufficiently inhibit clostridial spores that entered the processed container through leaks.

Table 9 attempts to summarize the contribution of nitrite to the protection of various meat products from the formation of botulinum toxin. It is evident that nitrite use has little if any protective effect in the meats involved in clinical cases of botulism. Improperly cured large hams become toxic in the interior before sufficient NaCl has penetrated, and in blood-containing sausages, nitrite has been almost completely transformed to compounds with no antimicrobial activity (nitrate, nitroso-hemochromogen). A comparable situation prevails in liver sausage (Hauschild et al., 1982; Lücke, 1985) where the heat process appears to liberate nitrite-binding iron from iron-storage proteins. In fermented sausages, *C. botulinum* is controlled by other factors. In most meat products, omission of

Table 9 Effect of Nitrite in Meat Products on Possible Botulism

Product	Heat treatment	Unrefrigerated storage	Effectiveness of added nitrite[a]	Recent cases of botulism
Ham, injection cured	mild	rare	marked	–
Ham, injection cured, canned	high	frequent	marked	–
Ham, bacon, injection cured	none	rare	marked	–
Ham, cured without brine injection	none	frequent	low or absent	+
Bologna-type sausages, luncheon meat, frankfurters	mild	rare	marked	–
Bologna-type sausages, luncheon meat, frankfurters	high	frequent	marked	–
Liver sausage	mild	rare	low	(+)
Liver sausage, canned	high	frequent	absent	+
Blood sausage	mild	rare	absent	+
Blood sausage, canned	high	frequent	absent	+
Raw sausage, soft	none	rare	[b]	–
Raw sausage, dry	none	frequent	[b]	–

[a]As determined in challenge studies.
[b]No growth of *C. botulinum*, irrespective of nitrite addition.

nitrite would diminish their protection by one or two log cycles (Hauschild, 1982). However, such products would still be better protected than certain meat products already on the market. Vacuum-packed bacon, for example, is well protected even if no nitrite is added (Hauschild, 1982), and the recontaminant spoilage microorganisms, refrigerated storage, and frying before eating would further diminish the risk. The wealth of investigations into the effect of nitrite and nitrite-replacing chemicals on the fate of *C. botulinum* in bacon can only be explained by remembering the formation of nitrosamines is a true (albeit controllable) hazard when that bacon is fried, and the pressure on the meat industry to stop using nitrite for bacon curing was particularly high.

In the early 1980s there was a trend to reduce the input of nitrite and to eliminate nitrate from many curing mixtures. In Norway, for example, only cured meats pasteurized in impervious containers may receive up to 150 mg of sodium nitrite/kg because they are considered products at risk with respect to botulism. The use of nitrite in the manufacture of pasteurized sliced meat products is limited to 60 mg/kg of sodium nitrite "as a colour fixative," while nitrite is not permitted in other meat products. Various other countries (e.g., Sweden, Denmark) have adopted the German regulations insofar as they only permit the use of nitrite as a defined mixture with common salt (nitrite curing salt, containing up to 0.6% sodium nitrite), a practice that very effectively eliminates the risk of adding excess nitrite.

In 1981, West Germany lowered the nitrite content in the nitrite curing salt from 0.5–0.6 to 0.4–0.5%, and thereby also the addition of nitrite to meats to generally 70–100 mg $NaNO_2$/kg (for fermented sausages: up to 150 mg/kg). This did not interfere with quality, microbiological stability, or safety of meat products. Since 1981, there was actually a decrease rather than an increase in outbreaks of botulism. However, any further reduction of the ingoing amount of nitrite requires caution and may have to be compensated for by other inhibitory factors such as lower a_w and/or a more intensive heating process.

REFERENCES

Anusz, Z. (1975). Epidemiologia zatruc toksyną botulinową w polsce. *Przeglad Epidemiol. 29*: 393.

Anusz, Z. (1980). Zatruc toksyną botulinową. *Przeglad Epidemiol. 34*: 87.

Baird-Parker, A. C., and Freame, B. (1967). Combined effect of water activity, pH and temperature on the growth of *Clostridium botulinum* from spore and vegetative cell inocula. *J. Appl. Bacteriol. 30*: 420.

Benedict, R. C. (1980). Biochemical basis for nitrite-inhibition of *Clostridium botulinum* in cured meat. *J. Food Prot. 43*: 877.

Berndt, S. F. (1978). Botulismus. *Med. Klin. 73*: 879.

Binkerd, E. F., and Kolari, O. E. (1975). The history and use of nitrate and nitrite in the curing of meat. *Food Cosmet. Toxicol. 13*: 655.

Blanche-Koelensmid, W. A. A., and van Rhee, R. (1968). Intrinsic factors in meat products counteracting botulinogenic conditions. *Antonie van Leeuwenhoek J. Microbiol. Serol. 34*: 287.

Buckley, J., Morrissey, P. A., and Daley, M. (1976). Aseptic techniques for obtaining sterile beef tissue. *J. Food Technol. 11*: 427.

Buhl, H., and Hertel, V. (1971). Botulismus—Bericht über neun Fälle. *Med. Welt. 22*: 2033.

Cassens, R. G., Greaser, M. L., Ito, T., and Lee, M. (1979). Reactions of nitrite in meat. *Food Technol. 33*(7): 46.

Castellani, A. G., and Niven, C. F. (1955). Factors affecting the bacteriostatic action of sodium nitrite. *Appl. Microbiol. 3*: 154.

Cerveny, J. G. (1980). Effects of changes in the production and marketing of cured meats on the risk of botulism. *Food Technol. 34*(5): 240.

Christiansen, L. N., Johnston, R. W., Kautter, D. A., Howard, J. W., and Aunan, W. J. (1973). Effect of nitrite and nitrate on toxin production by *Clostridium botulinum* and on nitrosamine formation in perishable canned comminuted cured meat. *Appl. Microbiol 25*: 357.

Christiansen, L. N., Tompkin, R. B., Shaparis, A. B., Johnston, R. W., and Kautter, D. A. (1975). Effect of sodium nitrite and nitrate on *Clostridium botulinum* growth and toxin production in a summer style sausage. *J. Food Sci. 40*(3): 488.

Christiansen, L. N., Tompkin, R. B., and Shaparis, A. B. (1978). Fate of *Clostridium botulinum* in perishable canned cured meat at abuse temperature. *J. Food Prot. 41*: 354.

Dainty, R. H. (1982). Biochemistry of undesirable effects attributed to microbial growth in proteinaceous foods stored at chill temperatures. *Food Chem. 9*: 103.

Dainty, R. H. (1985). Bacterial growth in food,—a nutrient-rich environment. In *Bacteria in Their Natural Environments*, Society for General Microbiology Special Publication No. 16 (Fletcher, M. M. and Floodgate, G. D., eds.) Cambridge University Press, p. 171.

Dainty, R. H., Shaw, B. G., and Roberts, T. A. (1983). Microbial and chemical changes in chill-stored red meats. In *Food Microbiology: Advances and Prospects* (T. A. Roberts and F. A. Skinner, eds.), Society for Applied Bacteriology Symposium Series No. 11, Academic Press, London, p. 151.

Delényi, M., Csiba, A., and Koritsansky, Z. (1981). Behaviour of *Clostridium botulinum* type B in nitrate vs. nitrite cured raw ham. *Proceedings, 27. European Congress of Meat Research Workers*, Vienna, p. 645.

Duncan, C. L., and Foster, E. M. (1968). Role of curing agents in the preservation of shelf stable canned meat products. *Appl. Microbiol. 16*: 401.

Esty, J. R., and Meyer, K. F. (1922). The heat resistance of spores of *B. botulinus* and allied anaerobes. *J. Infect. Dis. 31*: 650.

Gill, C. O. (1979). Intrinsic bacteria in meat—a review. *J. Appl. Bacteriol. 47*: 367.

Gill, C. O., and Newton, K. G. (1980). Growth of bacteria on meat at room temperatures. *J. Appl. Bacteriol. 49*: 315.

Gill, C. O., Penney, N., and Nottingham, P. M. (1978). Tissue sterility in uneviscerated carcasses. *Appl. Environ. Microbiol. 36*: 356.

Gill, C. O., Penney, N., and Wauters, A. M. (1981). Survival of clostridial spores in animal tissues. *Appl. Environ. Microbiol. 41*: 90.

Hannan, R. S. (1981). Use of nitrite as a preservative in British meat products. *Proc. Inst. Food Sci. Technol. 14*: 8.

Hauschild, A. H. W. (1982). Assessment of botulism hazards from cured meat products. *Food Technol. 36*(12): 95.

Hauschild, A. H. W. (1989). *Clostridium botulinum*. In *Foodborne Bacterial Pathogens* (M. P. Doyle, ed.), Marcel Dekker, Inc., New York, p. 111.

Hauschild, A. H. W., and Simonsen, B. (1985). Safety of shelf-stable canned cured meats. *J. Food Prot. 48*: 997.

Hauschild, A. H. W., Hilsheimer, R., Jarvis, G., and Raymond, D. P. (1982). Contribution of nitrite to the control of *Clostridium botulinum* in liver sausage. *J. Food Prot. 45*: 500.

Hauschild, A. H. W., Poste, L. M., and Hilsheimer, R. (1985). Toxin production by *Clostridium botulinum* and organoleptic changes in vacuum-packaged raw beef. *J. Food Prot. 48*: 712.

Incze, K., and Delényi, M. (1979). Influence of additives and ripening parameters on growth and toxin production of *Clostridium botulinum*. *Proceedings, 25th European Congress of Meat Research Workers*, Budapest, p. 879.

Ingram, M., and Roberts, T. A. (1971). Application of the 'D-concept' to heat treatments involving curing salts. *J. Food Technol. 6*: 21.

Jarvis, B., Rhodes, A. C., King, S. E., and Patel, M. (1976). Sensitization of heat-damaged spores of *Clostridium botulinum* type B to sodium chloride and sodium nitrite. *J. Food Technol. 11*: 41.

Jones, B., Nilsson, T., and Sörqvist, S. (1984). Contamination of pig carcasses with scalding water. *Fleischwirtsch. 64*: 857.

Krispien, K., Rödel, W., and Leistner, L. (1978). Vorschlag zur Berechnung der Wasseraktivität (aw-Wert) von Fleischerzeugnissen aus dem Gehalt von Wasser und Kochsalz. *Fleischwirtsch. 59*: 1173.

Krol, B., and Tinbergen, B. J. (1974). *Proceedings of the International Symposium on Nitrite in Meat Products*, PUDOC. Wageningen, The Netherlands.

Kueper, T. V., and Trelease, R. D. (1974). Variables affecting botulinum toxin development and nitrosamine formation in fermented sausage. *Proceedings, Meat Industry Research Conference*, American Meat Institute Foundation, Washington, DC, p. 69.

Lawrie, R. A. (1985). *Meat Science*, 4th ed., Pergamon Press, Oxford.

Lechovich, R. V., Graham, P. P., Pierson, M. D., Kelly, R. F., and Balcock, J. D. (1977). The effect of curing methods upon the quality and stability of country hams. Report to Nitrite Safety Council, Washington, DC.

Lecour, H., Ramos, M. H., Almeida, B., and Barbosa, R. (1988). Food-borne botulism—a review of 13 outbreaks. *Arch. Int. Med. 148*: 578.

Leistner, L. (1985). Hurdle technology applied to meat products of the shelf stable product and intermediate moisture food types. In *Properties of Water in Foods* (D. Simatos and J. L. Multon, eds.), Nijhoff, Dordrecht, p. 309.

Leistner, L. (1987). Shelf stable products and intermediate moisture food based on meat. In *Water Activity: Theory and Applications to Food* (L. B. Rockland and L. R. Beuchat, eds.), Marcel Dekker, Inc., New York, p. 295.

Leistner, L., Vuković, I., and Dresel, J. (1980). SSP: Meat products with minimal nitrite addition, storable without refrigeration. *Proceedings, 26th European Meeting Meat Research Workers*, Colorado Springs, p. 230.

Lücke, F.-K. (1983). *Wirkung von Hypophosphit auf Clostridien in Fleischerzeugnissen.* Jahresbericht, Bundesanstalt für Fleischforschung, Kulmbach, C21.

Lücke, F.-K. (1985). Heat inactivation and injury of *Clostridium botulinum* spores in sausage mixtures. In *Fundamental and Applied Aspects of Bacterial Spores* (G. J. Dring, G. W. Gould, and D. W. Ellar, eds.), Academic Press, London, p. 409.

Lücke, F.-K., and Hechelmann, H. G. (1986). Assessment of botulism hazards from German-type shelf-stable pasteurized meat products. *Proceedings of the 2nd World Congress Foodborne Infections and Intoxications*, 26–30 May, 1986, West Berlin, Vol. I, p. 578.

Lücke, F.-K., Hechelmann, H., and Leistner, L. (1983). Fate of *Clostridium botulinum* in fermented sausages processed with or without nitrite. *Proceedings of the XXIX European Meeting of Meat Research Workers*, Parma, I C/2.3, p. 403.

Lücke, F.-K., Hechelmann, H., and Leistner, L. (1982). Botulismus nach Verzehr von Rohschinken—Experimentelle Untersuchungen. *Fleischwirtsch. 62*: 203.

Lücke, F.-K., Hechelmann, H., and Leistner, L. (1981). The relevance to meat products of psychrotrophic strains of *Clostridium botulinum*. In *Psychrotrophic Microorganisms in Spoilage and Pathogenicity*, (T. A. Roberts, G. Hobbs, J. H. B. Christian, and N. Skovgaard, eds.), Academic Press, New York, p. 491.

Meyer, K. F. (1928). Botulismus. In *Handbuch der pathogenen Mikroorganismen*, 3rd ed., Vol. IV (W. Kolle, R. Kraus, and P. Uhlenhuth, eds.), Fischer + Urban and Schwarzenberg, Berlin/Jena, p. 1269.

Meyer, K. F., and Gunnison, J. B. (1929). European strains of *Clostridium botulinum*. *J. Infect. Dis. 45*: 96.

Mierzejewski, J., and Palec, W. (1981). Organoleptic changes and toxin production in canned meats infected artificially with *Clostridium botulinum* B. *Proceedings, 27. European Congress of Meat Research Workers*, Vienna, p. 655.

Narayan, K. G. (1966). Studies on the clostridia incidence in the beef cattle. *Acta Vet. Acad. Sci. Hung. 16*: 355.

NAS/NRC (National Academic of Sciences/National Research Council). (1981). The health effects of nitrate, nitrite and N-nitroso compounds. Part 1 of a 2-part study by the Committee on Nitrite and Alternative Curing Agents in Foods. Assembly of Life Sciences. National Academy Press, Washington, DC.

NAS/NRC (National Academy of Sciences/National Research Council). (1982). The alternatives to the current use of nitrite in foods. Part 2 of a 2-part study by the Committee on Nitrite and Alternative Curing Agents in Foods. Assembly of Life Sciences, National Academy Press, Washington, DC.

Nordal, J., and Gudding, R. (1975). The inhibition of *Clostridium botulinum* type B and E in salami sausage. *Acta Veterinaria Scandinavica 16*(4): 537.

Nowicki, L. (1976). Effect of post-transport fatigue of slaughter pigs on bacterial contamination of carcass and internal organs. *Med. Weterynaryjna 32*: 229.

Ockerman, H. W., Cahill, V. R., Weiser, H. H., Davies, C. E., and Siefker, J. R. (1969). Comparison of sterile and inoculated beef tissue. *J. Food Sci. 34*: 93.

Odlaug, T. E., and Pflug, I. J. (1977). Thermal destruction of *Clostridium botulinum* spores suspended in tomato juice in aluminum thermal death time tubes. *Appl. Environ. Microbiol. 34*: 23.

Odlaug, T. E., and Pflug, I. J. (1978). *Clostridium botulinum* and acid foods. *J. Food Prot. 41*: 566.

Overview. (1980). An assessment of nitrite for the prevention of botulism. *Food Technol. 34*(5): 228.

Overview. (1982). Current problems in botulism of interest to the food industry. *Food Technol. 36*(12): 87.

Overview. (1983). Bacterial spore resistance in food systems. *Food Technol. 37*(11): 87.

Pierson, M. D., and Smoot, L. A. (1982). Nitrite, nitrite alternatives, and the control of *Clostridium botulinum* in cured meats. *CRC Crit. Rev. Food Sci. Nutr. 17*: 141.

Pivnick, H., and Bird. H. (1965). Toxinogenesis by *Clostridium botulinum* types A and E in perishable cooked meats vacuum-packed in plastic pouches. *Food Technol. 19*(5): 132.

Pivnick, H., and Thacker, C. (1970). Effect of sodium chloride and pH on initiation of growth by heat-damaged spores of *Clostridium botulinum. Can. Inst. Food Technol. J. 3*(2): 70.

Pivnick, H., Barnett, H. W., Nordin, H. R., and Rubin, L. J. (1969). Factors affecting the safety of canned, cured, shelf-stable luncheon meat inoculated with *Clostridium botulinum. Can. Inst. Food Technol. J.* 2(3): 141.

Pusztai, S. (1970). Enterococcus occurrence in organs and muscles of slaughtery swine. *Act. Vet. Acad. Sci. Hung. 20*: 391.

Ralovich, B., and Barna, K. (1966). Durch toxikologische und bakteriologische Untersuchungen erwiesener Typ B-Botulismus. *Zbl. Bakt. I. Orig. B 200*: 509.

Rhodes, A. C., and Jarvis, B. (1976). A pork slurry system for studying inhibition of *Clostridium botulinum* by curing salts. *J. Food Technol. 11*: 13.

Riemann, H. (1963). Safe heat processing of canned cured meats with regard to bacterial spores. *Food Technol. 17*(1): 39.

Riemann, H., Lee, W. H., and Genigeorgis, C. (1972). Control of *Clostridium botulinum* and *Staphylococcus aureus* in semi-preserved meat products. *J. Milk Food Technol. 35*: 514.

Roberts, T. A., and Dainty, R. H. (1991). Nitrite and nitrate as food additives; Rationale and mode of action. In *Nitrate and Nitrite in Food and Water* (M. J. Hill, ed.), Ellis Horwood, London, p. 113.

Roberts, T. A., and Gibson, A. M. (1986). Chemical methods for controlling *Clostridium botulinum* in processed meats. *Food Technol. 40*: 163.

Roberts, T. A., and Hitchens, A. D. (1969). Resistance of spores. In *The Bacterial Spore* (G. W. Gould and A. Hurst, eds.), Academic Press, New York, p. 611.

Roberts, T. A., and Ingram, M. (1966). The effects of NaCl, KNO_3 and $NaNO_2$ on recovery of heated bacterial spores. *J. Food Technol. 1*: 147.

Roberts, T. A., and Ingram, M. (1973). Inhibition of growth of *Clostridium botulinum* at different pH values by sodium chloride and sodium nitrite. *J. Food Technol. 8*: 467.

Roberts, T. A., and Mead, G. C. (1985). Involvement of intestinal anaerobes in the spoilage of red meats, poultry and fish. In *Anaerobic Bacteria in Habitats Other than*

Man (E. M. Barnes and G. C. Mead, eds.), Society for Applied Bacteriology Symposium Series No. 13, Academic Press, London, p. 333.

Roberts, T. A., and Smart, J. L. (1974). Inhibition of spores of *Clostridium* spp. by sodium nitrite. *J. Appl. Bacteriol. 37*: 261.

Roberts, T. A., and Smart, J. L. (1976). The occurrence and growth of *Clostridium* spp. in vacuum-packed bacon with particular reference to *Clostridium perfringens* (*welchii*) and *Clostridium botulinum*. *J. Food Technol. 11*: 229.

Roberts, T. A., Gibson, A. M., and Robinson, A. (1981a). Factors controlling the growth of *Clostridium botulinum* types A and B in pasteurized, cured meats. I. Growth in pork slurries prepared from ''low'' pH meat (pH range 5.5–6.3). *J. Food Technol. 16*: 239.

Roberts, T. A., Gibson, A. M., and Robinson, A. (1981b). Factors controlling the growth of *Clostridium botulinum* types A and B in pasteurized, cured meats. II. Growth in pork slurries prepared from ''high'' pH meat (pH range 6.3–6.8). *J. Food Technol. 16*: 267.

Roberts, T. A., Gibson, A. M., and Robinson, A. (1981c). Prediction of toxin production by *Clostridium botulinum* in pasteurized pork slurry. *J. Food Technol. 16*: 337.

Roberts, T. A., Gibson, A. M., and Robinson, A. (1982). Factors controlling the growth of *Clostridium botulinum* types A and B in pasteurized, cured meats. III. The effect of potassium sorbate. *J. Food Technol. 17*: 307.

Roberts, T. A., Woods, L. F. J., Payne, M. J., and Cammack, R. (1991). Nitrite. *Food Preservatives* (G. W. Gould and N. J. Russell, eds.), Blackie, London, p. 89.

Robinson, A., Gibson, A. M., and Roberts, T. A. (1982). Factors controlling the growth of *Clostridium botulinum* types A and B in pasteurized meats. V. Prediction of toxin production: Non-linear effects of storage temperature and salt concentration. *J. Food Technol. 17*: 727.

Rowley, D. B., Firstenberg-Eden, R., Powers, E. M., Shattuck, G. E., Wasserman, A. E., and Wierbicki, E. (1983). Effect of irradiation on the inhibition of *Clostridium botulinum* toxin production and the microbial flora in bacon. *J. Food Sci. 48*: 1016.

Schulze, K., and Funke, R. (1981). Ein Fall von Botulismus unter dem besonderen Aspekt der Diagnose am Lebensmittel. *Arch. Lebensmittelhyg. 32*: 83.

Scott, V. N., and Bernard, D. T. (1982). Heat resistance of spores of non-proteolytic type B *Clostridium botulinum*. *J. Food Prot. 45*: 909.

Scott, V. N., and Bernard, D. T. (1985). The effect of lysozyme on the apparent heat resistance of nonproteolytic type B *Clostridium botulinum*. *J. Food Safety 7*: 145.

Sebald, M., Billon, J., Cassaigne, R., Rosset, R., and Poumeyrol, G. (1980). Le botulisme en France—incidence, mortalité, aliments responsables avec étude des foyers dûs à un aliment qui n'est pas de préparation familiale. *Méd. Nut. 16*: 262.

Sebald, M., Ionesco, H., and Prévot, A. R. (1972). Germination lzp-dépendantes spores de *Clostridium botulinum* type E. *Comptes Rend. Acad. Sci.*, (Série D) *275*: 2175.

Sebranek, J. G., and Cassens, R. E. (1973). Nitrosamines: A review. *J. Milk Food Technol. 36*: 76.

Simunovic, J., Oblinger, J. L., and Adams, J. P. (1985). Potential for growth of nonproteolytic types of *Clostridium botulinum* in pasteurized restructured meat products: A review. *J. Food Prot. 48*: 265.

Skoczek, A., and Mierzejewski, J. (1979). Izolowanie szczepów *Clostridium botulinum* B z konserw miesnych zbombazowanych. *Med. Weterynaryjna 35*: 736.

Sofos, J. N., and Busta, F. F. (1980). Alternatives to the use of nitrite as an antibotulinal agent. *Food Technol. 34*(5): 244.

Sofos, J. N., and Busta, F. F. (1983). Sorbate. In *Antimicrobials in Foods* (A. L. Branen and P. M. Davidson, eds.), Marcel Dekker, Inc., New York, p. 141.

Sofos, J. N., Busta, F. F., and Allen, C. E. (1979). Botulism control by nitrite and sorbate in cured meats: A review. *J. Food Prot. 42*: 739.

Tanaka, N., Meske, L., Doyle, M. P., Traisman, E., Thayer, D. W., and Johnston, R. W. (1985). Plant trials of bacon made with lactic acid bacteria, sucrose and lower sodium nitrite. *J. Food Prot. 48*: 679.

Tinbergen, B. J., and Krol, B. (1977). *Proceedings of the Second International Symposium on Nitrite in Meat Products*. PUDOC. Wageningen, The Netherlands.

Tompkin, R. B. (1980). Botulism from meat and poultry products—a historical perspective. *Food Technol. 34*: 229.

Tompkin, R. B. (1983). Nitrite. In *Antimicrobials in Foods* (A. L. Branen and P. M. Davidson, eds.), Marcel Dekker, Inc., New York, p. 205.

Tompkin, R. B. (1984). Indirect antimicrobial effect in foods: Phosphates. *J. Food Safety 6*: 13.

Tompkin, R. B., Christiansen, L. N., and Shaparis, A. B. (1978a). Enhancing nitrite inhibition of *Clostridium botulinum* with isoascorbate in perishable canned cured meat. *Appl. Environ. Microbiol. 35*: 59.

Tompkin, R. B., Christiansen, L. N., and Shaparis, A. B. (1978b). Antibotulinal role of isoascorbate in cured meat. *J. Food Sci. 43*: 1368.

Tompkin, R. B., Christiansen, L. N., and Shaparis, A. B. (1978c). Causes of variation in botulinal inhibition in perishable canned cured meat. *Appl. Environ. Microbiol. 35*: 886.

Tompkin, R. B., Christiansen, L. N., and Shaparis, A. B. (1979). Isoascorbate level and botulinal inhibition in perishable canned cured meat. *J. Food Sci. 44*: 1147.

Troeger, K., and Woltersdorf, W. (1987). Mikrobielle Kontamination von Schweineschlachtkörpern durch Brühwasser über das Gefäßsystem. *Fleischwirtsch. 67*: 857.

van Ermengem, E. (1897). Ueber einen neuen anaeroben *Bacillus* und seine Beziehungen zum Botulismus. *Zeitschr. Hyg. Infekt. 26*: 1.

Zeller, M. (1959). Ein Beitrag zur Botulismusdiagnose. *Arch. Lebensmittelhyg. 10*: 265.

8

Control in Fishery Products

Mel W. Eklund

Northwest Fisheries Science Center, Seattle, Washington

I. INTRODUCTION

Clostridium botulinum includes a very heterogeneous group of strains that have been divided into types A through G on the basis of the antigenic specificity of the neurotoxins that are produced. Recent studies have shown that other *Clostridium* species can also produce botulinum neurotoxins (see Chapters 1 and 4). When biochemical, physiological, serological, and DNA homologies are used to characterize the different strains, this species can be separated into: Group I, proteolytic types A, B, and F and subtypes AB, AF, BA, and BF; Group II, nonproteolytic types B, E, and F; Group III, nonproteolytic types C and D; and Group IV, proteolytic type G.

The majority of foodborne botulism outbreaks in humans have been caused by *C. botulinum* types A, B, E, and F. Types C and D are usually involved in animal and avian botulism outbreaks. Type G has not been involved in any botulism outbreaks, but has been isolated from soils in Argentina and from human autopsies in Switzerland.

In order for foodborne botulism to occur, the following conditions must be met: (1) the food must be contaminated with *C. botulinum* spores or vegetative cells from the environment or during further processing; (2) the processing treatment must be inadequate to inactivate *C. botulinum* spores, or the product must be contaminated after processing; (3) the food must support growth and toxin production by *C. botulinum* when storage temperatures exceed 38°F (3.3°C); and

(4) the food must be acceptable and consumed without cooking or after insufficient heating to inactivate the preformed toxin.

Condition 1. Based upon numerous investigations of the geographical distribution of *C. botulinum*, type E is the most prevalent type in most freshwater and marine environments (Hauschild, 1989). However, proteolytic types A, B, and F (Group I) and nonproteolytic types B and F (Group II) have also been isolated from these environments. In some areas, they are the predominant types (Eklund and Poysky, 1967). The incidence of *C. botulinum* in fish and crab has been shown to be closely correlated with the contamination of sediments from the same area (Cann et al., 1975; Eklund and Poysky, 1968; Huss et al., 1974; Sugiyama et al., 1970). In light of this evidence, we do not have any control of the contamination of raw products from nature. We must therefore always assume that, under normal conditions, *C. botulinum* can be present in any raw product.

Conditions 2 and 3. In preparing and preserving food products, we can prevent botulism outbreaks by concentrating on controlling conditions 2 and 3. Group II organisms, the most prevalent on fish, are the most heat-sensitive types, and are inhibited by 5–6% water-phase NaCl. They cannot grow in foods with an a_w below 0.96, and under most conditions they are inhibited at pH 5.0 and lower (Riemann, 1969). Group II organisms have the unique property of being nonproteolytic and growing at temperatures as low as 3.3°C (Eklund et al., 1967a,b; Schmidt et al., 1961). Because of their nonprotoelytic characteristics, their growth in foods cannot be detected by off-odors and off-flavors. The lack of correlation between spoilage and toxicity of food was observed during a type E botulism outbreak involving 16 people in 1963 (Anonymous, 1963). None of the victims detected any off-odors in the implicated food, and only three observed any unusual flavors. The optimum growth temperature for Group I strains is between 35 and 40°C. All strains grow and produce toxin at 15°c, but none of the strains will grow at 10°C or below. They are inhibited in foods below pH 4.6 (Ohye and Scott, 1953; Riemann, 1969). These strains are the most heat resistant and will grow and produce toxin in foods containing up to 8–9% water-phase NaCl. Foods with an a_w below 0.93 are inhibitory to their growth (Riemann, 1969; Smith and Sugiyama, 1988). The growth and proteolytic activity of this group are usually, but not always, accompanied by off-odors.

Condition 4. The acceptability of the food causing botulism depends upon the degree of organoleptic changes caused by spoilage microorganisms and the degree of spoilage that is acceptable to the consumer. This acceptability is also influenced by spices, acidulants, smoke, and other ingredients that can camouflage off-odors and off-flavors. When the insidious nonproteolytic types B, E, and F are involved and the spoilage microorganisms are inactivated or inhibited, the products are often acceptable to the victim, even though they can contain preformed botulinum neurotoxin. Condition (4) therefore cannot be controlled, especially when ready-to-eat products are involved. In addition, it is inadvisable to rely on the consumer to determine whether or not a food product is safe to consume.

II. EXAMPLES OF BOTULISM OUTBREAKS FROM FISHERY PRODUCTS

In order to set the course of our thinking on how to prevent botulism, we need to learn about the causes of previous foodborne botulism outbreaks.

A. Thermal-Processed Products

Botulism outbreaks have occurred from thermally processed fishery products prepared commercially and in the home. Outbreaks from commercially prepared fishery products have been associated with both underprocessing and with the postprocess contamination. Postprocess contamination has been associated with defects in the can or improperly sealed closures which have allowed entry of *C. botulinum*. Home-canned products are usually associated with underprocessing.

B. Fermented Products

Numerous botulism outbreaks have occurred from the consumption of fermented products. Japanese outbreaks have been associated with type E, and the incriminated foods were Izushi and Kirikomi (Nakamura et al., 1956; Nakao and Kodama, 1970; Sakaguchi, 1979). In Canada and Alaska, fermented salmon eggs, seal flippers, whale blubber and meat, and other meat products have been frequently associated with botulism outbreaks (Dolman, 1964; Hauschild, 1989). In Europe, trout and a number of other fish have been involved. In most cases, the products were produced by natural fermentation at ambient temperatures. Type E and occasionally nonproteolytic type B were associated with these outbreaks. In these natural fermentation processes, there is a race between the drop in pH resulting from fermentation of endogenous carbohydrate by the naturally occurring spoilage microorganisms (usually lactic acid bacteria) and the growth of *C. botulinum* and neurotoxin production. The carbohydrate source in fish is often inadequate, and the pH of the product does not reach inhibitory levels. In addition, the fermentation process is relatively slow and depends upon the microorganisms that contaminate the product. Neurotoxin produced under these conditions can be stable for months, and the fermented products are usually consumed without cooking.

C. Salt-Cured, Air-Dried, Uneviscerated Fish

Kapchunka, also marketed under the names Rybetz and Rostov, is prepared by layering raw, uneviscerated fish and salt. After curing for a minimum of 25 days, the fish are rinsed and air-dried at ambient temperatures for 3–7 days. After drying, the fish are marketed under refrigeration and are often consumed uncooked. In 1981, 1985, and 1987, three different botulism outbreaks were caused by this product (Anonymous, 1987; Slater et al., 1989; Telzak et al., 1990). Nonproteolytic type B was involved in the 1981 outbreak and type E in the other two. In the first two outbreaks, the flesh of the fish contained only 1.8 and 3.4% water-phase NaCl, totally inadequate for a product that is exposed to nonrefrigerated temperatures. In the 1987 outbreak, the NaCl level ranged from 18.8 to 24.0%. In spite of the inhibitory NaCl concentration, the product could have become toxic in one of two ways: (1) growth and toxin production occurred during drying at ambient temperatures before the NaCl had time to penetrate, or (2) the fish were dried, found to contain insufficient NaCl, and were rebrined and redried to meet the regulations of New York State. Experiments in the author's laboratory have shown that toxic fish can be brined and dried without affecting the neurotoxin titer (unpublished data). In addition, when spores are inoculated into the fish intestines and the flesh, the toxin titers are always 10–100 times greater in the intestinal area (Eklund et al., 1984; Huss et al., 1979).

D. Smoked Products

The botulism outbreaks from smoked fishery products in the early 1960s were caused by a combination of factors: (1) the products were inadequately processed and, in most cases, the water-phase NaCl concentrations were lower than 1%; (2) smoked products were vacuum packaged; and (3) they were grossly temperature abused during distribution. The vacuum packaging inhibited many of the spoilage bacteria and molds, thus extending the shelf life, but the NaCl concentrations were not inhibitory to *C. botulinum* type E.

E. Relationship of Preservation Procedures to Botulism

Figure 1 summarizes how processing and storage can create potential botulism problems. Botulism outbreaks may occur when we attempt to preserve a food product, prepare new products, extend their shelf life, and in the process inactivate or inhibit the natural spoilage microorganisms, but not *C. botulinum*. On the left of the figure, raw eviscerated fish or fish portions remain in the safe zone. These products have never been implicated in human botulism. Spoilage bacteria play a significant role causing deterioration and spoilage before the products become toxic. These bacteria outnumber *C. botulinum*, grow rapidly at 0–3°C, compete with *C. botulinum* for nutrients, and may be inhibitory to *C. botulinum*. In some cases, proteolytic enzymes can even inactivate preformed botulinum neurotoxin. In addition, raw products are often adequately cooked before consumption, which would inactivate any preformed toxin.

On the right side of Figure 1, sterilized products in which the spores are inactivated by heat, irradiation, or other means can be safely stored at ambient

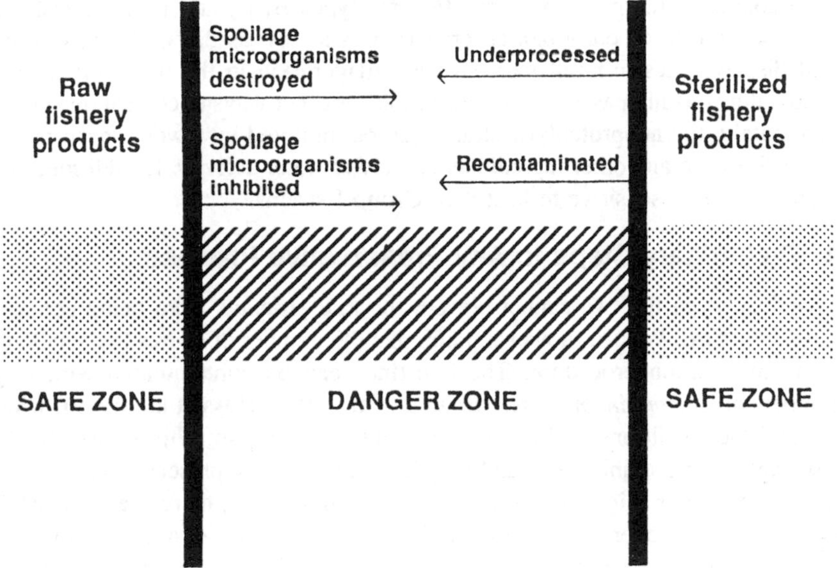

Figure 1 Relationship of preservation procedures to botulism problems in fishery products stored above 3.3°C.

temperatures. This is documented by the fact that the commercial canning industry has had an excellent record of producing over a trillion cans of thermally processed products with only a few outbreaks. The foods enter the danger zone of Figure 1 when they are underprocessed or when *C. botulinum* contaminates the products after processing. The danger zone is also entered when we destroy or inhibit the spoilage microorganisms to extend the shelf life of a product and the preservation procedure is inadequate to inhibit or inactivate *C. botulinum*.

III. CONTROL OF *C. BOTULINUM* IN FISHERY PRODUCTS BY INHIBITION OF GROWTH

Spores of proteolytic types A, B, and F (Group I) are very heat resistant and can grow and produce toxin in food products containing 8–9% water-phase NaCl. If processing procedures had to be based upon these two characteristics, all products would either have to be sterilized, which in many cases would markedly change the desired characteristics of some products, or the concentration of NaCl would be totally unacceptable to most consumers. The proteolytic strains, however, have a limiting characteristic in that they cannot grow at 10°C or below. We can minimize the risk from this group, therefore, by emphasizing refrigeration during distribution and storage.

In comparison, spores of nonproteolytic types B, E, and F can germinate, grow, and produce neurotoxin at temperatures as low as 3.3°C. Since refrigerated display cases in retail markets and refrigeration in the home often reach temperatures as high as 10°C, we must look for other ways to control this group. Fortunately, the nonproteolytic strains can be inhibited by lower concentrations of NaCl, sugar, and other ingredients that lower water activity. In addition, their spores are more sensitive to heat than Group I strains.

A. Preliminary Steps to Any Preservative Process

As indicated earlier, we cannot prevent the raw products from being contaminated in nature. We can, however, minimize the number of spores that challenge each preservation procedure. The intestines can be contaminated with large numbers of *C. botulinum* spores (Eklund et al., 1984; Huss et al., 1979), which under abuse conditions could grow and produce neurotoxin. This toxin could diffuse rapidly into edible flesh and may be unaffected by preservation measures such as salting or acidification. It is extremely important, therefore, that fish be stored in ice or under refrigeration and be eviscerated as soon as possible. The fish should be thoroughly washed before they are filleted and the preservation process is started. As it is unavoidable for raw products to be contaminated, cross-contamination from raw products to the finished product must be prevented and controlled.

B. Shelf Life of Raw Products

Raw fishery products have a very limited shelf life because of the rapid growth of psychrotrophic gram-negative bacteria. Even at 0°C, eviscerated fish and fish fillets usually spoil within 7–10 days of storage. For each 5°C increase in temperature, the shelf life of the product is halved because of the increased growth of the bacteria and their enzymatic activity. Spoilage therefore precedes growth and neurotoxin production by *C. botulinum*. This, in conjunction with cooking, has contributed to the absence of botulism from raw fish. This is further evidenced by the millions of pounds of uncooked fish consumed annually in Asian countries without any report of botulism.

C. Effect of Modified Atmospheres

Since the development of modern refrigeration systems, there have been many attempts to increase the shelf life of fresh fishery products and to extend their distribution from the points of harvesting and processing. The storage of fresh fish in modified atmospheres at refrigerated temperatures has been shown to inhibit the psychrotrophic bacteria, such as *Pseudomonas*, *Moraxella*, *Acinetobacter*, and *Flavobacterium*, and thereby markedly delay the spoilage of the product (Lee, 1982).

The use of modified atmospheres requires that the package or container be evacuated and replaced with mixtures of CO_2, N_2, and O_2. The gaseous mixtures recommended vary with the fishery product. Of the different gases, CO_2 inhibits the growth of spoilage bacteria, molds, and yeasts. N_2 is an inert gas and is used primarily to reduce the concentration of CO_2. O_2 was initially introduced to retard the growth of anaerobic bacteria (Ogrydziak and Brown, 1982) and, in some cases, to retain a desirable product color (Genigeorgis, 1985). Subsequent studies, however, have indicated that the O_2 and modified atmospheres have little effect on growth of *C. botulinum* in raw fishery products (Eklund, 1982). Modified atmosphere storage, therefore, could increase potential botulism problems if products are subjected to temperature abuse or prolonged refrigerated storage. This increased risk is created by the inhibition of the gram-negative spoilage bacteria, extension of shelf life, and absence of obvious organoleptic changes, which are safety signals to the consumer. The gram-positive bacteria, on the other hand, are more tolerant to the modified atmosphere gases and often become predominant. Growth of these gram-positive organisms is relatively slow and is usually accompanied by milder organoleptic changes, which are not as objectionable to the consumer.

To evaluate the potential health hazards from modified atmosphere preservation, numerous studies have been undertaken with different products and various gas mixtures. The main purpose of these studies was to determine whether botulinum toxin would be produced while the product remained organoleptically acceptable. It is not the intent of this chapter to review all of these studies but instead to highlight and summarize some of the results.

Raw Product Stored Above 3.3°C in CO_2

The results of numerous studies vary in their conclusions about whether *C. botulinum* can grow and produce neurotoxin in products stored in modified atmospheres before they spoil and become unacceptable to the consumer (Baker and Genigeorgis, 1990; Cann et al., 1983; Eyles and Warth, 1981; Post et al., 1985). These differences can be attributed to the initial quality of the product, the population of spoilage bacteria at the beginning of the experiment, the size of *C. botulinum* spore inoculum, the fishery product used, storage temperatures, and the procedures used for botulinum toxin assay.

The assay procedure plays an important role in determining whether or not a product becomes toxic before overt spoilage is detected. In some studies, the sample supernatant fluids were assayed in the untrypsinized state or only trypsinized samples were assayed, whereas others used both assay procedures. In some samples, especially after short storage periods or at refrigerated temperatures, the neurotoxins of nonproteolytic *C. botulinum* can be detected only after trypsin treatment. In other cases, the enzymes of other bacteria growing on the product can activate the neurotoxin, and further trypsin treatment may inactivate the toxin (unpublished data). It is extremely important that samples be assayed both in the trypsinized and untrypsinized state. In most cases, the samples can be assayed in the untrypsinized state first and the remainder of the sample refrigerated. Nontoxic samples can be assayed again after treatment with trypsin.

It is generally agreed that storage of fishery products in modified atmospheres at temperatures below 5°C does not present a potential botulism problem. In studies where storage temperatures of 10°C or above were used, the potential danger for development of *C. botulinum* increased.

In the author's laboratory (unpublished data), salmon samples were inoculated with type E spores and stored in the following gaseous atmospheres: (1) 60% CO_2, 25% O_2, and 15% N_2; (2) 90% CO_2 and 10% N_2; or (3) air. When the products were stored for 7 days at 10°C, CO_2 concentrations of 60% and 90% delayed neurotoxin production. Neurotoxin was produced in 100-g samples stored in air with an inoculum of 10^2 spores. In comparison, an inoculum of 10^3 spores was required for neurotoxin production in 60% CO_2 atmospheres, and 10^4 spores per sample were required in 90% CO_2 atmospheres. This inhibitory effect was overcome after 10 days of storage, and toxin was produced in all samples, irrespective of the spore inoculum (Table 1). In addition, O_2 in the 60% CO_2 mixture did not delay toxin production. Overt spoilage was evident in all samples stored in air for 4 days at 10°C, but was delayed until after 10 days when samples were stored in CO_2 atmospheres (Table 1).

When abuse storage temperatures of 25°C were used, samples inoculated with 10^3 spores per 100 g became toxic after 24 hours of storage. Samples inoculated with 10^2 spores were nontoxic after 24 hours but contained neurotoxin after 48 hours. All of these samples required trypsin treatment before neurotoxin could be detected. Salmon samples held in air were considered unacceptable within 24 hours of storage at 25°C, whereas spoilage in CO_2 atmospheres were delayed until after 48 hours.

Table 1 Relationship of Spoilage to Neurotoxin Production by *C. botulinum* Type E in Fresh Salmon in Modified atmospheres for 10 Days at 10°C[a]

Atmosphere[b]	Spoilage[c]	No. samples toxic/No. samples tested
Air	+	6/6
60% CO_2	−	6/6
90% CO_2	−	6/6

[a] 10^2 spores per 100-g sample.
[b] 60% CO_2 mixture = 60% CO_2, 25% O_2, 15% N_2; 90% CO_2 mixture = 90% CO_2 and 10% N_2.
[c] + = spoiled; − = not spoiled.

In the same studies, salmon samples inoculated with 10^3 and 10^4 type A spores per 100 g remained nontoxic over 21 days of storage at 10°C. At 25°C, toxin was not detectable until after 48 hours of storage.

These and other studies (Baker and Genigeorgis, 1990; Cann et al., 1983; Lindsay, 1982; Post et al., 1985; Stier et al., 1981) show that CO_2 does not increase the rate of toxin production by *C. botulinum* type E, but it does inhibit the spoilage microorganisms and delays spoilage at 10 and 25°C. Botulinum type E neurotoxin therefore can be produced in products stored in modified atmospheres before overt spoilage occurs.

In other experiments (unpublished data), the salmon were held on ice for 6 days prior to storage in modified atmospheres. During this period, the bacterial population increased rapidly and released enzymes and metabolites, but the fish were not spoiled. When the fish were inoculated with *C. botulinum* type E and stored at 10°C in CO_2 atmospheres, spoilage occurred before toxin production.

To obtain the full benefit from modified atmospheres and the best product quality, fishery products should be placed in this environment immediately after harvesting and processing (Bell, 1980). These products will have low populations of spoilage bacteria and a longer shelf life. Unfortunately, under these conditions, *C. botulinum* has a greater opportunity to grow and produce toxin before spoilage is evident if the products are stored at temperatures above 5°C. This emphasizes the importance of the initial quality and bacterial population in evaluating the potential hazards of CO_2 atmosphere storage.

Products Stored in CO_2 Atmospheres and Then Transferred to Air

Preservation of fresh fishery products by CO_2 modified atmospheres has a significant advantage over many other preservation procedures in that the gram-negative bacteria are inhibited but not inactivated. When products are removed from these atmospheres, the gram-negative bacteria will again grow and cause spoilage. The rate of growth of these bacteria is influenced by the concentration and exposure time to CO_2. Products therefore can be stored in CO_2 atmospheres when temperature controls are assured and removed from the atmosphere for further distribution. This practice was used in the late 1970s to transport several million pounds of fresh salmon to Seattle from Alaska in CO_2 atmosphere vans, while the temperature was monitored continually. In Seattle, the salmon was removed from CO_2 atmospheres, washed, and distributed fresh or frozen. Some of these fish were used in inoculated pack studies to determine potential public health problems related to *C. botulinum*. Salmon stored in 60% and 90% CO_2 at 2°C for 9 days during transit and then transferred to air and stored at 10°C spoiled before they became toxic (unpublished data). When salmon were grossly abused in air after CO_2 atmosphere storage, they also were spoiled when toxin was first detected.

Similar procedures have been proposed using impermeable master containers, which would hold products packaged in O_2-permeable films. At the distribution point, the packaged product would be removed from CO_2 and marketed in air. This would permit growth of the gram-negative bacteria to resume, and spoilage would be expected before toxin production. In comparison, the distribution of fish in CO_2 in consumer-size packages at the retail level poses a greater hazard from *C. botulinum* (Genigeorgis, 1985; Post et al., 1985).

D. Effect of Vacuum Packaging

When fishery products are stored in packaging films with low gas permeability, a self-generated CO_2 modified atmosphere system develops in a relatively short time. During vacuum packaging, much of the air is removed. The residual O_2 is rapidly decreased by the metabolism of the spoilage bacteria, and CO_2 is generated. The concentration of CO_2 does not approach that of most CO_2 modified atmosphere systems, but sufficient CO_2 accumulates to increase the shelf life of raw fishery products.

At abuse temperatures, botulinum neurotoxin production can occur in vacuum-packaged fishery products before or at the same time as overt spoilage. Increased shelf life will again depend upon the initial bacterial population, the time interval between harvesting, processing, and packaging, and the temperature throughout (unpublished data).

E. Effect of Ionizing Radiation on Shelf Life

The use of low doses of radiation (radurization process) has also been investigated as a process to increase the shelf life of fishery products. This method of preservation differs from CO_2 modified atmospheres in that the spoilage microorganisms are inactivated. With low doses of radiation, over 99% of the microorganisms are inactivated, and the refrigerated shelf life can be increased two or three times (Miyauchi et al., 1964). The main disadvantage of low-dose radiation is that it has little or no effect on the spores of *C. botulinum* (Eklund and Poysky, 1970).

Numerous studies in the 1960s and 1970s (Eklund and Poysky, 1970; Hobbs, 1977; Segner and Schmidt, 1968) showed that *C. botulinum* would not be a problem in radurized fishery products if the products were stored below 5°C. Neurotoxin production, however, did occur before spoilage if the products were irradiated at levels of 2 kGy or higher and stored at temperatures above 7°C. When products were irradiated at lower doses such as 1 kGy, some of the spoilage microorganisms survived the process, and the shelf life was reduced. When these lower doses were used, spoilage occurred before neurotoxin production (Eklund and Poysky, 1967).

The species of fish also has a marked effect on the rate of neurotoxin production by *C. botulinum*. Species of fish such as petrale sole are excellent substrates for type E, whereas others are relatively poor substrates (Eklund and Poysky, 1970; Strasdine and Kelly, 1967). In addition, detectable levels of neurotoxin are produced earlier in irradiated fish than in unirradiated fish. This, in part, can be attributed to the availability of nutrients such as ribose and glucose in fish, which increase after death during autolysis. These carbohydrates are utilized rapidly by the spoilage bacteria in unirradiated products, but are available for longer periods of storage in irradiated fish where the spoilage bacteria are inactivated. Ribose and glucose are readily used by *C. botulinum* types B, E, and F at refrigerated temperatures and increase growth and toxin production (Eklund and Poysky, 1970).

The permeability of packaging materials also has a marked effect on toxin production by type E in some fish species. In O_2-permeable films, neurotoxin was not produced by 5°C in petrale sole fillets, but was produced in similar samples packaged in high-barrier films after 42 days of storage (Eklund and Poysky, 1970).

Results from inoculated pack studies indicate that, in addition to the characteristics of different *C. botulinum* types, the size of spore inoculum, species of fish, packaging material, irradiation dose, and storage temperatures are very important in evaluating the potential public health problems of the radurization process (Eklund, 1982). These and other studies (Eklund and Poysky, 1970; Hobbs, 1977; Hussain et al., 1977; Segner and Schmidt, 1968) have shown that the radurization process increases the potential hazards from *C. botulinum* if irradiation doses over 1 kGy are employed and products are stored above 3.3°C. Doses below 1 kGy appear more favorable because of the rapid spoilage rate of the products. However, the results varied with the initial bacterial population prior to irradiation.

Because of regulatory requirements and organoleptic altering properties, most preservatives cannot be employed to control *C. botulinum* in fresh fish. Storage of the products below 3.3°C must therefore be relied upon to prevent *C. botulinum* growth.

F. Preservation by Salting

The use of NaCl for the preservation of fish is probably the oldest method of preservation and has been used for over 12 centuries in some areas of the world (Cutting, 1955). In the past, the products contained high concentrations of NaCl, and botulism hazards were avoided if the processes were carried out correctly. In recent years, consumers have been concerned about NaCl consumption, and, with increased use of mechanical refrigeration, some products now contain less NaCl.

Botulism outbreaks have occurred from salted fish, salted and dried fish such as Kapchunka, and also from salted fish eggs (Hauschild, 1989). In these cases, the products contained noninhibitory concentrations of NaCl, or neurotoxin production occurred before inhibitory levels of NaCl had penetrated into the intenstines (in cases of uneviscerated fish) or flesh.

To prevent botulism outbreaks, the water-phase NaCl content of the product must be above 10% in order to inhibit the proteolytic types of *C. botulinum* at ambient temperatures. At these higher temperatures, the water-phase NaCl concentrations, however, must be greater than 16% and the a_w less than 0.85 in order to inhibit growth and toxin production by *Staphylococcus aureus*. If the products are refrigerated *S. aureus* and the proteolytic *C. botulinum* types A, B, and F are inhibited, and 5% water-phase NaCl would be inhibitory to the nonproteolytic group.

Salting should be carried out at refrigerated temperatures whenever possible, and sufficient NaCl must be used to prevent growth and toxin production by *C. botulinum*. In addition, fish should be eviscerated to lower the spore load. It is also difficult to obtain inhibitory concentrations of NaCl in the visceral area. If products are to be salted and then dried, the NaCl concentration must be at inhibitory levels before the drying process is started, or the drying process must be completed at a higher temperature over shorter periods of time. For jerky products, 6 to 8 hour drying periods at relatively high temperatures are preferable to a drying period of days at ambient temperature. In the latter case, we are again confronted with a race between *C. botulinum* growth and the removal of moisture until inhibitory NaCl concentrations are reached. In addition, all the samples must be uniformly salted. Some outbreaks have occurred because the products varied widely in NaCl content.

The amount of NaCl required for inhibition is also influenced by pH (Segner et al., 1966). Hauschild and Hilsheimer (1979) surveyed different commercially prepared caviar and found that the product pH varied from 4.6 to 6.8. In studies with lumpfish caviar, a combined effect of NaCl and pH was demonstrated. The following water-phase NaCl and pH combinations were inhibitory to *C. botulinum*: (1) 5.6% NaCl at pH 5.6, (2) 4.7% NaCl at pH 5.4, and (3) 4% NaCl at pH 5.0.

G. Preservation by pH Control

Lowering of the pH in fish can be achieved by fermentation or by pickling. Products from both processes have been involved in botulism outbreaks.

A wide variety of fish have been preserved by fermentation. Most products have low levels of NaCl (Iida, 1970; Nakano and Kodama, 1970) in order to permit growth of resident bacteria. As mentioned earlier, the natural fermentation process is slow and there are usually insufficient carbohydrates available to lower the pH to inhibitory levels (pH ≤ 5.0 for nonproteolytic strains and pH ≤ 4.6 for proteolytic strains). This period, therefore, may be adequate for nonproteolytic types to grow and produce neurotoxin. The safety of these products relies upon addition of sufficient NaCl or fermentation at low temperatures until the required pH is obtained. Both low temperatures and higher concentrations of NaCl, however, can be inhibitory to the growth of lactic acid bacteria, which are responsible for the fermentation process.

The pickling process depends upon NaCl and adequate pickling solution to lower the pH of the product to inhibitory levels. A number of investigations have demonstrated that the interaction of pH and NaCl are important in the inhibition of *C. botulinum* (Baird-Parker and Freame, 1967; Ohye and Christian, 1967; Riemann et al., 1972; Roberts and Ingram, 1973). Pickled herring involved in outbreaks in Canada was shown to have a low pH inhibitory to type E at the time of the outbreak (Dolman, 1961). In this case, either the botulinum toxin was formed prior to brining or the time required for acid to penetrate was too long and type E neurotoxin was produced before the pH dropped to inhibitory levels. In order to prevent botulism, it is important that the process be carried out below 3°C and that a low pH be quickly obtained.

In some acidified products, it is also important to preclude the growth of yeasts and molds. These microorganisms have the ability to grow and utilize organic acids and raise the pH in localized regions of the product. This enables *C. botulinum* to grow and produce toxin in the microenvironments (Odlaug and Pflug, 1978).

Refrigeration of pickled products is advisable because the inhibitory effects of NaCl and pH are more effective at temperatures below the optimum for *C. botulinum* growth. These temperatures also inhibit growth of other microorganisms that could raise the pH.

H. Preservation by Smoking

The preservation of foods by smoking varies considerably in different areas of the world. Even though we have had considerable experience with this process, botulism outbreaks from smoked products are occasionally encountered. Many of the outbreaks have been caused by a lack of understanding that smoked fishery products may become hazardous if they are improperly processed or stored at abuse temperatures. Part of this misunderstanding is related to the changes in the processes that have occurred in recent years as a result of modern refrigeration systems. In order to become more acceptable to the consumer, the products now contain less NaCl and smoke and have higher concentrations of moisture. In addition, packaging systems with high barrier films have been introduced in the marketing of these products. Unfortunately, the requirement for smoked fish to be continuously refrigerated was not understood in the early

1960s. Consequently, the industry encountered botulism outbreaks from vacuum-packaged, lightly salted, hot-smoked products that were grossly temperature abused during distribution (Anonymous, 1963).

These outbreaks stimulated numerous investigations on the growth and toxin production of *C. botulinum* in different smoked fishery products. In inoculated pack studies, the spore inoculum varied from 10^3 to 10^5 per sample. Previous studies on the incidence of *C. botulinum* in raw uneviscerated fish indicated natural contamination levels of one to two spores per 100 g (Hobbs, 1976; Pace and Krumbeigel, 1973). If fish are eviscerated and thoroughly rinsed, the 10^3-spore level per 100-g sample used in experimentally contaminated samples should allow an adequate margin of safety. In addition, the bulk of naturally contaminating spores is located on the surface of the fish or fish fillets (Eklund et al., 1982) where they are affected the most by smoke and other processing variables.

In most studies, inhibition of *C. botulinum* by NaCl was examined at different storage times and temperatures. In recent studies (unpublished data), the processing steps and ingredients used in the smoking operation were examined individually and in combination to determine their inhibitory and lethal effects on nonproteolytic *C. botulinum*. These included NaCl, KCl, sodium nitrite, generated smoke, liquid smoke, processing time and temperature, storage time and temperature, and packaging materials.

In most experiments, the products were stored at either 10 or 25°C to simulate distribution at high refrigerated temperatures or severe temperature abuse. In general, the concentration of NaCl, NaCl plus nitrite, or other preservatives necessary to inhibit growth and toxin production were the same at 25 and 10°C. The main difference was that time for toxin production was markedly delayed at 10°C.

I. Comparison of O_2-Permeable and O_2-Impermeable Films

The results from experiments comparing the effect of packaging permeability are summarized in Table 2. In these and other experiments, higher concentrations of NaCl were always required for inhibition of *C. botulinum* in vacuum-packaged (O_2-impermeable film) products than in the air-packaged (O_2-permeable) products. The titer of toxin was always higher in the vacuum-packaged products. As discussed earlier in this text, this is in part related to the spoilage population that develops in these packaging systems. In vacuum-packaged products, the spoilage microorganisms are suppressed by the absence of O_2, thereby decreasing the competition with *C. botulinum* type E. Products packaged in O_2-permeable films always had bacterial populations 10–100 times greater than the products packaged in O_2-impermeable films. This resulted in rapid spoilage of O_2-permeable packaged products and delayed spoilage of the O_2-impermeable packaged products. The data emphasize the need to separate the air-packaged

Table 2 Toxin Production by *C. botulinum* Type E (10^3 spores/100 g) in Hot-Process Whitefish Steaks Vacuum Packaged in O_2-Permeable Film or O_2-Impermeable Film[a]

		Packaging film	
Storage at 25°C (days)	Water-phase salt (%)	O_2-permeable (MLD/g)[b]	O_2-impermeable (MLD/g)[b]
3	1.8	50	2,500
	2.6	0	500
	3.4	0	5
5	1.8	250	50,000
	2.8	0	2,500
	3.5	0	0
7	1.8	2,500	50,000
	2.8	0	5,000
	3.5	0	50

[a]Smoke applied during first half of smokehouse operation.
[b]Botulinum type E neurotoxin titer.

products from the vacuum-packaged products in the development of guidelines for processed smoked fishery products.

The hot-smoked process for fishery products thermally inactivates the spoilage microorganisms and/or inhibits their growth by NaCl, smoke, dehydration of the surface, sodium nitrite, or other preservatives used in smoked fish processing. These processing parameters eliminate or modify the type of spoilage typical of raw products.

Temperatures used to prepare hot-smoked products (60–80°C) are inadequate to inactivate the spores of even the most heat-sensitive nonproteolytic *C. botulinum* type E spores (Christiansen et al., 1968; Hobbs et al., 1969; Kautter, 1964; Pelroy et al., 1982). Because these temperatures markedly reduce or inactivate the spoilage microorganisms and do not eliminate *C. botulinum* spores, they place the smoked fish in the danger zone of Figure 1. By understanding the different steps and ingredients used in the process and their inhibitory effects on *C. botulinum*, safe products can be produced. In general, this knowledge, and an emphasis on critical control points, have contributed to the excellent record of the fish industry in recent years.

J. Smoked Fish Packaged in O_2-Impermeable Films

Table 3 summarizes the data on the effect of NaCl on outgrowth and toxin production of *C. botulinum* type E in hot-processed (smoke was not applied),

Table 3 Effect of NaCl on Toxin Production by *C. botulinum* Type E (10^3 spores/100 g) in Hot-Process Whitefish Steaks Vacuum (without smoke) Packaged in O_2 Impermeable Film and Stored at 25°C

Days of Storage	Maximum concentration of water-phase NaCl (%) permitting growth and toxin production
2	<1.8
3	2.7
4	3.0
5	3.3
6	3.7
7	3.6

vacuum-packaged whitefish during various storage periods at an abuse temperature of 25°C (unpublished data). Toxin was not produced during the first 2 days of storage. The minimum concentration of water-phase NaCl in any of the samples was 1.8%. The inhibitory level of NaCl increased with storage time until 6 days of storage when the maximum concentration of 3.6% was reached. In similar studies, the outgrowth and toxin production was studied in hot-processed salmon and carp stored at both 10 and 25°C (unpublished data). Water-phase NaCl greater than 3.8% was inhibitory to type E in both fish species stored at 10°C for 42 days. At 25°C, water-phase NaCl levels inhibitory to type E were greater than 3.8% in salmon and greater than 4.1% in carp.

In another set of experiments, generated or liquid smoke was applied to the products (unpublished data). Generated smoke was applied to the products during the first half, the second half, or during the entire process in the smokehouse. The efficacy of generated smoke as an inhibitor for *C. botulinum* type E was greatest when applied during the first half of the process or during the entire process. Processing products to an internal temperature of 63°C or higher and holding them at this temperature for 30 minutes enhanced the inhibitory effect of smoke and NaCl. Comparable results were obtained when liquid smoke was applied to the product (Table 4). Liquid smoke, however, was inhibitory even when it was applied during the last half of the process. Water-phase NaCl concentrations of 3.0% and higher inhibited toxin production for 7 days at 25°C and for 42 days at 10°C when liquid or generated smoke was used and the products were processed to an internal temperature of 63°C or higher and held at this temperature for 30 minutes or more.

The inhibitory effect of NaCl and sodium nitrite was also studied in salmon (Pelroy et al., 1985). In these studies, smoke was not used because the experiments were designed to determine the efficacy of NaCl and sodium nitrite. So-

Table 4 Effect of Liquid and Generated Smoke on Inhibition of *C. botulinum* Type E (10^3 spores/100 g) in Hot-Process Whitefish Steaks Vacuum Packaged in O_2-Impermeable Film and Stored at 25°C for 7 Days

Treatment[a]	Inhibitory water-phase NaCl (%)
No smoke	3.7
Liquid smoke	2.9
Generatred smoke[b]	3.0

[a]Product processed to internal temperatures of 62–77°C for 30 min.
[b]Smoke applied during first half of smokehouse operation.

dium nitrite at concentrations of 50 ppm in combination with NaCl was no more effective than NaCl alone. When the sodium nitrite concentration was increased to between 100 and 200 ppm, only 2.5% NaCl was required for inhibition.

In recent years, consumers have become concerned about the amount of sodium in their diet because of the possible association between sodium intake and hypertension. This has created a dilemma for the smoked fished industry where relatively high concentrations of NaCl are needed, not only for traditional flavor and texture, but also to assure microbiological safety during distribution. Studies were therefore undertaken to determine whether KCl could be substituted in part for NaCl. Products containing NaCl, KCl, or equimolar concentrations of both were compared for their inhibitory effects on *C. botulinum* type E (Pelroy et al., 1985). Products were inoculated intramuscularly with 10 or 100 type E spores per g and hot-processed to an internal temperature of 62–76°C for 30 minutes of a 2- to 3-hour process, vacuum packaged, and stored for 7 days at 25°C. Toxin production was inhibited in whitefish steaks containing more than 0.66 ionic strength NaCl, 0.64 KCl, or 0.71 equimolar NaCl and KCl. The results indicate that KCl can be substituted for NaCl in hot-process smoked fish for inhibition of outgrowth and toxin production by type E. From a practical standpoint, the amount of KCl that can be used in products will depend mainly on its effect on flavor. Flavor changes become objectionable when more than one third to one half of NaCl is replaced with KCl.

Based upon these and many other studies, 3.5% water-phase NaCl would be adequate to inhibit *C. botulinum* type E in vacuum-packaged products that are smoked at least during the first half of the process and heated to an internal temperature of 63°C or above for a minimum of 30 minutes. If 100–200 ppm of sodium nitrite is used in the process (in the United States permitted only in salmon, sablefish, chubs, and shad), then the water-phase NaCl level can be lowered to 3%.

K. Smoked Fish Packaged in O_2-Permeable Films

Inoculated salmon samples prepared without smoke, processed to an internal temperature of 65–77°C, and held at this temperature for 30 minutes were packaged in O_2-permeable film (unpublished data). Toxin was not observed until after 21 days of storage at 10°C, and only one sample, with a water-phase NaCl concentration of 1.5%, was toxic. All samples were spoiled from both bacterial and mold growth after 14 days of storage. When the samples were challenged with an inoculum of 10^4 spores per sample, botulinum toxin was produced after 14 days in one sample with 1.9% water-phase NaCl. After 21 days, the NaCl requirement increased to above 2.5%. These samples, however, were obviously spoiled after 14 days.

In another set of experiments, samples were processed to internal temperatures of 48–77°C and stored at 10°C (unpublished data). Smoke was applied to the products during the first half, the second half, or over the entire processing period. Toxin was detected in only two of the 160 samples. One sample contained 1.8% and the other 1.7% water-phase NaCl. In an additional set of experiments, type E was replaced with the more NaCl resistant nonproteolytic type B. Only one of the 80 samples became toxic during the 35-day storage period, and this sample contained only 2.0% water-phase NaCl. The combination of temperature and smoke also increased the inhibitory effects of NaCl for *C. botulinum* type E in smoked fish products packaged in O_2-permeable films. Based upon these data, water-phase NaCl concentrations of 2.5% and above would be adequate to inhibit type E in O_2-permeable packaged smoked fish at refrigerated temperatures. Similar findings have been reported by Cann and Taylor (1979).

Refrigeration is the most important factor in the safe distribution of smoked fishery products. Storage of products at temperatures below 3°C should be emphasized at all stages of marketing. This not only improves the shelf life but also ensures the safe distribution of products that occasionally may have lower NaCl concentrations.

IV. CONTROL OF *C. BOTULINUM* IN FISHERY PRODUCTS BY INACTIVATION

A. Inactivation of Nonproteolytic *C. botulinum* by Heat Pasteurization

In the previous section, emphasis was placed on controlling *C. botulinum* by inhibiting outgrowth and neurotoxin production. In this section, the use of heat pasteurization will be considered as a method of inactivating the spores of the more heat-sensitive strains of nonproteolytic *C. botulinum* in products to be marked at refrigerated temperatures.

Pasteurized Crab Meat

The pasteurization of blue crab meat in hermetically sealed cans was introduced to the industry on the East Coast of the United States to increase refrigerated shelf life. Pasteurized crab meat has a refrigerated shelf life of at least 6 months. During 34 years of production, there have been no reported illnesses.

A thermoprocess of 1 minute at an internal temperature of 85°C has been followed by industry as recommended by a Tri-State Seafood Committee (1971). Based upon the studies of Lynt et al. (1977), a 12D process would require 3 minutes at 85°C for type E spores. In earlier studies by Cockey and Tatro (1974), a process of 1 minute at 85°C reduced 10^8 viable spores per 100 g of crab meat to 6 or less, and the pasteurized meat remained nontoxic for 6 months at 4–6°C.

It is generally agreed that a 4–6D process is adequate for refrigerated products. In addition, crab meat has been shown to be a relatively poor substrate for growth and toxin production by type E, especially when the spores have been heat injured (Lynt et al., 1977). Solomon et al. (1976) also indicated that crab meat was a poor substrate and that unheated or heated type E spores would not produce neurotoxin at storage temperatures of 4 or 8°C.

Pasteurized Smoked Fish

Although the spores of nonproteolytic *C. botulinum* are relatively heat sensitive as compared to proteolytic strains, attempts to inactivate them at high temperatures in the smokehouse operation have been successful because of the adverse effects on the products and loss of consumer acceptability. Eklund et al. (1988) demonstrated the feasibility of a heat pasteurization process for inactivation of Group II spores in hot-smoked fishery products. The products were heat-processed in the smokehouse to an internal temperature of 62–77°C. Smoke was not applied to the sample because of possible variable inhibitory effects. After heat processing, the products were cooled, inoculated with 10^6 type B or E spores, and vacuum packaged. Packaged products were pasteurized in a water bath at 85, 89, and 92°C. A total of 75, 65, and 55 minutes, respectively, in the water bath inactivated 10^6 type E spores in 180 to 220-g samples. Longer times—175, 85, and 65 minutes—were required to inactivate type B spores. This process not only increased the refrigerated shelf life of the product, but also increased the safety of smoked fish without jeopardizing product quality.

The process is applicable to products that are prepared as fillets and steaks, but somewhat less applicable to fish such as trout, chubs, etc., which are processed in the eviscerated form. When the latter species were used, the flesh separated from the backbone.

The data from these studies were presented as total time in the water bath. Additional studies are in progress so that results can be extrapolated to products of different sizes and configurations in a commercial process.

Previous studies (Eklund et al., 1982) have shown that smoke used in combination with heat increases the inhibitory effect of NaCl on spores that occur on the product surface. Whether heat pasteurization would have increased effectiveness against spores introduced after the smoking process remains to be investigated. Even though *C. botulinum* spores of the nonproteolytic Group II are inactivated by this process, the move heat-resistant strains of Group I are not inactivated and will grow if the pasteurized products are temperature abused. The packages therefore must carry labels such as "Continuous refrigeration required. Store below 3°C."

V. CONCLUSIONS

It is generally agreed that fish cannot be protected from natural contamination with *C. botulinum*. It is essential, therefore, to control processing and storage so that outgrowth and neurotoxin production can be prevented. This process starts by eviscerating the fish as soon as possible to remove the reservoir of spores in the intestines and thoroughly washing the fish before further processing. This step minimizes the spore load and increases the effectiveness of any preservation method.

Unprocessed eviscerated fish, fish fillets, or other forms have had an excellent safety record with regard to problems from *C. botulinum*. When preservation procedures are introduced to increase shelf life, i.e., by preparing smoked, salted, marinated, fermented, dried, or other product forms, then the potential problems from *C. botulinum* are increased. These processes share a common denominator in that they selectively destroy or inhibit the spoilage bacteria common to raw fishery products, but often have little or no inhibitory effect on *C. botulinum* spores.

Of the preservation methods discussed in this chapter, smoking, marinating, salting, and drying have the advantage in that preservatives can be used in conjunction with refrigeration to inhibit *C. botulinum*. In comparison, the extension of the shelf life of raw fishery products by modified atmospheres, vacuum packaging, irradiation, or other methods must rely strictly on refrigerated storage below 3.3°C for inhibition. Heat pasteurization procedures can also be used to inactivate spores of the more heat-sensitive nonproteolytic *C. botulinum* types B, E, and F, but the products must also be stores at refrigerated temperatures in order to inhibit growth of the more heat-resistant proteolytic types A, B, and F.

To minimize the risks from food poisoning, characteristics of the nonproteolytic and proteolytic strains of *C. botulinum* and other bacterial pathogens and the effects of process and product ingredients on the survival and growth of these microorganisms must be thoroughly understood. In products preserved short of sterilization, storage at refrigerated temperatures below 3°C is probably the most important parameter to preclude potential food-poisoning problems. The requirement for refrigerated storage therefore should be clearly labeled on all packages and master cartons.

REFERENCES

Anonymous. (1963). Botulism. *Morbid. Mortal. Weekly Rep. 12* (41): 337.

Anonymous. (1987). International outbreak of type E botulism associated with ungutted, salted whitefish. Centers for Disease Control. *Monthly and Morbidity Report 35* (49): 812.

Baird-Parker, A. C., and Freame, B. (1967). Combined effect of water activity, pH and temperature on the growth of *Clostridium botulinum* from spore and vegetative cell inocula. *J. Appl. Bact. 30*: 420.

Baker, D. A., and Genigeorgis, C. (1990). Predicting the safe storage of fresh fish under modified atmospheres with respect to *Clostridium botulinum* toxigenesis by modelling length of the lag phase of growth. *J. Food Prot. 53*: 131.

Bell, L. (1980). Gas in trucks, containers, helps packages keep foods fresh. *Package Eng. 25*: 72.

Cann, D. C., and Taylor, L. Y. (1979). The control of the botulism hazard in hot smoked trout and mackerel. *J. Food Technol. 14*: 124.

Cann, D. C., Smith, G. L., and Houston, N. G. (1983). Further studies on marine fish stored under modified atmosphere packaging. Technical Report, Torry Research Station, Aberdeen, United Kingdom.

Cann, D. C., Taylor, L. Y., and Hobbs, G. (1975). Incidence of *C. botulinum* in farmed trout raised in Great Britain. *J. Appl. Bacteriol. 39*: 331.

Christiansen, L. N., Deffner, J., Foster, E. M., and Sugiyama, H. (1968). Survival and outgrowth of *C. botulinum* type E spores in smoked fish. *Appl. Microbiol. 16*: 133.

Cockey, R. R., and Tatro, M. C. (1974). Survival studies with spores of *C. botulinum* type E in pasteurized meat in the blue crab *Callinectes sapidus*. *Appl. Microbiol. 27*: 629.

Cutting, C. L. (1955). *Fish Saving. A History of Fish Processing from Ancient to Modern Times*. Leonard Hill, London.

Dolman, C. E. (1964). Botulism as a world health problem. In *Botulism 1964* (K. H. Lewis and K. Cassel, Jr., ed.). U.S. Department of Health, Education, and Welfare, Public Health Service, Cincinnati, Ohio, p. 5.

Dolman, C. E. (1961). Further outbreaks of botulism in Canada. *Can. Med. Assoc. J. 84*: 191.

Eklund, M. W. (1982). Significance of *Clostridium botulinum* in fishery products preserved short of sterilization. *Food Technol. 36*: 107.

Eklund, M. W., and Poysky, F. T. (1970). The significance of nonproteolytic *Clostridum. botulinum* types B, E, and F in the development of radiation-pasturized fishery products. *Proceedings of Preservation of Fish by Irradiation*, Vienna, Austria. p. 125.

Eklund, M. W., and Poysky, F. T. (1968). Distribution of *Clostridium botulinum* on the Pacific Coast of the United States. *Proceedings of the 1st U.S.-Japan Conf. on Toxic Microorganisms*, Honolulu, HI, p. 304.

Eklund, M. W., and Poysky, F. T. (1967). Incidence of *Clostridium botulinum* on the Pacific Coast of the United States. In *Botulism 1966* (M. Ingram and T. A. Roberts, eds.), Chapman and Hall Ltd., London, England, p. 49.

Eklund, M. W., Pelroy, G. A., Paranjpye, R., Peterson, M. E., and Teeny, F. M. (1982). Inhibition of *Clostridium botulinum* types A and E toxin production by liquid smoke and NaCl in hot-process smoke-flavored fish. *J. Food Prot.*, *45*: 935.

Eklund, M. W., Peterson, M. E., Paranjpye, R. N., and Pelroy, G. A. (1988). Feasibility of a heat pasteurization process for the inactivation of nonproteolytic *Clostridium botulinum* types B and E in vacuum-packaged, hot-process (smoked) fish. *J. Food Prot.*, *51*: 720.

Eklund, M. W., Poysky, F. T., Peterson, M. E., Peck. L. W., and Brunson, W. D. (1984). Type E botulism in salmonids and conditions contributing to outbreaks. *Aquaculture 41*: 293.

Eklund, M. W., Poysky, F. T., and Weiler, D. I. (1967a). Characteristics of *Clostridium botulinum* type F isolated from the Pacific Coast of the United States. *Appl. Microbiol. 15*: 1316.

Eklund, M. W., Weiler, D. I., and Poysky, F. T. (1967b). Outgrowth and toxin production of nonproteolytic type B *Clostridium botulinum* at 3.3 to 5.6°C. *J. Bacteriol. 93*: 1461.

Eyles, M. J., and Warth, A. D. (1981). Assessment of the risk of botulism from vacuum-packaged raw fish: a review. *Food Technol. Aust. 33*: 574.

Genigeorgis, C. (1985). Microbial and safety implications of the use of modified atmospheres to extend the storage life of fresh meat and fish. *Int. J. Food Microbiol 1*: 237.

Hauschild, A. H. W. (1989). *Clostridium botulinum*. In *Foodborne Bacterial Pathogens* (M. P. Doyle, ed.) Marcel Dekker, Inc., New York, p. 111.

Hauschild, A. H. W., and Hilsheimer, R. (1979). Effect of salt content and pH on toxigenesis by *Clostridium botulinum* in caviar. *J. Food Protect. 42*: 245.

Hobbs, G. (1977). *Clostridium botulinum* in irradiated fish. *Food Irradiation Info. No. 7*: 39.

Hobbs, G. (1976). *Clostridium botulinum* and its importance in fishery products. In *Advances in Food Research* (C. O. Chichester, ed.), Vol. 22, Academic Press, London, p. 135.

Hobbs, G., Cann, D. C., and Wilson, B. B. (1969). An evaluation of the botulism hazard in vacuum-packed smoked fish. *J. Food Technol. 4*: 185.

Huss, H. H., and Larsen, A. (1979). Changes in the oxidation-reduction potential (Eh) of smoked and salted fish during storage. *Lebensmittelwiss. Technol. 13*: 40.

Huss, H. H., Pedersen, A., and Cann, D. C. (1974). The incidence of *C. botulinum* in Danish trout farms. I: Distribution in fish and their environment. *J. Food Technol. 9*: 443.

Huss, H. H., Schaeffer, I., Rye Petersen, E., and Cann, D. C. (1979). Toxin production by *Clostridium botulinum* type E in fresh herring in relation to the measured oxidation-reduction potential. *Nord. Vet. Med. 31*: 81.

Hussain, A. M., Ehlermann, D., and Diehl, J. F. (1977). Comparison of toxin production by *Clostridium botulinum* type E in irradiated and unirradiated vacuum-packed trout (*Salmon gairdneri*). *Arch. Lebensmittelhyg. 28*: 23.

Iida, H. (1970). Epidemiological and clinical observations of botulism outbreaks in Japan. In *Toxic Microorganisms* (M. Herzberg, ed.) Department of the Interior, Washington, DC, p. 357.

Kautter, D. A. (1964). *Clostridium botulinum* type E in smoked fish. *J. Food Sci. 29*: 843.

Lee, J. S. (1982). Selection and growth of seafood microorganisms under carbon dioxide treatment. *Proceedings of the 1st National Conference on the Modified and Controlled Atmosphere Packaging of Seafood Products* (R. E. Martin, ed.), National Fisheries Institute, Washington, DC, p. 30.

Lindsay, R. C. (1982). Controlled atmosphere packaging. *Proceedings of the 4th Annual International Seafood Conference*, Washington, DC, p. 80.

Lynt, R. K., Solomon, H. M., Lilly, T., and Kautter, D. A. (1977). Thermal death time of *Clostridium botulinum* type E in meat of the blue crab. *J. Food Sci. 42*: 1022.

Miyauchi, D., Eklund, M., Spinelli, J., and Stroll, N. (1964). Irradiation preservation of Pacific Coast shellfish. *Food Technol. 18*: 138.

Nakamura, Y., Iida, H., Saeki, K., Kanazawa, K., and Karashimada, T. (1956). Type E botulism in Hokkaido, Japan. *Jpn. J. Med. Sci. Biol. 9*: 45.

Nakano, W., and Kodama, E. (1970). On the reality of "izushi," the casual food of botulism, and on its folklore meaning. In *Toxic Microorganisms* (M. Herzberg, ed.), U.S. Department of Interior, Washington, DC, p. 388.

Odlaug, T. E., and Pflug, I. J. (1978). *Clostridium botulinum* and acid foods. *J. Food Prot. 41*: 566.

Ogrydziak, D. M., and Brown, W. D. (1982). Temperature effects in modified atmosphere storage of seafood. *Food Technol. 35*: 86.

Ohye, D. F., and Christian, J. H. B. (1967). Combined effects of temperature, pH, and water activity on growth and toxin production by *Clostridium* types A, B, and E. In *Botulism 1966* (M. Ingram and T. A. Roberts, eds.), Chapman and Hall Ltd., London, p. 217.

Ohye, D. F., and Scott, W. J. (1953). The temperature relations of *Clostridium botulinum* types A and B. *J. Biol. Sci. 6*: 178.

Pace, P. J., and Krumbeigel, E. R. (1973). *Clostridium botulinum* and smoked fish production 1963–1972. *J. Milk Food Technol. 36*: 42.

Pelroy, G. A., Scherer, A., Peterson, M. E., Paranjpye, R., and Eklund, M. W. (1985). Inhibition of *Clostridium botulinum* type E toxin formation by potassium chloride and sodium chloride in hot-process (smoked) whitefish (*Coregonus clupeaformis*). *J. Food Prot. 48*: 971.

Pelroy, G. A., Eklund, M. W., Paranjpye, R., Suzuki, E. M., and Peterson, M. E. (1982). Inhibition of *Clostridium botulinum* types A and E production by sodium nitrite and sodium chloride in hot-processed (smoked) salmon. *J. Food Prot. 45*: 833.

Post, L. S., Lee, D. A., Solberg, M., Furang, D., Specchio, J., and Graham, C. (1985). Development of botulinal toxin and sensory deterioration during storage of vacuum- and modified-atmosphere-packaged fish fillets. *J. Food Sci. 50*: 990.

Riemann, H. (1969). Botulism—Types A, B, and F. In *Food-Borne Infections and Intoxications* (H. Riemann, ed.), Academic Press, New York, p. 291.

Riemann, H., Lee, W. H. E., and Genigeorgis, C. (1972). Control of *Clostridium botulinum* and *Staphylococcus aureus* in semi-preserved meat products. *J. Milk Food Technol. 35*: 514.

Roberts, T. A., and Ingram, M. (1973). Inhibition of growth of *C. botulinum* at different pH values by sodium nitrite. *J. Food Technol. 8*: 467.

Sakaguchi, G. (1979). Botulism. In *Food-Borne Infections and Intoxications* (H. Riemann, ed.), Academic Press, London, p. 390.

Schmidt, C. F., Lechowich, R. V., and Folinazzo, J. F. (1961). Growth and toxin production by type E *Clostridium botulinum* below 4°C *J. Food Sci. 25*: 626.

Segner, W. P., and Schmidt, C. F. (1968). Inoculated pack studies on *Clostridium botulinum* type E in unirradiated and irradiated haddock. *Proceedings of the 1st U.S.-Japan Conference on Toxic Microorganisms*. U.S. Government Printing Office, Washington, DC, p. 483.

Segner, W. P., Schmidt, C. F., and Boltz, J. K. (1966). Effect of sodium chloride and pH on the outgrowth of spores of type E *Clostridium botulinum* at optimal and suboptimal temperatures. *Appl. Microbiol. 14*: 49.

Slater, P. E., Addiss, D. G., Cohen, A., Leventhal, A., Chassis, G., Zehavi, H., Bashari, A., and Costin, C. (1989). Foodborne botulism: An international outbreak. *Int. J. Epidemiol. 18*: 693.

Smith, L. D. S., and Sugiyama, H. 1988. *Botulism*. Charles C Thomas, Springfield, IL.

Solomon, H. M., Lynt, R. K., Lilly, T., Jr., and Kautter, D. A. (1976). Effect of low temperature on growth of *Clostridium botulinum* spores in meat of blue crab. *J. Food Prot. 40*: 5.

Stier, R. F., Bell, L., Ito, K. A., Shafer, B. O., Brown, K. A., Seeger, M. L., Allen, B. H., Porcuna, M. N., and Lerke, P. A. (1981). Effect of modified atmosphere storage on *Clostridium botulinum* toxigenesis and the spoilage microflora of salmon fillets. *J. Food Sci. 41*: 1639.

Strasdine, G. A., and Kelly, J. M. (1967). Germination of spores of *C. botulinum* type E in fish and shellfish extracts. *J. Fish. Res. Board Canada 24*: 1833.

Sugiyama, H. T., Bott, L., and Foster, E. M. (1970). *Clostridium botulinum* type E in the Inland Bay (Green Bay of Lake Michigan). *Proceedings of the 1st U.S.-Japan Conference on Toxic Microorganisms*, U.S. Government Printing Office, Washington, DC, p. 287.

Tanaka, N. (1982). Toxin production by *Clostridium botulinum* in media at pH lower than 4.6. *J. Food Prot. 45*:234.

Telzak, E. E., Bell, E. P., Kautter, D. A., Crowell, L.., Budnick, L. D., Morse, D. L., and Schultz, S. (1990). An international outbreak of Type E botulism due to uneviscerated fish. *J. Infect. Dis. 161*:340.

Tri-State Seafood Commission. (1971). Rule governing the pasteurization of crabmeat. Maryland, North Carolina, and Virginia State Health Departments and Universities.

Tsang, N., Post, L. S., and Solberg, M. (1985). Growth and toxin production by *Clostridium botulinum* in model acidified systems. *J. Food Sci. 40*: 961.

9

Control in Fruits and Vegetables

S. H. W. Notermans

National Institute of Public Health and Environmental Protection, Bilthoven, The Netherlands

I. INTRODUCTION

The "classic" form of botulism is caused by preformed toxin produced by *Clostridium botulinum* in food. The types of food involved in recorded outbreaks of foodborne botulism are summarized in Table 1. The results presented in this table are a summary of records reviewed by Hauschild (1989). Of the recorded outbreaks, 24.1% are attributed to contaminated fruit and vegetables. Countries such as China, Spain, Italy, the United States and Argentina have recorded relatively frequent outbreaks caused by such products. These countries accounted for 71% of the recorded outbreaks caused by fruit and vegetables, and in almost all cases the incriminated products were homemade.

In China, and especially the northwestern province Xinjiang (Sinkiang), fermented bean curd was the prevailing type of food involved in the outbreaks (Ying and Shuan, 1986). In the recorded outbreaks in China, botulinum toxin type A dominated (94%). In Argentina, mostly home-preserved vegetables such as string beans, peppers, and tomatoes are implicated in outbreaks (Ciccarelli and Giménez, 1981), and *C. botulinum* type A is the most common causative organism. This organism is also the most common form in the Argentinian soil environment (Ciccarelli and Giménez, 1981). Also in Spain and Italy, home-preserved vegetables are the prevailing causative foods (Aureli et al., 1984), but the responsible *C. botulinum* type is usually B. As in Argentina, botulism in the United States is commonly associated with home-canned vegetables and

Table 1 Recorded Outbreaks of Foodborne Botulism

Type of food	Number of outbreaks recorded
Meats	693 (39.3%)
Fish	575 (32.5%)
Fruit and vegetables	425 (24.1%)
Others[a]	72 (4.1)
Total	1765 (100%)

[a]Including mixed vehicles.
Source: Hauschild, 1989.

C. botulinum type A. Most of the incidents are small and limited to families. Of the 215 outbreaks between 1945 and 1973 in which the toxin type was diagnosed, 150 (68.5%) were caused by vegetables and fruits (Centers for Disease Control, 1974). Of the incriminated vegetables and fruits, string beans caused 98 outbreaks and involved 246 cases and 189 deaths.

In contrast, botulism outbreaks caused by fruit and vegetables are only rarely observed in some European countries such as Denmark, Sweden, Germany, and Poland (Hauschild, 1989). In Poland, for example, vegetables accounted only for 1.5% of the 541 botulism cases recorded in 1984 (Anusz, 1985). Most cases of botulism in Poland were associated with meats and were caused by botulinum toxin type B (95.8%).

II. SOME EXAMPLES OF CASES OF BOTULISM CAUSED BY FRUIT AND VEGETABLES

Carré et al. (1987) described an outbreak of botulism in France. Spanish canned asparagus containing type B toxin was implicated as the source. Studies at the factory where the asparagus was canned revealed that the heat process was insufficient and that process control was inadequate.

Outbreaks of botulism in Chang-kwa county, Taiwan, in 1986 were traced to botulinum type A toxin in peanuts packed in glass jars (Chen et al., 1987). Twenty-three of 90 bottles of the same lot number in the manufacturer's warehouse also contained botulinum toxin. Further investigations revealed that the manufacturing plant was not properly equipped to process low-acid canned foods and that the jars were underprocessed.

An outbreak of botulism involving approximately 10 persons in the U.S.S.R. was reported by Podreshetnikova et al. (1970). The causative agent was *C. botulinum* type B toxin, which was isolated from blown cans of beetroot. The cause of the defect was traced to lack of a heat-regulating device in the canning stage.

Commercial chopped garlic in soybean oil used to make garlic butter served on sandwiches in a restaurant was implicated as the vehicle for transmission in a large outbreak of type B botulism in Vancouver, Canada (St. Louis et al., 1988). Thirty-six cases were recorded. The implicated garlic was an aqueous mixture with a pH above 4.6 and was stored at room temperature. According to the label, the product was to be kept refrigerated.

In November 1978, several cases of type A botulism occurred in persons who had eaten potato salad in a restaurant in Colorado (Seals et al., 1981). It appears that *C. botulinum* grew in baked potatoes wrapped in aluminum foil while they were stored at ambient temperatures for up to 5 days before they were used to prepare the salad.

A case of type B botulism following consumption of commercially marinated mushrooms occurred in Canada in 1973 (Todd et al., 1974). All 8656 bottles recovered from the market were examined, and 10 additional bottles were found to contain type B toxin. The product was to be preserved by acidification below pH 4.6 and subsequent pasteurization. The most likely cause of the outbreak was failure to sufficiently acidify the mushrooms prior to packing them in oil.

Haller et al. (1969) described a case of botulism in Kentucky from home-bottled tomato juice. The symptoms—nausea, vomiting, and cramping abdominal pain—indicated that the tomato juice was possibly contaminated by several types of microorganisms. In all probability the bottled tomato juice was underprocessed.

In 1977, 59 individuals in Michigan developed type B botulism. All ill persons had eaten at the same Mexican restaurant, and all had consumed a hot sauce made with improperly home-canned jalapeno peppers (Terranova et al., 1978).

A rare *C. botulinum* type A outbreak occurred in Japan in 1984. Eleven (31%) of 36 patients from 14 different areas died after eating commercially vacuum-packaged lotus-rhizome with mustard. A total of 42 packages of the food collected from different districts were examined. Thirteen of these (31%) were contaminated with *C. botulinum* type A, and in 11 (26%) a small amount of toxin A had been produced (Otofugi et al., 1987).

A case of botulism in California caused by home-canned olives has been reported by Casillas et al. (1978). The home-canned olives contained type A toxin.

This anthology of examples of human cases of botulism shows that a wide variety of fruits and vegetables such as peppers, peanuts, tomatoes, mushrooms, asparagus, garlic, etc. have been involved in outbreaks. When botulism outbreaks from fruits and vegetables occur, the products have generally been stored for long periods at ambient temperatures under almost anaerobic conditions. Furthermore, the products have not been treated adequately to destroy or to inhibit spores of *C. botulinum* present in the raw materials. It also becomes clear that in all reported cases of botulism involving fruits and vegetables, either toxin of type A or B is involved. Type E and F botulism have never been observed from consumption of fruits and vegetables, except for a single episode of type E botulism due to canned mushrooms that contained some other ingredients (Geiger, 1941). These findings indicate that proteolytic types of *C. botulinum* are primarily involved in outbreaks of botulism in which fruits and vegetables are implicated.

III. PRESENCE OF *C. BOTULINUM* IN RAW FRUIT AND VEGETABLES

C. botulinum is a ubiquitous organism found principally in soil and aquatic environments. However, sewage sludge and other organic fertilizers may also contain spores of *C. botulinum*. Microorganisms present on fruit and vegetables will mainly originate from the soil of the production fields and from organic fertilizers. Therefore, the contamination of raw fruit and vegetables depends largely on the contamination of the soil and on agricultural practices. The occurrence of *C. botulinum* in soil at different locations is presented in Table 2. In soil specimens originating from all parts of the United States, Smith (1978) found *C. botulinum* in 23.8% of the 260 samples tested. Type A strains were almost all from the portion of the United States west of the rise of the Rocky Mountains out of the great plains. Type B strains were, however, not so sharply localized as type A. Type A also predominates in Argentina (Ciccarelli and Giménez, 1981). In contrast to this, *C. botulinum* type A is only rarely detected in the soils of western European countries. A survey by Haagsma (1973) in the Netherlands failed to detect *C. botulinum* type A, whereas types B and E were frequently found. The absence of *C. botulinum* type A in Dutch soil samples was also demonstrated by Notermans et al. (1979, 1980a). In a study on the presence of

Table 2 Occurrence of *Clostridium botulinum* in Environmental Soil Samples

Location	No. of samples	No. with *C. botulinum*	Type %[a]				Ref[b]
			A	B	C/D	E	
United States	260	62	42	35	13	10	1
Argentina	722	244	67	20	0	0	2
The Netherlands	135	39	0	24	2	74	3
Italy	144	1	0	100	0	0	4
Britain	174	10	0	100	0	0	5
Denmark	215	29	0	93	7	0	6
U.S.S.R.	4242	445	8	28	2	62	7

[a]As percentage of the types identified.

[b]1) Smith, 1977a; 2) Ciccarelli and Giménez, 1981; 3) Haagsma, 1973; 4) Barbuti et al., 1980; 5) Smith and Young, 1980; 6) Huss, 1980; 7) Kravchenko and Shishulina, 1967.

C. botulinum in soils cultivated for vegetables, Barbuti et al. (1980) tested 144 soil samples in Italy, and only one was shown to contain *C. botulinum* type B. Smith and Young (1980) tested 174 soil samples from various parts of Britain, and only 10 contained *C. botulinum*, all of type B. *C. botulinum* type B also predominates in Danish soils (Huss, 1980). In U.S.S.R. survey (Kravchenko and Shishulina, 1967) more than 10% of the soil specimens contained *C. botulinum*. Types A, B, and E were all found. Rozanova et al. (1972) detected *C. botulinum* in samples of soil originating from fields used for growing vegetables. Seventy-nine of 300 samples (26.3%) contained *C. botulinum* spores and $>$ 1 spore was present per 20 g of soil. *C. botulinum* type A was dominant (see Chapter 2).

From the above results it can be concluded that *C. botulinum* is a ubiquitous organism and that soil is its principal habitat. The origin of *C. botulinum* spores in soil is not completely clear. However, organic fertilizers of animal origin contribute to the contamination of soil with *C. botulinum*. This was demonstrated by Notermans et al. (1981) on farms involved in botulism outbreak. Spreading of cow manure increased the contamination of soil considerably. An increase in the contamination of soil with *C. botulinum* was also observed after repeated fertilization with sewage sludge (Notermans et al., 1985). *C. botulinum* may also be present in large numbers in freshly drained land. In clay soil from such land, Notermans et al. (1985b) detected an average 450 *C. botulinum* type E per 100 g of soil.

As a consequence of the contamination of soil and organic fertilizers (manure, sewage, sludge) with *C. botulinum*, both fruit and vegetables become contaminated with these organisms. When fruit and vegetables are harvested, *C. botulinum* may be transferred from the soil to the processing plant. In wash waters from carrots grown in the U.S.S.R., *C. botulinum* was found in 6% of the samples; type A dominated (Prokhorovich et al., 1975). In the analysis of wash waters from red peppers and apricots, Prokhorovich et al. (1975) detected *C. botulinum* in 9 and 7% of the samples, respectively. These results demonstrate that *C. botulinum* is present on the raw product. The presence of *C. botulinum* on raw products is summarized in Table 3. Rozanova et al., (1972) examined samples of cabbage, carrot, onion, tomato, potato, horseradish leaves, green dill, and green parsley taken 1–7 days after harvesting. Thirteen of the 30 samples tested contained *C. botulinum* spores; type A spores were presented in 11 samples and type B spores in 7 samples. Insalata et al. (1969) determined the incidence of *C. botulinum* spores in 50 samples each of cut green beans and chopped spinach vacuum packaged in butter sauce and frozen. None of the bean samples contained *C. botulinum* spores, whereas six of the spinach samples contained spores of type A or B. Solomon and Kautter (1986) detected *C. botulinum* type A in 5 out of 75 raw onions obtained from a restaurant involved in an outbreak of botulism from sauteed onions. Vergieva and Incze (1979) did not detect

Table 3 Occurrence of *Clostridium botulinum* in Raw Vegetables

Type of product	Country	No. of samples	No. with *C. botulinum*	MPN/ 100 g	Types	Ref[a]
Different vegetables	U.S.S.R.	30	13	—	A,B	1
Chopped spinach	United States	50	6	—	A,B	2
Cut green beans	United States	50	0	—	—	2
Onions	United States	75	5	—	A	3
Potatoes	Hungary	26	0	—	—	4
Carrots	Hungary	18	0	—	—	4
Peeled washed potatoes	Germany	72	0	—	—	5
Mushrooms	Canada	—	—	41	B	6
Mushrooms	The Netherlands	50	—	< 0.08–0.16	—	7

[a]1) Rozonova et al., 1972; 2) Insalata et al., 1969; 3) Solomon and Kauter, 1986; 4) Vergieva and Incze, 1979; 5) Baumgart, 1987; 6) Hauschild et al., 1975; 7) Notermans et al., 1989.

C. botulinum in any of 26 samples of potatoes and 18 samples of carrots. Also, Baumgart (1987) did not detect *C. botulinum* in 72 samples of raw potatoes.

Hauschild et al. (1975) estimated the incidence of *C. botulinum* in Canadian mushrooms as 41 type B spores per 100 g. The presence of *C. botulinum* in mushrooms grown in the Netherlands was intensively tested using an MPN method (Notermans et al., 1989). *C. botulinum* was not detected in any of the samples, indicating numbers of less than one per 100 g. Craig (cited by Kautter et al., 1978) found *C. botulinum* type A spores in all samples of mushrooms compost (a mixture of straw and horse feces) and casing soil of mushrooms production beds. In contrast to this, the contamination of casing soil originating from Dutch farms was very low and varied from less than 1.3 to 1.6 spores of *C. botulinum* per 100 g (Notermans et al., 1989). These results suggest that the contamination of mushrooms can be minimized by using clean casing soil.

From the above it is apparent that *C. botulinum* is frequently present in soil and organic fertilizers and that, due to the close contact of fruit and vegetables with these, they in turn become contaminated with *C. botulinum*. Although good agricultural and horticultural practices may reduce the contamination, the presence of *C. botulinum* on raw fruit and vegetables must always be considered. Therefore, during processing of fruit and vegetables, factors have to be built in to prevent outgrowth and toxin formation by *C. botulinum*.

IV. GROWTH OF *C. BOTULINUM* IN FRUIT AND VEGETABLES

According to *Bergey's Manual of Systematic Bacteriology* (Sneath et al., 1986), all clostridia that produce any of the characteristic botulinal neurotoxins, A to G, are combined in the species *Clostridium botulinum*. The species is subdivided into four groups (Smith, 1977). Group I strains, including the so-called proteolytic types A, B, and F, differ from group II strains, including the so-called nonproteolytic types B, E, and F, by their temperature range of growth and by the heat resistance of their spores. The temperature range of growth is 10–48°C for group I organisms and 3.3–45°C for group II organisms. The decimal reduction time at 100°C (D_{100}) of spores of group I is approximately 25 minutes; for group II spores it is < 0.1 minute (see chapter 1).

Essentially all outbreaks of botulism involving fruit and vegetables have been caused by type A and B organisms, probably of group I. There are several reasons for this assumption. Because of their heat sensitivity, spores of group II organisms are rapidly destroyed during heat processing of fruit and vegetables. Second, fruit and vegetables may be relatively poor substrates for the growth of *C. botulinum* group II. Due to nutrient insufficiency, organisms of this group fail to grow in many fruits and vegetables (Notermans et al., 1979; 1981). Third, fruit and vegetables may contain inhibitory factors preventing outgrowth of spores of group II organisms. The suitability of some vegetables for growth and toxin production by *C. botulinum* is presented in Table 4. These results show that group I organisms grow well on potatoes. This was also observed by Dodds (1989), Dignon (1985), and Sugiyama et al. (1981). The nonproteolytic type B *C. botulinum* strains tested did not grow on potatoes, and growth of *C. botulinum* type E strains was poor. Carrots are also an excellent medium for *C. botulinum* of group I, and high quantities of toxin are produced. In contrast, group II *C. botulinum* did not grow on carrots. Similar results were obtained with red beets as substrate. However, the quantity of toxin produced by group I organisms was low.

In a study by Sugii and Sakaguchi (1977), growth of *C. botulinum* was tested in string beans. All types of *C. botulinum* produced measurable quantities of toxin, with the exception of *C. botulinum* type E. This finding again underlines that vegetables are less supportive media for *C. botulinum* group II organisms. Peas and mushrooms, however, support growth and toxin production by both *C. botulinum* type I and type II organisms (Gola and Mannino, 1985; Sugii and Sakaguchi, 1977).

Considering the above-mentioned findings, it appears appropriate that studies on toxin production by *C. botulinum* in fruit and vegetables have been carried out mainly with group I organisms. Rosanova (1974) tested growth of

Table 4 Toxin Production by *Clostridium botulinum* in Different ''Vegetable'' Media

C. botulinum		Potatoes (pH 6.0)		Carrots (pH 6.2)		Red beets (pH 5.7)		Grass (pH 5.8)	
Type	Strain	x/y[a]	Toxin titer[b]	x/y	Toxin titer[c]	x/y	Toxin titer[d]	x/y	Toxin titer[e]
A	62A	5/5	6.6	2/2	6.2	3/3	2.0	1/1	6.1
	141A	5/5	6.4	2/2	NT				
	73A			2/2	NT	3/3	2.7		
B, proteolytic	CDI3	4/4	5.8	2/2	NT			1/1	5.3
	CD14	4/4	5.8	2/2	NT			1/1	4.7
	SNB77	4/4	6.2	2/2	5.3	3/3	2.1	1/1	5.8
	Okra	4/4	6.2	2/2	5.9	3/3	2.2	1/1	5.8
B, nonproteolytic	CDI1	0/4	—	0/2	—			0/2	—
	CDI2	0/3	—	0/2	—	0/3	—	0/2	—
C	C_a	3/5	4.3	0/2	—	1/3	2.0	0/2	—
E	RIV1	4/5	4.3	0/2	—			0/2	—
	German sprats	2/5	2.7	0/2	—				
	Beluga	5/5	3.6	0/2	—	0/3	—		

[a]x = number of tubes with toxin; y = number of tubes tested.
[b]Highest toxin titer in culture fluid (log_{10} i.p. mouse (LD_{50}/ml) after 8 days at 30°C.
[c]Highest toxin titer in culture fluid (log_{10} i.p. mouse (LD_{50}/ml) after 5 days at 30°C.
[d]Highest toxin titer in culture fluid (log_{10} i.p. mouse (LD_{50}/ml) after 10 days at 30°C.
[e]Highest toxin titer in culture fluid (log_{10} i.p. mouse (LD_{50}/ml) after 5 days at 37°C.

C. botulinum in commercially produced tomato juice and observed that *C. botulinum* grew if the pH was ≥ 4.7. Growth in pumpkin purée was observed as pH ≥ 4.45. Mordvinova et al. (1984) showed that *C. botulinum* grew and produced toxin in tomato purée at pH > 4.4, and Mazokhina and Levshenko (1983) observed growth at pH values ≥ 4.2 in gherkins. Growth of *C. botulinum* has also been observed in puréed figs (Ito et al., 1978), but it only occurred at pH > 4.9. Botulinum toxin was not produced in vacuum-packed celery (Johnson, 1979) unless it was fortified with nutrient broth. Spores of certain *C. botulinum* type A strains were able to produce high toxin titers within 48 hours in experimentally inoculated sautéed onions (Solomon and Kautter, 1986), although onion oil has been shown to be inhibitory to *C. botulinum* type A organisms (de Wit et al., 1979).

Different forms of toxin are produced by *C. botulinum*, depending on the type and the medium (Hauschild, 1989). Sugii and Sakaguchi (1977) observed that type A organisms produced both the 16 S and 19 S molecular-sized toxins in vegetables, whereas meat and fish products contained the 12 S and 16 S forms or the 12 S form alone. This may explain the relatively high oral toxicity of *C. botulinum* type A in vegetables, since the stability of botulinum toxins in the intestine increases with increased molecular size (Sugii et al., 1977).

V. INHIBITION OF GROWTH OF *C. BOTULINUM*

From the foregoing it is clear that the presence of *C. botulinum*, especially group I, is of concern in the processing of fruit and vegetables. Therefore, for the production of safe food, either outgrowth of *C. botulinum* spores has to be prevented or the spores have to be completely destroyed. The main factors limiting growth of *C. botulinum* in fruit and vegetables are pH, water activity (a_w), redox potential (E_h), and competing microorganisms. These factors, as well as added preservatives, are included as intrinsic safety factors. Besides intrinsic safety factors, extrinsic factors such as storage temperature, storage time, and method of packaging (vacuum or gas packaging) are also important in limiting growth of *C. botulinum* in fruit and vegetables.

A. Intrinsic Factors

pH

The minimum pH requirement for growth of strains of group I organisms is in the range of 4.6–4.8, though for many strains it may be well over 5.0 (Ingram and Robinson, 1951; Ohye and Christian, 1967). Most fruits and some vegetables, such as rhubarb, are sufficiently acidic to inhibit *C. botulinum* completely by their pH alone. Other products have to be acidified to inhibit outgrowth of *C. botulinum*. Some reports mentioned growth of *C. botulinum* at pH levels below

4.6. Rozanova (1974) reported toxin production in pumpkin in tomato sauce at pH 4.45. Mordvinova et al. (1984) showed that tomato purée with pH > 4.4 could cause botulism after prolonged storage. The studies of Prokhorovich et al. (1976), also suggest that the pH limit of 4.6 does not ensure inhibition of spore outgrowth and toxin formation in canned apricot juice and compote. These findings, however, have never been confirmed by others. In acidified fruit and vegetables, development of *C. botulinum* may also result from excessively slow pH equilibration. Studies by Quast et al. (1975) on pH changes at various points in canned acidified palm hearts showed that pH equilibration took several weeks. To prevent growth of *C. botulinum*, the rate of acidification could be accelerated by cutting the palm hearts into smaller sections approximately 3 cm long. Stroup et al. (1984) determined the time required for the pH to decrease to 4.8 or less inside mushrooms, pearl onions, and cherry peppers. Blanched products were packed into home canning jars, covered with acidified brine, pasteurized in steam at 100°C, and subsequently stored at 25°C. The time required for the pH decrease was a function of acid type, initial acid concentration, and the product being acidified. With sufficient acidulant to achieve an equilibrium pH of 4.6 or less, the time required was 7 days or less in all instances.

Regulations for acidified foods generally set the upper pH limit at 4.6. Notwithstanding reports of toxigenesis below pH 4.6, adherence to this level has ensured the production of safe acidified fruits and vegetables in the past.

Water Activity (a_w)

The minimum a_w for growth of *C. botulinum* group I is 0.94. The water activity of fruit and vegetables can be reduced to this level either by drying or by the addition of humectants. Salt (NaCl) has usually been used as a humectant in the preservation of vegetables, and sugar in the preservation of fruits. Under otherwise optimal conditions, the highest concentrations of salt tested allowing growth of *C. botulinum* type A, and proteolytic *C. botulinum* type B were 8 and 9.4%, respectively (Ohye and Christian, 1967). For *C. botulinum* type E the limit was 5.1% (Ohye and Christian, 1967). Salt is usually used in combination with acids to prevent outgrowth of *C. botulinum*, e.g., in fermented products such as sauerkraut.

Redox Potential (E_h)

Pasteur (1861) reported the existence of microorganisms in food that could not survive in the presence of oxygen and postulated that oxygen was toxic to them. This concept has been supported by others (Holland et al., 1969; Knaysi and Dutky, 1936). However, not all investigators believe that oxygen is toxic to anaerobes (Hanke and Katz, 1943; Quaste and Stephenson, 1926). Instead, it has been suggested that upper E_h limits exist, above which many anaerobes will not grow (Smith, 1975; Vennesland and Hanke, 1940; Barnes and Ingram, 1956;

Hanke and Bailey, 1945). In general, rapid growth of *C. botulinum* will take place in a suitable medium when the E_h is between −6 and −436 mV (Morris and O'Brien, 1971). Among the substrates present in fruit and vegetables that help to maintain reducing conditions are acids such as ascorbic acid and reducing sugars. Due to the relatively high content of reducing agents in fruit and vegetables, the E_h of these products is always in the range that allows growth of *C. botulinum*. Smith (1975) reported E_h values of −399 mV in canned tomato sauce, −309 mV in tomato juice, −296 mV in peeled potatoes, and −194 mV in tomato bisque soupe. Montville and Conway (1982) determined the E_h of a wide variety of canned foods. All types of canned fruit and vegetables tested had E_h values ranging from −18 to −438 mV. Foods packed in glass had higher redox potentials than foods packed in cans. Mushrooms packed in glass had E_h values ranging from −194 to −201 mV. In cans, the values ranged from −466 to −509 mV. Inoculated canned mushrooms, whole corn, asparagus, and different types of beans supported toxin production by *C. botulinum* (Montville and Conway, 1982).

Vegetables and fruits are sold increasingly as fresh products. Some of them, such as mushrooms, are often packed in plastic containers and sealed with polyvinylchloride (PVC) film. Respiration of the vegetables decreases the available O_2, produces CO_2, and results in a less aerobic atmosphere. This atmosphere extends the shelf life of the product (Nichols and Hammond, 1973), but favors the growth of *C. botulinum*. Sugiyama and Yang (1975) showed that *C. botulinum* can grow and produce toxin in the PVC-wrapped packs before obvious spoilage occurs.

Associating Microorganisms

Microorganisms other than *C. botulinum* have a significant role in the development of *C. botulinum*. On the one hand, they can inhibit growth of *C. botulinum*. On the other, they can promote growth. Furthermore, associating microorganisms can have a protective function by causing spoilage that would make a toxic product less likely to be consumed. Growth of *C. botulinum* can be inhibited by lactic acid bacteria, e.g., due to an early pH drop. Growth can be promoted when accompanying microorganisms decrease the redox potential or increase the pH of the food. Promotion of growth by accompanying organisms is known as metabiosis. Since tomato products are frequently implicated in botulism outbreaks, much research has been carried out on metabiosis, especially with respect to the pH-increasing effect of microorganisms in tomato products that would allow growth of *C. botulinum*. Yeast and molds surviving the heating process have the ability to raise the pH during growth, at least in their vicinity, to a level that would allow *C. botulinum* to develop, while the overall pH may remain well below 4.6. Tanner et al. (1970), De Lagarde and Beerens (1979), and Huhtanen et al. (1976) were able to detect botulinum toxin in acid foods inoculated with *C. botulinum* spores where molds such as *Penicillium* spp., *Mycoderma* spp., *Trichosporon* spp., and *Cladosporium* spp. had grown and raised the pH locally to above 4.6. The possibility of growth of *C. botulinum* spores, in combination with molds, in tomato juice with a pH of 4.0–4.3 was shown by Gola and Casolari (1980). Odlaug and Pflug (1979) demonstrated that *Aspergillus gracilis*, inoculated along with *C. botulinum* spores into tomato juice with pH 4.2, developed a pH gradient under the mycelial mat that resulted in growth of *C. botulinum* and toxin production. In a hermetic unit, however, mold growth was reduced, and no pH gradient was detected. *C. botulinum* also grew in fresh tomatoes and produced toxin in a metabolic relationship with certain fungi (Draughon et al., 1988). Infection of fresh tomatoes with *Alternaria*, *Rhizoctonia*, and *Fusarium* spp. increased with pH of tomato tissue to 8.0. However, not all mold species capable of increasing the pH of tomato products will allow growing of *C. botulinum*. Various molds inhibit growth of *C. botulinum*, presumably by the production of antibiotic metabolites (Gola and Casolari, 1980). Bacteria are also able to promote growth of *C. botulinum* in tomato products. Montville (1982) showed that *Bacillus licheniformis* elevated the initial pH of 4.4 of canned tomatoes sufficiently to allow *C. botulinum* to grow and produce toxin. Toxin production was also observed when spores of both species were co-inoculated at levels as low as 10 spores/ml. Gola and Casolari (1980) found the same metabiotic effect with *Serratia marcescens* and *Enterobacter aerogenes*.

Preservatives

Factors such as pH, a_w, incubation temperature, etc., are the main parameters in preventing botulism. Besides these, preservatives are often incorporated as additional safety factors to control *C. botulinum*. Commonly used preservatives include sodium nitrite, usually combined with sodium chloride (Roberts et al., 1981; Sofos et al., 1989), potassium sorbate (Ivey et al., 1978; Seward et al., 1982), and sulfites (Ingram, 1972; Robach, 1980). The disadvantages of nitrite, a potential precursor of carcinogenic nitrosamines as a preservative, are well documented (Krol and Tinbergen, 1974; Proc. U.S. Department of Agriculture, 1978). The use of other preservatives is often limited because their taste becomes noticeable (Robach, 1980) or because of legal limitation, as in the case of sulfites.

Preservatives for fruits and vegetables are almost all acidulants, e.g., citric acid, acetic acid, and ascorbic acid. These acids are natural components of many foods and are generally regarded as safe to the consumers. The ability of bacteria to grow in foods or culture media in acidic conditions depends on both the pH and the nature of the acidulant. Graham and Lund (1986) demon strated that if the pH of a medium was adjusted to 5.2 with citric acid, the growth of *C. botulinum* (both spores and vegetative cells) was strongly inhibited. However, if the pH of the medium was adjusted to 5.2 with HCl, no such inhibition was observed. The inhibitory effect of citric acid was prevented by adding Ca^{2+} and Mg^{2+}. The effect of citric acid, in combination with ascorbic acid, on outgrowth and toxin formation by *C. botulinum* on potatoes was demonstrated by Notermans et al. (1985a). Dipping potatoes in a solution of citric acid (1% v/v) and ascorbic acid (2% v/v) before vacuum packing and cooking inhibited growth and toxin production by *C. botulinum* type B (group I) at an incubation temperature of 15°C for 70 days. The preservative treatment also resulted in an organoleptically acceptable product with a prolonged shelf life. Without the preservative treatment, the shelf life of the potatoes at 15°C was only 12–13 days. In general, ready-to-eat products without safety factors such as low pH or a_w levels should be stored at temperatures below 4°C to prevent growth of *C. botulinum*. The above-mentioned treatment adds an additional safety factor to the potato product. The nature of the antibotulinal effect of the ascorbic/citric acid treatment is not fully understood. It cannot be explained by the pH reduction alone because the treatment decreased the pH of the potato surface only from pH 5.9–6.0 to pH 5.4–5.8 (Notermans et al., 1985a). A similar extension of the shelf life of cooked, vacuum-packed potatoes by decreasing the pH was shown by Dodds (1989).

Acetic acid is also effective in delaying growth of *C. botulinum*. Baumgart (1987) dipped pasteurized potatoes in 0.5% acetic acid solution for 15 minutes and did not observe toxin formation until after 21 days of incubation at 20°C. However, due to the acetic acid treatment, the pH of the potatoes was reduced to 4.8. Ito et al. (1976b) demonstrated that addition of various amounts of acetic acid to puréed cucumbers inhibited outgrowth of *C. botulinum* at pH 4.8 but not at pH 5.0.

In fruit and vegetables, especially in spices, components may be present that inhibit growth and toxin production by *C. botulinum*. Ismaiel (1988) tested essential oils of clove, thyme, black pepper, pimento, origanum, garlic, onion, and cinnamon for their effects on germination, growth, and toxin formation of *C. botulinum*. Garlic oil was the most potent inhibitor of germination, while the oils of black pepper, clove, cinnamon, and origanum were the strongest inhibitors of vegetative growth. Huhtanen (1980) showed that alcoholic extracts of mace and achiote were strongly inhibitive, whereas extracts of bay leaf, white pepper, and nutmeg were moderately inhibitive to *C. botulinum*. Extracts of mace, bay leaf, and nutmeg also inhibited toxin formation by *C. botulinum* (Hall and Maurer, 1986). As well, alcoholic extracts of cloves, oregano, rosemary, sage, and thyme were inhibitory to *C. botulinum* (Ueda et al., 1982). Of vegetables, *Alllium* spp. have been found to contain components inhibiting the growth of *C. botulinum* and other microorganisms (Dankert et al., 1979; Giménez et al., 1988; Mantis et al., 1979). Oil of garlic (*A. sativum*) or onion (*A. sepa*) at concentrations of $\geq$ 1500 μg/g inhibited toxin production by spores of *C. botulinum* type A. Inhibition, however, was not complete. Toxin production by *C. botulinum* type B (group I) and type E was not inhibited (Wit et al., 1979).

Whether the inhibitory components present in spices and some type of vegetables are of practical value is not yet clear and has to be tested further. The different outbreaks of botulism caused by botulinum toxin–containing garlic and onions (St. Louis et al., 1988; Solomon and Kautter, 1986) demonstrate that these intrinsic factors alone are of little or no value as inhibitory factors.

B. Extrinsic Factors

Temperature

Temperature is an important safety factor in the control of growth and toxin production by *C. botulinum* in all types of food products. The established lower limits are approximately 10°C for group I organisms (Sperber, 1982; Smelt and Haas, 1978) and 3.5°C for group II organisms (Eklund et al., 1967; Roberts and Hobbs, 1968; Schmidt et al., 1961; Simunovic et al., 1985). These limits have been determined under otherwise optimal growth conditions.

In a study by Notermans et al. (1981b) toxin production by group I organisms in vacuum-packed cooked potatoes was observed at an incubation temperature of 10°C. However, the quantity of toxin was less than 10 i.p. mouse (LD_{50}/g). The maximal quantities of toxin after incubation at 15 and 20°C amounted to 1.2×10^4 i.p. mouse (LD_{50}/g and 2×10^6 i.p. mouse (LD_{50}/g), respectively. *C. botulinum* type E (group II) produced 2.5×10^3 i.p. mouse LD_{50}/g at an incubation temperature of 10°C. At 4°C, neither group I nor group II organisms were found to produce toxin. Gola and Mannino (1985) observed no production of toxin by *C. botulinum* group I in canned peas stored at 8°C.

Storage Time

Toxigenesis by *C. botulinum* at the lower end of the growth range takes several weeks and is influenced by suboptimal temperature, a_w, pH, and inoculum size (Eklund et al., 1967; Smelt and Haas, 1978; Solomon et al., 1977). Smelt and Haas (1978) observed that at an incubation temperature of 12°C it takes 3–4 weeks before group I organisms produce toxin. At 15°C, toxin formation could be observed within 1 week, and at 20°C, within 2–3 days. For their growth experiments, Smelt and Haas (1978) used cooked meat medium. In experiments with vacuum-packed cooked potatoes stored at 15°C, Notermans et al. (1981b) observed toxin production within only 4 days. Notermans et al. (1970) observed that by pretreatment of spores of *C. botulinum* type B of group II at 3°C, toxinogenesis during subsequent incubation at 8°C was stimulated. Similar results have been observed by Smelt and Haas (1978) for group I organisms. Incubation of spores of group I at 10°C prior to incubation at 20°C reduced the lag time to toxin detection.

In cucumber purée inoculated with group I organisms, no toxin was produced at pH 4.8, even after prolonged incubation. However, toxin production was observed at pH 5.0 after 18 days and at pH 5.5 within 4 days of incubation (Ito et al. 1976a).

The water activity (a_w) is another factor determining the time of toxigenesis. In experiments with cooked, vacuum-packed potatoes with a pH of 5.5 and an a_w of 0.96, Dodds (1989) found no toxin production by *C. botulinum* of group I over 35 days of incubation. At a_w levels of 0.97 and 0.98, toxin was detected in the potatoes after incubation times of 14 days and 7 days, respectively.

In addition to temperature, pH, and a_w, inoculum size also influences the time in which toxin is produced. In cucumber puree with a pH of 5.2 inoculated with 10^2 spores, production of toxin was observed after 12 days of incubation. If 10^6 spores were inoculated, production of toxin was observed after 8 days. At pH 5.0, growth and toxin production were observed only if $\geq 10^6$ spores were added (Ito et al., 1976a). The study of Sugiyama et al. (1981) with foil-wrapped baked potatoes also shows that toxin was produced in an inoculum dose-dependent manner. Times to first detectable toxin ranged from 3 days with an inoculum of 10^5 spores to 6–7 days with an inoculum of 10 spores. Rosanova (1974) found that in tomato juice with a pH of 4.7, toxin was produced only if the spore inoculum was larger than 1450/ml. Production of toxin was not observed with lower numbers, even after 13 months of incubation. In experiments by Notermans et al. (1990), toxin production at 8°C by nonproteolytic *C. botulinum* was tested in fortified egg-meat medium inoculated with different numbers of spores. If low numbers of spores (5 spores/30 ml) were added, no toxin production occurred within approximately 3 weeks of incubation. With higher numbers of spores (5000 spores/30 ml), toxin production was observed after 12 days of incubation.

Storage Conditions and Packaging

Storage conditions and the method of packaging strongly influence the shelf life of fruit and vegetables. Recontamination of prepacked cooked vegetables does not occur during storage. However, due to the low redox potential obtained after cooking, surviving spores of *C. botulinum* may grow and produce toxin. This has been demonstrated by Dodds (1989) and by Notermans et al. (1981b). Kautter et al. (1981) evaluated the botulism hazard from nitrogen-packed sandwiches. They found that these became toxic after 4 days of incubation at room temperature, while remaining fully acceptable organoleptically. Use of modified atmosphere for packaging of precooked foods may reduce growth of *C. botulinum*, especially when gas mixtures with a certain percentage of oxygen are chosen.

Modified-atmosphere packaging can extend high-quality storage life for a range of fresh fruits and vegetables (Priepke et al., 1976; McLachlan and Stark, 1985). This is due to a reduction in the rates of metabolic reactions and the retardation of ripening, senescence, and associated loss of acceptability. The optimal O_2 and CO_2 concentrations to achieve these effects are dependent on the individual product and its storage temperature (Smock, 1979; Isenberg, 1979). In a sealed package containing fresh produce, a modified atmosphere is produced as a result of respiration. If the film enclosing the product is insufficiently permeable, anaerobic conditions soon prevail. Studies in which mushrooms were artificially contaminated with spores of *C. botulinum* showed that botulinum toxin can be produced in such packs before obvious spoilage (Sugiyama and Yang, 1975; Sugiyama and Rutledge, 1978). However, if the film is permeable to O_2 and CO_2, an equilibrium concentration of both gases is established when the rate of gas transmission through the package is equal to the rate of respiration. The production of botulinum toxin in the packaged mushrooms was avoided by allowing entry of sufficient oxygen into the package (Sugiyama and Rutledge, 1978). This can be achieved by placing small air holes in the packages or by use of permeable packaging materials (Ballantyne et al., 1988). The equilibrium-modified atmosphere attained within a sealed package depends further upon product rate of respiration, fill weight, and surface area of gas exchange (Henig, 1972). The rate of respiration is incubation temperature dependent. Due to these factors it is difficult to predict the oxygen concentration in the package and whether or not *C. botulinum* will develop and produce toxin. Therefore, more research is needed to test the mode of packaging on growth of *C. botulinum*.

A common method of packaging fresh fruit and vegetables is by vacuum. Vacuum packaging removes almost all atmospheric gases from both product and package and is distinct from hypobaric storage conditions (Brecht, 1980). The effects of conditions within a vacuum package on senescence of fruits and vegetables have received little research attention. The anaerobic conditions prevailing in vacuum packages may result in off-flavor development in some product (McLachlan and Stark, 1985), including prepeeled potatoes (Keybets, 1981), and would facilitate growth and toxin production by *C. botulinum* (Tamminga et al., 1978). Johnson (1979) did not observe production of botulinum toxin in vacuum-packed fresh celery, probably because of nutrient deficiency. If, however, the product was fortified with nutrient broth, toxin production was observed after 8 weeks at 25°C.

Controlled atmosphere storage of fruit and vegetables is maintained by a continuous flow of appropriate mixtures of O_2, CO_2, N_2, etc. The optimal gas combination and storage temperature are product dependent (Sing et al., 1972). The beneficial effects of controlled atmosphere storage have been attributed to the inhibition of the respiration rate by low O_2 flow combined with low storage temperature (Karnik et al., 1970; Sing et al., 1970). Storage temperature below 3.3°C does not allow *C. botulinum* to grow.

The preservative effects of gas exchange procedures have been examined by Dignon (1985) using diced potatoes. Their results show that neither SO_2, CO_2, nor CO had any sporicidal effect on *C. botulinum* spores, and that toxigenesis occurred without the usual accompanying signs of spoilage. A small sporicidal effect of SO_2 was observed by Kaffezaki et al. (1969). Their study showed that gas exchange of fresh apple and potato pieces with pure SO_2 resulted in partial destruction of *C. botulinum* spores. Treatment with pure CO or N_2 had no appreciable antimicrobial effect.

VI. DESTRUCTION OF SPORES OF *C. BOTULINUM*

For production of safe food it must be assumed that spores of *C. botulinum* are always present on raw food materials. Products such as fruit and vegetables that have been in close contact with soil are the most likely to be contaminated. Therefore, during processing of raw food materials, factors have to be built in to prevent toxin formation by *C. botulinum*. If intrinsic factors (pH, a_w, etc.) and extrinsic factors (temperature, time, etc.) are not adequate to prevent *C. botulinum* spores from developing, the spores must be destroyed. Spores present in food are stable and survive for long periods, even under adverse conditions. Odlaug and Pflug (1977a) reported that *C. botulinum* type A and B spores stored in tomato juice (ph 4.2) and citric acid phosphate buffer (pH 4.2) at 4, 22, and 32°C showed no significant decrease in their numbers over a period of 180 days. Spores of *C. botulinum* also survived in apricot purée with a pH of 3.7 (Flaumenbaum et al., 1970). Fruit and vegetables packed in hermetically sealed containers without effective intrinsic and extrinsic safety factors, therefore, must be processed at a temperature and for a time sufficient for complete microbial spore destruction. The processing industry normally assures safety by adhering to the so-called 12D concept (Anonymous, 1965), unless preservation is assured by other safety factors. Besides heat, irradiation and chemical destruction are used as sterilizing agents in food processing.

Heating is the most widely applied technique for destruction of bacterial spores. The kinetics of spore destruction is discussed in detail in Chapter 6. As mentioned in this chapter, the resistance of spores can be expressed in decimal reduction time (D-value). D-values depend on the heating environment. Hutchings (cited by Odlaug and Pflug, 1978) extensively investigated the effect of pH on the heat resistance of *C. botulinum* spores in food and reported a significant decrease in the spore D-value as the food became more acidic. The extrapolated $D_{100°C}$ for spores of *C. botulinum* type A (strain 62A) in tomato sauce product was approximately 6 minutes at pH 4.0 and 42 minutes at pH 7.0. Odlaug and Pflug (1977b) and Mordvinova et al. (1980) found that the heat resistance of *C. botulinum* spores in tomato juice (pH 4.2) was three times lower than in neutral buffer. These results all show that *C. botulinum* spores have lower heat resistance in acid foods than in low-acid foods.

Stumbo et al., (1950) determined the $D_{105°C}$-value of *C. botulinum* spores in different media. In puréed peas the $D_{105°C}$-value was 10.2 minutes. For strained squash, phosphate buffer (pH 7.2) and distilled water the D-values were 6.9,

10.0, and 5.5 minutes, respectively. Notermans et al. (1981) showed that the thermal inactivation of *C. botulinum* spores in vacuum-packed potatoes was slower than in phosphate buffer. The $D_{95°C}$-values for spores of *C. botulinum* type A (strain 62A) and *C. botulinum* type B (strain Okra) were 65 and 35 minutes, respectively. In phosphate buffer with an identical pH as the potatoes, the $D_{95°C}$-values were 38 and 23 minutes, respectively. Odlaug et al. (1978) found marked differences between D-values for spores heated in buffer and in mushroom purée. At 110°C, the mean D-value for spores was 0.78 minutes in mushroom purée and 1.17 minutes in buffer. The a_w also affects the D-value. In general, D increases with decreasing a_w (Alderman et al., 1972; Gombas, 1983). Fats and proteins are also protective (Lücke, 1983).

Double pasteurization has also been tested for destruction of *C. botulinum* spores. Lund et al. (1988) examined this process for the preservation of vacuum-packed potatoes. After the first pasteurization, potatoes were vacuum packed, stored at 25–35°C for up to 24 hours to allow germination of bacterial spores, and then pasteurized again. However, the double pasteurization process was not effective because a high proportion of inoculated spores did not germinate and remained viable after the second pasteurization. This resulted in growth and formation of toxin within 5–9 days at 25°C.

VII. CONCLUSIONS AND OUTLOOK TO THE FUTURE

Fruit and vegetables are frequently involved in human cases of botulism. To reduce the botulism risk, certain safety factors must be in place during production and processing and/or subsequent storage. Relevant factors in the safety of fruit and vegetables are summarized in Table 5. A major factor contributing to the botulism risk is the frequent presence of *C. botulinum* spores in raw products. Major contamination sources of raw fruit and vegetables are soil and organic fertilizers. An increasing quantity of fruit and vegetables are grown under increasingly controlled conditions, e.g., tomatoes in greenhouses using aquaculture. Mushroom production is another example of a well-controlled growing system requiring high hygienic standards. As a consequence of such horticultural practices, contamination of some raw fruits and vegetables with spores of *C. botulinum* can be held to a minimum. Nevertheless, even though the spore incidence may be low, some potential for botulism outbreaks always exists. Not all fruits and vegetables are good substrates for growth of *C. botulinum*. Growth of *C. botulinum* organisms of group II (the nonproteolytic strains) is generally poor, whereas growth of group I organisms (the proteolytic strains) is usually excellent. This may be the main reason why botulism outbreaks involving fruit and vegetables are commonly caused by group I organisms.

To prevent outgrowth and toxin formation by *C. botulinum* in fruit and vegetables, either the growth of the organism must be inhibited or the organism has

Table 5 Factors Affecting Botulinal Safety of Fruit and Vegetables

C. botulinum spore incidence	Contact
	Agricultural practices
Fruit and vegetables as substrate	Suitable for group I organisms
	Less suited for group II organisms
Inhibition of growth	
Intrinsic factors	
pH	No growth at pH $\leq$ 4.6
	PH equilibration time
a_w	At a_w < 0.94 no growth with NaCl as humectant
E_h	Growth can be initiated at high E_h (+200 mV)
	Optimal growth from − 6 mV to − 500 mV
Associating microorganisms	Growth inhibition
	Spoilage
	Metabiotic effect
Preservatives	Additives
	Natural components
Extrinsic factors	
Temperature	No growth of group I organisms at < 10°C
	No growth of group II organisms at < 3.5°C
Storage time	Related to intrinsic factors, storage temperatures, and spore incidence
Storage/packaging	Modified atmosphere packaging of fresh produce
	Controlled atmosphere storage
	Vacuum packaging of heat processed vegetables.
Destruction of spores	Heat destruction (12D concept)
	Irradiation
	Chemical destruction

to be destroyed. The main intrinsic factors inhibitory to growth are pH, a_w, E_h, associating microorganisms, and preservatives. The minimum pH requirement for growth of *C. botulinum* is in the range of 4.6–4.8, although Russian researchers reported growth of *C. botulinum* in tomato sauce, apricot juice, and compote at pH < 4.6. However, these reports gave no detailed information about the adequacy of the pasteurization process and the potential for fungal growth with development of *C. botulinum* adjacent to the fungal layer. Therefore, additional research should be carried out to confirm these findings. This is of interest since it does appear that under specialized circumstances toxin production at pH < 4.6 can occur (Raatjes and Smelt, 1979; Tanaka, 1982). In addition to a minimum pH level for growth inhibition, attention has to be paid to pH equilibrium of large pieces of fruit and vegetables if they are preserved by low pH. In media or foods with NaCl as the main a_w depressant, the growth-limiting water activity is about 0.94. However, the minimum a_w at which growth initiates depends on both the pH and temperature (Baird-Parker and Freame, 1967). Ascorbic acid and reducing sugars present in fruit and vegetables, as well as respiratory activity of the raw product, all contribute to a low E_h level that is conducive to growth of *C. botulinum*. Associating microorganisms present in a product generally inhibit the growth of *C. botulinum*. Furthermore, associating microorganisms have a protective function by causing spoilage. In some cases, however, a metabiotic effect of associating microorganisms enabling growth of *C. botulinum* in acid products is observed. Preservatives are important in the inhibition of growth of *C. botulinum*, but the social acceptance of certain preservatives is decreasing. Therefore, the preservative effects of acidulants that are natural components of many foods, e.g., citric and ascorbic acids, and that are regarded as safe need further research. The preservative effect of some natural acids cannot be ascribed to their pH-lowering effect alone. Ascorbic acid prolongs the shelf life of processed vegetables due to its reducing capacity.

Many studies have been carried out to determine the antibotulinal capacities of extracts of spices, garlic, onions, etc. However, their practical value is limited, due to the relatively high concentrations needed for an effective preservation.

Extrinsic factors limiting growth of *C. botulinum* include storage temperature and time. In general, no growth of group I and group II organisms is observed at temperatures $< 10°C$ and $< 3.5°C$, respectively, but these limits have been determined under otherwise optimal conditions, and long periods are required before toxin production occurs. The inoculum size also influences the time needed for toxin to be produced. Suboptimal storage temperature, in combination with suboptimal intrinsic factors such as pH, a_w, and preservatives and a low contamination rate makes is possible to produce safe products for a limited storage time. Prediction of product safety requires experimental support with artificially contaminated products, and safe processing is generally assured by a code of production and by the so-called hazard analysis critical control point (HACCP) concept.

Modified atmosphere packaging is a new development and has a range of applications in the preservation of fruit and vegetables. These are largely based on reducing the respiration rate of the produce, which slows down the process of physiological aging. Mushrooms packed under modified atmosphere conditions allow toxin production to occur. Therefore, more research is needed to improve the botulinal safety of this method of packaging. Vacuum packaging of pasteurized fruit and vegetables is increasing. These products have a limited shelf life, and the control of *C. botulinum* depends largely on a low storage temperature. Research is needed to ensure safety of such products when the shelf life is extended. Controlled atmosphere storage is commonly used in combination with low temperatures to extend the shelf life of fruit and vegetables. In all probability, this type of storage can be applied to pasteurized foods as well. If well-selected gas mixtures are chosen, growth of *C. botulinum* may be avoided.

Finally, botulinal safety can be obtained by spore destruction. A heat treatment of F_o = 3.0 minutes has traditionally been shown to adequately control botulism (Riemann, 1963; Tompkin and Christiansen, 1976). Presently, irradiation cannot be considered as a satisfactory alternative to the heat destruction of *C. botulinum* spores.

REFERENCES

Alderman, G. G., King, G. J., and Sugiyama, H. (1972). Factors in survival of *Clostridium botulinum* type E spores through the fish smoking process. *J. Milk Food Technol.* *35*:162.

Anellis, A., Grecz, N., and Berkowitz, D. (1965). Survival of *Clostridium botulinum* spores. *Appl. Microbiol.* *13*:397.

Anusz, Z. (1986). Zatrucia jadem kielbasianym. *Przeg. Epidemiol.* *40*:89.

Anon. (1965). The technical basis for legislation of irradiated food. Report of a Joint FAO/IAEA/WHO Expert Committee. WHO Tech. Rept. Series 316, Appendix 8.

Aureli, P., Fenicia, L., and Ferrini, A. M. (1984). Intossicazione alimentare da *Clostridium botulinum* tipo E. *Boll. 1st. Sieroter Milan* *63*:553.

Baird-Parker, A. C., and Freame, B. (1967). Combined effect of water activity, pH and temperature on the growth of *Clostridium botulinum* from spore and vegetative cell inocula. *J. Appl. Bacteriol.* 30:420.

Ballantyne, A., Stark, K., and Selman, J. D. (1988). Modified atmosphere packaging of shredded lettuce. *Int. J. Food Sci. Technol.* *23*:267.

Barbuti, S., Quarto, M., Ricciardi, G., and Armenise, E. (1980). Research on the presence of *Clostridium botulinum* in soils cultivated for vegetables. *Igiene Moderna* *73*:3.

Barnes, E. M., and Ingram, M. (1956). The effect of redox potential on the growth of *Clostridium welchii* strains isolated from horse muscle. *J. Appl. Bacteriol.* *19*:117.

Baumgart, J. (1987). Occurrence and growth of *Clostridium botulinum* in vacuum-packed raw and pasteurized potatoes and potato salad. *Chem. Mikrobiol. Techn. Lebensmitt.* *11*:74.

Brecht, P. E. (1980). Computer assessment of the variables affecting respiration and quality of produce. Ph.D. thesis, Rutgers University, Brunswick, NJ.

Carré, H., Gledel, J., Poumeyrol, M., Sebald, M., Thomas, G., and Veit, P. (1987). Enquête sur un foyer de botulisme. Nécessité du respect des bonnes pratiques professionnelles. *Med. Nutr.* *23*:393.

Casillas, D., Moncado, R., Raffer, P. K., Vista, C., Ramras, D G., Redmond, R. B., Renger, G., Townsend, W. A., Vera, A. G., Casteneda, R., Cota, T., Midura, T., and Werner, S. B. (1978). Botulism—California. *Morbid. Mortal. Weekly Rep.* *27*:501.

Chen, L. H., Cheng, C. M., Chang, R. M., Du, S. J., and Wang, C. C. (1987). Report of botulism tests from outbreak caused by peanuts packaged in a glass jar. *Food Science, China* *14*:92.

Ciccarelli, A. S., and Giménez, D. F. (1981). Clinical and epidemiological aspects of botulism in Argentina. In *Biomedical Aspects of Botulism* (S. E. Lewis, ed.), Academic Press, New York, p. 291.

Dankert, J., Tromp, Th. F. J., De Vries, H., and Klasen, H. J. (1979). Antimicrobial activity of crude juices of *Allium ascalonicum*, *Allium sepa* and *Allium sativum*. *Zbl. Bakt. Hyg. I. Abt. Orig. A 245*:229.

De Lagarde, A., and Beerens, H. (1979). Contribution à l'étude de la formation de toxine botulique dans les conserves de fruits. *Ann. Inst. Pasteur Lille 21*:231.

Dignon, E. D. (1985). Evaluation of the botulism hazard from diced raw potatoes preserved by gas exchange. *Diss. Abstr. Int. B*; 46, 1011 University of Maryland.

Dodds, K. L. (1989). Combined effect of water activity and pH on inhibition of toxin production by *Clostridium botulinum* in cooked, vacuum-packed potatoes. *Appl. Environ. Microbiol. 55*:656.

Draughon, F. A., Chen, S., and Mundt, J. O. (1988). Metabiotic association of *Fusarium*, *Alternaria* and *Rhizoctonia* with *Clostridium botulinum* in fresh tomatoes. *J. Food Sci. 53*:120.

Dymickey, M., and Trenchand, H. (1982). Inhibition of *Clostridium botulinum* 62A by saturated N-aliphatic acids, N-alkyl formates, acetates, propionates and butyrates. *J. Food Sci. 45*:1117.

Eklund, M. W., Wieler, D. I., and Poysky, F. T. (1967). Outgrowth and toxin production of non-proteolytic type B *Clostridium botulinum* at 3.3 to 5.6°C. *J. Bacteriol. 93*: 1461.

Flaumenbaum, B. I., Fedotov, N. I., Mordivinova, S. A., and Raevskaya, M. V. (1970). A new method for sterilizing apricot purée. *Konservnaya i Ovoshchesushil' naya Promyshlennost 9*:18.

Gaze, J. E., and Brown, K. L. (1988). The heat resistance of spores of *Clostridium botulinum* 213B over the temperature range 120 to 140°C. *Int. J. Food Sci. Technol. 23*:373.

Geiger, J. C. (1941). An outbreak of botulism. *JAMA 117*:22.

Giménez, M. A., Solanes, R. E., and Giménez, D. F. (1988). Growth of *Clostridium botulinum* in media with garlic (*Allium sativum*). *Rev. Arg. Microbiol. 20*:17.

Gola, S., and Casolari, A. (1980). Possibility of growth of *Clostridium botulinum* in tomato juice. *Industria Conserva 55*:294.

Gola, S., and Mannino, S. (1985). Toxin production by some *Clostridium botulinum* strains in culture and food media. *Industria Conserva 60*:39.

Gombas, D. E. (1983). Bacterial spore resistance to heat. *Food Technol. 37*:105.

Graham, A. F., and Lund, B. M. (1986). The effect of citric acid on growth of proteolytic strains of *Clostridium botulinum*. *J. Appl. Bacteriol. 61*:39.

Haagsma, J. (1973). The etiology and epidemiology of waterfowl botulism in The Netherlands. Thesis, Rijks Universiteit Utrecht, The Netherlands.

Hall, M. A., and Maurer, A. J. (1986). Spice extracts, lauricidin, and propylene glycol as inhibitors of *Clostridium botulinum* in turkey frankfurter slurries. *Poultry Sci. 65*:1167.

Haller, H. D., May, R. T., and Roth, R. L. (1969). Botulism--a case report. *J. Kentucky Med. Assoc. 61*:820.

Hanke, M. E., and Bailey, J. H. (1945). Oxidation-reduction potential requirements of *Cl. welchii* and other clostridia. *Proc. Soc. Exp. Biol. Med. 59*:163.

Hanke, M. E., and Katz, Y. J. (1943). An electrolytic method for controlling oxidation-reduction potential and its application in the study of anaerobiosis. *Arch. Biochem. Biophys. 2*:183.

Hansen, P. I. E. (1966). A consumer survey for acceptance evaluation of cured ham treated by a combination of heat and irradiation. *Food Technol., Champaign 20*:99.

Hauschild, A. H. W., Aris, B. J., and Hilsheimer, R. (1975). *Clostridium botulinum* in marinated products. *Can. Inst. Food Sci. Technol. J. 8*:84.

Hauschild, A. H. W. (1989). *Clostridium botulinum*. In *Foodborne Bacterial Pathogens* (M. P. Doyle, ed.), Marcel Dekker, New York, p. 111.

Henig, Y. (1972). Computer assessment of the variables affecting respiration and quality of produce. Ph.D. thesis, Rutgers University, Brunswick, NJ.

Holland, D., Barker, A. N., and Wolf, J. (1969). Factors affecting germination of clostridia. In *Spores IV* (Campell, L. L., ed.), American Society Microbiol., Bethesda, MD., p. 317.

Huhtanen, C. N. (1980). Inhibition of *Clostridium botulinum* by spice extracts and aliphatic alcohols. *J. Food Prot. 43*:195.

Huhtanen, C. N., Naghski, J., Custer, C. S., and Russell, R. W. (1976). Growth and toxin production by *Clostridium botulinum* in molded tomato juice. *Appl. Environ. Microbiol. 32*:711.

Huss, H. H. (1980). Distribution of *Clostridium botulinum*. *Appl. Environ. Microbiol. 39*:764.

Ingram, M. (1972). Meat preservation-past, present and future. *Roy. Soc. Health J. 92*:121.

Ingram, M., and Robinson, R. H. M. (1951). The growth of *Clostridium botulinum* in acid bread media. *Proc. Soc. Appl. Bacteriol. 14*:62.

Insalata, N. F., Witzeman, J. S., and Berman, J. H. (1969). The problems and results of an incidence study of the spores of *Clostridium botulinum* in convenience foods. *Health Lab. Sci. 7*:54.

Isenberg, F. M. R. (1979). Controlled atmosphere storage of vegetables. *Horticult. Rev. 1*:337.

Ismaiel, A. A. (1988). The inhibition of *Clostridium botulinum* growth and toxin production by essential oils of spices. *Diss. Abstr. Int.*, B *48*, 2834.

Ito, K. A., Chen, J. K., Lerke, P. A., Seeger, M. L., and Unverferth, J. A. (1976). Effect of acid and salt concentration in fresh-pack pickles on the growth of *Clostridium botulinum* spores. *Appl. Environ. Microbiol. 32*:121.

Ito, K. A., Chen, J. K., Seeger, M. L., Unverferth, J. A., and Kimball, R. N. (1978). Effect of pH on the growth of *Clostridium botulinum* in canned figs. *J. Food Sci. 43*:1634.

Ivey, F. J., Shaver, K. J., Christiansen, L. N., and Tompkin, R. B. (1978). Effect of potassium sorbate on toxinogenesis by *Clostridium botulinum* in bacon. *J. Food Prot. 41*:621.

Johnson, C. E. (1979). Behavior of *Clostridium botulinum* in vaccum-packed fresh celery. *J. Food Prot. 42*:49.

Kaffezaki, J. G., Palmer, S. J., and Kramer, A. (1969). Microbiology of fresh apple and potato plugs preserved by gas exchange. *J. Food Sci. 34*:426.

Karnik, V. V., Salunkhe, D. K., Olson, L. E., and Post, F. J. (1970). Physiochemical and microbiological effect of controlled atmosphere on sugar beet. *J. Am. Soc. Sugar Beet Technol. 16*:156.

Kautter, D. A., Lynt, R. K., Lily, T., and Solomon, H. M. (1981). Evaluation of the botulism hazard from nitrogen-packed sandwiches. *J. Food Prot. 44*:59.

Keybets, M. J. H. (1981). Vacuum and gas packaging, an alternative for sulphite in prepeeled potatoes. *Proc. 8th Triennial Conf. Eur. Ass. Potato Res.*, Munich, p. 263.

Knaysi, G., and Dutky, S. R. (1936). The growth of a butanol clostridium in relation to the oxidation-reduction potential and oxygen content of the medium. *J. Bacteriol. 31*:137.

Kravchenko, A. T., and Shishulina, L. M. (1967). Distribution of *Cl. botulinum* in soil and water in the U.S.S.R. In *Botulism 1966* (M. Ingram and T. A. Roberts, eds.), Chapman and Hall, London, p. 13.

Krol, B., and Tinbergen, B. J., eds. (1974). *Proceedings of the International Symposium on Nitrite in Meat Products*, Pudoc, Wageningen, The Netherlands.

Lücke, F. K. (1983). Botulismus in Europa—epidemiologische und experimentelle Befunde, neue Erkenntnisse über die Erreger mikrobieller Lebensmittel-Vergiftungen. *Schweiz. Ges. Lebensmittelhyg. 13*:30.

Lund, B. M., Graham, A. F., and George, S. M. (1988). Growth and formation of toxin by *Clostridium botulinum* in peeled, inoculated, vacuum-packed potatoes after a double pasteurization and storage at 25°C. *J. Appl. Bacteriol. 64*:241.

Mantis, A. J., Koides, P. A., Karaioannoglou, P. G., and Panetsos, A. G. (1979). Effect of garlic extract on food poisoning bacteria. *Lebensm.-Wiss. Technol. 12*:330.

Mazokhina, N. N., and Levshenko, M. T. (1983). High active acidity a basic factor for the safety of gherkin preserves. *Konservnaya i Ovoshchesushil' Naya Promyshlemmost' 4*:44.

McLachlan, A., and Stark, R. (1985). Modified atmosphere packaging of selected prepared vegetables. Technical Memorandum No. 412, Campden Food Preservation Research Association, Chipping Campden, United Kingdom.

Montville, T. J. (1982). Metabiotic effect of *Bacillus licheniformis* on *Clostridium botulinum*: Implications for home-canned tomatoes. *Appl. Environ. Microbiol. 44*:334.

Montville, T. J., and Conway, L. K. (1982). Oxidation-reduction potentials of canned foods and their ability to support *Clostridium botulinum* toxigenesis. *J. Food Sci. 47*:1879.

Mordvinova, S. A., Belonsova, M. V., and Flaumenbaum, B. L. (1984). Determining the possible development of the factors responsible for botulism in concentrated tomato products. *Konservnaya i Ovoshchesushil' Naya Promyshlennost 8*:32.

Mordvinova, S. A., Belousova, M. V., and Titarenko, I. O. (1980). Formation of toxins of botulism-producing spores and their heat stability in juice from machine harvested tomatoes. *Konservnaya i Ovoshchesuchil' Naya Promyshlennost; 4*:37.

Morris, J. G., and O'Brien, R. W. (1971). Oxygen and clostridia; a review. In *Spore Research—1971* (A. N. Barker, G. S. Gould, and J. Wolf, eds.), Academic Press, London, p. 1.

Nicols, R., and Hammond, J. B. W. (1973). Storage of mushrooms in pre-packs; the effect of changes in carbon dioxide and oxygen on quality. *J. Sci. Food Agric. 24*:1371.

Notermans, S., Dufrenne, J., and Oosterom, J. (1981a). Persistence of *Clostridium botulinum* type B on a cattle farm after an outbreak of botulism. *Appl. Environ. Microbiol. 41*:179.

Notermans, S., Dufrenne, J., and Keybets, M. J. H. (1981b). Vacuum-packed, cooked potatoes: Toxin production by *Clostridium botulinum* and shelf life. *J. Food Prot. 44*:572.

Notermans, S., Dufrenne, J., and Keybets, M. J. H. (1985a). Use of preservatives to delay toxin formation by *Clostridium botulinum* (type B, strain Okra) in vacuum-packed, cooked potatoes. *J. Food Prot. 48*:851.

Notermans, S., Dufrenne, J., and Gerrits, J. P. G. (1989). Natural occurrence of *Clostridium botulinum* on fresh mushrooms (*Agaricus bisporus*). *J. Food Prot. 52*:737.

Notermans S., Dufrenne, J., and van Schothorst, M. (1979). Recovery of *Clostridium botulinum* from mud samples incubated at different temperatures. *Eur. J. Appl. Microbiol. Biotechnol. 6*:403.

Notermans, S., Dufrenne, J., and Lund, B. M. (1990). Botulism risk of refrigerated, processed foods of extended durability. *J. Food Prot. 53*:1020.

Notermans, S., Havelaar, A. H., Dufrenne, J. B., and Oosterom, J. (1985b). Voorkomen van *Clostridium botulinum* op rundveehouderijbedrijven. *Tijdschr. Diergeneesk 110*:175.

Notermans, S., Havelaar, A. H., and Schellart, J. (1980a). The occurrence of *Clostridium botulinum* in raw-water storage areas and their elimination in water treatment plants. *Water Res. 14*:1631.

Notermans, S., Kozaki, S., Dufrenne. J., and van Schothorst, M. (1980b). *In vitro* inactivation of *Clostridium botulinum* toxins types B, C and E by digestive juices of man and ducks. *Jap. J. Med. Sci. Biol. 33*:255.

Odlaug, T. E., and Pflug, I. J. (1977a). Effect of storage time and temperature on the survival of *Clostridium botulinum* spores in acid media. *Appl. Environ. Microbiol. 34*:30.

Odlaug, T. E., Pflug, I. J. (1977b). Thermal destruction of *Clostridium botulinum* spores suspended in tomato juice in aluminium thermal death time tubes. *Appl. Environ. Microbiol. 34*:23.

Odlaug, T. E., and Pflug, I. J. (1978). *Clostridium botulinum* in acid foods. *J. Food Prot. 41*:566.

Odlaug, T. E., and Pflug, I. J. (1979). *Clostridium botulinum* growth and toxin production in tomato juice containing *Aspergillus gracilis*. *Appl. Environ. Microbiol. 37*:496.

Odlaug, T. E., Pflug, I. J., and Kautter D. A. (1978). Heat resistance of *Clostridium botulinum* type B spores grown from isolates from commercially canned mushrooms. *J. Food Prot. 41*:351.

Ohye, D. F., and Christian, J. H. B. (1967). Combined effects of temperature, pH and water activity on growth and toxin production by *Clostridium botulinum* types A, B and E. In *Botulism 1966* (M. Ingram and T. A. Roberts, eds.), Chapman and Hall Ltd. London, p. 217.

Otofugi, T., Tokiwa, H., and Takahashi, K. (1987). A food poisoning incident caused by *Clostridium botulinum* toxin A in Japan. *Epidemiol. Inf. 99*:167.

Pasteur, L. (1861). Animalcules infusoires vivant sans gas oxygène libre et déterminant des fermentations. *C. R. Acad. Sci. 52*:344.

Podreshetnikova, N. A., Shapiro, B. M., Steblyanko, S. N., and Tryasina, S. G. (1970). Case of botulism due to consumption of canned vegetables. *Voprosy Pitaniya 29*:89.

Powers, J. J. (1976). Effect of acidification of canned tomatoes on quality and shelf life. *CRC Rev. Food Sci. Nutr. 7*:371.

Priepke, P. E., Wei, L. S., and Nelson, A. I. (1976). Refrigerated storage of prepacked salad vegetables. *J. Food Sci. 41*:379.

Proc. U.S. Department of Agriculture (1979). Nitrites: Availability of dietary nitrite in an animal study. *Fed. Ref.* (August 18) *43*:36697.

Prokhorovich, L. E., Saltykova, L. A., Shenderovskaya, L. M., Gritsko, L. P., and Klepa, N. F. (1976). Possibility of development of *Clostridium botulinum* in apricot juice and compote. *Konservnaya i Ovoshchesushil' Naya Promyshlennost' 1*:35.

Prokhorovich, L. E., Saltykova, L. A., Malkina, Z. M., Gritsko, L. P., and Shenderouskaya, L. M. (1975). Botulism contamination of canned apricots and red peppers. *Konservnaya i Ovoshchesushil' Naya Promyshlennost' 6*:38.

Quast, D. G., Zapata, M. N., and Bernhardt, L. W. (1975). Preliminary study on penetration of acidity into canned palm hearts. *Coletanea do Instituto de Technologia de Alimentos 6*:341.

Quastel, J. H., and Stephenson, M. (1926). Experiments on "strict" anaerobes. *Biochem. J. 20*:1125.

Raatjes, G. J. M., and Smelth, J. P. P. M. (1979). *Clostridium botulinum* can grow and form toxin at pH lower than 4.6. *Nature* (London) *281*:398.

Riemann, H. (1963). Safe heat processing of canned cured meats with regard to bacterial spores. *Food Technol. 17*:39.

Robach, M. C. (1980). Use of preservatives to control microorganisms in food. *Food Technol. 34*:81.

Roberts, T. A., and Hobbs, G. (1968). Low temperature growth characteristics of clostridia. *J. Appl. Bacteriol. 31*:75.

Roberts, T. A., Gibson, A. M., and Robinson, A. (1981). Factors controlling the growth of *Clostridium botulinum* types A and B in pasteurized, cured meats. *J. Food Technol. 16*:239.

Rosanova, L. I. (1974). The development of microorganisms causing botulism in preserved tomato products. *Trudy, Vsesoyuznyi Nauchno- issledovatel' skii Institut Konservnoi i Ovoshchesushil' Noi Promyshlennosti 21*:54.

Rozanova, L. I., Zemlyakov, V. L., and Mazokhina, N. N. (1972). Factors affecting *Clostridium botulinum* contamination of vegetables intended for preservation and of materials used. *Gigiena i Sanitariya 37*:102.

Schmidt, C. F., Lechowich, R. V., and Folinazzo, J. F. (1961). Growth and toxin production by type E *Clostridium botulinum* below 40°F. *J. Food Sci. 26*:626.

Seals, J. E., Snijder, J. E., Edell, T. A., Hatheway, C. L., Johnson, C. J., Swanson, R. C., and Hughes, J. M. (1981). Restaurant-associated type A botulism: Transmission by potato salad. *Am. J. Epidemiol. 113*:436.

Seward, R. A., Deibel, R. H., and Lindsay, R. C. (1982). Effects of potassium sorbate and other antibotulinal agents on germination and outgrowth of *Clostridium botulinum* type E spores in microcultures. *Appl. Environ. Microbiol. 44*:1212.

Simunovic, J., Oblinger, J. L., and Adams, J. P. (1985). Potential for growth of non-proteolytic *Clostridium botulinum* in pasteurized restructured meat products; a review. *J. Food Prot. 48*:265.

Sing, B., Littlefield, N. A., and Salumkhe, D. K. (1970). Effect of CA storage on amino acids, organic acids, sugar and rate of respiration of 'Lambert' sweet cherry fruit. *J. Am. Soc. Hort. Sci. 95*:458.

Sing, B., Yang, C. C., and Salunkhe, D. K. (1972). Controlled atmosphere storage of lettuce. 1. Effects on quality and respiration rate on lettuce heads. *J. Food Sci. 37*:48.

Smelt, J. P. P. M., and Haas, H. (1978). Behaviour of proteolytic *Clostridium botulinum* types A and B near the lower temperature limits of growth. *Eur. J. Appl. Microbiol. Biotechnol. 5*:143.

Smith, L. D. S. (1978). The occurrence of *Clostridium botulinum* and *Clostridium tetani* in the soil of the United States. *Health Lab. Sci. 15*:74.

Smith, L. D. S. (1977). *Botulism. The Organism, Its Toxins, the Disease*. Charles C Thomas, Springfield, IL.

Smith, G. R., and Young, A. M. (1980). *Clostridium botulinum* in British soil. *J. Hyg. 85*:271.

Smith M. W. (1975). The effect of oxidation-reduction potential on the outgrowth and chemical inhibition of *Clostridium botulinum* type E spores. Ph.D. thesis Virginia Polytechnic Institute, Blacksburg, VA.

Smock, R. M. (1979). Recent advances in controlled atmosphere storage of fruits. *Horticult. Rev. 1*:301.

Sneath, P. H. A., Mair, N. S., Sharpe, M. E., and Hilt, J. G. eds. (1986). *Bergey's Manual of Systematic Bacteriology*, Vol. 2, Williams and Wilkins, Baltimore.

Sofos, J. N., Busta, F. F., and Allen, C. E. (1979). Botulism control by nitrite and sorbate in cured meats: A review. *J. Food Prot. 42*:739.

Solomon, H. M., and Kautter, D. A. (1986). Growth and toxin production by *Clostridium botulinum* in sauteed onions. *J. Food Prot. 49*:618.

Solomon, H. M., Lynt, R. K., Lilly, T., and Kautter, D. A. (1977). Effect of low temperatures on growth of *Clostridium botulinum* spores in meat of the blue crab. *J. Food Prot. 40*:5.

Sperber, W. H. (1982). Requirements of *Clostridium botulinum* for growth and toxin production. *Food Technol. 36*(12):89.

St Louis, M. E., Peck, S. H. S., Bowering, D., Morgan, G. B., Blatherwick, J., Bannerjee, S., Kettyls, G. D. M., Black, W. A., Milling, M. E., Hauschild, A. H. W., Tauxe, R. U., and Blake, P. A. (1988). Botulism from chopped garlic: Delayed recognition of a major outbreak. *Ann. Int. Med. 108*:363.

Stroup, W. H., Dickerson, R. W., and Johnston, M. R. (1985). Acid equilibrium development in mushrooms, pearl onion, and cherry peppers. *J. Food Prot. 48*:590.

Stumbo, C. R., Murphey, J. R., and Cochran, J. (1950). Nature of thermal death time curves for P. A. 3675 and *Clostridium botulinum*. *Food Technol. 4*:321.

Sugii, S., Ohishi, I., and Sakaguchi, G. (1977). Correlation between oral toxicity and *in vitro* stability of *Clostridium botulinum* type A and B toxins of different molecular sizes. *Infect. Immun. 16*:910.

Sugii, S., and Sakaguchi, G. (1977). Botulogenic properties of vegetables with special reference to the molecular size of the toxin in them. *J. Food Safety* 2:53.

Sugiyama, H., and Yang, K. H. (1975). Growth potential of *Clostridium botulinum* in fresh mushrooms packaged in semi-permeable plastic film. *Appl. Microbiol. 30*:964.

Sugiyama, H., and Rutledge, K. S. (1978). Failure of *Clostridium botulinum* to grow in fresh mushrooms packaged in plastic film overwraps with holes. *J. Food Prot. 41*:348.

Sugiyama, H., Woodburn, M., Yang, K. H., and Movroydis, C. (1981). Production of botulinum toxin in inoculated pack studies of foil-wrapped baked potatoes. *J. Food Prot. 44*:896.

Tamminga, S. K., Beumer, R. R., Keybets, M. J. H., and Kampelmacher, E. H. (1978). Microbial spoilage and development of food poisoning bacteria in peeled, completely or partly cooked vacuum packed potatoes. *Arch. Lebensmitt. Hyg.* 2:215.

Tanaka, N. (1982). Toxin production by *Clostridium botulinum* in media at pH lower than 4.6. *J. Food Prot. 45*:234.

Tanner, F., Beamer, P. R., and Ricker, C. J. (1940). Further studies on development of *Clostridium botulinum* in refrigerated foods. *Food Res. 5*:323.

Terranova, W., Breman, J. G., Locey, R. P., and Speck, S. (1978). Botulism type B: Epidemiological aspects of an extensive outbreak. *Am. J. Epidemiol. 108*:150.

Todd, E., Chang, P. C., Hauschild, A., Sharpe, A., Park, C., and Pivnick, H. (1974). Botulism from marinated mushrooms. *Proceeding IV Int. Congress Food Sci. Technol. 3*:182.

Tompkin, R. B., and Christiansen, L. N. (1976). *Clostridium botulinum*. In *Food Microbiology: Public Health and Spoilage Aspects* (M. P. Defigueredo and D. F. Splitstoeser, eds.), AVS Publishing Co., Inc., Westport, CT, p. 156.

Tompkin, R. B., Christiansen, L. N., and Shaparis, A. B. (1980). Antibotulinal efficacy of sulfur dioxide in meat. *Appl. Environ. Microbiol. 39*:1096.

Ueda, S., Yamashita, H., and Kuwabata, Y. (1982). Inhibition of *Clostridium botulinum* and *Bacillus* spp. by spices and flavouring compounds. *J. Jap. Soc. Food Sci. Technol. 29*:389.

Vessesland, B., and Hanke, M. E. (1940). The oxidation-reduction potential requirements of a non-spore-forming, obligate anaerobe. *J. Bacteriol. 39*:139.

Vergieva, V., and Incze, K. (1979). The ecology of *Clostridium botulinum*. *Husipar 28*:79.

Wit, J. C. de, Notermans, S., Gorin, N., and Kampelmacher, E. H. (1979). Effect of garlic oil and onion oil on toxin production by *Clostridium botulinum* in meat slurry. *J. Food Prot. 42*:222.

Ying, S., and Shuan, C. (1986). Botulism in China. *Rev. Infect. Dis. 8*:984.

10

Control in Dairy Products

David L. Collins-Thompson

Westreco, Inc., New Milford, Connecticut

Diane S. Wood

University of Guelph, Guelph, Ontario, Canada

I. INTRODUCTION

Clostridia are present in low numbers in most milk samples. Evidence for common species such as *Clostridium perfringens* suggest spore levels of less than 1 spore per ml of milk (Jones and Langlois, 1977). Studies by Bergere (1969) with *Clostridium tyrobutyricum* show that this species is normally present in milk at less than 0.2 spores/ml. Though outbreaks of botulism are not usually associated with dairy products, *Clostridium botulinum* may enter the milk chain as an environmental contaminant (e.g., in soil, feces, or contaminated ingredient). Information relating to levels of *C. botulinum* spores in milk is sparse. Studies at the University of Guelph indicate less than 1 *C. botulinum* spore per liter of raw milk (unpublished data). The low incidence of such spores is confirmed by surveys of dairy products. In a study by Taclindo et al. (1967) with vacuum-packed products such as cheese, the few products examined were negative for spores of *C. botulinum*. A more extensive study of convenience foods including dairy products was undertaken by Insalata et al. (1969). None of 40 samples of Edam and cheddar or of 10 samples of cheese spreads were found to contain spores of *C. botulinum*.

The low incidence of botulism outbreaks from dairy products must also relate to the question of whether these products are able to support outgrowth of spores of *C. botulinum* when they are present. The answer to this question will depend on the intrinsic and extrinsic properties of the dairy product in question. Such controlling influences are in part the subject of this chapter.

II. OUTBREAKS OF *BOTULISM* INVOLVING DAIRY PRODUCTS

Outbreaks of botulism involving dairy products are listed in Table 1. It is obvious that the involvement of *C. botulinum* in outbreaks from such products is rare, accounting for less than 1% of the total number of foodborne botulism outbreaks recorded since 1899. Only two serotypes have been associated with these outbreaks: type A and, more predominantly, type B. Both home and commercially prepared products appear to be equally responsible for the outbreaks. Accounts of such outbreaks have not always been fully discussed in the literature.

The outbreak in 1951 (Meyer and Eddie, 1951) involved a 5-ounce jar of Liederkranz brand cheese (Limburger) containing toxin estimated at 50 MLD (mouse lethal dose) per gram of cheese. Approximately 70 g of the cheese was eaten by the victim, a 53-year-old man, who experienced the characteristic symptoms of botulism within 20 hours. Death ensued 3 days after eating the cheese. The cheese involved was found to have a pH of 5.9 and the odor of butyric acid. This outbreak was confined to one person. Examination of other jars of cheese of the same batch showed some with type B *C. botulinum* spores but

Table 1 Outbreaks of Botulism Involving Milk or Dairy Products (1912–1990)

Year	Toxin type	Product	Location	Outbreaks	Cases	Deaths	Ref.[a]
1912	—	Cottage cheese	California	1	7	2	1
1914	—	Neufchatel cheese	California	1	2	2	1
1914	B	Cottage cheese	New York	1	3	3	1
1931	—	Milk	California	1	1	0	1
1935	—	Curd cheese	California	1	3	0	1
1939	A	Cottage cheese	New York	1	3	0	1
1951	B	Cheese spread (Liederkranz)	California	1	1	1	1
1973	B	Brie	Marseille, France	1	22	—	2
1973	B	Brie	Lausanne, Switzerland	1	43	—	2
1974	A	Cheese spread	Buenos Aires, Argentina	1	6	3	3
1978	B	Soft cheese	St. Etienne, France	—	—	—	4
1989	B	Hazelnut yoghurt	Wales and England	1	27	1	5

[a]1) Meyer and Eddie, 1965; 2) Sebald et al., 1974; 3) de Lágarde, 1974; 4) Billon et al., 1980; 5) Critchley et al., 1989.

without preformed toxin. A wide range of contaminating anaerobic microorganisms were also present in these jars.

The French and Swiss outbreaks involving *C. botulinum* type B introduced an interesting aspect to both the epidemiology of botulism outbreaks and to cheese technology. In 1973, ripened Brie cheese was responsible for two simultaneous outbreaks of botulism in Marseille, France (32 cases), and Lausanne, Switzerland (43 cases) (Sebald et al., 1974). Since the incriminated cheese was not available for analysis, the diagnosis was based on the detection of the toxin in the blood of 37.5% of the patients. The ensuing epidemiological investigation showed the involvement of straw on which the cheeses were allowed to ripen. One lot of straw, sold to cheese manufacturers in France and Switzerland, was found to be contaminated with animal waste. It was believed that the spores of *C. botulinum* type B entered the cheese via the straw and were able to germinate. This possible route was established by studies of Billon et al. (1980). The storage of soft cheese on inoculated straw (1000 spores of type B/cm^2) led to the production of toxin in the rind but not in the body of the cheese. The toxin was also found to be unstable and disappeared in the latter phases of ripening.

A year later, in 1974, an outbreak of *C. botulinum* type A involving a commercial cheese spread with onions occurred in Buenos Aires, Argentina (de Lágarde, 1974). Six cases were reported and three people died. Certain intrinsic parameters such as a_w and pH in the commercial spread were shown to support growth and toxin production by *C. botulinum* type A. The a_w value found in the contaminated spread was 0.97. In a following survey, cheese spreads found in Argentinian markets had a pH range of 5.6–6.1 and a_w values from 0.97 to 0.98 (Magrini et al., 1983). Studies by Briozzo et al. (1983) have shown that in media with pH values similar to those of process cheese spreads, *C. botulinum* can grow and produce toxin at an a_w as low as 0.965. When the a_w was lowered to 0.949, growth and toxin were absent. Similar results have been reported by Ohye and Christian (1966), who found that the minimal a_w for *C. botulinum* type A growth was 0.95. Baird-Parker and Freame (1967) reported that the minimal a_w for growth of type A spores is 0.96 at pH 7.0 and 0.97 at pH 5.5.

Perhaps the most unusual outbreak involving *C. botulinum* in a dairy product occurred in northwest England and Wales in 1989 (Critchley et al., 1989). It was the largest recorded outbreak of foodborne botulism in the United Kingdom and involved a yoghurt product containing hazelnut conserve. Type B toxin was found in a blown can of hazelnut conserve and in cartons of yoghurt. *C. botulinum* was also cultured from the yoghurt. During the investigation it was determined that the process used for the hazelnut conserve was a heat treatment of 90°C for 10 minutes, followed by heating in a sealed can for 20 minutes in boiling water. This process was inadequate for the destruction of *C. botulinum* spores, and the low acidity of the conserve (pH between 5.0 and 5.5) and high a_w allowed *C. botulinum* growth and toxin production. While the hazelnut conserve had been previously made using sugar, the producer had recently switched to using aspartame. No processing changes were made to compensate for a final product with a much higher a_w. A total of 27 people were affected by the outbreak, and 8 were classed as serious. One 74-year-old patient died. The level of toxin found in the conserve was 600–1800 MLD/ml, while the yoghurt was limited to 14–30 MLD/ml (O'Mahoney et al., 1990).

III. CONTROL OF *C. BOTULINUM* IN MILK

Raw milk is processed using a number of time/temperature combinations. These range from vat pasteurization (62.8°C for 30 min), to the high temperature, short time process (HTST, 71.5°C for 15 sec), and up to 137.8°C for 2 seconds for the ultra high temperature process (UHT). The vat pasteurization and HTST processes allow survival of spores (Olson and Mocquot, 1980). Studies conducted at the University of Guelph have shown that heat-activated type A and B spores inoculated into retail HTST milk do not germinate and grow. However, spores inoculated into milk sterilized by autoclaving (121°C for 15 min) do germinate and produce toxin. This suggests that the microorganisms surviving in HTST milk may play a competitive role against *C. botulinum*. This control mechanism has been demonstrated in other products (Hauschild, 1989). Although milk contains a number of antibacterial substances such as lysozyme, lactoferrin, and lactenin, such defense systems tend to operate against gram-negative bacteria. It has been indicated that these antibacterial substances are minor mechanisms and do little to inhibit the growth of pathogens in milk (Ayres et al., 1980).

Studies at the University of Guelph have shown that a minimum of 125°C for 5 seconds is necessary to destroy *C. botulinum* spores in milk. When processing conditions of 116°C for 3 seconds were used, milk inoculated with 1000 heat-activated spores of types A and B per ml showed outgrowth and toxin production within 1 week at 20°C. After processing at 122°C for 4 seconds, growth and toxin production occurred in the milk after 2 weeks of incubation at 20°C.

Growth and toxin production of six strains of *C. botulinum* type E were studied in commercially sterile whole milk at different storage temperatures (Read et al., 1970). The lowest storage temperature that allowed toxin production was 7.2°C. The "Tenno" type E strain produced toxin after 70 days of storage at this low temperature. Four of the six strains in milk stored at 10°C grew and produced toxin at times ranging from 21 to 56 days. At 20°C storage, all of the six type E strains produced toxin in milk within 28 days. These authors also examined the minimum number of *C. botulinum* type E cells required in milk before toxin could be detected. An initial inoculation level of 2.1×10^4 cells/ml milk produced detectable toxin. The milk also had a normal appearance when toxin was first detected. However, strains of types A and B require a higher storage temperature to produce toxin in sterile milk. Toxin was not detected after 60 days of storage at 10°C when inoculated with 1×10^4 heat-activated spores (University of Guelph, unpublished data).

One of the possible factors controlling growth of *C. botulinum* is the oxidation reduction potential (Eh) of milk. Initiation of spore outgrowth in foods has been recorded at positive Eh values (Hauschild, 1989). This has also been found for milk (Kaufmann and Marshall, 1965). The effect of heating on the development of *C. botulinum* 62A in milk was examined. Using a sterilizing temperature of 121°C and heating for 18 and 30 minutes, the Eh of each of two samples was measured. The sample heated for 18 minutes had an Eh value of +234 mV, while the 30-minute sample had a value of +192 mV. Outgrowth of 5×10^5 heat-activated spores occurred faster in the milk with the lower Eh. The 30-minute sample spores had a lag period of 5 days compared to 7 days for the 18-minute sample. Increasing the heating period of the milk to 45 minutes, however, increased the lag phase, despite a lower Eh value. These data suggest that factors other than Eh are involved. It was also noted that the age of the milk influenced the rate of spore germination. The lag phase of spores in milk held for 3 days between heating and inoculation, was shorter than milk inoculated 2 hours after sterilizing. The data further indicated that the Eh of sterile, uninoculated milk decreased on storage, suggesting the presence of active reducing compounds. This "autoreductive" property of milk may play a role in the outgrowth of *C. botulinum* in milk.

Studies using UHT milk and *C. botulinum* at the University of Guelph present a different picture concerning the possible role of an autoreductive process. The autoreductive properties of milk do not appear to be displayed by UHT processed milk. The Eh values of uninoculated milk (as measured using an oxidation/reduction electrode) did not change over a period of 120 days when incubated at 20°C. The values remained at +177 mV. As a consequence, the age of the milk had little influence on the rate of outgrowth of spores of *C. botulinum* types A and B (Table 2). In this study, heat-activated spores were added 4 and 120 days after UHT processing. The 120-day-old inoculated milk showed a marginal reduction in the number of positive samples when compared to the 4-day-old samples. A more significant influence on outgrowth of *C. botulinum* in UHT milk was associated with the spore inoculum level (Table 2). At a low spore level, fewer samples allowed spore outgrowth. This was also shown in a comparison study with other processed milks (Table 3). It appears that UHT milk is more resistant to the outgrowth of *C. botulinum* at low spore inoculum levels than HTST milk. Since Eh values were not monitored in these experiments, it is not known whether these results might be related to the autoreductive process observed by Kaufmann and Marshall (1965). This process could be operating in HTST milk but not in UHT milk. The fat level (2% vs. 3%) in HTST milk did not appear to influence spore outgrowth either. It has been observed that the

Table 2 Spore Load and Outgrowth of *C. botulinum* Types A and B in Different Aged UHT Milk

Inoculation time (days after processing)	Spores/ml	Samples with growth and toxin present/samples inoculated at			
		7 days[a]	15 days	34 days	60 days
4	3×10^2	1/5	3/5	5/5	5/5
	3×10^1	0/5	0/5	2/5	4/5
120	3×10^2	0/5	4/5	5/5	5/5
	3×10^1	0/5	0/5	1/5	1/5

[a]Incubation was at 20°C.

addition of heat-activated spores of *C. botulinum* to milk leads to an initial drop in spore levels with fewer viable cells (Kaufmann and Marshall, 1965). This reduction, however, has not been reported by other investigators (Read et al., 1970).

The role of pH in the control of *C. botulinum* in skimmed milk was evaluated by Smelt et al. (1982). Using *C. botulinum* types A and B spores, growth and toxin production were initiated at pH 4.4. The different acids used in lowering the pH of milk samples were found to be important. Samples acidified to pH 4.4 with hydrochloric or citric acids resulted in toxin production after 4 weeks of incubation at 30°C. Samples adjusted to pH 4.4 with lactic or acetic acids did not show signs of toxin production until after 12–14 weeks of incubation. Thus, the nature of the acid and not the pH value alone appears to be important in controlling the growth of *C. botulinum* in milk.

Table 3 Growth of *C. botulinum* Types A and B in Processed Milk Stored for 30 Days at 20°C

Number of spores/ml	Samples with growth and toxin present/samples inoculated in		
	UHT (2%)[a]	HTST (2%)	HTST (3%)
2×10^3	5/5	5/5	5/5
2×10^2	4/5	5/5	5/5
2×10^1	1/5	5/5	5/5
2×10^0	0/5	4/5	4/5

UHT = Ultra high temperature;
HTST = high temperature, short time.
[a]Percent butterfat.

The growth of *C. botulinum* in milk thus appears to be related to a number of factors. The few studies indicate that control of growth can be linked to the biological makeup (autoreductive properties) of the milk, the degree of processing, the level of spore contamination, the pH, and, to some extent, Eh values. There also appears to be a role for the contaminating microorganisms in providing some degree of control.

IV. CONTROL OF *C. BOTULINUM* IN PROCESS CHEESE PRODUCTS

It is generally recognized that there are three classes of process cheese products: pasteurized process cheese, pasteurized process cheese food, and pasteurized process cheese spread. In pasteurized process cheese, nearly all the solids (90%) are contributed by the cheese. Other ingredients include emulsifying agents, salt, acid, and cream. The moisture, a_w, and fat content are usually similar to natural cheese. In pasteurized process cheese food, some of the solids may be replaced with skim milk, cheese whey, or cream. The moisture content is usually about 40% and the a_w around 0.95. The pH values range from 5.1 to 5.6. In pasteurized process cheese spreads, the moisture level is about 50–60%. The ingredients are similar to the process cheese products but may also contain sweetening agents and gums. The pH values range from 4.1 to 6.0 and the a_w from 0.93 to 0.95. Process cheese foods and spreads may also contain delta-gluconolactone to assist in lowering the pH.

Process cheese products have had an excellent safety record with respect to *C. botulinum*. The two cases on record where toxin has been found in commercially made cheese spreads must be placed in perspective considering the millions of tons of these products made throughout the world (de Lágarde, 1974; Meyer and Eddie, 1965). These two outbreaks have initiated a number of studies regarding the controlling factors against *C. botulinum* in process cheese products. Karahadian et al. (1985) evaluated the potential for botulinal toxigenesis in cheese foods and spreads processed with reduced sodium. Cheese products were inoculated with 1000 spores of types A, B, and E per gram and stored at 30°C for 84 days. When delta-gluconolactone was added to these products, toxigenesis did not occur. The pH of such samples was 5.26 or less. In the absence of this compound, the pH values ranged from 5.4 to 5.6, and toxin was detected in the food. The a_w of both study groups was 0.96, emphasizing the critical role of pH. Several emulsifiers were screened for antibotulinal activity, including various phosphates and citrates at 2.5%. Most allowed toxin formation, but some inhibition was observed when disodium phosphate and, to a lesser extent, trisodium citrate were included in the formulation. This confirmed the findings by Tanaka et al. (1979) that phosphate emulsifiers have a greater antibotulinal effect than citrates in cheese spreads.

A study by Karahadian et al. (1985) also evaluated moisture levels associated with cheese spreads and their relation to toxin production. Spores (1000/g) of types A and B were added during the processing of cheese spreads made with different levels of water, salt, and emulsifier. The a_w of the product was maintained between 0.95 and 0.96. Cheese spreads made with 52 and 54% moisture and disodium phosphate (2%) were stable, and toxin production was not observed. At 58% moisture, however, toxin was produced. Toxin was also produced in cheese spread samples made with 2% sodium citrate at moisture levels of 52, 54, and 58%.

Kautter et al. (1979) evaluated the hazards from *C. botulinum* in shelf-stable pasteurized process cheese spreads. Using commercially available products, they selected samples with an a_w range from 0.930 to 0.953 and a pH range of 5.05–6.32. Five strains of *C. botulinum* types A and B spores at levels of 460 and 24,000/jar were used. The incubation period was for 6 months at 35°C. Two commercial products, cheese with bacon and Limburger spread, were found to support *C. botulinum* growth and toxin production. A significant portion of the 50 jars inoculated with the higher spore inoculum were positive for toxin. At an inoculum level of 460 spores/jar, only one jar of cheese with bacon spread became positive. This indicates that an inoculum of only 460 spores is capable of toxin production. It is of interest to note that the Limburger product had the highest pH and a_w values (pH 6.3, a_w 0.952).

One important aspect of a second study by Kautter et al. (1981) was that it partially refuted the claim by Tanaka et al. (1979) that commercial process cheese products had a substantial margin of safety against toxin production by *C. botulinum*. This was clearly not the case with the cheese and bacon and the Limburger spreads. In a further study by Tanaka (1982), a comparison was made between two published methods of inoculating spores of *C. botulinum* into cheese spreads. In the earlier Tanaka study (1979), spores of *C. botulinum* were added by the "in-process" method, i.e., to the product as it was being processed. The Kautter et al. (1981) study used the postprocess or "cold" inoculation method. In this method, heat-activated spores were directly added to the finished product in 0.1 ml of diluent (0.85% NaCl). Using similar products, the two inoculation methods were compared. Some products challenged with 10^3 spores by "cold" inoculation became toxic, whereas samples inoculated by the "in-process" method remained toxin-free. Tanaka (1982) suggested that localized pockets of added water in the "cold" inoculation process could account for the discrepancies between the two studies, since all other parameters such as cheese formulation, a_w, and pH were identical.

One of the most complete studies on factors involved in antibotulinal properties in process cheese spreads was conducted by Tanaka et al. (1986). Response surface methodology was used in the experimental design, and logistic regression was used to produce predictive models for botulinal safety in cheese spreads. A number of relationships were examined including pH, a_w, sodium chloride levels, disodium phosphate, and lactic acid addition. The spore levels used were 1000 per gram and consisted of a mixture of five strains of *C. botulinum* types A and B. Samples were incubated at 30°C and toxin levels determined after 4 weeks and up to 42 weeks. In the pH range tested (up to 6.1), no toxin was produced at or below an a_w of 0.944. With increasing a_w, the critical pH decreased, e.g., a pH of 5.8 or below was required to prevent toxigenesis at an a_w of 0.946.

A_w measurements in cheese systems are not very reliable. Tanaka et al. (1986) therefore related pH requirement to moisture content and concentration of NaCl and disodium phosphate. Since these two salts were found to be similarly effective and additive in the range studied, they were combined as "NaCl + disodium phosphate" concentration. As the moisture content increased from 51 to 60%, the pH required to inhibit toxin production decreased from 5.8 to 5.2.

Another interesting aspect of this study was the effect of added 0.25% lactic acid. This amount of acid reduced the pH by an average of approximately 0.17 pH units, yet the inhibitory effect on botulinum toxin production was "beyond its effect" due to pH reduction alone. This result links up with the observation by Smelt et al. (1982) relating the nature of the acid as more important than the actual pH change in *C. botulinum* studies. The interrelationship of pH and a_w shown in the above study agreed well with data reported by Briozzo et al. (1983), who demonstrated that a pH value of 5.8 and an a_w of 0.948 were required to prevent growth and toxin production in the cheese spreads under study.

In all studies involving pasteurized process cheese spreads or foods, it is apparent that no single parameter is responsible for the safety from *C. botulinum*.

V. USE OF NISIN TO CONTROL *C. BOTULINUM* IN PASTEURIZED PROCESSED CHEESE AND CHEESE SPREADS

Nisin is a bacteriocin produced by certain strains of *Lactococcus lactis*. It has been used in a variety of foods including dairy products such as cheese, processed cheese, pasteurized processed cheese, pasteurized flavored and long life milks, and clotted cream (Anonymous, 1989). Nisin has been included in formulations for approximately 40 years and is permitted for use in 46 countries.

Nisin's most effective use has been in the formulation of pasteurized, processed cheese and pasteurized, processed cheese spreads. Tanaka et al. (1986) conducted a series of experiments determining factors influencing the antibotulinum properties in this type of cheese product. The factors examined included pH, moisture, sodium chloride, and disodium phosphate levels. The results of this study showed that if the pH was increased to 5.7 and the moisture level to >54%, *C. botulinum* would be expected to grow. Using various combinations of the above-mentioned factors, a cheese spread was formulated containing 60% moisture. If the sodium chloride and disodium phosphate levels were high enough and the pH low enough, *C. botulinum* would not grow. Any small alteration in the formulation, though, could lead to unsafe conditions.

To produce a pasteurized processed cheese spread with high moisture, yet low in sodium chloride, and keep it free of *C. botulinum* toxigenesis, nisin can be added. Somers and Taylor (1987) formulated a cheese spread with no added sodium chloride, 14% added disodium phosphate, and a moisture level of 55–57%. Nisin added to a level of 100–200 ppm prevented *C. botulinum* spore outgrowth over a 48-week incubation period at 30°C. Higher nisin levels were required for products formulated with higher moisture levels. The addition of nisin allows for a greater formulation range in this type of product.

The addition of nisin to other dairy products, including dairy desserts (Anonymous, 1985), fluid milk (Anonymous 1988, 1989), sterilized milk (Wajid and Kalra, 1976), and canned evaporated milk (Gregory et al., 1964), was shown to extend shelf life. Challenge studies with *C. botulinum* were not performed.

Reviews of nisin, its properties, biosynthesis, mode of action, and application can be found in Hurst (1981, 1983), Delves-Broughton (1990), and Lipinska (1977).

VI. CONTROL OF *C. BOTULINUM* IN SURFACE-RIPENED CHEESES

The most definitive studies with surface-ripened cheeses and *C. botulinum* are those of Wagenaar and Dack (1958a,b,c). Strictly speaking, these studies were performed in cheese slurries rather than in the actual product. In a series of papers, the relationships between brine concentration (%NaCl × 100/moisture content), pH, stabilizer concentration (sterile locust bean gum) and age of cheese, and possible growth and toxin production by *C. botulinum* types A and B were examined. Three cheese types were studied: type I (microbial population composed chiefly of yeasts and bacteria), type II (microbial population composed chiefly of molds), and type III (microbial population composed mainly of bacteria). Samples for these studies were prepared by adding distilled water (10–15%, depending on the cheese moisture level) to the cheese type. Sodium chloride was added to some samples to yield a series of cheese slurries with 3–8% salt. The cheese preparations were heated for 10 minutes at 90°C, cooled, and inoculated with 10^2–10^4 spores/g, depending on the experiment. As a result of the various experimental parameters, certain trends were observed for all cheese types.

The inoculum level did not affect toxin production; toxin was produced at the same salt level, regardless of inoculum size. Type A spores of *C. botulinum* grew and produced toxin at higher brine concentrations than type B. Since type B showed poor growth in cheese type III, only type A spores were used. The responses to changes in pH and growth/toxin production depended on the cheese type. In cheese type I, toxin production appeared to be independent of pH in the range tested (5.8–6.2). In contrast, when type II cheese samples were adjusted upward to the same pH range, higher brine concentrations were required to inhibit toxin production. At the normal pH values of 5.5, toxin production was inhibited at all brine concentrations tested. These data suggested that the pH was more critical in type II cheese than in type I cheese. In type III cheese, any increase in pH above 7.0 caused a decrease in growth and toxin production by *C. botulinum*. Some type III cheeses had initial pH values of 7.0 and higher. The maximum brine concentration allowing toxin production varied with the cheese type. Using type A spores, it was the lowest in type III cheese (6.5%). For types I and II cheese, the critical concentrations were 7% and 8%, respectively. The addition of the locust bean gum stabilizer at levels of 0.2–0.4% had little effect on growth and toxin production.

Wagenaar and Dack (1953, 1954) also explored the effect of disodium phosphate and sodium citrate added to slurries of surface-ripened cheeses inoculated with spores of *C. botulinum* types A and B. They found that 2.5% disodium phosphate reduced the amount of sodium chloride required to inhibit *C. botulinum*. Sodium citrate at a level of 2.5% was less inhibitory. These results were confirmed by Tanaka et al. (1979) and Karahadian et al. (1985). A general conclusion from these studies was that well-ripened cheese was more inhibitory to the growth of *C. botulinum* than fresh, unaged cheese. The inhibitory nature of aged surface-ripened cheese was explored by Grecz et al. (1959a). These authors suggested several factors that might contribute to this phenomenon, including the fatty acid composition of the cheese, pH, and growth of competing microorganisms. This last suggestion was based on the changes in the ability of the cheese to support *C. botulinum* during ripening. Cheese inoculated after storage for 4–6 weeks at 2–4°C permitted growth of *C. botulinum*. After 8 weeks, however, growth was inhibited. Therefore, inhibitory compounds, possibly antibioticlike, were formed at refrigerated temperatures and were linked to an increase in the level of competing microorganisms. It was also noted that the inhibitory agents appeared to form at the surface of the cheese, slowly penetrating towards the center, and that their formation was dependent on aerobic conditions. The role of pH in aged cheese was shown to be limited.

In a separate paper, Grecz et al. (1959b) explored the relationship of fatty acids to inhibition of *C. botulinum* in aged cheese. Using type I cheese as a model, the production of fatty acids during ripening was followed. The total free fatty acids increased from 0 to 1.10–2.16%, depending on the lot of cheese tested. The development of propionic and butyric acids was found to correlate with inhibition of growth and toxin production. Inhibition due to increasing brine concentrations were required for inhibition in young cheese than in older, aged cheese. Since rancidity was not observed in 8-week-old cheese, oxidation products did not appear to be part of the inhibitory process. In contrast, Roth and

Halvorson (1952) reported that oxidized samples of oleic, linolenic, and linolenic acids and their methyl esters caused inhibition of germination of *C. botulinum* spores when present in culture medium at a level of 0.01% and that nonrancid samples, at the same concentration, had little or no effect on spore germination. The difference between the conclusions of these two papers may lie in the concentration of nonoxidized vs. oxidized fatty acids used. It is likely that higher levels of nonoxidized fatty acids (1.7–2.2%) are required for inhibition, whereas a lower level (1%) of oxidized fatty acid inhibits spore germination. The inhibition by such acids may also be linked to the increasing acidity in the cheese during ripening. Interestingly, the fat fraction taken from 8-week ripened cheese was also capable of inhibiting *C. botulinum* when added to fresh cheese (Grecz et al., 1959b).

Studies by Billon et al. (1980) on toxin formation by *C. botulinum* during ripening of soft cheese demonstrated the complexity of the cheese environment. Soft cheese made from milk and inoculated with 1500 spores/ml of *C. botulinum* type B did not develop toxin after 11 weeks of storage at 20°C. Uninoculated cheese stored on straw contaminated with 1000 spores/cm^2 led to the production of toxin in the cheese rind. The toxin, however, was unstable and disappeared one month after the maximum level had been attained. It would have been interesting to carry out these experiments with spores of *C. botulinum* type A, as other studies (Wagenaar and Dack 1958a,b) suggest that this type is more likely to grow in soft cheeses.

The question of toxin and spore survival in aged cheeses was studied by Grecz et al. (1965). Using the three types of cheeses described by Wagenaar and Dack (1958a,b,c) as a model, toxin and spore survival of *C. botulinum* 62A were evaluated. The results indicated a two- to fivefold increase in toxin titer in inoculated samples after a 12-month storage period at 2–4°C. The toxin titer remained constant for 2–4 years and then declined. After 6 years of storage, low levels of toxin were still detectable. If toxic filtrates were added to the three types of cheeses and stored for 60 days at 30°C, some decline in toxin levels was observed. The decline varied with the cheese type, being most pronounced in cheese types II and III. Type I cheese showed no decline in toxin levels. Viable spores of *C. botulinum* 62A could be recovered from all cheese types, even after 5 years of storage at 2–4°C. For cheese type III, spores were recovered after 6 years with levels similar to that of the initial inoculum (10^5 spores/g).

VII. IMITATION CHEESE

In imitation cheese products, the milk components are replaced with vegetable oils and proteins. These products may also contain milk derivatives such as whey or modified whey ingredients. Cheeses on the market include imitation mozzarella, cream, and similar soft cheese products. Such products were evaluated by

Kautter et al. (1981) for botulism hazards. Eleven commercial products were exposed to heat-activated spores of *C. botulinum* types A and B, at an inoculum level of approx. 800–1930 spores per 150- to 250-g jar. The cheese products were inoculated by the "cold process" (spores in 0.85% NaCl added to the final product) and incubated at 26°C for up to about 300 days. Of the 11 cheeses tested, only one product was shown to support growth and toxin production by *C. botulinum*, although viable spores were found in most of the products after the incubation period. This was a mozzarella cheese substitute, which became toxic after 26–28 days at 25°C. In a retest of this product with a spore load of 240 spores per jar, the cheese again became toxic, but only after 70 days of incubation. The substitute mozzarella had a pH of 5.86 and an a_w of 0.973. This latter value was the highest a_w of all imitation cheeses tested (range 0.942–0.973). The authors believed that the a_w value may have been the key to the product becoming toxic.

It would have been interesting to evaluate the same products using the "in-processing" method of spore addition suggested by Tanaka (1982), which avoids the possibility of localized water pockets in the product.

VIII. LOW-SODIUM CHEESE PRODUCTS

Trends towards reduced sodium levels in food products have raised the concern about the microbiological stability and safety of dairy products. Studies by Ibrahim (1982) and Koenig and Marth (1982) suggest that a reduction of the sodium concentration does not necessarily compromise the microbial safety of the product. Karahadian et al. (1985) evaluated the potential for botulinal toxigenesis in reduced-sodium processed cheese foods and spreads. Inoculated pack studies showed that highly reduced sodium products were resistant to *C. botulinum*, especially when delta-gluconolactone was added as an acidulant. Replacing sodium-based emulsifiers with the equivalent concentration (2.5%) of potassium emulsifiers allowed toxin formation. This suggests that sodium and potassium ions are not equivalent in inhibiting *C. botulinum*. These researchers, however, did not account for differences in ionic strength between the two ions. In summary, the effect of sodium concentrations on toxin formation by *C. botulinum* in processed cheeses is not fully understood.

IX. CONTROL OF *C. BOTULINUM* IN OTHER DAIRY PRODUCTS

Many dairy products have not attracted much attention with respect to the potential hazards from *C. botulinum*. This is partly due to the lack of epidemiological evidence indicating such hazards. It is remarkable that products such as cheddar cheese have not been involved in any recorded outbreaks, considering the volume of the product manufactured around the world. Although cheddar is a relatively low-moisture, low-pH product, with high levels of lactic acid bacteria, it is still prone to spoilage by other clostridial species. This suggests that *C. botulinum* spores must be rare in milk supplies.

Products such as butter are essentially made from pasteurized cream, which reduces some risk from pathogens. In the case of *C. botulinum*, one of the controlling factors in this product is the salt concentration. The salt level in butter rarely exceeds 2.5%, but this translates to about 15% salt in the water phase. Such levels would control the outgrowth of spores of *C. botulinum*. Storage temperatures of 5°C also contribute to the control, especially in low-sodium butter. As yet unexplored is the possible migration of spores to the lipid phase in high-fat products. Spores of *Clostridium* sp. are hydrophobic, due to the presence of hydrocarbons in the spore coat (Murrell, 1981). Under such conditions, germination requirements such as a_w may not be met. Little is also known about the partition of spores in dairy products with varying degrees of lipid material and the consequences on the control of *C. botulinum*. The work by Grecz et al. (1959b) with fatty acids may be related to this partition phenomenon. In addition, there is no work relating the influence of ions such as Na^+ to the hydrophobicity of clostridial spores.

The excellent record of fermented milks, with reference to outbreaks of botulism, attest to the important role of pH in controlling this organism. Basically, these products have been acidified by a lactic acid fermentation, producing pH values of 4.0 or less. During such fermentations, small quantities of organic acids and other inhibitory compounds are produced which provide an unfavorable environment for organisms such as *C. botulinum*. In the recent outbreak of botulism associated with yoghurt, the toxin was preformed in one of the ingredients (Critchley et al., 1989). Problems can also arise when ingredients are added which interfere with the fermentation, e.g., too much sugar (Olson and Mocquot, 1980). Other controlling factors include the antagonistic properties of several *Lactobacillus* (e.g., *L. acidophilus*) and *Streptococcus* strains to pathogens, including clostridial species (Gilliland and Speck, 1977).

It must be remembered that dairy products are generally stored at low temperatures and that frozen products such as ice cream and dairy desserts are well protected against both spoilage and pathogens.

ACKNOWLEDGMENT

The authors wish to thank Jeanne Hogeterp for her typing skills and patience.

REFERENCES

Anonymous. (1985). Nisin preservation of chilled desserts. *Dairy Ind. Int. 50*: 41.

Anonymous. (1988). Preservation of milk and milk products with nisaplin. Tech. Info. Leaflet No. 11/88, Aplin and Barrett Ltd., Dorset, England.

Anonymous. (1989). International acceptance of nisin as a food preservative. Tech. Info. Leaflet No. 4/89/11, Aplin and Barrett Ltd., Dorset, England.

Ayres, J. C., Mundt, J. O., and Sandine, W. E. (1980). Dairy products. In *Microbiology of Foods* (B. S. Schweigert, ed.), W. H. Freeman and Company, San Francisco, p. 341.

Baird-Parker, A. C., and Freame, B. (1967). Combined effect of water activity, pH and temperature on the growth of *Clostridium botulinum* from spore and vegetative cell inocula. *J. Appl. Bacteriol. 30*: 420.

Bergere, J. L. (1969). La bactofugation du lait et l' élimination des spores de *Clostridium tyrobutyricum*. *Le Lait 488*: 507.

Billon, J., Guérin, J., and Sébald, M. (1980). Etude de la toxinogénese de *Clostridium botulinum* type B au cours de la maturation de fromages a pate molle. *Le Lait 60*: 392.

Briozzo, J., de Lágarde, E. A., Chirife, J., and Parada, J. L. (1983). *Clostridium botulinum* type A growth and toxin production in media and process cheese spread. *Appl. Environ. Microbiol. 45*: 1150.

Critchley, E. M. R., Hayes, P. J., and Isaacs, P. E. T. (1989). Outbreak of botulism in north west England and Wales, June 1989. *Lancet ii*: 849.

de Lágarde, E. A. (1974). Boletin informativo del Centre Paramericano de Zoonosis, Vol. 1. Centro Paramericano de Zoonosis, Buenos Aires, Argentina.

Delves-Broughton, J. (1990). Nisin and its uses as a food preservative. *Food Technol. 44*: 100.

Gilliland, S. E., and Speck, M. L. (1977). Antagonistic action of *Lactobacillus acidophilus* toward intestinal and food-borne pathogens in associative culture. *J. Food Prot. 40*: 820.

Gould, G. W. (1964). Effect of food preservatives on the growth of bacteria from spores. In *Microbial Inhibitors in Food* (N. Molin, ed.), Almqvist and Wiksell, Stockholm, p. 17.

Grecz, N., Wagenaar, R. O., and Dack, G. M. (1959a). Inhibition of *Clostridium botulinum* and molds in aged, surface ripened cheese. *Appl. Microbiol. 7*: 33.

Grecz, N., Wagenaar, R. O., and Dack, G. M. (1959b). Relation of fatty acids to the inhibition of *Clostridium botulinum* in aged surface ripened cheese. *Appl. Microbiol. 7*: 228.

Grecz, N., Wagenaar, R. O., and Dack, G. M. (1965). Storage stability of *Clostridium botulinum* toxin and spores in processed cheese. *Appl. Microbiol. 13*: 1014.

Gregory, M. E., Henry, K., and Kon, S. K. (1964). Nutritive properties of freshly prepared and stored evaporated milks manufactured by normal commercial procedure or by reduced thermal processes in the presence of nisin. *J. Dairy Res. 31*: 113.

Hauschild, A. H. W. (1989). *Clostridium botulinum*. In *Foodborne Bacterial Pathogens* (M. P. Doyle, ed.), Marcel Dekker, Inc., New York, p. 111.

Hirsch, A., and Wheater, D. M. (1951). The production of antibiotics by streptococci. *J. Dairy Res. 18*: 193.

Hurst, A. (1981). Nisin. *Adv. Appl. Microbiol. 27*: 85.

Hurst, A. (1983). Nisin and other inhibitory substances from lactic acid bacteria. In *Antimicrobials in Foods* (A. L. Branen and P. M. Davidson, ed.), Marcel Dekker, Inc., New York, p. 327.

Ibrahim, G. F. (1982). Effect of non-salting on cheddar cheese made with induced starter failure on growth of clostridia and keeping quality of process cheese. *J. Food Prot. 45*: 356.

Insalata, N. F., Witzeman, S. J., Fredericks, G. J., and Sunga, F. C. A. (1969). Incidence study of spores of *Clostridium botulinum* in convenience foods. *Appl. Microbiol. 17*: 542.

Jones, F. T., and Langlois, B. E. (1977). Microflora of retail fluid milk products. *J. Food Prot. 40*: 693.

Karahadian, C., Lindsay, R. C., Dillman, L. L., and Deibel, R. H. (1985). Evaluation of the potential for botulinal toxigenesis in reduced-sodium processed American cheese foods and spreads. *J. Food Prot. 48*: 63.

Kaufmann, O. W., and Marshall, R. S. (1965). Factors affecting the development of *Clostridium botulinum* in whole milk. *Appl. Microbiol. 13*: 521.

Kautter, D. A., Lilly, T., Jr., Lynt, R. K., and Solomon, H. M. (1979). Toxin production by *Clostridium botulinum* in shelf-stable pasteurized process cheese spreads. *J. Food. Prot. 42*: 784.

Kautter, D. A., Lynt, R. K., Lilly, T., Jr., and Solomon, H. M. (1981). Evaluation of the botulism hazard from imitation cheeses. *J. Food. Sci. 46*: 749.

Koenig, S., and Marth, E. M. (1982). Behavior of *Staphylococcus aureus* in cheddar cheese made with sodium chloride or a mixture of sodium chloride and potassium chloride. *J. Food Prot. 45*: 996.

Lipinska, E. (1977). Nisin and its applications. In *Antibiotics and Antibiosis in Agriculture* (M. Woodbine, ed.), Butterworths, London, p. 103.

Magrini, R. C., Chirife, J., and Parada, J. L. (1983). A study of *Staphylococcus aureus* growth in model systems and processed cheese. *J. Food Sci. 48*: 882.

McClintock, M. L., Serres, L., Marzolf, J. J., Hirsch, A., and Mocquot, G. (1952). Action inhibitrice des streptocoques producteurs de nisine sur le developpement des sporules anaerobies dans le formage de Gruyere fondu. *J. Dairy Res. 19*: 187.

Meyer, K. F., and Eddie, B. (1951). Perspectives concerning botulism. *Z. Hyg. Infektionskr. 133*: 255.

Meyer, K. F., and Eddie, B. (1965). Sixty-five years of human botulism in the United States and Canada. Epidemiology and tabulations of reported cases 1899 through 1964, University of California Printing Department, San Francisco Medical Center.

Murrell, W. G. 1981. Biophysical studies on the molecular mechanisms of spore heat resistance and dormancy. In *Sporulation and Germination* (H. S. Levinson, A. L. Sonenshein, and D. J. Tripper, eds.), American Society of Microbiology, Washington, DC, p. 64.

Ohye, D. F., and Christian, J. H. B. (1966). Combined effects of temperature, pH and water activity on growth and toxin production by *Clostridium botulinum* types A, B and E. *Proceedings of the Fifth International Symposium on Food Microbiology*, Moscow, p. 217.

Olson, J. C., and Mocquot, G. (1980). Milk and mild products. In *Microbial Ecology of Foods*. Vol. 2, *Food Commodities* (The International Commission on Microbiological Specifications for Foods), Academic Press, New York, p. 470.

O'Mahoney, M., Mitchell, E., Gilbert, R. J., Hutchinson, D. N., Begg, N. T., Rodhouse, J. C., and Morris, J. E. (1990). An outbreak of foodborne botulism associated with contaminated hazelnut yoghurt. *Epidemiol. Infect. 104*: 389.

Read, R. B., Jr., Bradshaw, J. G., and Francis, D. W. (1970). Growth and toxin production of *Clostridium botulinum* type E in milk. *J. Dairy Sci. 53*: 1183.

Rogers, L. A. (1928). The inhibiting effect of *Streptococcus lactis* on *Lactobacillus bulgaricus. J. Bacteriol. 16*: 321.

Roth, N., and Halvorson, H. O. (1952). The effect of oxidative rancidity in unsaturated fatty acids on the germination of bacterial spores. *J. Bacteriol. 63*: 429.

Scott, V. N., and Taylor, S. L. (1981). Effect of nisin on the outgrowth of *Clostridium botulinum* spores. *J. Food Sci. 46*: 117.

Sébald, M., Jouglard, J., and Gilles, G. (1974). Botulisme humain de type B aprés ingestion de fromage. *Ann. Microbiol.* (Inst. Pasteur) *125A*: 349.

Smelt, J. P. P. M., Raatjes, G. J. M., Crowther, J. S., and Verrips, C. T. (1982). Growth and toxin formation by *Clostridium botulinum* at low pH values. *J. Appl. Bacteriol. 52*: 75.

Somers, E. B., and Taylor, S. L. (1987). Antibotulinal effectiveness of nisin in pasteurized processed cheese spreads. *J. Food Prot. 50*: 842.

Taclindo, C., Jr., Midura, T., Nygaard, G. S., and Bodily, H. L. (1967). Examination of prepared foods in plastic packages for *Clostridium botulinum. Appl. Microbiol. 15*: 426.

Tanaka, N., Goepfert, J. M., Traisman, E., and Hoffbeck, W. M. (1979). A challenge of pasteurized process cheese spread with *Clostridium botulinum* spores. *J. Food Prot. 42*: 787.

Tanaka, N. (1982). Challenge of pasteurized process cheese spreads with *Clostridium botulinum* using in-process and post-process inoculation. *J. Food Prot. 45*: 1044.

Tanaka, N., Traisman, E., Plantinga, P., Finn, L., Flom, W., Meske, L., and Guggisberg, J. (1986). Evaluation of factors involved in antibotulinal properties of pasteurized process cheese spreads. *J. Food Prot. 49*: 526.

Wagenaar, R. O., and Dack, G. M. (1953). Some factors influencing the growth and toxin production of *Clostridium botulinum* experimentally inoculated into surface ripened cheese. *J. Dairy Sci. 36* 563.

Wagenaar, R. O., and Dack, G. M. (1954). The effect of emulsifier on growth and toxin production of *Clostridium botulinum* experimentally inoculated into surface-ripened cheese. *J. Dairy Sci. 37*: 640.

Wagenaar, R. O., and Dack, G. M. (1958a). Factors influencing growth and toxin production in cheese inoculated with spores of *Clostridium botulinum* types A and B. I. Studies with surface-ripened cheese type III. *J. Dairy Sci. 41*: 1182

Wagenaar, R. O., and Dack, G. M. (1958b). Factors influencing growth and toxin production in cheese inoculated with spores of *Clostridium botulinum* types A and B. II. Studies with surface-ripened cheese type II. *J. Dairy Sci. 41*: 1191.

Wagenaar, R. O., and Dack, G. M. (1958c). Factors influencing growth and toxin production in cheese inoculated with spores of *Clostridium botulinum* types A and B. III. Studies with surface-ripened cheeses type 1. *J. Dairy Sci. 41*: 1196.

Wajid, H. R. A., and Kalra, M. S. (1976). Nisin as an aid for extending the shelf life of sterilized milk. *J. Food Sci. Technol.* (*Mysore*) *13*: 6.

11

Potential Hazards Associated with REPFEDS

Barbara M. Lund

Agricultural and Food Research Council Institute of Food Research, Norwich Laboratory, Norwich, United Kingdom

S. H. W. Notermans

National Institute of Public Health and Environmental Protection, Bilthoven, The Netherlands

I. THE DEVELOPMENT OF REPFEDS

The use of precooked, refrigerated meals has increased greatly in recent years. Guidelines in the United Kingdom in relation to the production of precooked refrigerated meals by catering operations (Department of Health and Social Security, 1980; Department of Health, 1989) specify that these meals should be stored between 0 and 3°C for no longer than 5 days. The ''cook-chill'' system of catering based on these guidelines has been introduced into many hospitals and catering units in the United Kingdom (Wilkinson, 1990). Close control is used to ensure that these foods are stored in the range of temperature specified, and their short storage period limits the microbiological risks associated with any exposure to slightly higher temperatures.

There is a trend towards the production of similar precooked, refrigerated foods for catering or retail sale with an extended shelf life at refrigeration temperatures. For some of these, a shelf life of between 3 and 6 weeks at refrigeration temperatures is proposed (Mossel and Thomas, 1988). Foods of this type produced in catering establishments or for retail sale have been described as ''refrigerated, processed foods of extended durability'' (REPFEDS) (Mossel and Thomas, 1988). These foods are among the range of products that have been referred to as ''new generation refrigerated foods'' (Conner et al., 1989), which include sauces, soups, entrees, salads including pasta, seafood and meat salads, fresh pasta, and complete meals. Examples of complete meals are canneloni with meat or vegetables, chili con carne, lasagna, moussaka, and tagliatelle with pork or mushrooms or chicken (Notermans et al., 1990). Other examples of REPFEDS are canned, pasteurized crab meat, canned, pasteurized ham, and vacuum-packed, hot-smoked fish products. According to Glew (1990), the key to extending the shelf life from a nutritional and sensory point of view is to package the food in the absence of O_2.

The REPFEDS now being produced share the following characteristics (NACMCF, 1990): (1) precooked, packaged products intended to receive little or no additional heat treatment prior to consumption; (2) extended refrigerated shelf life; (3) not usually protected by conventional preservation systems, such as a low water activity or a low pH; and (4) marketed under vacuum, modified atmosphere packaging (MAP) and/or in sealed containers.

REPFEDS are commonly prepared by one of the following methods (Rosset and Poumeyrol, 1986): (1) foods are packaged then cooked, e.g., *sous vide* foods; (2) foods are cooked, then packaged; (3) foods are cooked, then packaged, then heated again.

Sous vide foods were first introduced in France over 20 years ago (Leadbetter, 1989) and represent one of the most common types of REPFEDS today. While the shelf life specified for these foods in France was originally not more than 6 days, this could be extended by a ministerial decision based on tests of microbiological criteria and sensory properties in specified laboratories. In *sous vide* processing, the food is packaged in heat-stable, air-impermeable bags under vacuum, the bags are sealed, and the product is cooked (pasteurized), cooled, and then stored under refrigeration.

While it is intended that REPFEDS should be stored at temperatures below 3–5°C, there is a risk that they will be maintained at somewhat higher temperatures. From the point of view of microbiological safety, the main concern is the risk of survival and growth of foodborne pathogens able to multiply at refrigeration temperatures, in particular *Listeria monocytogenes* and nonproteolytic strains of *Clostridium botulinum*. The heat treatments specified in some of the regulations and guidelines for production of these foods (see Sec. IV) are aimed at the destruction of both of these organisms, whereas the treatments in others are aimed only at vegetative bacteria. Because of the difficulty in ensuring that REPFEDS are maintained at temperatures below 3–5°C, it has been recommended that additional inhibitory factors be incorporated so that the safety of these foods in relation to *C. botulinum* is not dependent solely on refrigeration (Conner et al., 1989; Agriculture Canada, 1990). The heat treatments are aimed both at elimination of pathogenic bacteria and at extension of shelf life. Any surviving spoilage microorganisms should be controlled at a storage temperature of 0–4°C. Extension of shelf life depends on the core temperature of the food during cooking. If recontamination of the product after cooking can be prevented and the storage temperature can be maintained lower than 4°C, the shelf life of the product will be a function of the severity of the heat treatment during cooking (MAFR, 1988; Beaufort and Rosset, 1989).

II. POTENTIAL HAZARDS DUE TO *C. BOTULINUM*

A. Heat Resistance of Spores of Nonproteolytic *C. botulinum*

Information regarding the heat resistance of spores of *C. botulinum* and the ability of heated spores to result in growth at refrigeration temperatures is of critical importance in the control of this bacterium in REPFEDS. Basic determinations of the heat resistance of spores of *C. botulinum* have generally been made using spores suspended in phosphate buffer as a reference medium, and this has been followed by determinations of the heat resistance in specific foods, which often show protective effects. The minimum heat processing used in the canning of low-acid foods is 2.45 minutes at 121°C, which is sufficient to give a 10^{12} reduction in the numbers of the most heat-resistant spores of proteolytic *C. botulinum* types A and B (Stumbo, 1973). The heat resistance of proteolytic type F strains falls within the range of that of the proteolytic type A and B strains (Lynt et al., 1981). The heat treatments for REPFEDS are not designed to cause destruction of heat-resistant spores of these proteolytic strains of *C. botulinum* because these strains are not able to multiply at or below 10°C. Their temperature range for growth, therefore, is substantially higher than the storage temperatures specified for REPFEDS.

Strains of nonproteolytic *C. botulinum* types B, E, and F can multiply at temperatures as low as 3.3°C under otherwise optimal conditions (Ohye and Scott, 1957; Schmidt et al., 1961; Eklund, 1967a,b). If REPFEDS are not maintained at temperatures below 4°C, but are exposed to storage temperatures between 4 and 10°C, any surviving spores of these strains may multiply and form toxin before the foods show obvious spoilage. The heat resistance of spores of nonproteolytic strains of *C. botulinum* is much lower than that of proteolytic strains and has been less investigated. In order to determine the heat treatments necessary to give a specified lethality for spores of nonproteolytic strains, as suggested by the Sous Vide Advisory Committee (SVAC, 1991) in the United Kingdom and by the Draft Code for Holland and Belgium (Anonymous, 1992), further research is required.

Reports of the heat resistance of spores of nonproteolytic strains of *C. botulinum* in reference media are shown in Table 1, where the D-value is the time for a 10-fold reduction in viable numbers and the z-value is the change in temperature resulting in a 10-fold change in D-value. The results of Scott and Bernard (1982) indicated that spores of nonproteolytic type B strains were more resistant than those of types E and nonproteolytic F. Many of these studies used fraction-negative methods, which involved subculturing in an optimal medium

Table 1 Heat Resistance of Spores of Nonproteolytic *C. botulinum* in Reference Media

Number of strains tested[a]	Most resistant strain	Heating medium	Temp. range (°C)	D-value[b] (min) 82.2°C	z-value[c] (°C)	Ref.[d]
4 (E)	Saratoga	M/30 PO_4 pH 7.0	77	0.15	—	1
1 (E)	Saratoga	M/15 PO_4 pH 7.0	77–100	0.33	4.8	2
4 (E)	Saratoga	M/60 PO_4 pH 7.0	70–80	0.48	5.6	3
9 (E)	1957/61	water	70–80	0.54	6.0	4
4 (E)	Saratoga	M/15 PO_4 pH 7.0	70–80	0.69	7.2	3
1 (E)	Minneapolis	M/15 PO_4 pH 7.0	74–86	1.25	8.3	5
2 (E)	Nanaimo	M/15 PO_4 pH 7.0	70–80	1.93	9.6	6
4 (E)	Alaska	M/15 PO_4 pH 7.0	82.2	>4.9	—	2
3 (F)	610	M/15 PO_4 pH 7.0	71–85	0.84	6.3	7
4 (B)	2129B	M/15 pH 7.0	80–100	32.3	9.7	2

[a]Type in parentheses.
[b]D-Value = time for 10-fold reduction in viable numbers.
[c]z-value = change in temperature resulting in a 10-fold change in D-value.
[d]1) Ito et al., 1967; 2) Scott and Bernard, 1982; 3) Bohrer et al., 1975; 4) Roberts et al., 1967; 5) Schmidt, 1964; 6) Ohye and Scott, 1957; 7) Lynt et al., 1979. *Source*: Simunovic et al., 1985.

after a series of heating times, and recording growth or no growth after incubation. In these methods, the shape of the curve relating the number of survivors to heating time is not known, and the heat resistance reported may be determined by a low proportion of very resistant spores. In none of the above studies was lysozyme added to the medium used for the recovery of heated spores. All of these reports must be evaluated, therefore, in the light of the observations, first published in the early 1970s, that the use of a recovery medium containing lysozyme for the enumeration of surviving, heated spores of nonproteolytic strains of *C. botulinum* increased the recovery of viable, sublethally damaged spores. Sebald et al. (1972) showed that when spores of *C. botulinum* type E were heated at 80°C for 10 minutes, surviving spores could be detected on medium with lysozyme but not in the absence of lysozyme. The percentage of spores that germinated and produced colonies was dependent on the concentration of lysozyme added to the medium in the range 0.1–5.0 μg/ml, but higher concentrations gave no increase in the number of surviving spores. The proteolytic enzymes trypsin, pronase, and protease also increased the percentage of spores that germinated and formed colonies, but they were much less effective than lysozyme. Alderton et al. (1974) reported that when spores of a strain of type E were heated in phosphate buffer at 79.5°C and enumerated on a medium without lysozyme, the estimated D-value was 1.3 minutes. When lysozyme (5 μg/ml) was included in the recovery medium, the heating temperature had to be raised in order to obtain measurable destruction; at 195°F (90.5°C) and 200°F (93.3°C) the D-values were 13.5 and 3.8 minutes, respectively. These D-values were considerably higher than the values reported from many previous studies. The heat resistance of spores of nonproteolytic type B and E strains was investigated by Smelt (1980), who enumerated the surviving spores on a trypticase-peptone-glucose-thioglycollate medium containing 5% egg yolk (egg yolk emulsion, Oxoid), which probably contained some lysozyme. When spores were heated in phosphate buffer, M/15, pH 7.0, at temperatures in the range of 77–92.5°C, biphasic survival curves were obtained, indicating the presence of a heat-sensitive and a heat-resistant fraction. The heat-resistant fraction appeared to form between 0.1 and 1% of the total number of viable spores. The estimated D-values for the heat-sensitive and the heat-resistant fractions are shown in Table 2. Scott and Bernard (1982) used a quantal method to determine the heat resistance of spores of nonproteolytic type B *C. botulinum*. The spores were heated in phosphate buffer, M/15, pH 7.0, and subcultured into a pork-pea broth in order to recover surviving viable spores. At 82.2°C, D-values between 1.49 and 32.3 minutes were reported, compared with D-values of 0.33 and >4.9 minutes for two strains of type E. A similar method was used to determine the effect of lysozyme on the estimated heat resistance of spores of nonproteolytic type B and E strains (Scott and Bernard, 1985). After heat treatment in phosphate buffer, the spores were transferred into pork-pea broth containing 5 μg

Table 2 Heat Resistance of Spores of Nonproteolytic Strains of *C. botulinum* in Phosphate Buffer, M/15, pH 7.0

Type[a]	Temp. (°C)	Fraction of the population	D-value (min)	Standard deviation of D-value	Number of observations
B	77.5	S	3.98	0.48	6
(B17)		R	103	27.6	6
	85	S	2.53	0.37	5
		R	51	10.6	6
	87.5	S	1.5	—	2
		R	24	2.2	9
	90.0	S	0.4	—	2
		R	8.3	17.08	7
E	77.5	S	1.52	0.12	6
(Beluga)		R	38	2.87	6
	80.0	S	1.24	0.58	5
		R	36	10.66	6
	82.5	S	0.53	0.16	4
		R	23.6	3.74	5
	85.0	S	0.30	0.147	3
		R	10.40	1.73	5
	87.5	S	0.15	—	2
		R	6.1	1.42	3

[a] Strain in parentheses.
S, Sensitive fraction; R, resistant fraction.
Source: Smelt, 1980.

lysozyme (Sigma, 41,000 units/mg)/ml. The presence of lysozyme resulted in an increase of up to 160-fold in the estimated heat resistance of spores of nonproteolytic type B strains, giving D-values at 82.2°C between 28.2 and 2224 minutes, and a 73-fold increase in the estimated heat-resistance of type E strain Saratoga, to give a D-value at 82.2°C of 24.2 minutes (Table 3). Because this work was based on an endpoint method for the determination of heat resistance, the shape of the curve relating number of survivors to the heating time was not determined. The D-values reported by Smelt (1980) for the resistant fraction of the spore populations were of the same order as the D-values reported by Scott and Bernard (1985).

It is clear that lysozyme in the recovery medium enables the detection of surviving spores that, in the absence of lysozyme or an alternative enzyme, would have remained dormant. The effect of lysozyme probably results from its ability to substitute for a germination mechanism that has been inactivated by the heat treatment (Gould, 1984). Lysozyme in the recovery medium enables a

Table 3 Effect of Lysozyme in the Recovery Medium on the Estimated Heat Resistance of Spores of Nonproteolytic *C. botulinum* in Phosphate Buffer, M/15, pH 7.0

		D-value (min) at 82.2°C	
Strain	Type	PPB	PPB + L[a]
CBW 25	nonproteolytic B	1.49	73.6
ATCC 17844	nonproteolytic B	4.17	245
17 B (crop 1)	nonproteolytic B	0.83	134
17 B (crop 2)	nonproteolytic B	16.7	28.2
2129 B	nonproteolytic B	32.3	2224
Saratoga	E	0.33	24.2

[a] Recovery media: PPB, pork pea broth; L, lysozyme.
Source: Scott and Bernard, 1985.

much improved assessment of the numbers of spores that have survived heat treatment. It may be argued, however, that unless comparable concentrations of lysozyme are likely to be present in heat-treated foods, it is unnecessary to take account of heat-damaged spores that require lysozyme in order to germinate and initiate growth.

In their review, Proctor and Cunningham (1988) reported that lysozyme has been found in several foods (Table 4). Many of the vegetables that contained lysozyme were Brassicas, and the enzyme was not detected in 24 other types of vegetables and fruits tested. Lysozyme has also been reported at the following concentrations in the organs of cattle: spleen, 50–160 μg/g; thymus, 80 μg/g; pancreas, 20–35 μg/g; liver, kidneys, and lungs, trace amounts (Proctor and Cunningham, 1988). Low concentrations of lysozyme have also been reported in some fish (Lie et al., 1989) and shellfish (Mochizuki and Matsumiya, 1983). The enzyme is relatively heat-stable, particularly in acidic conditions.

In view of the facts that lysozyme may be present in a range of foods and that it may survive pasteurization processes, the effect of this enzyme on the germination of heat-damaged *C. botulinum* spores must be considered in assessing the heat treatments required to ensure the safety of REPFEDS.

Estimates of the heat resistance of bacterial spores in foods are often higher than those obtained in phosphate buffer. Type E spores added to foods which were heated in thermal-death-time (TDT) cans, cooled, and incubated at 29.4°C for up to 12 months had higher D-values than those of spores in heated phosphate buffer, subcultured into nutrient medium and incubated at the same temperature for a minimum of 3 weeks (Bohrer et al., 1973) (Table 5). This result was probably due, in part, to the longer incubation time used to recover spores from the cans. Nevertheless, there is evidence that foods may protect spores from heat inactivation. The most protective foods in the study by Bohrer et al.

Table 4 Reported Concentrations of Lysozyme in Foods

Food	Concentration of lysozyme (μg/ml)
Hen egg albumen (fresh)	2250–3270
Hen egg albumen (fresh after storage of eggs for 2 months at 20°C)	~3000
Bovine milk	0.16–0.32
Juice of:	
cauliflower	27.6
rutabaga	8.9
papaya	7.9
cabbage	2.3
kohlrabi	3.3
red radish	4.8
white radish	4.5
turnip	1.8
parsnip	1.6
broccoli	8.1

Source: Proctor and Cunningham, 1988.

Table 5 Decimal Reduction Times (D-Values) and z-Values for Spores of *C. botulinum* Type E Strain Saratoga in Phosphate Buffer and in Foods

	D-value (min)at		
Heating medium	76.7°C	80°C	z-value (°C)[a]
Phosphate buffer M/15, pH 7.0	4.1	1.4	7.2
Tuna in oil	40.9	10.5	6.1
Crabmeat		4.4	6.3
Corn (W. K. in brine)	11.2	3.2	6.0
Milk, evaporated	6.4	1.9	6.3
Sardines in tomato sauce	20.0	4.5	6.3
Shrimp		1.9	
Peas	18.6	3.3	4.4
Salmon		2.9	

[a]Over approximate temperature range 70–82.2°C.
Source: Bohrer et al., 1973.

Table 6 Decimal Reduction Times (D) and z-Values of Spores of Nonproteolytic Type F *C. botulinum* in Phosphate Buffer, M/15, pH 7.0, and in Crabmeat

Heating medium	D-value (min) at 82.2°C	z-value[a] (°C)
Phosphate buffer	0.25	5.83
Crabmeat	1.16	7.5

[a]Over approx. temperature range 76–85°C
Source: Lynt et al., 1979.

(1973) were those with the highest content of fat and the lowest content of water, i.e., tuna in oil and sardines in tomato sauce. The authors implied that to ensure safety from *C. botulinum* type E, heat processes for specific foods should be based on the actual D-values for strain Saratoga in those foods. In experiments using a fraction-negative method with subculture of heated samples into a nutrient medium, spores of a nonproteolytic type F strain showed higher D-values in crabmeat than in phosphate buffer (Lynt et al., 1979) (Table 6). Other reports of the heat resistance of spores of nonproteolytic *C. botulinum* in foods are summarized in Table 7. In none of these experiments was lysozyme added to the recovery medium.

Table 7 Heat Resistance of Spores of Nonproteolytic Strains of *C. botulinum* in Foodstuffs

Number of strains tested	Most resistant strain	Heating medium	Temp. range (°C)	D-value (min) 82.2°C	z-value (°C)	Ref.[a]
5	Beluga (E)	blue crabmeat	74–85	0.74	6.5	1
1	Saratoga (E)	evaporated milk	77–82.2	0.8	6.3	2
1	Saratoga (E)	corn in brine	77–82.2	1.3	6.0	2
1	Saratoga (E)	ground crabmeat	80–82.2	1.9	7.6	2
5	Alaska (E)	whitefish chubbs	74–85	2.21	7.6	3
1	Saratoga (E)	sardines in tomato sauce	77-82.2	2.9	6.3	2
1	Saratoga (E)	tuna in oil	72–82.2	6.6	6.1	2
1	Beluga (E)	oyster	50–80	0.78[b]	7.6	4
1	202 (F)	crabmeat	76.6–85	1.16	6.38	5

[a]1) Lynt et al., 1977; 2) Bohrer et al., 1973; 3) Crisley et al., 1968; 4) Bucknavage et al., 1990; 5) Lynt et al., 1979.
[b]At 80°C
Source: Simunovic et al., 1985.

When chub (*Leucichthys hoyi*) was inoculated with 10^6 spores of nonproteolytic type B per fish, heated at 82.2°C for 30 minutes, vacuum-packaged, and incubated at 20–25°C, spores survived and led to toxin formation unless the concentration of brine in the fish was higher than 2.75% (Christiansen et al., 1968).

Lynt et al. (1977) investigated the survival of spores of type E *C. botulinum* in meat of the blue crab during pasteurization processes. Spores were heated in the crabmeat and cultured in a nutrient medium at 26°C. D-values at 82.2°C for the most resistant strains are shown in Table 7. The authors concluded that the pasteurization process used by industry, the equivalent of 180°F (82.2°C) for 18 minutes, was well above a 12-D treatment for type E spores in crabmeat. As their recovery medium did not contain added lysozyme, however, their results may have overestimated the lethality of the process by a factor of 70 (Scott and Bernard, 1985). In a subsequent paper, Lynt et al. (1979) reported similar results with a nonproteolytic type F strain (Table 7); lysozyme was not added to the recovery medium.

B. Ability of Nonproteolytic *C. botulinum* to Grow at Low Temperature

Average doubling times of 10 strains of *C. botulinum* type E in culture medium at 5, 10, and 15°C were 42, 7.19, and 3.06 hours, respectively (Ohye and Scott, 1957). For a single strain of nonproteolytic type B, mean doubling times at 5, 10, and 20°C were 30.3, 6.1, and 1.5 hours (Graham and Lund, 1991). These doubling times mean that at 10°C, for example, about 5 days would be required for one nonproteolytic type B *C. botulinum* per gram of food to multiply to 10^6/g, giving a significant amount of toxin.

Jensen et al. (1987) investigated the effect of incubation temperature on growth of nonproteolytic strains of types B, E, and F in a model broth system. For the strain that was best able to initiate growth at each temperature, the probabilities of growth from a single spore in 28 days were 0.1% at 4°C, 10% at 8°C, and 100% at 12°C. The probabilities of growth from a vegetative cell in 28 days were 6.3% at 4°C and 100% at 8 or 12°C. For a mixture of four nonproteolytic type B strains, the probabilities of growth from a single vegetative bacterium were 10–100% at 6°C after 20 days or at 8°C after 11 days and 100% at 20°C after 2 days (Lund et al., 1990). None of these studies used spores that had survived a heat treatment, but they illustrate the short incubation times and the low numbers of spores that can result in formation of toxin by nonproteolytic *C. botulinum* in foods stored at temperatures in the range of 4–12°C.

The ability of unheated and heated type E and nonproteolytic type B and F spores to grow and form toxin in blue crabmeat at low temperatures was studied by Solomon et al. (1977,1982). At temperatures of 12°C and below, there was a greater incidence of growth and toxin formation in broth than in crabmeat, and only unheated spores of type E gave growth and produced toxin in crabmeat (Table 8). Neither unheated nor heated spores produced growth in crabmeat at 4 or 8°C. Both unheated and heated spores of type E and unheated spores of types B and F gave growth and toxin in broth at 4°C, but heated spores of types B and F failed to give growth and toxin in broth after 180 days; lysozyme was not added to the broth used in these experiments. According to these results, (1) unheated and heated spore inocula were less able to give growth in crabmeat than in broth at temperatures of 12°C and below, (2) heated spores of type E were less able than unheated spores to result in growth in crabmeat at 12°C, and (3) heated spores of nonproteolytic types B and F were less likely to result in growth in broth at 4°C than unheated spores.

Both the rate of growth and the probability of growth of *C. botulinum* in culture media and in foods can be reduced by the presence of combinations of preservative factors (Lund et al., 1990), but many REPFEDS are likely to be produced without any preservative factors.

III. CONTROL OF *C. BOTULINUM* IN REPFEDS

The minimum heat treatment given to low-acid, canned foods is designed, in general, to reduce by a factor of 10^{12} the number of surviving spores of proteolytic strains of *C. botulinum* with the maximum heat resistance (see Chapter 6). In many foods, the control of *C. botulinum* relies on a combination of destruction by heat and inhibition by chemical preservative factors. The safety factor or the degree of protection (Pr) can be expressed as

$$\mathrm{Pr} = \mathrm{Ds} + \mathrm{In}$$

where Ds is the decimal reduction of *C. botulinum* spores by heat and In is the decimal inhibition (Hauschild and Simonsen, 1986). The thermal process given to the majority of shelf-stable, canned, cured meats may range from 0 to $1.5F_0$, but is usually in the range of $0.05–0.6F_0$ (Hauschild, 1989), which, at most, would reduce the number of viable spores of proteolytic strains of *C. botulinum* by a factor of 10^3. Additional protection is given by inhibition of growth from the surviving spores, primarily by the presence of NaCl and nitrite.

The degree of protection can also be expressed as (Hauschild, 1982)

$$\mathrm{Pr} = \log \frac{1}{\mathrm{P}}$$

where P is the probability that individual spores would survive any heat process and grow and form toxin in the product. This probability can be calculated from the result of experiments in which *C. botulinum* spores are incorporated into the product that is then processed, stored, and tested for toxin. For a canning

Table 8 The effect of Incubation Temperature on Growth and Formation of Toxin by Spores of Nonproteolytic *C. botulinum* in Broth and in Crabmeat

	Unheated spores, types						Heated spores[a], types					
	E		B		F		E		B		F	
Incubation temperature (°C)	Broth[b]	CM[b]	Broth[c]	CM[c]	Broth[c]	CM[c]	Broth[d]	CM[e]	Broth[c]	Cm[c]	Broth[c]	Cm[c]
26	GT3[f]	GT3	GT3	GT3	GT3	G3T10	GT6	GT6	GT3	GT3	GT3	G3T10
12	GT14	GT14	GT6	—[g]	GT6	—	GT20	—	GT6	—	GT10	—
8	GT14	—	GT15	—	G15T20	—	GT30	—	GT15	—	GT20	—
4	GT52	—	GT20	—	GT20	—	GT30	—	—	—	—	—

[a]Type E spores heated in the medium at 82.3°C for 6 min, type B and F spore suspensions heated, before inoculation, sufficiently to give 10-fold lethality.
[b]2 × 10^3 viable spores/ml of broth and /g of crabmeat (CM).
[c]Approx. 10^3 viable spores/ml of broth or g of crabmeat.
[d]2 × 10 viable spores/ml of broth.
[e]2 × 10^3 viable spores/g of crabmeat.
[f]Results are shown as time (days) when growth (G) and toxin (T) were detected.
[g]— = No growth or toxin detected.
Source: For type E, Solomon et al., 1977; for types B and F, Solomon et al., 1982.

process given to a low-acid food with a 12-D heat treatment (F_0 = 3.0), the degree of protection would be 12. For commercial shelf-stable, canned, cured meats, the degree of protection against *C. botulinum* has been estimated as equivalent to not more than 8 (Hauschild, 1989). Under conditions of temperature abuse (the equivalent of one week at 27°C), the degree of protection for vacuum-packed bacon was between 5 and 7, and for vacuum-packed turkey between 3 and 4, (Hauschild, 1982). Control of *C. botulinum* in this last product relies largely on refrigeration.

Similar alternatives—heat treatment alone or a combination of heat treatment and inhibition—can be applied to control nonproteolytic *C. botulinum* in REPFEDS. The first approach is to ensure that the heat treatment alone is sufficient to give the required safety factor against spores of nonproteolytic strains. For this approach it is necessary to determine the greatest intrinsic heat resistance likely to be shown by spores of nonproteolytic strains and the degree to which specific foods protect the spores against heat (see Sec. II.A). The application of this heat treatment to give the required number of decimal reductions of these spores would assure the safety of REPFEDS packed before heat treatment, provided that the storage temperature always remained below 10°C. In the case of products packed after heat treatment, there would be some risk of contamination with nonproteolytic *C. botulinum* after cooking, but this risk should be minimized by the use of "high care" conditions during packing (Chilled Food Association, 1989; Anonymous, 1992).

If the heat treatment required for the first approach is not practicable for sensory reasons, an alternative approach is to set up challenge tests to determine whether spores of nonproteolytic strains of *C. botulinum* would both survive the heat-processing and result in growth and formation of toxin in the product under conditions to which it may be exposed between production and consumption. Tests could be used in which foods are inoculated with a mixture of heat-resistant spores of nonproteolytic strains of *C. botulinum* (for example, about 10^6 spores per sample) and subjected to the conditions used for processing. Samples would then be incubated at several temperatures for suitable times and tested for growth of *C. botulinum* and formation of toxin. If such tests show consistently that surviving spores fail to result in growth and formation of toxin during incubation at temperatures up to 10°C and for a period exceeding the shelf life of the food, it may be concluded that the combination of the heat processing and the nutrient/inhibitory properties of the food gives a safety factor of at least 6. The view has been expressed that the safety of REPFEDS should not rely entirely on refrigeration, but that attempts should be made to add other barriers to microbial growth (Conner et al., 1989; Agriculture Canada, 1990). Recommendations in the United States require that for inoculated pack tests with nonproteolytic *C. botulinum*, the packs are incubated at 12.8°C for one and one-half times the product's shelf life, or up to the time when the product is overtly spoiled. Tests with proteolytic *C. botulinum* are recommended at 27°C for up to the time when the product is overtly spoiled or for one-half the refrigerated shelf life if the product does not appear spoiled by that time (NACMCF, 1990). If the product contains multiple components, then testing should terminate 1 day beyond the time that the last component of the product is overtly spoiled. The limitation of this empirical approach is that it relies on the production of heat-resistant spores for the challenge testing of each individual food and on repeated testing when changes are made to recipes.

Safety could also be achieved by a combination of mild heat processing and additional barriers to growth of *C. botulinum*, such as low pH, acidulants, NaCl or other humectants, or preservatives such as nitrite, sorbic acid, or nisin. (Lund et al., 1990). Inhibitory systems could be devised by experiments in culture media, where these systems can best be controlled and their effects analyzed. Individual foods may contain additional inhibitors or protective factors, which means that the effects of inhibitory systems (predicted on the basis of experiments in culture media) should be confirmed in foods.

REPFEDS have been referred to as "new-generation refrigerated foods." It seems likely that they will be used and sold widely in developed countries, and it is essential that rational methods are agreed upon internationally for the control of microbiological risks. In adopting such methods, it is important to note that, on the basis of experience over many years, several similar foods are considered safe, although the basis for that safety may not have been thoroughly analyzed.

Meat of the blue crab is highly prized in the United States. It cannot be heat processed to give a shelf-stable product, because the required heat treatment results in discoloration and off-flavors, nor can it be frozen satisfactorily. Since the 1950s, blue crabmeat has been sold as a pasteurized, refrigerated product with a shelf life of up to 6 months. No case of botulism has been attributed to this product, but the presence of *C. botulinum* has been demonstrated in the blue crab and its environment, and the organism has also been found in commercially packed crabmeat (Lynt et al., 1977). The recommended minimum pasteurization process by a water-bath method was heating hermetically sealed containers at 190–192°F to an internal meat temperature of 185°F (85°C) and holding at that temperature for 1 minute (Cockey and Tatro, 1974). This process was reported to result in a lethality of approximately 8D for spores of type E (Cockey and Tatro, 1974), but the surviving spores were enumerated in a medium without added lysozyme. Cans of crabmeat (1 lb) that had been inoculated with 10^7–10^8 spores of type E per 100 g meat and pasteurized by the above procedure remained nontoxic when stored at 4.4°C for up to 6 months, but the effect of storage at temperatures up to 10°C was not reported. In these experiments, the core temperature was above 180°F (82.2°C) for at least 20 minutes. Therefore, the come-up time to the pasteurization temperature would have contributed significantly to the lethality of the pasteurization. Pasteurized crabmeat has a good safety record, and in over 30 years of production it has not been implicated in botulism (Hauschild, 1989). Nevertheless, there is a need to understand the basis of this safety and the extent to which the protection relies on refrigeration.

In hot-smoked fish products, surviving nonproteolytic *C. botulinum* spores are inhibited by a combination of NaCl (or NaCl plus nitrite), smoke, and refrigerated storage below 38°F (3.3°C). The interest in vacuum-packaging smoked fish products and longer refrigerated storage life, together with consumer preference for reduced levels of NaCl, nitrite, and other additives, have increased the extent to which the safety of these products relies on refrigeration (see Chapter 8).

It has been proposed that minimally processed oysters should be heated to an internal temperature of at least 55°C for 10 minutes (Bucknavage et al., 1990). The shelf life of the heated product at refrigeration temperature was claimed to be 45 days, but could be increased to 60 days by the incorporation of 1.0% NaCl and 0.13% potassium sorbate (Bucknavage et al., 1990). Spores of type E strain Beluga were added to a homogenate of oysters at the required temperature, and the surviving spores were enumerated after a series of heating times (Bucknavage et al., 1990). The D-value in oyster homogenate is shown in Table 7. The authors concluded that the heating times achieved by the proposed process, 27.7–42.6 minutes at 55°C (Bucknavage et al., 1990), would result in little, if any, inactivation of type E spores, and that a heat treatment at 80°C for 9.36 minutes would be required to achieve a 12-D process. This would give the oysters a cooked appearance. Since the survival of spores of *C. botulinum* was highly probable in the proposed process, Bucknavage et al. (1990) concluded that an investigation of the conditions needed for growth and toxin formation in minimally processed oysters would be required before such a product could be marketed. In this study, lysozyme was not added to the recovery medium, and counts were made after incubation for only 2 days. Thus, the number of surviving spores was probably greatly underestimated, and no investigation was made of the survival of the more heat-resistant spores of nonproteolytic type B strains.

The potential for preservation of restructured meat products by vacuum-packaging, heat-pasteurization, and subsequent refrigerated storage was reviewed by Simunovic et al. (1985). These authors suggested that the most important factor controlling growth of *C. botulinum* in foods preserved short of sterilization is storage of the finished product at temperatures below 3.3–5°C, but that temperature abuse of the products could occur frequently. While the interaction of factors such as salt, nitrites, liquid smoke, and other ingredients might enhance the inhibition at refrigeration temperatures, there was a lack of quantitative information regarding the magnitude of these effects.

Proposals have been made, particularly in Japan, to add lysozyme as a preservative to foods including cheese, meat, fish, vegetables, and fruit (Proctor and Cunningham, 1988). There has been considerable interest in the use of low concentrations of the enzyme in cheese to prevent "late-blowing" due to butyric acid bacteria, in particular *Clostridium tyrobutyricum*. Its use for this purpose has been permitted in Germany, Italy, Denmark, France, and Australia (Crawford, 1987). Roberts and Kruger (1984) proposed its use in prepared foods, and Scott (1988–1989) suggested that the addition of lysozyme may permit the extension of shelf life for refrigerated foods and milder processing conditions for heat-processed, low-acid foods. In an investigation of the potential use of lysozyme as a preservative for foods, Hughey and Johnson (1987) reported that four strains of nonproteolytic *C. botulinum* types B and E grew in culture medium at 30 and 37°C with a lysozyme concentration of 200 μg/ml, equivalent to 10,000 units/ml. However, the combination of 1 mM EDTA and 20 μg lysozyme/ml prevented growth of both proteolytic and nonproteolytic strains of *C. botulinum* and nongrowing cells of a type E strain were lysed by lysozyme or, more actively, by lysozyme plus EDTA at 10 and 4°C.

The effect of preservative factors and of foods on the probability of growth of *C. botulinum* may be estimated by predictive microbiology (see Chapter 14). In studies of this type, particularly in miniaturized systems at low temperature, it is important to ensure that traces of O_2 are eliminated and that redox potentials are maintained at very low levels (Lund et al., 1984), in order that the effects of storage temperature and of the chemical composition of the food can be measured.

IV. REGULATIONS AND GUIDELINES

Regulations for the hygienic production, distribution, and sale of *sous vide* foods and other REPFEDS in France and for the extension of their shelf life have been published by the French Ministry of Agriculture (MAFR, 1974, 1988). The storage temperature specified for precooked, ready-to-eat foods is between 0 and 3°C.

The relationship between heat treatment and shelf life of the product is shown in Table 9. In calculating the "pasteurizing value," the target organism selected in France by the Veterinary Service of Food Hygiene (S. V. H. A.) was a strain of *Enterococcus faecalis* (*Streptococcus faecalis*) for which the D-value at 70°C was 2.95 minutes and the z-value was 10°C (Rosset and Poumeyrol, 1986; Mossel and Thomas, 1988). The pasteurizing value is the time (in minutes) of heating at 70°C that has the same destructive effect on *E. faecalis* as the heat treatment applied to the food. The relationship between core temperature, time at temperature, and pasteurizing value is shown in Table 10. The manufacturer must ensure that the products remain in accordance with specified microbiological criteria throughout the shelf life plus an additional 48 hours at 3°C. Products with a 21-day shelf life must remain in accordance with these criteria after 14 days at 4°C, followed by 7 days at 8°C; and products with a 42-day shelf life

Table 9 Relation Between Core Temperature During Cooking, Pasteurizing Value, and Shelf Life of *Sous Vide* Foods

Core temperature (°C)	Pasteurizing value[a] (Pt_{70}^{10})	Shelf life
57–65°C	Pasteurization value and shelf life are the subject of a technical file that is submitted to an approved laboratory.	
≥65°C	≥100	The responsibility of the manufacturer up to 21 days
≥70°C	≥1000	The responsibility of the manufacturer up to 42 days

[a]Total pasteurizing value, Pt_T^z, where T = the reference temperature for heat treatment, 70°C, and z = change in temperature, in °C, required to change the decimal reduction time by a factor of 10, or one log cycle. The pasteurizing value is expressed in minutes at the reference processing temperature, 70°C, and is calculated using a z-value of 10 for a strain of *Enterococcus faecalis*.
Source: MAFR, 1988.

must comply with these criteria after storage for 28 days at 3°C, followed by 14 days at 8°C.

In the United Kingdom, the Chilled Food Association (1989) issued guidelines for good hygienic practice in the manufacture of chilled foods other than those produced by catering operations. Four categories of prepared, chilled food are covered: (1) those prepared from raw components; (2) those prepared from cooked and raw components processed to extend the safe shelf life; (3) those prepared from only cooked components; and (4) those cooked in their own packaging prior to distribution. The guidelines are intended to ensure that the cooking processes are sufficient to destroy vegetative pathogens, in particular *L. monocytogenes*. In order to achieve this, all parts of the food or food component

Table 10 Relationship Between Core Temperature, Heating Time, and Pasteurizing Value of Precooked Foods

Temperature (°C)	Time at temperature (min)	Pasteurizing value (Pt_{70}^{10})
69 ± 2[a]	50	25–63
71 ± 2	50	40–100
74 ± 2	40	63–159
75.3 ± 2	295	631–1584

[a]A difference of 2°C in core temperature results in a large change in pasteurizing value (Beaufort and Rosset, 1989).
Source: MAFR, 1988.

should have a treatment at least equivalent to maintenance at 70°C for 2 minutes. If foods in categories 3 and 4 have an extended shelf life, it must be demonstrated that the heating process is sufficient to destroy the spores of psychrotrophic strains of *C. botulinum*, unless other preservation systems are present.

The Sous Vide Advisory Committee of the United Kingdom has produced a Code of Practice for *sous vide* catering systems (SVAC, 1991). These guidelines relate specifically to the use of *sous vide*, cook-chill foods with a shelf life of not greater than 8 days, which is regarded as the maximum for nonindustrially produced *sous vide* foods. The basic principles specify that the vacuum-packing and heat treatment must ensure the destruction of the vegetative stages of any pathogenic microorganisms and significantly reduce the number of psychrotrophic *C. botulinum* (types B and E). The recommended minimum heat treatments, expressed as core temperatures, are shown in Table 11. Any products or recipes that cannot be processed according to these specified heat treatments are regarded as unsuited to *sous vide* production and should not be produced by this process. After heat treatment, the food must be cooled rapidly and stored at 0–3°C. Should the temperature of the food rise above 5°C but not 10°C, it must be used within 12 hours; should the temperature rise above 10°C the food must be destroyed immediately.

American recommendations for REPFEDS containing cooked, uncured meat or poultry products (NACMCF, 1990) divide the foods into the following three categories: (1) products that are assembled, then packed and cooked, including *sous vide* and cook-in-the-bag foods; (2) products that are cooked, then assembled into the final package, with no heat treatment applied after the final packaging; and (3) products with components cooked individually, combined with raw ingredients, then assembled and packaged. It is specified that the final products should be maintained at a temperature not exceeding 40°F (4.4°C) and that the shelf life of the product is to be determined and documented by the manufacturer. The heat processing must be sufficient to give a minimum 4-D kill of *L. monocytogenes*. The recommended minimum processing times at specified temperatures are shown in Table 12. The recommendations also specify that it is

Table 11 Recommended Minimum Heat Treatments for *Sous Vide* Cook-Chill Foods with a Shelf Life of Not Greater than 8 Days at 0–3°C

Core temperature (°C)	Heating time (min)
80	26
85	11
90	4.5
95	2

Source: SVAC, 1991.

Table 12 Recommended Minimum Processing Times at Specified Temperatures for Refrigerated Foods Containing Cooked, Cured, or Uncured Meat or Poultry Products Packaged for Extended Refrigerated Shelf Life and Ready to Eat or Prepared with Minimal Additional Heat Treatments

Core temperature		Time
°F	°C	(min)
125	51.6	244
135	57.2	31.6
140	60.0	11.4
149	65.0	1.82
158	70.0	0.29
160	71.1	0.19

Source: NACMCF, 1989.

incumbent on the processor to verify that production of botulinum toxin in these products can be prevented from the time of production to consumption, and guidelines are provided for inoculated pack studies to determine the potential for the production of botulinum toxin in these foods. The recommended conditions for inoculated pack studies with nonproteolytic *C. botulinum* are incubation at 55°F (12.8°C) for one and one-half times the product's shelf life, or until the time when the product becomes overtly spoiled (unfit for human consumption). Studies with proteolytic *C. botulinum* are recommended at 80.6°F (27°C) for up to the time when the product is overtly spoiled or for one-half the refrigerated shelf life if the product does not appear spoiled by that time. If the product contains multiple components, then testing should terminate 1 day beyond the time the last component of the product is overtly spoiled. If the product supports the formation of botulinum toxin within the time frame described in the guidelines or before the food is unfit for human consumption, the processor must either reformulate the product to prevent botulinum toxin production or design a strategy to ensure that the consumer can recognize temperature abuse before botulinum toxin may be produced.

The Canadian Code of Recommended Manufacturing Practices for REPFEDS (Agriculture Canada, 1990) deals with products that are pasteurized, packaged under modified atmospheres (vacuum-packaging, gas flushing) in hermetically sealed containers, prior to or immediately after heat treatment, and require refrigeration (−1 to 4°C) throughout their shelf life. The focus of the code is on products whose microbiological safety relies heavily on proper control of the above process. The code states that refrigeration by itself could be insufficient to guarantee microbiological safety of the pasteurized/MAP/refrigerated foods, since at some point during distribution or consumer handling, temperature abuse is likely to occur. For this reason, the use of additional barriers to the growth of microorganisms, such as acidification, the addition of preservatives or the reduction of water activity is ''strongly encouraged.'' The stated objective of the pasteurization process is the same as that in the French recommendations (above), and a similar relationship between ''pasteurizing value'' and shelf life is adopted.

The draft Code of Practice for the production, distribution and sale of REPFEDS in Holland and Belgium (Anonymous, 1992) covers meals intended for consumption after heating by the consumer, and with a storage life of 11 days–6 weeks at 0–5°C. The Code covers meals that have been prepared by one of three types of process: (1) foods packed before processing and given a pasteurization treatment equivalent to heating at 90°C for 10 minutes (Table 13); (2) foods given a heat treatment equivalent to 90°C for 10 minutes, assembled and packed in high-care conditions, and repasteurized using a heat treatment equivalent to 70°C for 2 minutes (Table 13); and (3) foods assembled, given a heat treatment equivalent to 90°C for 10 minutes, cooled, and packed in high-care conditions. Heating at a core temperature of 70°C for 2 minutes is designed to ensure the destruction of *L. monocytogenes* (Chilled Food Association, 1989; Department of Health, 1989). Heat treatment at 90°C for 10 minutes is designed to reduce the number of viable spores of nonproteolytic *C. botulinum* by six log cycles. The Code states that there is some uncertainty about the z-value of spores of nonproteolytic type B *C. botulinum* (the most heat-resistant of the three nonproteolytic types). For calculations of the lethal value of temperatures above 90°C, a z-value of 10 is used, whereas a z-value of 7 has been adopted for temperatures below 90°C. The official French guidelines for chilled meals with a shelf life of 6 weeks also require a pasteurization equivalent to 10 minutes at 90°C (Anonymous, 1992). The products are to be tested after storage for two-thirds of their intended storage life at 3°C and for one-third at 8°C. For a meal with a storage life of 6 weeks, this would mean 4 weeks at 3°C and 2 weeks at 8°C. After this time, vegetative pathogenic or toxigenic bacteria and microbial toxins ''must be absent,'' and the count of aerobic bacteria present must not exceed 10^6/g.

Table 13 Recommended Heat Treatment for Chilled, Long-Life Pasteurized Foods with a Shelf Life of Up to 6 Weeks

Target organism	Temperature/time	z-value (°C)
L. monocytogenes	70°C/2 min	7.5
Nonproteolytic *C. botulinum*	90°C/10 min	<90°C, 7 >90°C, 10

Source: Anonymous, 1992.

The French regulations for REPFEDS (MAFR, 1988) and the codes of practice for Canada (Agriculture Canada, 1990), the draft code of practice for Holland and Belgium (Anonymous, 1992), the American Recommendations (NACMCF, 1990) and the code of practice for *sous vide* catering systems (SVAC, 1991) specify that storage temperatures should not exceed 3–5°C. Under the guidelines of the Chilled Food Association (1989), the storage temperatures that apply are those given in the Food Hygiene (Amendment) Regulations (1990). These specify that cooked products should be stored at or below 8°C. As of April 1, 1993, cooked products that have been prepared for consumption without the necessity for further cooking or reheating must be stored at or below 5°C. The difficulty of ensuring this temperature during commercial transport, in retail display cabinets, and in home refrigerators has been documented (Conner et al., 1989; Smith et al., 1990; James and Evans, 1990). Surveys of current retail cabinets for wrapped and unwrapped refrigerated products showed that maintenance of a set product temperature requires an air temperature reduction of approximately 7°C during passage through the evaporator (James and Evans, 1990). Thus, to ensure a maximum product temperature of 5°C, which is specified for many foods in the United Kingdom by the Food Hygiene (Amendment) Regulations (1990), air must enter the cabinet at −2°C. In many existing cabinets this would result in products near the air inlets being frozen. In chilled food compartments of refrigerators in 21 households, the overall mean temperatures ranged from 3 to 10.4°C, and over 85% of the appliances operated above the recommended 5°C (James and Evans, 1990). Because of the difficulty of ensuring that REPFEDS are maintained at a low temperature, the draft Code for Holland and Belgium specified that for a product with a maximum shelf life of 6 weeks, not more than 3 weeks of this time should be taken up by the distribution/sales channel, and not more than 1 week should elapse after sale. Time-temperature integrators are available (Selman, 1990), and their use to monitor the time-temperature experienced by refrigerated foods in general is under consideration. In the United Kingdom, a report of research into consumer attitudes to time-temperature indicators has been published (National Consumer Council, 1991).

REFERENCES

Agriculture Canada. (1990). Canadian code of recommended manufacturing practices for pasteurized/modified atmosphere packaged/refrigerated food. Agri-Food Safety Division, Ottawa, Ontario, Canada.

Alderton, G., Chen, J. K., and Ito, K. A. (1974). Effect of lysozyme on the recovery of heated *Clostridium botulinum* spores. *Appl. Microbiol. 27*: 613.

Ando, Y., and Iida, H. (1970). Factors affecting the germination of spores of *Clostridium botulinum* type E. *Jpn. J. Microbiol. 14*: 361.

Anonymous. (1992). Draft code for the production, distribution and sale of chilled, long-life, pasteurized meals. Belgian-Dutch Chilled Meals Working Group, The Netherlands. (In Dutch).

Beaufort, A., and Rosset, R. (1989). Durée de vie des plats cuisinés sous vide réfrigérés. Adaptation de la réglementation française. *Actualités des Industries Alimentaires et Agro-Industrielles 106*: 475.

Bohrer, C. W., Denny, C. B., and Yao, M. G. (1973). Thermal destruction of type E *Clostridium botulinum*. Final Report on RF 4603, National Canners Association Research Foundation, Washington, DC.

Bucknavage, M. W., Pierson, M. D., Hackney, C. R., and Bishop, J. R. (1990). Thermal inactivation of *Clostridium botulinum* type E spores in oyster homogenates at minimal processing temperature. *J. Food Sci. 55*: 372.

Chilled Food Association. (1989). Guidelines for good hygienic practice in the manufacture of chilled foods. Chilled Food Association, London.

Christiansen, L. N., Deffner, J., Foster, E. M., and Sugiyama, H. (1968). Survival and outgrowth of *Clostridium botulinum* type E spores in smoked fish. *Appl. Microbiol. 16*:133.

Cockey, R. R., and Tatro, M. C. (1974). Survival studies with spores of *Clostridium botulinum* type E in pasteurized meat of the blue crab *Callinectes sapidus. Appl. Microbiol. 27*: 629.

Conner, D. E., Scott, V. N., Bernard, D. T., and Kautter, D. A. (1989). Potential *Clostridium botulinum* hazards associated with extended shelf-life refrigerated foods: A review. *J. Food Safety 10*: 131.

Crawford, R. J. M. (1987). The use of lysozyme in the prevention of late-blowing in cheese. *Bull. International Dairy Fed.* No. 216.

Crisley, F. D., Peeler, J. T., Angelotti, R., and Hall, H. E. (1968). Thermal resistance of spores of five strains of *Clostridium botulinum* type E in ground whitefish chubs. *J. Food Sci. 33*: 411.

Department of Health. (1989). Chilled and frozen. Guidelines on cook-chill and cook-freeze catering systems. Her Majesty's Stationery Office, London.

Department of Health and Social Security. (1980). Guidelines on precooked chilled foods. Her Majesty's Stationery Office, London.

Eklund, M. W., Poysky, F. T., and Wieler, D. I. (1967a). Characteristics of *Clostridium botulinum* type F isolated from the Pacific Coast of the United States. *Appl. Microbiol. 15*: 1316.

Eklund, M. W., Wieler, D. I., and Poysky, F. T. (1967b). Outgrowth and toxin production of non-proteolytic type B *Clostridium botulinum* at 3.3° to 5.6°C. *J. Bacteriol. 93*: 1461.

Food Hygiene (Amendment) Regulations. (1990). Statutory instruments no. 1431. Her Majesty's Stationery Office, London.

Glew, G. (1990). Precooked chilled foods in catering. In *Processing and Quality of Foods*, Vol 3, *Chilled Food: The Revolution in Freshness* (P. Zeuthen, J. C. C. Cheftel, C. Eriksson, T. R. Gormley, P. Linko, and K. Paulus, eds.), Elsevier Applied Science, London, p. 3.31.

Gould, G. W. (1984). Injury and repair mechanisms in bacterial spores. In *The Revival of Injured Microbes* (M. H. E. Andrew and A. D. Russell, eds.), Academic Press, London, p. 199.

Graham, A. F., and Lund, B. M. (1991). The effect of temperature on the growth of non-proteolytic, type B *Clostridium botulinum* (abstract). *J. Appl. Bacteriol. 71*: xxi.

Hauschild, A. H. W. (1982). Assessment of botulism hazards from cured meat products. *Food Technol. 36*(12): 95.

Hauschild, A. H. W. (1989). *Clostridium botulinum*. In *Foodborne Bacterial Pathogens*. (M. P. Doyle, ed.), Marcel Dekker, Inc., New York, p. 111.

Hauschild, A. H. W., and Simonsen, B. (1986). Safety assessment for shelf-stable canned cured meats—an unconventional approach. *Food Technol. 40*(4): 155.

Hughey, V. L., and Johnson, E. A. (1987). Antimicrobial activity of lysozyme against bacteria involved in food spoilage and food-borne disease. *Appl. Environ. Microbiol. 53*: 2165.

Ito, K. A., Seslar, D. J., Mercer, W. A., and Meyer, K. F. (1967). The thermal and chlorine resistance of *C. botulinum* types A,B and E spores. In *Botulism 1966* (M. Ingram, and T. A. Roberts, eds.), Chapman and Hall, London, p. 108.

James, S., and Evans, J. (1990). Temperatures in the retail and domestic chilled chain. In *Processing and Quality of Foods*, Vol. 3, *Chilled Foods: The Revolution in Freshness* (P. Zeuthen, J. C. C. Cheftel, C. Eriksson, T. R. Gormley, P. Linko, and K. Paulus, eds.), Elsevier Applied Science, London, p. 3.273.

Jensen, M. J., Genigeorgis, C., and Lindroth, S. (1987). Probability of growth of *Clostridium botulinum* as affected by strain, cell and serologic type, inoculum size and temperature and time of incubation in a model broth system. *J. Food Safety 8*: 109.

Leadbetter, S. (1989). Sous-vide—A technology guide. Food Focus. The British Food Manufacturing Industries Research Association. Leatherhead, Surrey, United Kingdom.

Lie, O., Evensen, O., Sorensen, A., and Froysadal, E. (1989). Study on lysozyme activity in some fish species. *Dis. Aquatic Organisms 6*: 1.

Lund, B. M., Knox, M. R., and Sims, A. P. (1984). The effect of oxygen and redox potential on growth of *Clostridium botulinum* type E from spore inocula. *Food Microbiol. 1*: 277.

Lund, B. M., Graham, A. F., George, S. M., and Brown D. (1990). The combined effect of incubation temperature, pH and sorbic acid on the probability of growth of non-proteolytic, type B *Clostridium botulinum*. *J. Appl. Bacteriol. 69*: 481.

Lynt, R. K., Solomon, H. M., Lilly, T., Jr., and Kautter, D. A. (1977). Thermal death time of *Clostridium botulinum* type E in meat of the blue crab. *J. Food Sci. 42*: 1022.

Lynt, R. K., Kautter, D. A., and Solomon, H. M. (1979). Heat resistance of nonproteolytic *Clostridium botulinum* type F in phosphate buffer and crabmeat. *J. Food Sci. 44*: 108.

Lynt, R. K., Kautter, D. A., and Solomon, H. M. (1981). Heat resistance of proteolytic *Clostridium botulinum* type F in phosphate buffer and crabmeat. *J. Food Sci. 47*: 204.

MAFR (Ministry of Agriculture, French Republic). (1974). Regulations for the hygienic conditions concerning preparation, preservation, distribution and sale of ready to eat meals. *Journal Officiel de la Rèpublique Française*. Order of the 26th June, 1974. Paris.

MAFR (Ministry of Agriculture, French Republic). (1988). Extension of the shelflife of ready to eat meals, alteration to the protocol for obtaining authorization. Memorandum DGAL/SVHA/N88/No. 8106, 31st May, 1988. Paris.

Mochizuki, A., and Matsumiya, M. (1983). Lysozyme activity in shellfishes. *Bull. Jpn. Soc. Sci. Fish. 49*: 131.

Mossel, D. A. A., and Thomas, G. (1988). Microbiological safety of refrigerated meals: Recommendations for risk analysis, design and monitoring of processing. *Microbiologie-Aliments-Nutrition 6*: 289.

NACMCF (National Advisory Committee on Microbiological Criteria for Foods). (1990). Recommendations for refrigerated foods containing cooked, uncured meat or poultry products that are packed for extended, refrigerated shelf life and that are ready-to-eat or prepared with little or no additional heat treatment. Adopted January 31, 1990. Washington, D.C.

National Consumer Council. (1991). Time-temperature indicators: Research into consumer attitudes and behavior. Food Safety Directorate, Ministry of Agriculture, Fisheries and Food. London.

Notermans, S., Dufrenne, J., and Lund, B. M. (1990). Botulism risk of refrigerated, processed foods of extended durability. *J. Food Protect. 53*: 1020.

Ohye, D. F., and Scott, W. J. (1957). Studies in the physiology of *Clostridium botulinum* type E. *Aust. J. Biol. Sci. 10*: 85.

Proctor, V. A., and Cunningham, F. E. (1988). The chemistry of lysozyme and its use as a food preservative and a pharmaceutical. *CRC Crit. Rev. Food Sci. Nutr. 26*: 359.

Roberts, J. J., and Kruger, J. (1984). Lysozyme and its effects on the microflora of prepared food. *Food Rev. 11*: 73.

Roberts, T. A., Ingram, M., and Skulberg, A. (1965). The resistance of *Clostridium botulinum* type E to heat and radiation. *J. Appl. Bacteriol. 28*: 125.

Rosset, R., and Poumeyrol, G. (1986). Modern processes for the preparation of ready-to-eat meals by cooking before or after sous-vide packaging. *Sci. Aliments* 6 H. S. VI. 161–167.

Schmidt, C. F. (1964). Spores of *C. botulinum*: Formation, resistance, germination. *In Botulism 1964* (K. H. Lewis and K. Cassel, Jr., eds.) U. S. Public Health Service, Cincinnati, OH, p. 69.

Schmidt, C. F., Lechowich, R. V., and Folinazzo, J. F. (1961). Growth and toxin production by type E *Clostridium botulinum* below 40°F. *J. Food Sci. 26*: 626.

Scott, D. (1988–1989). Antimicrobial enzymes. *Food Biotechnol. 2*: 119.

Scott, V. N., and Bernard, D. T. (1982). Heat resistance of spores of non-proteolytic type B *Clostridium botulinum. J. Food Safety 45*: 909.

Scott, V. N., and Bernard, D. T. (1985). The effect of lysozyme on the apparent heat resistance of non-proteolytic type B *Clostridium botulinum. J. Food Safety 7*: 145.

Sebald, M., Ionesco, H., and Prévot, A. R. (1972). Germination lz-dépendante des spores de *Clostridium botulinum* type E. *C. R. Acad. Sci Paris(Serie D) 275*: 2175.

Selman, J. D. (1990). Time/temp. indicators: How they work. *Food Manufacture 65*(8): 30.

Simunovic, J., Oblinger, J. L., and Adams, J. P. (1985). Potential for growth of nonproteolytic types of *Clostridium botulinum* in pasteurized, restructured meat products: A review. *J. Food Protect. 48*: 265.

Smelt, J. P. P. M. (1980). Heat resistance of *Clostridium botulinum* in acid ingredients and its significance for the safety of chilled foods. Doctoral thesis, University of Utrecht, Utrecht, The Netherlands.

Smith, J. P., Toupin, C., Gagnon, B., Voyer, R., Fiset, P. P., and Simpson, V. (1990). A hazard analysis critical control point approach (HACCP) to ensure the microbiological safety of sous vide processed meat/pasta product. *Food Microbiol 7*: 177.

Solomon, H. M., Lynt, R. K., Lilly, T., Jr., and Kautter, D. A. (1977). Effect of low temperatures on growth of *Clostridium botulinum* spores in meat of the blue crab. *J. Food Protect. 40*: 5.

Solomon, H. M., Kautter, D. A., and Lynt, R. K. (1982). Effect of low temperatures on the growth of non-proteolytic *Clostridium botulinum* types B, and F and proteolytic type G in crabmeat and broth. *J. Food Protect. 45*: 516.

Stumbo, C. R. (1973). *Thermobacteriology in Food Processing*, 2nd ed. Academic Press, New York, p. 113.

SVAC (Sous Vide Advisory Committee) (1991). Code of Practice for Sous Vide Catering Systems. Sous Vide Advisory Committee, Tetbury, Glos. United Kingdom.

Wilkinson, P. (1990). Cook-chill in perspective. *British Food J. 92*(4): 37.

12

Hazards from Northern Native Foods

Robert B. Wainwright

Arctic Investigations Program, Centers for Disease Control, Anchorage, Alaska

I. INTRODUCTION

For many years outbreaks of illness associated with the consumption of traditionally prepared and preserved foods among northern populations have been reported. Clusters of deaths in families and villages were described by early explorers in the Arctic. These early outbreaks were attributed to "ptomaine" poisoning or trichinosis. However, many of these clusters of illness resemble outbreaks of foodborne botulism.

Botulism among northern native people is likely an old problem and is currently a disease of great concern in these groups. In spite of changing dietary patterns and the influx of Western foods and food preparation traditions, individuals in all age groups currently consume the traditionally prepared foods associated with botulism outbreaks, particularly the "fermented"* foods. Changes in traditional food preparation methods may be increasing the incidence of foodborne botulism among these populations.

It has been 30 years since Dolman (1960) highlighted and reviewed type E botulism as "a hazard of the north." Since that report, the problem of botulism in northern native populations has remained. Dolman (1960) and Smith (1977)

*Though fermentation is the term commonly used to describe these traditional foods, the level of fermentable carbohydrate is too low to foster a decrease in pH. A more appropriate term to describe the process occurring in these foods would be putrefaction.

have summarized outbreaks of botulism occurring in northern Europe and the former U.S.S.R., which have been few in number and probably not completely reported. This chapter will discuss the occurrence of well-reported botulism outbreaks associated with traditional foods prepared and consumed by the Alaska Natives, the Canadian Inuit, and the Greenland Eskimos.

II. HISTORICAL ACCOUNTS OF BOTULISM IN THE ARCTIC

Arctic explorers and scientists have referred to outbreaks of fatal disease occurring among northern native people (Stefansson, 1914). Ethnographers have described food storage conditions and cultural traditions regarding food preparation that would allow for the production of botulinum toxin (Nelson, 1971). Stefansson (1914) describes a group of eight Eskimos he met on a visit to an area at the head of the Mackenzie River delta. Upon returning 2 years later, he found that they had all died soon after consuming meat of a freshly killed white whale. He reports that while the Eskimos attributed these deaths to the breaking of taboos regarding whaling, the nonnative whalers operating in the Arctic waters attributed cases of this kind to ''ptomaine'' poisoning. Though the frequency of these clusters of deaths following consumption of sea mammal meat is not known, a review of all reports suggests that such outbreaks were not uncommon. Stefansson speculated that these deaths are ''rather the nature of trichinosis than of the type of ptomaines.''

Other scientists have also attributed the clusters of deaths to trichinosis. Parnell (1934) speculated that these cases may be related to *Trichina spiralis* infestation. Later Connell (1949) noted that *T. spiralis* occurs in several animal hosts that serve as direct foods for the northern native people—polar bear, walrus, and white whale—and in other animals that serve only sporadically as food—sled dogs and the arctic fox. There is ample evidence, both from the distribution of cases of trichinosis and from documented outbreaks, that trichinosis is and has been a risk for northern native people. However, Nelson's (1971) description of northern native food-handling practices also raises the possibility that the outbreaks described in the Arctic explorers' journals were due to botulism. The descriptions of potentially dangerous storage practices and traditional fermentation of fish and sea mammals indicate that conditions were present that would allow botulinum toxin to develop in these foods.

III. EPIDEMIOLOGY OF BOTULISM IN THE ARCTIC

Though botulism occurs in four different forms—foodborne, infant, wound, and unspecified—virtually all botulism reported in the Arctic is foodborne. No cases of wound botulism have been reported from Alaska, Greenland, or the Canadian north. Only two cases of infant botulism have been reported from Alaska and four cases from all of Canada (McCurdy et al., 1981; Hauschild and Gauvreau, 1986, 1987). No cases cases of infant botulism have been reported in northern indigenous people.

Information on foodborne botulism outbreaks among indigenous people in Arctic areas is available from some areas, but is incomplete or unavailable from other areas. Numerous outbreaks in the Canadian north since the early 1900s have been reported and discussed (Dolman, 1974; Hauschild and Gauvreau, 1985, 1986, 1987; Hauschild et al., 1988; Dodds et al., 1989, 1990). Outbreaks in Alaska have been reported since the mid-1900s (Rabeau, 1959; Barrett et al., 1977; Wainwright et al., 1988; CDC, personal communication; Epidemiology Office, State of Alaska, personal communication). The first laboratory-confirmed outbreaks in the Arctic occurred in Canada in 1945 (Dolman and Kerr, 1947) and in Alaska in 1947 (Rabeau, 1959). The first laboratory-confirmed outbreak reported in Greenland Eskimos occurred in 1967 (Muller and Thomsen, 1968), though unconfirmed prior outbreaks had been reported (Kern Hansen and Bennike, 1981), and others have since been reported (J. Misfeldt, personal communication). Reports of outbreaks among the Siberian Eskimos in the former U.S.S.R. are generally unavailable.

Since the early 1900s, 200 outbreaks of foodborne botulism have been recorded among the native populations of the Arctic (Table 1). This includes both laboratory-confirmed outbreaks (154) and outbreaks suspected on the basis of clinical findings and a history of consumption of foods frequently associated with botulism outbreaks. One hundred and two outbreaks were reported from Canada from 1919 through 1989; 73 outbreaks were reported in Alaska between 1947 and 1989; and 25 outbreaks were reported in Greenland Eskimos from 1967 through 1989. Approximately 505 individuals have been involved in these outbreaks.

Table 1 Reported Foodborne Botulism Outbreaks Among Northern Native People, 1919–1989

Location	No. of outbreaks	Toxin type				No. of cases	No. of deaths	Ref[a]
		A	B	E	Unknown			
Canada	102	1	3	84	14	232	76	1–7
Alaska	73	6	6	46	15	200	19	8–10
Greenland	25	0	0	8	17	70–75	15	11, 12
Total	200	7	9	138	46	502–507	110	

[a]1) Dolman, 1974; 2) Hauschild and Gauvreau, 1985; 3) Hauschild and Gauvreau, 1986; 4) Hauschild and Gauvreau, 1987; 5) Hauschild et al., 1988; 6) Dodds et al., 1989; 7) Dodds, 1990; 8) Wainwright et al., 1988; 9) EOB, 1990; 10) EO, 1990; 11) Muller and Thomsen, 1965; 12) Misfeldt, 1990.

Among the northern native people, a preponderance of Eskimos/Inuit are involved in foodborne botulism outbreaks. Of the approximately 485 individuals involved in outbreaks whose ethnicity is specified, 387 (80%) were Eskimo/Inuit; 94 (19%) were Indian; and 4 (1%) were Aleut (Table 2). The ethnic distribution parallels the geographic distribution of outbreaks. The majority of outbreaks occur in the coastal and riverine areas of Alaska, Canada, and Greenland—the geographic areas inhabited mainly by Alaskan Eskimos or Canadian Inuit (Fig. 1). However, there may be some cultural factors related to the methods of preparation of traditional foods which also influence the distribution of foodborne botulism outbreaks.

From 1919 through 1989, a total of 110 foodborne botulism-associated deaths have been recorded in the northern native population (Table 1). Most deaths have been associated with outbreaks in Canada, followed by Alaska and Greenland. Overall, the case:mortality ratio in recorded outbreaks from 1919 through 1989 is 22%. In the Canadian outbreaks, the case:mortality ratio is 33%. In the Alaska Native and Greenland outbreaks, the ratio is 10% and 21%, respectively.

The botulism type accounting for the great majority (86%) of the laboratory-proven outbreaks in these northern areas is type E (Table 1). Where the botulinum type is known, all Greenland outbreaks were caused by type E, as were 84 (95%) of the Canadian outbreaks. In 46 (78%) of the Alaska Native outbreaks, botulism type E was identified. Overall, type A was identified in 7 (4%) outbreaks and type B in 9 (5%).

The foods involved in foodborne botulism outbreaks among northern native people are primarily fish and sea mammals; occasionally, land mammals are implicated (Table 3). The fish most commonly involved is salmon, including salmon eggs. Other implicated fish include whitefish, trout, cod, and Arctic char. Sea mammals involved in outbreaks include whale, seal, and walrus. Of the reported foodborne botulism outbreaks in northern native people in which a specific food could be implicated, 61% were associated with sea mammals, 35% with fish, and 5% were land mammals.

Table 2 Ethnicity of Persons Involved in Foodborne Botulism Outbreaks Among Northern Native Populations by Country

	Ethnicity		
Country	Eskimo/Inuit	Indian	Aleut
Canada	174	57	0
Greenland	49–54	0	0
Alaska	159	37	4
Total	382–387	94	4

Source: see Refs. for Table 1.

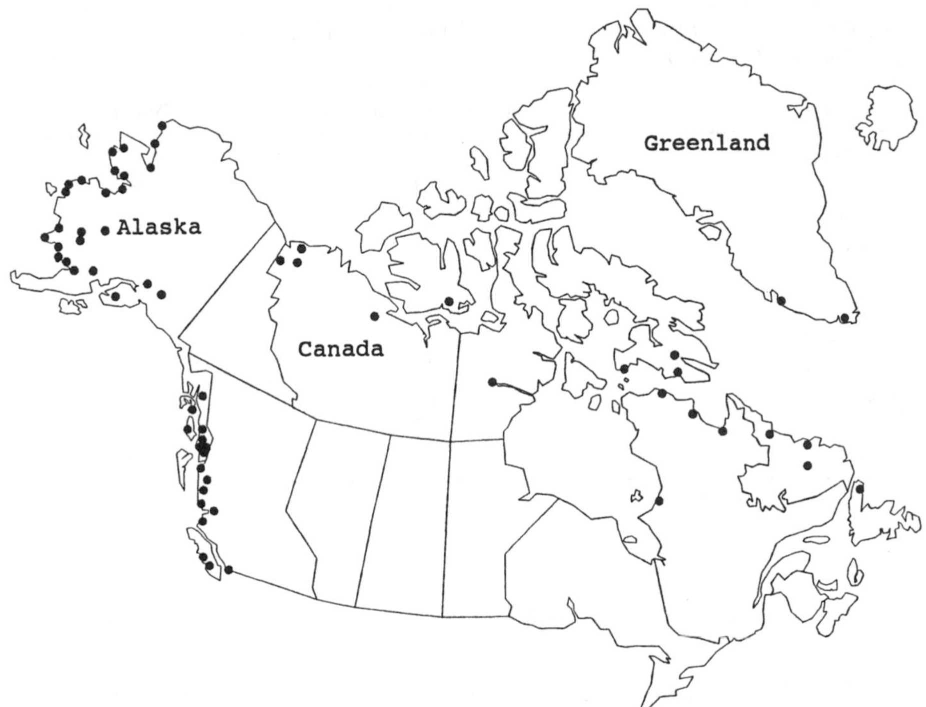

Figure 1 Geographic distribution of foodborne botulism outbreaks in northern native people. *Note*: Each dot represents one or more outbreaks. In many cases, particularly in eastern Canada, multiple outbreaks have occurred at a single site.

The handling of foods before consumption is a critical factor in the etiology of botulism among the northern native people. Putrefaction (or fermentation) of foods, whether intentional, as in preparing traditional "stink" foods, or unintentional, as in poorly stored and preserved fish and mammals, is the principal cause of foodborne botulism in these people. In addition, certain foods (whale, seal, beaver tail) that are found dead in various states of decay are eaten by northern native people and have been implicated in botulism outbreaks.

In an ethnographic report of Eskimo food handling in the late 1800s, Nelson (1971) noted,

> Meat is frequently kept for a considerable length of time and sometimes until it becomes semiputrid. At Point Barrow, in the middle of August, 1881, the people still had the carcasses of deer which had been killed the preceding winter and spring. This meat was kept in small underground pits, which the frozen subsoil rendered cold, but not cold enough to prevent a bluish fungus

Table 3 Foods Associated with Reported Botulism Outbreaks Among Northern Native People, 1919–1989

Food type	Number of outbreaks	Percent
Sea Mammals		
Seal	54	45
Whale	15	13
Walrus	4	3
Fish		
Salmon	3	3
Salmon eggs	27	23
Other fish	11	9
Land Mammals		
Caribou	3	3
Beaver tail	3	3

Source: see Refs. for Table 1.

growth which completely covered the carcasses of the animals and the walls of the storerooms.

Freshly killed and stored foods are often eaten with fermented condiments such as *micerak* (fermented sea mammal fat) or *ogsuk* (fermented seal oil). Dried foods have also been implicated in botulism outbreaks.

Of the botulism outbreaks among the Canadian native people from 1971 to 1984, 11 of 54 (20%) were associated with traditionally fermented foods. However, since 1984, 9 of 14 (64%) outbreaks were associated with fermented foods. In the botulism outbreaks in Alaska Natives since 1947, 62% were associated with fermented foods. Slightly less than a quarter of all reported Alaskan outbreaks have been associated with dried fish or sea mammals.

Meaningful rates of foodborne botulism in northern native people are difficult to determine because the exact population truly at risk is not known. The Inuit of Canada have a high rate of botulism with an annual rate of 30 cases/100,000 (Smith, 1977). In Alaska, from 1976 to 1984, the annual incidence of foodborne botulism was 8.6 cases/100,000 population (MacDonald et al., 1986). In the United States, the state with the next highest incidence during the same period of time was Washington, with an annual incidence of 0.43 cases/100,000 population, about 1/20 the Alaska rate. However, all reported cases of foodborne botulism in Alaska have occurred in Alaska Natives. If the population at risk is considered to be only the Alaska Natives, the annual incidence rate would increase severalfold. However, it is difficult to determine what the true risk of eating the traditional foods may be. Because not all northern native groups uniformly prepare and consume the traditional fermented foods, it is likely that the annual incidence of botulism in the groups who regularly prepare and consume these foods is greater than the 30 cases/100,000 population seen in the northern natives as a whole.

IV. CLINICAL PRESENTATION OF BOTULISM IN NORTHERN NATIVE PEOPLE

The clinical course of foodborne botulism occurring among northern native people is similar to that seen elsewhere. Symptoms, usually gastrointestinal, occur 12–36 hours after ingesting foods containing botulinum toxins of types A, B, or E. The usual initial symptoms are nausea and/or vomiting. Neurological symptoms are typically apparent within 72 hours of ingesting contaminated foods. Visual symptoms such as blurred vision and diplopia, as well as dysphagia, dysarthria, and dry mouth, are common early neurological symptoms.

In a series of 156 Alaska native botulism cases reported from 1947 through 1985, dry mouth was found in 90% of patients (Wainwright et al., 1988). One or more eye signs (diplopia, blurred vision, ptosis, or dilated and fixed pupils) occurred in 93% of patients with laboratory-proven botulism. Three or more signs or symptoms of the pentad of dry mouth, nausea/vomiting, dysphagia, diplopia, or dilated fixed pupils also occurred in 93% of the patients with laboratory-confirmed botulism.

Of 35 patients with foodborne botulism that were hospitalized at the Alaska Native Medical Center in Anchorage from 1973 through 1986, D. H. Barrett (personal communication) reports that in 33 patients where information is available, 27 (82%) presented with nausea, 26 (79%) with vomiting, 26 (79%) with extremity weakness, 26 (79%) with dry mouth, and 20 (61%) with double vision. On physical examination of this group, 28/31 (90%) were found to have dry oral membranes; 27/32 (84%) weakness; 26/33 (79%) pupils poorly reactive to light; 25/32 (78%) hypoactive bowel sounds; 19/25 (76%) ptosis; and 13/20 (65%) decreased gag reflex.

The clinical course of individuals with botulism intoxication varies greatly. Of the 35 hospitalized Alaskan patients, 23 (66%) eventually required ventilatory assistance because of respiratory muscle paralysis. Patients that required a tracheostomy needed respiratory assistance for a mean of 15.5 days, while those that required nasotracheal intubation needed assistance for a mean of 8.6 days (D. H. Barrett, personal communication). Other individuals in confirmed foodborne botulism outbreaks developed dry mouth and/or eye signs, but the symptoms did not progress to respiratory paralysis. The role of factors such as toxin type, amount of toxin ingested, and timing of antitoxin administration in the course of the disease is unclear.

The initial diagnosis of foodborne botulism is always clinical, and the list of conditions in the differential diagnosis is lengthy (D. H. Barrett, personal communication; Sanders et al., 1983). Due to delays in obtaining appropriate specimens for laboratory analysis and the length of time required for the mouse assay laboratory test for botulinum toxin, clinicians must decide on treatment based upon history and physical findings. In Alaska, 25% of the botulism outbreaks from 1947 to 1989 were not laboratory confirmed, yet the histories, physical findings, and clinical courses of persons involved in these outbreaks are consistent with botulism. Clinicians in northern areas, particularly those treating native populations, must consider the possibility of botulism intoxication in any patient with gastrointestinal and neuromotor signs and symptoms. A patient with a history of consuming traditionally prepared or stored foods, nausea/vomiting, a dry mouth, with or without eye signs, must be treated as having botulism until proven otherwise.

V. BOTULISM AND FERMENTED FOODS IN OTHER POPULATIONS

Foodborne botulism associated with fermented fish and meats is not unique to northern native people. One of the most completely reported fermented foods associated with botulism outbreaks is *isushi*, a frequently home-prepared, traditional Japanese relish made with fermented fish (Iida et al., 1958). The pickled product is prepared by soaking the fish for 5 days, during which time the toxin is produced. The preformed toxin is stable in the acidic pickling step, which follows the soaking step. Most outbreaks of *isushi*-associated botulism are due to type E toxin (Smith, 1977).

Denmark experiences infrequent outbreaks of foodborne botulism associated with processed fish. Philipsen et al. (1986) summarized the reported cases of foodborne botulism in Denmark from 1833 through 1978. During that period, 17 outbreaks were reported, 14 associated with the consumption of fish. Seven of nine outbreaks in which the toxin type was identified were caused by type E.

In the former Soviet Union, traditional home-processed, dried, salted, smoked, or pickled fish are associated with type A and type E botulism (Smith, 1977). As seen in an outbreak among Russian immigrants to the United States, the fish in these traditional foods are often uneviscerated (Telzak et al., 1990). Since *C. botulinum* spores can concentrate in the viscera of fish (Ward et al., 1967), botulinum toxin can form if fish are not properly handled during the time from catch to processing. It has also been suggested that, even in the presence of salt, *C. botulinum* spores can germinate, grow, and produce toxin in the relatively protected environment of the fish viscera (Stuart et al., 1970). The formed toxin can then diffuse into the flesh of the fish.

Many outbreaks of botulism associated with fermented foods have occurred in China. From 1958 through 1983, 986 botulism outbreaks were reported in China, involving 4377 individuals with 548 deaths (Shih and Chao, 1986). Most of these outbreaks (824, 84%) were associated with fermented bean curd. Of the 733 outbreaks where the botulinum toxin type was determined, 685 (93%) were type A.

VI. USE AND PREPARATION OF TRADITIONAL FOODS BY NORTHERN NATIVE PEOPLE

With the arrival of Western foods into the villages during the early to mid-1900s, the introduction of Western culture through radio, television, and increased travel of northern native people themselves, changes in traditional dietary patterns were inevitable. What effect the Western cultural influences have had on the consumption of traditional fermented foods and on other traditional food processing practices is unclear, because no complete surveys of traditional dietary patterns have been made. However, whatever changes there may have been in dietary patterns and food-handling practices, current northern native cultures still include a number of practices associated with botulism intoxication.

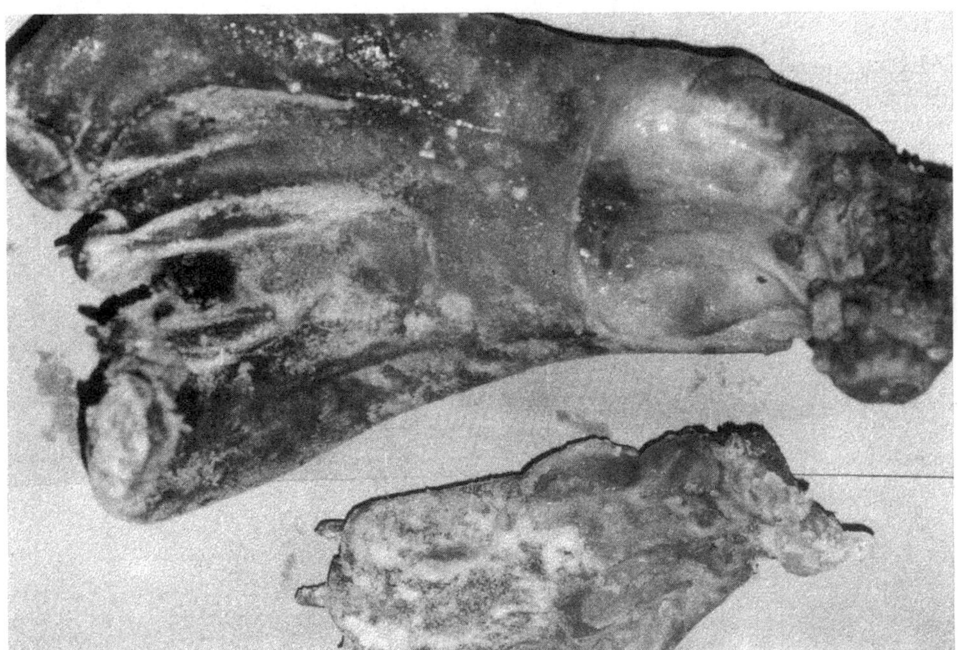

Figure 2 Fermented seal flipper.

In a survey of the diets of 308 Alaska Native adults in 11 communities, Nobmann (1989) found that 42% of those responding to the survey had consumed traditionally fermented foods sometime during their lifetime. The most commonly consumed fermented foods included salmon heads and eggs and other types of fish and their eggs. In addition, seal flipper (Fig. 2), beaver tail (Fig. 3), caribou, whale meat, and seal meat were mentioned by survey respondents as foods they had fermented and eaten.

In a survey of food-handling practices in southwest Alaska, Shaffer et al. (1990) found that the practice of preparing and eating fermented foods was widespread in this area populated by Yupik Eskimos, Aleuts, and Athapaskan Indians. In a survey of health aides in 20 of the 29 villages in the region, he found that fermented foods were prepared by at least one family in each of the Yupik Eskimo villages and three-fourths of the other villages (Indian and Aleut). The villages ranged in size from 50 persons to about 500 persons. However, the village health aides estimated that in two-thirds of the Yupik Eskimo villages and in all other villages, the majority of families did not prepare fermented foods. The health aides also estimated that less than 25% of the children eat fermented foods in these villages.

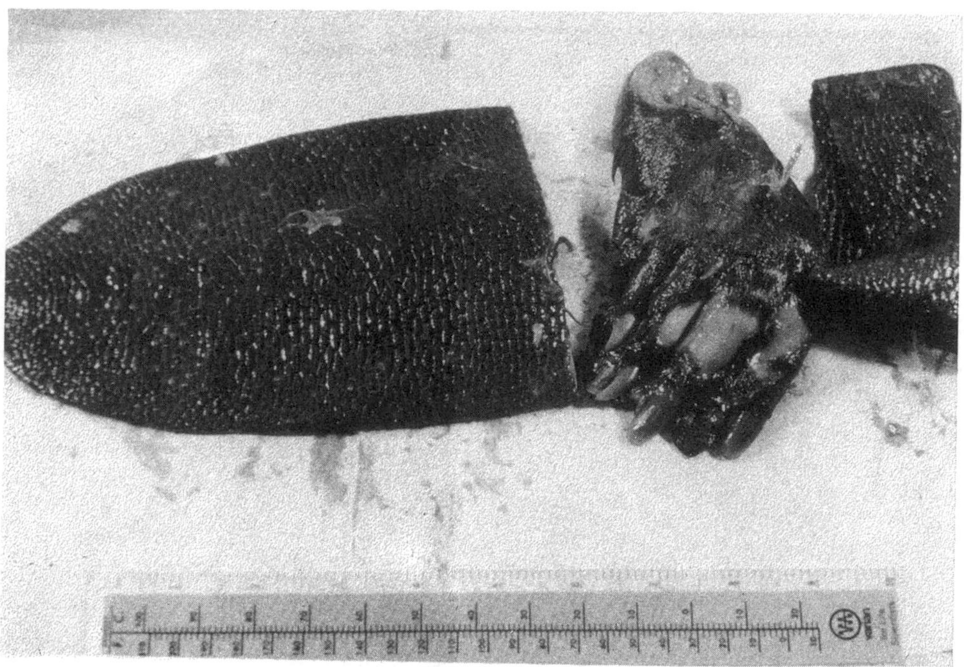

Figure 3 Fermented beaver tail and paw.

In a survey of 45 Alaska Native high school students in four Yupik Eskimo villages, Shaffer et al. (1990) found that only 15% of the students indicated that they regularly consumed fermented foods, compared to 71% of their parents and 80% of their grandparents.

Interviews with 24 individuals preparing fermented foods in six Yupik Eskimo villages indicated that about one-half fermented their foods aboveground in wooden barrels or buckets, while the other half used pits dug into the ground (Shaffer et al., 1990). Of those using pits for fermentation, most had a barrel buried in the ground. The older Yupik Eskimos who prepared fermented food indicated that, traditionally, the foods were fermented in clay pits with the family using the same pit each year. When wooden barrels became available in the villages, these were placed in the pits as fermentation vessels. Over the years, many families have stopped placing the barrels in the ground, but kept them in a cool, shady place aboveground (Fig. 4).

There are several methods of preparing traditional fermented foods. An Eskimo delicacy called "stinky heads" is a popular food prepared in regions with access to salmon. Nelson (1971) described the preparation of fermented fish heads, which he observed during his stay in the coastal villages of northwest Alaska in 1878–1881:

> In the district between the Yukon and Kuskokwim, the heads of king salmon, taken in the summer, are placed in small pits in the ground surrounded by straw and covered with turf. They are kept there during the summer and in the autumn have decayed until even the bones have become of the same consistency as the general mass. They are then taken out and kneaded in a wooden tray until they form a pasty compound and are eaten as a favorite dish by some of the people. The odor of the mess is almost unendurable to one not accustomed to it, and is even too strong for the stomachs of many of the Eskimo.

The method of preparing this food has apparently undergone some changes in the last generation or so. Currently, some villagers place the fish heads in a pit lined with moss and grass (Arnariak and Andrews, 1980a). Others place the foods to be fermented in a seal skin bag or "poke," which is then buried or hung up to ferment. In addition to fish heads, other fish parts such as sperm, heart, liver, and the blood vein are placed in the pit. The pit is then covered with more moss and grass and with soil or sand. It remains undisturbed for 1–3 weeks. When the pit is opened, it contents are consumed, often with seal oil, without being heated or cooked. There is some indication that families have traditional sites for preparing the fermented foods, and the same pit may be used year after year.

The more modern way of preparing the "stinky" foods is similar to the old way, except that other types of containers like wooden barrels, plastic bags, buckets, or glass jars (Fig. 5) are often used. These containers are frequently not buried but are left aboveground inside the home or in the food cache (a small,

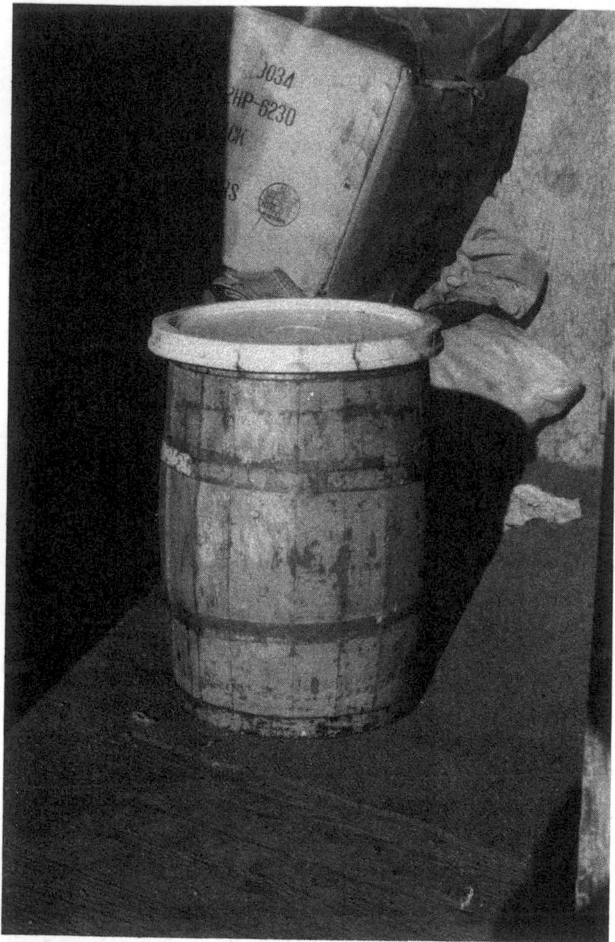

Figure 4 Wooden barrel used to ferment "stinkheads."

aboveground, enclosed storage shed). Fermentation times are considerably reduced if the containers are kept in the home.

Other ethnic groups have different ways of preparing the traditional fermented foods. The Athapaskan Indians of interior Alaska report preparing their fermented fish heads by stringing the heads together and leaving them in a running stream (Shaffer et al., 1990). When the heads are ready to eat, in about 2 weeks, they float to the surface.

The more northern groups also ferment seal meat, seal flipper, walrus flipper, walrus nose, whale flipper, or *muktuk* (whale skin and attached blubber) (Fig.

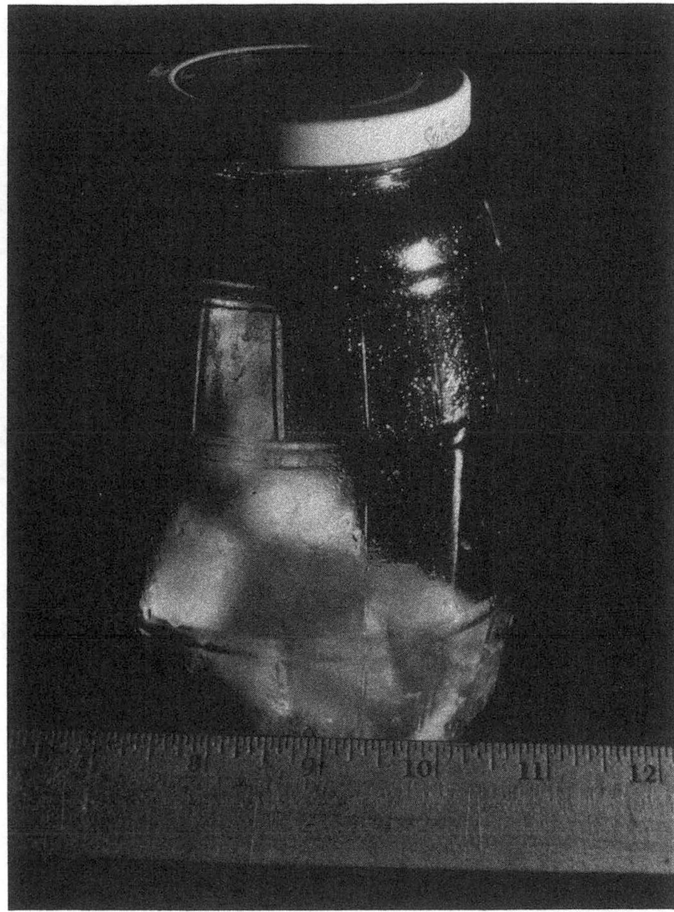

Figure 5 Glass jar used to ferment seal meat.

6). In one area of Alaska, beaver tails are fermented. With all of these foods, the fermentation method is similar to the method for "stinky heads." Seal oil is sometimes added to the foods as they are fermenting.

Fermented fish eggs are either liquid or made into a "cheese." For the former, the eggs are placed in a container, sometimes with seal oil, and left out for about a week. To prepare fish egg "cheese," the eggs are pressed firmly into a porous container, such as a fish skin bag, and hung out to dry, usually for months. When ready to eat, the product is quite firm or hard on the outside and soft on the inside (Lockuk, 1980). The "cheese" is cut into small pieces and eaten as is.

Figure 6 Fermented whale *muktuk*.

Fermented seal oil (*ogsuk*) and fat from seal, whale, and walrus are often used as condiments with other foods. *Micerak*, fermented fat of marine mammals, is prepared by placing the fat in containers, such as a plastic or glass jar or a plastic or skin bag, and letting it sit either indoors or outside for up to 4 months (Dodds et al., 1989). Meats of land and sea mammals, birds, and fish are dipped into the fermented oils as they are eaten.

Drying, with or without salting, is a common traditional form of preservation among northern native people. The most common dried foods are fish and sea mammals, particularly salmon and seal meat. Fish are usually split and filleted, with the two fillets joined at the tail. The skin is left on the fillets, and the thick flesh is cut crosswise to the skin. The filleted fish then may be placed in a brine solution of various compositions for several days. The fish are then hung up by the tail to dry. The drying time depends upon the thickness of the fish and the weather. If it rains while the fish are being dried, attempts are made to cover them. However, fish at times ''get ruined'' by rain (Fig. 7) (Arnariak and Andrews, 1980b). When dry, some fish are smoked and stacked in the food cache.

The degree of dryness and the amount of salt in the fish probably varies greatly. A preliminary study of drying/salting practices by Alaska Natives found

Figure 7 Fish "drying" in the rain.

that the salt levels are often too low and the water levels too high to inhibit the production of botulinum toxin in these foods (Zottala and Zoltai, 1981).

VIII. PREVENTION OF FOODBORNE BOTULISM IN NORTHERN NATIVE POPULATIONS

Prevention of foodborne botulism among the consumers of traditional northern native foods is difficult. Though a vaccine that will provide protection against botulinum toxin is available, its widespread use in northern native populations would not be cost-effective. Current programs to control botulism in northern native populations have focused on educating the population of the dangers of fermented foods and on education of health professionals on early detection and proper treatment.

Though cooking the traditional foods would inactivate the botulinum toxin, the change in texture and flavor of the foods brought about by cooking is not acceptable to the consumers. Prohibition and condemnation of the traditional foods would alienate the population and drive the practice underground. The ideal solution would be to develop food-preparation techniques that would eliminate the risk of botulinum toxin formation and still preserve the flavor and texture of the traditional foods. Until safe food preparation methods are developed and accepted, high rates of botulism associated with the consumption of traditional foods among northern native populations are likely to continue. In addition, cases of repeat botulism are now occurring among the northern native people (Beller and Middaugh, 1990). This not only brings the risk associated with a second episode of botulism, but the added risk from a second exposure to the equine trivalent antitoxin.

REFERENCES

Arnariak, C., and Andrews, J. (1980a). How to make stinkyhead. *Ak'a Tamaani 1*: 11.

Arnariak, C., and Andrews, J. (1980b). Dry fish for winter. *Ak'a Tamaani 1*: 18.

Barrett, D. H. (1990). Alaska Native Medical Center, Anchorage Alaska. Personal Communication.

Barrett, D. H., Eisenberg, M. S., Bender, T. R., Burks, J. M., Hatheway, C. L., and Dowell, V. R. (1977). Type A and type B botulism in the North: first reported cases due to toxin other than type E in Alaskan Inuit. *Can. Med. Assoc. J. 117*: 483.

Beller, M., and Middaugh, J. P. (1990). Repeated type E botulism in an Alaskan Eskimo. *N. Engl. J. Med. 322*: 855.

Connell, F. H. (1949). Trichinosis in the Arctic: A review. *Arctic 2*: 98.

Dodds, K. (1989). Botulism in Fort Chimo, Quebec. *Can. Dis. Wkly. Rep. 15*: 79.

Dodds, K. (1990). Health and Welfare Canada, Ottawa, Canada. Personal Communication.

Dodds, K., Hauschild, A., and Dubuc, B. (1989). Botulism in Canada: Summary for 1988. *Can. Dis. Wkly. Rep. 15*: 78.

Dolman, C. E. (1960). Type E botulism: A hazard of the North. *Arctic 13*: 230.

Dolman, C. E. (1974). Human botulism in Canada (1919–1973). *Can. Med. Assoc. J. 110*: 191.

Dolman, C. E., and Kerr, D. E. (1947). Botulism in Canada, with report of a type E outbreak at Nanaimo, B.C.. *Can. J. Pub. Health 38*: 48.

EDB, Enteric Diseases Branch (1990). Division of Bacterial Diseases, Center for Infectious Diseases, Centers for Disease Control, Atlanta, GA. Personal Communication.

EO, Epidemiology Office (1990). Division of Public Health, Department of Health and Social Services, State of Alaska, Anchorage, AK. Personal Communication.

Hauschild, A. H. W., and Gauvreau, L. (1985). Foodborne botulism in Canada, 1971–84. *Can. Med. Assoc. J. 133*: 1141.

Hauschild, A., and Gauvreau, L. (1986). Botulism in Canada: Summary for 1985. *Can. Dis. Wkly. Rep. 12*: 53.

Hauschild, A., and Gauvreau, L. (1987). Botulism in Canada: Summary for 1986. *Can. Dis. Wkly. Rep. 13*: 47.

Hauschild, A., Dodds, K., and Gauvreau, L. (1988). Botulism in Canada: Summary for 1987. *Can. Dis. Wkly. Rep. 14*: 41.

Iida, H., Nakamura, Y., Nakagawa, I., and Karashimada, T. (1958). Additional type E botulism outbreaks in Hokkaido, Japan. *Jpn J. Med. Sci. Biol. 11*: 215.

Kern Hansen, P., and Bennike, T. (1981). Botulismus in Greenland Eskimos. *Circumpolar Health 81: Proceedings of 5th International Symposium on Circumpolar Health*, Copenhagen, Denmark, p. 438.

Lockuk, M. (1980). Fish eggs. *Ak'a Tamaani 1*: 23.

MacDonald, K. L., Cohen, M. L., and Blake, P. A. (1986). The changing epidemiology of adult botulism in the United States. *Am. J. Epidemiol. 124*: 794.

McCurdy, D. M., Krishnan, C., and Hauschild, A. H. W. (1981). Infant botulism in Canada. *Can. Med. Assoc. J. 125*: 241.

Misfeldt, J. (1990). Chief Medical Officer, Landslaegeembedet I Gronland, Nuuk/Godthag, Greenland. Personal Communication.

Muller, J., and Thomsen, B. F. (1968). An outbreak of type E botulism in West-Greenland. *Nord. Vet. Med. 20*: 485.

Nelson, E. W. (1971). *The Eskimo About Bering Strait*. Johnson Reprint Corp., New York.

Nobmann, E. D. (1989). Assessment of current dietary intakes of Alaska Native adults: Final report. Alaska Area Native Health Service, Anchorage.

Parnell, I. W. (1934). Animal parasites of north-east Canada. *Can. Field-Naturalist 48*: 111.

Philipsen, E. K., Hartvig, T., and Andersen, L. W. (1986). A review of the cases of botulism treated during the period 1945–1983 in the Department of Epidemic Diseases, Blegdamshospitalet/Rigshospitalet, Copenhagen. *Ugeskr Laeger 148*: 313.

Rabeau, E. S. (1959). Botulism in Arctic Alaska. Report of 13 cases with 5 fatalities. *Alaska Med. 1*: 6.

Sanders, A. B., Seifert, S., and Kobernick, M. (1983). Botulism. *J. Fam. Pract. 16*: 987.

Shaffer, N., Wainwright, R. B., Middaugh, J. P., and Tauxe, R. V. (1990). Botulism among Alaska natives: The role of changing food preparation and consumption practices. *West J. Med. 153*: 390.

Shih, Y., and Chao, S. (1986). Botulism in China. *Rev. Infect. Dis. 8*: 984.

Smith, L. D. S. (1977). *Botulism: The Organism, Its Toxins, the Disease*. Charles C. Thomas, Springfield, IL, p. 183.

Stefansson, V. (1914). The Stefansson-Anderson Arctic expedition: Preliminary ethnological report. *Anthropol. Papers Am. Mus. Nat. Hist. 14*: 449.

Stuart, P. F., Wiebe, E. J., McElroy, R., Cameron, Douglas G., Todd, E. C. D., Erdman, I. E., Albalas, B., and Pivnick, Hilliard. (1970). Botulism among Cape Dorset Eskimos and suspected botulism at Frobisher Bay and Wakeham Bay. *Can. J. Pub. Health 61*: 509.

Telzak, E. E., Bell, E. P., Kautter, D. A., Crowell, L., Budnick, L. D., and Schultz, S. (1990). An international outbreak of type E botulism due to uneviscerated fish. *J. Infect. Dis. 161*: 340.

Wainwright, R. B., Heyward, W. L., Middaugh, J. P., Hatheway, C. L., Harpster, A. P., and Bender, T. R. (1988). Foodborne botulism in Alaska, 1947–1985: Epidemiology and clinical findings. *J. Infect. Dis. 157(6)*: 1158.

Ward, B. Q., Carroll, B. J., Garrett, E. S., and Reese, G. B. (1967). Survey of the U.S. Gulf Coast for the presence of *Clostridium botulinum*. *Appl. Microbiol. 15*: 629.

Zottola, E. A., and Zoltai, P. T. (1981). *A Preliminary Report on Research Concerning Native Alaska Foods, Methods of Preparation, Preservation and the Effect of These Methods on Their Nutritional Quality and Safety*. Department of Food Sciences and Nutrition, Agricultural Extension Service, University of Minnesota, St. Paul.

13

Destruction of Botulinum Toxins in Food and Water

Lynn S. Siegel

U.S. Army Medical Research Institute of Infectious Diseases, Fort Detrick, Frederick, Maryland

I. INTRODUCTION

There are seven types of *Clostridium botulinum*, designated types A through G, each producing an immunologically distinct protein neurotoxin. Human botulism is associated with types A, B, E, and F. Types C and D cause botulism in animals. Type G was isolated from soil in Argentina (Gimenez and Ciccarelli, 1970), and from autopsy samples obtained from humans whose cause of death was not known (Sonnabend et al., 1981), but has not yet been associated with botulism in any species.

When synthesized by *C. botulinum*, the neurotoxin is a single polypeptide chain with a molecular weight of approximately 150,000. It has little or no toxic activity, and the molecule must be activated to attain its high toxicity. In proteolytic cultures (all type A and some B and F strains), the toxin is activated to a dichain molecule endogenously by proteases produced by the organisms. This dichain is composed of a heavy (H) and a light (L) polypeptide chain, which are joined together by at least one disulfide bond. The molecular weight of the heavy chain is about 100,000; that of the light chain, about 50,000. In nonproteolytic strains (all type E and some B and F strains), such enzymes are either inhibited or are not present. However, these toxins are fully activated by proteolytic enzymes of other organisms or by experimental treatment with trypsin (Smith and Sugiyama, 1988).

The neurotoxins are sometimes referred to as 7S toxins, based on their sedimentation coefficient, or S toxin (small). In foods and in cultures, type A and B neurotoxins associate with one or two nontoxic proteins to form complexes of 12S (M toxin, medium) or 16S (L toxin, large). The 16S complex contains a nontoxic protein with hemagglutinating activity. Type A also occurs as a 19S complex (LL toxin, extra large), which is probably a dimer of 16S. Type E and F neurotoxins also associate with nontoxic proteins in culture to form a complex with a sedimentation coefficient of 12S for E and 10S for F. Types E and F lack hemagglutinating activity (Smith and Sugiyama, 1988).

Type A toxin was crystallized in the 1940s and believed to be pure because it did crystallize (Lamanna et al., 1946). However, the crystals were actually L toxin, as the neurotoxin remains associated with other proteins in acidic solutions. At a slightly alkaline pH, the neurotoxin can be separated from the other contaminating proteins using ion exchange chromatography (DasGupta et al., 1966). Purification procedures for the other neurotoxin types have also been developed.

Toxin levels are usually determined using a bioassay. Serial dilutions of the toxin-containing sample are injected into a susceptible species, usually mice, to determine lethality (after a 3- or 4-day holding period). Using the dilution factor, the quantity of toxin in a volume or weight of the original sample is calculated and expressed as minimum lethal doses (MLD) or as median lethal doses (LD_{50}). The minimum lethal dose, as the name implies, is the least amount of toxin that is determined to be fatal. Usually, toxin levels are presented as LD_{50}s, the quantity of toxin that kills 50% of the group of experimental animals receiving the same dilution.

Extensive research has addressed the problem of inactivating botulinum toxins in foods without destroying nutrients or the organoleptic qualities (taste, color, odor, texture) of the products. Processes that have been examined include heating, freezing, and irradiating (gamma or beta). Investigations to determine effective measures for the destruction of botulinum toxins in water have also been conducted.

II. DESTRUCTION OF BOTULINUM TOXINS IN FOODS

A. Heat Inactivation

The fact that botulinum toxins can be inactivated by heat has long been known. In 1897, van Ermengem described his isolation of the causative organism, which he called *Bacillus botulinus*, from cured ham implicated in an outbreak of human botulism. He also reported that extracts of the ham produced paralysis in laboratory animals that resembled botulism in humans, demonstrating the association of a toxin with the disease. In addition, van Ermengem noted that heating at 70°C for 1 hour, at 80°C for 30 minutes, or boiling for 5 minutes inactivated the toxin. Thus, the original paper describing the isolation of the causative organism from cases of botulism also noted the lability of the toxin to heating (van Ermengem, 1897). Subsequent investigators (Landmann, 1904; Orr, 1920) reported that botulinum toxins from various strains could be destroyed by temperatures of 65–80°C in times from 1 hour to 30 seconds. The numerous early investigations concerning heat inactivation of botulinum toxin were reviewed and extended by Schoenholz and Meyer (1924).

In 1932, Sommer and Sommer described their studies on factors affecting the heat inactivation of types A and C toxins in buffers and in fruit and vegetable juices. In that classic paper, they reported that (1) the rate of toxin inactivation by heat was not a linear function, (2) the toxins were more stable at or near pH 5.0, and (3) substances in food products protected toxin from heat inactivation. These findings have been confirmed by other investigators for each type of botulinum toxin causing human disease.

Type A

Sommer and Sommer (1932) demonstrated that the rate of thermal inactivation for type A toxin at 68°C was nonlinear (toxicity versus heating time), but appeared to be a second-order reaction. Using several preparations of type A toxin, they found that the toxin was most stable at pH 5 when subjected to heating at 68–70°C.

These results were confirmed by Scott and Stewart (1950) for type A toxin and extended to type B as well (see below). They examined the heat inactivation of toxins produced in several different canned vegetables. Since the vegetable liquors containing the toxin were centrifuged prior to heating, differences in heat inactivation of the toxin could not be attributed to differences in heat penetration. The time needed to destroy the toxin at 75°C depended more on the vegetable in which the toxin had been produced than on the initial concentration of the toxin. The authors also commented that the heat inactivation at 75 and 80°C was not a first-order reaction, but that the rate decreased significantly when the majority of the toxic activity had been eliminated. The rate of toxin inactivation was the same for six strains of type A tested. Although type A toxin was most stable at approximately pH 5.0, it was inactivated at different rates in different vegetable liquors having the same pH. Therefore, the rate of toxin destruction depended on the composition of the vegetable liquor, and certain substances present in the liquid presumably protected the toxin.

Scott (1950) demonstrated that divalent cations (Ca^{2+} and Mg^{2+}) and organic acid anions (malate and citrate) present in vegetable juice had a protective effect on the toxin when it was subjected to a temperature of 75°C. A solution of high ionic strength also made the toxin more stable to heat. The author concluded that the time required to inactivate a given amount of type A toxin at a given temperature can only be defined if the ions present in the toxin solution are known.

The heat inactivation of crude type A toxin, of crystalline toxin (neurotoxin plus hemagglutinin), and neurotoxin (crystalline toxin repeatedly absorbed with chicken red blood cells to eliminate the hemagglutinin) was examined by Cartwright and Lauffer (1958). Their study is somewhat unusual in that they assayed the LD_{50} of the toxin by titration in goldfish. They demonstrated that the rates of inactivation at 50 and 60°C were similar for all preparations of type A toxin, and that the curves (change in toxicity versus heating time) were biphasic. Approximately 95% of the toxic activity was rapidly inactivated, and the remainder was destroyed more slowly. Sedimentation diagrams from ultracentrifugation of toxin samples before and after heating suggested the presence of two types of toxin molecules, one being more susceptible to heat inactivation than the other.

Losikoff (1978) studied the heat inactivation at 80°C of type A toxin in beef broth (bouillon) adjusted to various pHs from 3.5 to 6.8, and confirmed maximum heat stability at pH 5.0. In beef broth adjusted to pH 5.0, the toxin was destroyed in 8 minutes at 80°C, whereas in broth at pH 6.8, the toxin was inactivated in only 3 minutes. The inactivation curve (temperature versus heating time required for inactivation) was linear, with a slope (z-value) of 4.6°C at pH 5.0, and 5.4°C at pH 6.8. Losikoff also compared the use of hydrochloric acid with citric, lactic, tartaric, acetic, and malic acids for pH adjustment, since the organic acids were often used to modify the pH of foods. However, in these studies, the type of acid used had little effect on the heat stability of the toxin.

To determine the effect of pH on heat stability, Woolford et al. (1978) added crystalline type A toxin to 0.05 M sodium acetate, citrate, or phosphate buffer at pH 4.1, 4.7, 6.1, and 6.8 supplemented with 1% gelatin, to 9×10^4 mouse LD_{50}/ml. At 60°C, the type A toxin was more stable at pH 4.1 and 4.7 than at 6.1 and 6.8.

Woodburn et al. (1979a) added type A toxin (as well as type B or E, see below) to canned salmon to discover the minimum heating time and temperature required to destroy the toxin (Fig. 1). Although boiling was of course known to inactivate the toxin, that treatment adversely affected the appearance and palatability of the food. Samples containing 2×10^5 mouse LD_{50}/g of salmon were heated at 68, 71, 74, 79, and 85°C for time periods of up to 20 minutes. Some toxicity remained after 20 minutes at 68 or 71°C, but toxin could not be detected (≤ 5 LD_{50} in their assay) after 2–5 minutes at 79 and 85°C. The inactivation curve at each temperature was not linear, and a small percentage of the toxin (0.5–4%) was more stable to heat. The relative purity of the type A toxin (crude to crystalline) added to the salmon did not influence the heat stability of the preparation, and added toxin was destroyed at a rate similar to that of toxin produced in the salmon. The high level of toxin used in the experiments would

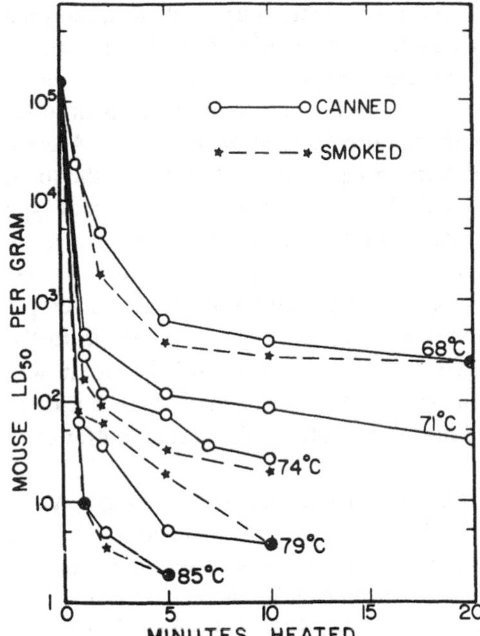

Figure 1 Thermal inactivation of type A botulinum toxin in salmon paste, heated at the temperatures indicated. (Reprinted with permission from Woodburn et al., 1979a.)

probably not occur in foods deemed to be edible, and employing such levels increased the margin of safety.

The rate of heat inactivation of type A (as well as types B, E, and F, see below) in foods and in buffers was examined by Woodburn et al. (1979b). Their goal was to determine the minimum heat treatment necessary to destroy the toxins and thereby circumvent the thermal destruction of nutrients and organoleptic qualities of the food. Products tested were tomato juice (pH 4.1) and tomato soup (pH 4.2) as representative of acid foods (pH $\leq$ 4.6), and canned string beans (pH 5.1), canned mushrooms (pH 5.8), canned corn (pH 6.2) and creamed corn (pH 6.3). The initial toxin concentration used was 10^5 LD_{50}/g or ml, since that amount of type A toxin had occasionally been found in foods. Results were expressed as the heating time at 74, 79, and 85°C needed to reduce the toxicity in the various foods by 1 log, 3 log, and to levels undetectable by the mouse bioassay. This clearly showed the nonlinearity of thermal inactivation for each type of toxin. To explain the small percentage of heat-resistant molecules, the authors suggested that the toxin molecules had aggregated or combined with the denatured toxin, or else changed in shape. They also demonstrated the effect of pH on heat inactivation for type A toxin. In tomato soup at pH 4.2, the time required for the destruction of toxin was 15 minutes at 79°C and 5 minutes at 85°C. In contrast, with the pH of the tomato soup adjusted to 6.2, times were 8 minutes at 79°C and 2 minutes at 85°C. The effect of other components in foods on toxin stability was examined. In canned corn (natural pH 6.2), type A toxin was destroyed in less time than in the tomato soup, and destruction of the toxin appeared to be exponential at both 79 and 85°C. The rate of inactivation of type A toxin in buffers was similar to the rate in food at a comparable pH. Thus, the rate of destruction of type A toxin in 0.05 M phosphate buffer, pH 6.8, was similar to that obtained in canned corn, and the rate in 0.05 M acetate buffer, pH 4.2, followed that in tomato juice and tomato soup. Adding 1% gelatin to the phosphate buffer tended to stabilize type A toxin to heating at 74°C (toxin was inactivated in 40 minutes with gelatin, and in 10 minutes without it), but there was little effect at 79 or 85°C. The addition of gelatin to the acetate buffer or to the citrate buffer was without effect.

Bradshaw et al. (1979) investigated the heat inactivation of crude types A and B toxins in 0.1 M acetate buffer (pH 5.0) and in 0.1 M phosphate buffer (pH 6.05) compared to beef and mushroom patties (pH 6.05) to assess the ability of the toxins to survive the heat processing of convenience foods. The initial concentration of toxin was 1.3×10^4 to 6.4×10^4 mouse LD_{50}/ml, which is approximately 10 times the maximum toxicity expected in convenience foods (Bradshaw et al., 1979). Heat-inactivation studies were conducted at 67.8–80.0°C. Even though the initial titers of crude toxin were comparable, they noted that the rates of heat inactivation of type A toxin prepared from three strains (62A, 73A, and V141) were different, in contrast to previous results by Scott and Stewart (1950). The authors (1979) suggested that the strain differences they noted might be more apparent due to advances in the techniques used for heat-inactivation studies. Confirming previous results, type A toxin was less heat resistant at pH 6.05 than at 5.0 and less stable in phosphate buffer than in beef and mushroom patties at the same pH (6.05), showing the protective effects of foods.

Like the numerous investigators noted above, Bradshaw et al. (1979) found that toxin inactivation rates were not linear functions and therefore could not be expressed by D-values, which are used to describe the logarithmic heat inactivation of viable spores of *C. botulinum*. The D-value is defined as the time required to reduce the viable count by one log (or by 90%) at a particular temperature. To describe toxin inactivation by heat, these authors proposed the F-value, defined as the time needed to reduce toxicity from a known initial concentration (C_0) in mouse LD_{50}/0.5 ml to one mouse LD_{50}/0.5 ml at a particular temperature. F-values determined for types A and B botulinum toxins varied from < 1 to about 8 minutes at 76.7°C (a process temperature commonly employed for such convenience food products), illustrating the variations imposed by differences in pH and food constituents on the heat inactivation of the toxin.

Type B

Scott and Stewart (1950) found that the rate of heat inactivation of type B toxin in canned vegetables was not linear, analogous to their findings for type A. Compared to type A, type B toxin appeared to be more readily destroyed, and the rate of toxin inactivation showed little variation among the nine strains tested. Woodburn et al. (1979a) also noted that the heat-inactivation curves for type B toxin in salmon paste were biphasic (Fig. 2), like those for types A (Fig. 1) and E (Fig. 3). Woodburn et al. (1979b) also demonstrated that heat inactivation of type B toxin in various foods was biphasic at 74, 79, and 85°C, in experiments as described above for type A. Type B toxin was more readily inactivated in canned corn (natural pH 6.2) than in tomato soup, and inactivation appeared exponential at temperatures of 79 and 85°C, as for types A and E. Adding 1% gelatin to phosphate buffer served to stabilize type B toxin to heating at 74°C, as noted above for type A. Bradshaw et al. (1979) demonstrated that the heat-inactivation rates of toxin from two type B strains (999B and Beans-B) were different. Type B toxin was less heat resistant at pH 6.05 than at 5.0 and

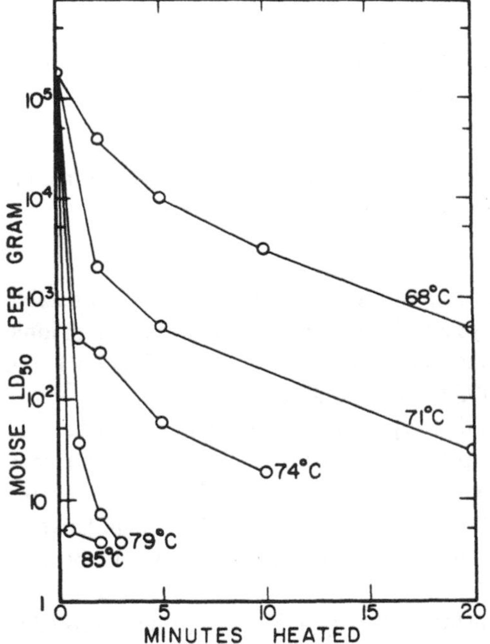

Figure 2 Thermal inactivation of type B botulinum toxin in salmon paste, heated at the temperatures indicated. (Reprinted with permission)

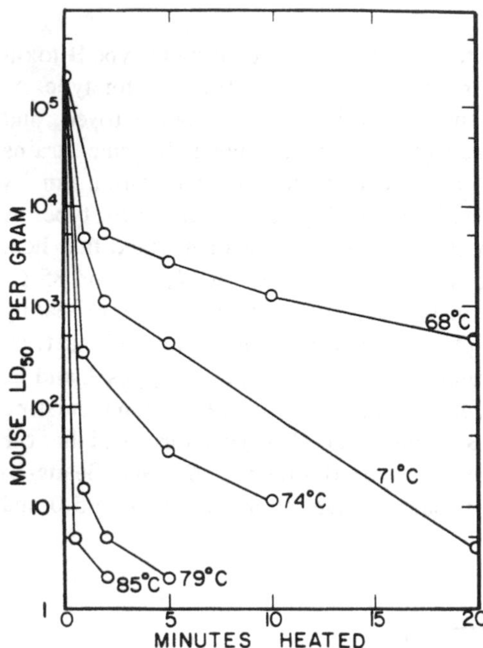

Figure 3 Thermal inactivation of type E botulinum toxin in salmon paste, heated at the temperatures indicated. (Reprinted with permission)

less stable in phosphate buffer than in beef and mushroom patties at the same pH (6.05), again showing the protective effects of foods, as for type A.

Type E

When type E toxin was heated at 65°C at various pHs, the toxin was most stable to heat inactivation at pH 4.5–5 (Ohye and Scott, 1957). A solution of high ionic strength did not increase the heat stability of type E toxin at 60°C and pH 5.0, in contrast to the results of Scott (1950) for type A toxin.

Abrahamsson et al. (1966) found that type E toxin was more readily heat inactivated in 0.1 N phosphate buffer, pH 7.2 (requiring 5 minutes at 58°C), than in the phosphate buffer plus 0.2% gelatin (10 minutes at 62°C). However, the toxin was more stable in meat broth (pH 6.2) and most stable in fish dialysate (pH 6.5), demonstrating that the nature of the food markedly affects the heat stability of type E toxin.

Licciardello et al. (1967a) demonstrated that type E toxin (trypsin-activated prior to heating) was most stable to heat at pH 5.5, and that increasing the pH increased the lability of the toxin. For type E toxin in haddock substrate (initially 1×10^4 mouse LD_{50}/ml), the heat-inactivation time curve (temperature versus heating time to inactivate the toxin) was linear for the range of temperatures studied (135–160°F, 57.2–71.1°C) and was very dependent on temperature (z-value of 7.5).

In a subsequent study, Licciardello et al. (1967b) showed that heat inactivation of type E toxin was comprised of an initial rapid rate of inactivation followed by a slower rate. This pattern was shown for (1) toxin in culture filtrate that was heated for different time periods and then injected into mice, (2) the prototoxin, which was heated for various times, trypsin-activated, and then injected into mice, and (3) activated toxin, which was trypsin-activated, heated for different times, and then injected. The rate of inactivation was similar for all three toxin preparations at 140°F (60°C). Using the prototoxin, further studies were conducted at 125°F (51.7°C), 135°F (57.2°C), and 145°F (62.8°C). Each curve had two segments: a rapid initial rate leading to about 99% destruction, followed by a second, slower rate. The authors proposed that this was due to a small number of unusually heat-resistant toxin molecules (1–4%) within the population tested. They further hypothesized that the more stable forms are actually aggregates of toxin molecules that are present initially, or else are formed by heating, or they could have a higher degree of hydrogen bonding and therefore require more heat to be denatured.

Woodburn et al. (1979a) also noted the nonlinearity of heat-inactivation curves for type E toxin in salmon paste (Fig. 3). They found that type E toxin was more stable to heat than previously reported by Ohye and Scott (1957), Licciardello et al. (1967a), or Yao et al. (1973).

Analogous to types A and B toxins, Woodburn et al. (1979b) demonstrated that heat inactivation of type E toxin in various foods was biphasic at 74, 79, and 85°C. However, the type E toxin seemed more readily heat-inactivated than type A or B in the foods and buffers tested.

Type F

The rate of thermal destruction of type F toxin in string beans (pH 5.1) was examined by Woodburn et al. (1979b) and found to be similar to those for types A, B, and E. The type F toxin was destroyed in 25 minutes at 74°C, in 10 minutes at 79°C, and in 1 minute at 85°C.

Bradshaw et al. (1981), using the same experimental design as for types A and B (Bradshaw et al., 1979), extended their study to the Wall 8 strain of type F and the G89 strain of type G. Prior to heating, the crude type G toxin was activated by trypsin. They found that the inactivation rates for types F and G toxins were also nonlinear and could not be expressed by D values. Therefore, as described above (Bradshaw et al., 1979), they reported their results as F-values. At 76.7°C, F-values for types F and G toxins were from 1 to 3+ minutes. Both F and G toxin were most stable at pH 5.0, as determined by heating each toxin at 71.1°C in 0.1 M acetate buffer adjusted to pH values ranging from 4.0 to 6.0.

The increase in stability at lower pH was more significant for type G toxin than for type F. As noted previously for types A and B (Bradshaw et al., 1979), types F and G toxins were more stable in the beef and mushroom patties than in phosphate buffer at the same pH. Thus, food components also protected the F and G toxins from heat inactivation.

B. Heat Inactivation After Frozen Storage

Yao et al. (1973) found that the titer of type E toxin, produced by the whitefish strain in canned salmon, remained constant at −15°C for up to 264 days. For heat-inactivation studies, the toxin-containing ampules were thawed and heated at various temperatures (149–162°F) for specified time intervals, cooled, and the contents trypsinized prior to injection into mice. Although the titer remained unchanged over the storage period, the heat stability of the toxin decreased with storage time. The same result was found for type E toxin produced in canned corn. However, the toxin from the canned corn (pH 5.2) required about 10-fold more time at a given temperature for inactivation than did toxin from canned salmon (pH 6.3).

Woolford et al. (1978) examined the thermal destruction of type A toxin in convenience foods after frozen storage. Crystalline type A toxin was added to an initial concentration of 6×10^4 mouse LD_{50}/g or ml to tomato bisque soup (pH 4.1), cream of mushroom soup (pH 6.2), 0.05 M phosphate buffer (pH 5.9), and beef pie filling (pH 5.9) that had been mixed in a blender. The foods and buffer containing the toxin were placed in ampules, frozen, and stored at −20°C. Ampules were removed from storage, thawed, and the contents assayed for toxicity at 1, 7, 21, 63, and 180 days, and the rate of heat inactivation of the toxin at 68°C determined. This temperature was chosen because in preliminary studies, the investigators found that the interior temperature of some convenience foods, when heated according to the manufacturer's directions, rarely attained more than 60–68°C. Frozen storage had no effect on the toxicity prior to heating. In contrast to results previously obtained with type E toxin (Yao et al., 1973), the rate of heat inactivation of type A toxin did not change with time of frozen storage. Woolford et al. (1978) noted that Yao et al. (1973) had used a temperature of −15°C for storage, whereas they had used −20°C. The authors also expected intrinsic differences between the two serotypes of toxin. The heat-inactivation curves for type A toxin in blended beef pie filling, tomato bisque soup, and phosphate buffer were biphasic, with an initial rapid decline in toxicity, followed by a slower rate for 1–4% of the toxin. There was little difference in the heat inactivation for the beef pie blend versus the phosphate buffer at the same pH. The food apparently exerted no protective effect.

C. Irradiation

Irradiation of food was investigated as an alternative to heat processing for the inactivation of botulinum toxins. Over time, the units used to measure the absorbed dose of ionizing radiation have changed. The rep (roentgen equivalent physical), now an obsolete unit, is approximately 0.084 joule per kilogram. The rad (short form of radiation) is equal to 0.01 joule per kilogram. The gray (Gy) is equal to 1 joule per kilogram. Thus, 1 rad equals 0.01 Gy. A megarad (Mrad) equals 10^6 rad, and a kilogray (kGy) equals 10^3 Gy.

Wagenaar and Dack (1956) found that spores of *C. botulinum* were more readily inactivated than was botulinum toxin by gamma irradiation from a cobalt 60 (^{60}Co) source. Type A spores innoculated on soft, surface-ripened cheese samples were inactivated by 8×10^5 to 1.1×10^6 rep, while type B spores required 5.5×10^5 rep. In contrast, the inactivation of crude type A toxin added to cheese required at least 7.8×10^6 rep to reduce the toxicity to less than 20 MLD from an initial level of above 1×10^3 MLD. Lower doses of irradiation were effective for the inactivation of toxin in broth culture. For crude type A toxin in trypticase broth or in beef heart broth, irradiation with less than 4.3×10^6 rep failed to reduce toxin levels from above 1×10^3 MLD to below 20 MLD, but 5.3×10^6 rep was usually effective. Similar results were obtained for type B toxin in broth. In these and subsequent investigations (Wagenaar and Dack, 1960), toxin was assayed at only three levels: 1×10^3, 200, and 20 MLD. Nonspecific deaths of mice from breakdown products precluded assay of lower levels in cheese products. Skulberg and Coleby (1960) noted that although the limitations of the toxin assay of Wagenaar and Dack (1956) did not permit an exact assessment of their inactivation curves, an exponential function was suggested. They calculated D-values of 1.47 to 6.15 Mrad from Wagenaar and Dack's (1956) data. Similarly, Miura et al. (1967) attributed D-values of 1.8–4.2 Mrad for type A or B toxin to Wagenaar and Dack (1956).

Wagenaar et al. (1959) extended their earlier studies using a more purified toxin preparation, crystalline type A toxin, and components of the crystalline toxin separated by ultracentrifugation. Previously, Wagman and Bateman (1953) had used ultracentrifugation to fractionate crystalline type A toxin into two portions, one with a molecular weight of 70,000, the second of 100,000. The inactivation patterns of the three preparations in 0.05 M sodium phosphate buffer, pH 7.5, using ^{60}Co irradiation were very similar. They noted that the preparations were reduced from high initial levels of toxicity (5×10^4 MLD for the crystalline toxin, 1×10^5 MLD for each of the fractions) to 20–100 MLD by 6×10^4 rad. However, further decreases in toxicity occurred more slowly, and doses of 7.3×10^5 rad (or 0.73 Mrad) were required to reduce toxicity to less than 5 MLD. In a second experiment, 3×10^3 rad reduced toxicity 80% from the initial level of 1 × 10^5 MLD to 2 × 10^4 MLD, but 9 × 10^5 rad (or 0.9 Mrad) was required to reduce the level 99.99%, to 10 MLD. Wagenaar and Dack (1956) had already demonstrated that large doses of ^{60}Co irradiation were required to inactivate crude type A toxin in broth. In fact, the crude type A toxin in trypticase broth or in cheese was about 100-fold more stable to gamma irradiation than the crystalline toxin in phosphate buffer. Therefore, they suggested that other materials in the crude toxin preparation served to protect the toxin from irradiation.

Wagenaar and Dack (1960) next used crude type A toxin at a lower initial level, 10^3–10^4 MLD, in trypticase broth and in heat-treated surface-ripened cheese. In cheese, 7.3 Mrad of ^{60}Co irradiation was required to reduce the toxicity from greater than 1 × 10^3 MLD to less than 20 MLD. In trypticase broth, at least 4–4.9 Mrad was necessary to reduce toxin levels from above 1 × 10^3 MLD to below 20 MLD. Using crystalline type A toxin in 0.05 M sodium phosphate buffer (pH 7.5), 3.4 × 10^5 rad (or 0.34 Mrad) decreased the toxicity from 2 × 10^4 MLD to less than 20 MLD, and only 2.9 × 10^4 rad reduced the activity from 200 MLD to less than 20 MLD. Parallel experiments with crystalline toxin added at 2 × 10^4 MLD to 0.05 M sodium phosphate buffer (pH 7.5), 5% trypticase broth (pH 7.1), and surface-ripened cheese (58% solids, pH 6.2) showed that the toxin in the phosphate buffer was approximately 100-fold less stable to inactivation by ^{60}Co irradiation than was the toxin in the trypticase broth, and that the cheese product protected the toxin somewhat more. Thus, the toxin was protected from damage when additional protein was present. Therefore, the authors concluded that irradiation was not a suitable replacement for heat inactivation of the toxins. Skulberg (1965) plotted the data of Wagenaar and Dack (1960) for crude type A toxin and determined a D-value of 4.1 Mrad. In contrast, for more purified (crystalline) type A toxin in 0.05 M phosphate buffer (pH 7.5), Miura et al. (1967) attributed a D-value of approximately 0.003–0.015 Mrad to Wagenaar and Dack (1960).

In an effort to determine the molecular weight of the toxins, Skulberg and Coleby (1960) examined the effect of ionizing radiation on crude type A and B toxins. Toxins were diluted in acetate buffer (pH 3.8 and 4.8), or in phosphate buffer (pH 6.2), containing 0.2% gelatin at about 2 × 10^6 LD_{50}/ml and subjected to 4 MeV electrons from a linear accelerator, at 1 Mrad/min as the average dose. They found an exponential relationship between the quantity of toxin inactivated and the dose of radiation absorbed, over an extended range of inactivation. The D-value for type A toxin at 15°C and at pH 3.8 or 4.8 was 1.25 Mrad, and 2.9 Mrad at pH 6.2. For type B toxin at 15°C and pH 6.2, the D-value was 0.65, but at −75°C it was 4.12. They concluded that the toxins were not readily destroyed by ionizing radiation.

Using an electron beam as a source of irradiation, Skulberg (1965) examined the effect of doses of 1, 2, 4, 6, and 8 Mrad on 6.4 × 10^3 MLD/ml of type E toxin. To prepare the toxin, *C. botulinum* type E was grown in Robertson's cooked meat medium, the fluid portion of the medium centrifuged, and the supernatant fluid irradiated. The D-value was determined to be 2.1 Mrad. Thus, the stability of type E toxin to irradiation was comparable to that previously determined for type A toxin (Skulberg and Coleby, 1960). Skulberg noted that the type E toxin, as well as types A and B, was so stable to ionizing radiation that practical quantities of radiation (approximately 5 Mrad) could not be depended upon to inactivate it. Furthermore, the variation in radiation resistance from one preparation of toxin to another for the same type precluded comparison of different types.

Miura et al. (1967) investigated the inactivation of type E toxin at different purities by irradiation with ^{60}Co. Three preparations of toxin were used, each containing approximately 5×10^4 LD_{50}/ml in 0.05 M acetate or 0.2 M phosphate buffer, pH 6.0. They reported D-values of about 0.04 Mrad for the most purified preparation, about 0.21 Mrad for the cell extract, and about 2.1 Mrad for the cell suspension. Activation of the toxin by trypsin did not alter the effect of irradiation, except for the cell suspension, which became more sensitive after trypsin treatment. The investigators assumed that the presence of other substances in the preparation of toxin played a significant role in the protection of toxin from radiation inactivation. They then demonstrated experimentally that addition of casein, serum albumin, DNA, or RNA to purified and activated toxin markedly increased the resistance of the toxin to irradiation, while addition of sugars or ascorbic acid had little effect. The authors concluded that type E toxin was as readily inactivated by irradiation as type A, but that the toxins were not likely to be destroyed at usual doses of irradiation due to the presence of protective substances.

Licciardello et al. (1969) used type E toxin from crude culture fluid that had been centrifuged and filtered, irradiated it with ^{60}Co, and then trypsinized it. Their D-value of 2.1 Mrad agreed exactly with the value published by Skulberg (1965).

The biological activity of chromatographically purified type A neurotoxin in gelatin phosphate buffer was completely destroyed by gamma irradiation at a dose of 2.5 kGy (Tranter et al., 1987). A concentration of toxin similar to those found in foods (10^4 LD_{50}/ml) was used. As the amount of beef mince was increased in the foods, the residual toxicity after irradiation with 25 kGy increased markedly, substantiating earlier reports that high doses of irradiation were required to inactivate botulinum toxins in foods.

Rose et al. (1988) found that the biological activity of purified type A neurotoxin (initially 10^4 LD_{50}/ml) in gelatin phosphate buffer, pH 6.5, was completely destroyed by 1.9 kGy, the minimum dose of ^{60}Co irradiation tested. This yielded a D-value of < 0.4 kGy (< 0.04 Mrad). For toxin in a 15% beef mince slurry (initially 10^4 LD_{50}/ml), a D-value of 12.99 kGy was determined. The stability of the type A neurotoxin to irradiation increased as the concentration of mince in the slurry increased.

III. DESTRUCTION OF BOTULINUM TOXINS IN WATER

A. Toxin Stability in Water

Brygoo (1953a) diluted crude culture fluid of types A, B, C, D, and E to 10 MLD/ml in tap water and incubated the preparations at room temperature and exposed to light. Samples were removed at 1-hour and at 24-hour intervals thereafter and tested for toxicity in mice (0.5 ml corresponding to 5 MLD/mouse, three mice/sample). The data were presented as a histogram of toxin type (three bars per type, one for each different preparation of toxin) versus the days required to observe a loss of more than 80% of the lethality. Type A remained toxic for 1–2 days, type B for 2–3 days, and E for 3–4 days. A footnote explained that the author used the French-Belgian names for the toxins and that his type D toxin corresponded to the South African and American type C and vice versa. Thus, his type D (South African type C) remained toxic for 2–4 days, but his type C (actually type D) remained toxic for 5–7 days. Toxins had been diluted at least 100-fold to achieve the initial level of 10 MLD/ml (type E), more for the other types (500-fold for type B and 1000-fold for type A), so the other components added with the toxin probably did not play a role in the inactivation.

The effect of pH on toxin stability in distilled water buffered at pH 6.0, 7.0, 8.0, 9.0, 10.0, and 11.0 was examined by Brazis et al. (1959). Using type A toxin at an initial level of 2×10^4 to 3×10^4 LD_{50}/ml at 25°C, they determined that increasing the pH above 7.0 decreased the stability of the toxin. The toxin was completely inactivated at pH 9.0 in 28 days, at pH 10.0 in 7 days, and at pH 11.0 in 3 minutes or less. Type E toxin was at least 50% inactivated in less than 1 hour at pH 8.0–11.0.

Type A botulinum toxin added to surface water obtained from a reservoir (pH 7.9) was inactivated by $\geq$ 99% in 3–6 days (Notermans and Havelaar, 1980). In drinking water sampled after slow sand filtration (pH 7.9), 9 days were needed for 99% reduction in the toxicity of type A. In sterile distilled water, pH 8.0, types A and B retained 56 and 63% of their toxicity, respectively, after 15 days.

B. Halogens

Brygoo (1953b) used the same stock preparations of toxin as before (Brygoo, 1953a) and diluted them in tap water to 10 MLD/ml, then added various concentrations of permanganate, hypochlorite, or Lugol's solution (iodine). After 15 and 30 minutes of incubation, three mice were injected as previously described (Brygoo, 1953a). The concentration of agent that inactivated at least

80% of the toxin (shown by the survival of all the injected mice) in 15 minutes was reported. For chlorine, type A toxin required 10 mg/l; type B or E, 500 mg/l; and type C or D, 50 mg/l. For iodine, type A and his type D (South African type C) needed 2.5 mg/l; type B or E, 5 mg/l; and his type C (actually D) 3 mg/l. He concluded that the toxins were each destroyed by concentrations of permanganate, chlorine, or iodine in the range of those usually used for water disinfection.

D'Arca and D'Arca Simonetti (1955) examined the effect of free chlorine on type A toxin added to spring water, assaying the residual toxicity in guinea pigs. Active chlorine at 0.1, 1, 2.5, 5, and 10 mg/l was incubated with toxin at levels of 1, 5, or 10 guinea pig MLD for 30 minutes, and assayed for residual toxicity. One MLD of toxin was destroyed by all concentrations of chlorine tested, 5 MLD was inactivated by concentrations > 0.1 mg/l, and 10 MLD by concentrations > 1 mg/l. Next, filtered liver broth culture medium was added to the toxin, to increase the content of organic materials, to a chlorine demand of 1.25 and 2.5 mg/l. Chlorine was added at 1, 2.5, 5, 10, 12.5, 15, and 17.5 mg/l (above the chlorine demand), and the mixtures incubated for 30 minutes. The authors reported that 15–17.5 mg/l of chlorine was needed to inactivate the toxin. Thus, the addition of culture medium to type A toxin protected the toxin from inactivation by chlorine. The amount of free active chlorine usually employed for the purification of drinking water inactivated the purified type A botulinum toxin. However, less purified toxin containing more organic materials was more resistant to chlorine inactivation.

Brazis et al. (1959) examined the effect of various concentrations of chlorine and found that toxin was quickly inactivated as long as there were free available chlorine residuals present. At pH 7.0 and 25°C, partially purified type A toxin was 99.99% inactivated in 28, 56, and 103 seconds by concentrations of 0.5, 0.33, and 0.22 mg/l of applied free available chlorine. Differences in sensitivity among the toxin types to free available chlorine at pH 7.0 and 25°C were A > D > C > B > E, with A the most sensitive to inactivation. In the range of 7.0–10.0, pH had little effect on the rate of toxin inactivation at 25°C by free available chlorine. However, decreasing the temperature from 25 to 5°C increased the time needed for toxin inactivation by a factor of three. Correlating the results of previous investigators was difficult, due to the problems encountered in measuring the chlorine concentration. They concluded that botulinum toxins can be readily inactivated by either free available chlorine or chlorine dioxide, but that chloramines were much less effective.

Notermans and Havelaar (1980) showed that chlorine, as a solution of hypochlorite, was effective in inactivating types B and E botulinum toxins in 0.005 M phosphate buffer, pH 7.8 (prepared from distilled water). The toxicity was reduced 99% in 30 seconds by a chlorine concentration of 0.3–0.5 mg/l. Since, according to the authors, chlorine is added to water to achieve a free chlorine concentration of 0.2–0.5 mg/l after a contact time of 30 minutes, any botulinum toxin present would be inactivated.

C. Ozone

Miller et al. (1957) added type A toxin to raw sewage to an initial level of 1×10^6 mouse MLD/ml. They found that the toxin was totally inactivated in 30 minutes of treatment with ozone. In three experiments, the range of ozone used was 210–290 ppm with an average of 250 ppm.

The effect of ozone on type E toxin (centrifuged culture fluid that had been trypsin-activated) was examined by Graikoski et al. (1985). The toxin, maintained at pH 5.6–5.8 during the treatment, was exposed to ozone gas for various times from 0 to 20 minutes. At a flow rate of 110 ml/min (8.5 ppm dissolved residual), 2×10^4 mouse units were totally inactivated in 12 minutes. At a 10-fold lower toxin concentration, which the authors believed more realistic under natural conditions, an ozone concentration of 0.81 ppm completely inactivated the toxin in less than 1 minute.

Notermans and Havelaar (1980) found that types A and B toxins were 99% inactivated by 1 mg/l ozone in 2 minutes. Since, according to the authors, 3 mg/l ozone are used in water treatment plants, that concentration would be adequate to inactivate botulinum toxins.

D. Other

Coagulation with $FeCl_3$ and absorption to activated powdered carbon failed to remove botulinum toxins from water (Notermans and Havelaar, 1980).

IV. SUMMARY

Developing processes to inactivate botulinum toxins in foods without destroying the nutritional value or the palatability of the product has been problematic. Heat inactivation is the method of choice. The time required to inactivate botulinum toxin is inversely proportional to the temperature. However, at any one temperature, the rate of toxin destruction by heat is not a linear function. The resistance of the toxin is also dependent on the pH and the presence of protective substances in the food.

Scott and Stewart (1950) recommended that foods that could contain botulinum toxin should be heated to temperatures greater than 80°C, where the rate of toxin inactivation is rapid. Based on the nonlinearity of toxin destruction and the effects of protective substances, they also advocated adding a large safety factor to any suggested procedure. Normal cooking procedures, such as frying, were found to be adequate to destroy type E toxin in haddock fillets (Licciardello et al., 1967a). Bradshaw et al. (1979) stated that 1 minute at 82°C (internal temperature) would inactivate types A and B botulinum toxin in convenience foods like beef and mushroom patties if the toxin concentration was less than 2×10^4 LD_{50}/ml. However, if the internal temperature was < 66°C, toxin probably would not be inactivated. For types F and G (Bradshaw et al., 1981), 1 minute at 78°C (internal temperature) would inactivate 2×10^4 LD_{50}/ml in the beef and mushroom patties. Woodburn et al. (1979b) noted that the effects of pH and protective substances in foods precluded recommending a single procedure that would be adequate to inactivate toxin in all foods. They demonstrated that 10^5 LD_{50} of type A, B, E, or F botulinum toxin was inactivated by heating at 79°C for 20 minutes, or at 85°C for 5 minutes, and suggested that these procedures could be used as a guideline.

Botulinum toxins are not adversely affected by freezing. The effect of freezing on subsequent heat inactivation has also been investigated. Although Yao et al. (1973) reported that type E toxin became more labile to heat inactivation with increased time of storage at −15°C, a more extensive study by Woolford et al. (1978) using type A toxin did not confirm that observation. Instead, Woolford et al. (1978) found that the rate of heat inactivation did not change with the time of frozen storage.

Irradiation of foods was examined as an alternative to heat processing. However, botulinum toxin was more resistant to gamma irradiation from a ^{60}Co source than were spores of *C. botulinum*. Substances in foods also exerted a significant protective effect. Botulinum toxins could not be reliably inactivated by doses of irradiation that preserved the organoleptic qualities of the food.

Botulinum toxins are relatively stable in water under laboratory conditions. However, the toxins are readily inactivated by the concentrations of agents normally used for water treatment and purification.

REFERENCES

Abrahamsson, K., Gullmar, B., and Molin, N. (1966). The effect of temperature on toxin formation and toxin stability of *Clostridium botulinum* type E in different environments. *Can. J. Microbiol. 12*: 385.

Bradshaw, J. G., Peller, J. T., and Twedt, R. M. (1979). Thermal inactivation of *Clostridium botulinum* toxins types A and B in buffer and beef and mushroom patties. *J. Food Sci. 44*: 1653.

Bradshaw, J. G., Peller, J. T., and Twedt, R. M. (1981). Thermal inactivation of *Clostridium botulinum* toxin types F and G in buffer and in beef and mushroom patties. *J. Food Sci. 46*: 688.

Brazis, A. R., Bryant, A. R., Leslie, J. E., Woodward, R. L., and Kabler, P. W. (1959). Effectiveness of halogens or halogen compounds in detoxifying *Clostridium botulinum* toxins. *J. Am. Waterworks Assoc. 51*: 902.

Brygoo, E. R. (1953a). Résistance des toxines botuliques en dilution dans l'eau. *Ann. Inst. Pasteur 84*: 1039.

Brygoo, E. R. (1953b). Action des agents épurateurs sur les toxines botuliques en dilution dans l'eau. *Ann. Inst. Pasteur 84*: 1040.

Cartwright, T. E., and Lauffer, M. A. (1950). Temperature effects on botulinum A toxin. *Proc. Soc. Exp. Biol. Med. 98*: 327.

D'Arca, S., and D'Arca Simonetti, A. (1955). Further studies on the detoxifying power of free active chlorine with respect to type A botulinum toxin. *Nuovi Ann. Igiene e Microbiol 4*: 355.

DasGupta, B. R., Boroff, D. A., and Rothstein, E. (1966). Chromatographic fractionation of the crystalline toxin of *Clostridium botulinum* type A. *Biochem. Biophys. Res. Commun. 22*: 750.

Gimenez, D. F., and Ciccarelli, A. S. (1970). Another type of *Clostridium botulinum. Zentralbl. Bacteriol. Parasitenkd. Infektionskr. Hyg. Abt. I Orig. 215*: 221.

Graikoski, J. T., Blogoslawski, W. J., and Choromanski, J. (1985). Ozone inactivation of botulinum type E toxin. *Ozone: Sci. Eng. 6*: 229.

Lamanna, C., McElroy, O. E., and Eklund, H. W. (1946). The purification and crystallization of *Clostridium botulinum* type A toxin. *Science 103*: 613.

Landmann, G. (1904). Ueber die Darmstadter Bohnenvergiftung. *Hyg. Rundschau 10*:449.

Licciardello, J. J., Nickerson, J. T. R., Ribich, C. A., and Goldblith, S. A. (1967a). Thermal inactivation of type E botulinum toxin. *Appl. Microbiol. 15*: 249.

Licciardello, J. J., Ribich, C. A., Nickerson, J. T. R., and Goldblith, S. A. (1967b). Kinetics of the thermal inactivation of type E *Clostridium botulinum* toxin. *Appl. Microbiol. 15*: 344.

Licciardello, J. J., Ribich, C. A., and Goldblith, S. A. (1969). Effect of irradiation temperature on inactivation of *Clostridium botulinum* toxin type E by gamma rays. *J. Appl. Bacteriol. 32*: 476.

Losikoff, M. E. (1978). Establishment of a heat inactivation curve for *Clostridium botulinum* 62A toxin in beef broth. *Appl. Environ. Microbiol. 36*: 386.

Miller, S., Burkhardt, B., Ehrlich, R. and Peterson, R. J. (1957). Disinfection and sterilization of sewage by ozone. *Adv. Chem. 21*: 381.

Miura, T., Sakaguchi, S., Sakaguchi, G., and Miyaki, K. (1967). Radiosensitivity of type E botulinus toxin and its protection by proteins, nucleic acids and some related substances. In *Microbiological Problems in Food Preservation by Irradiation. Report of a Panel*. International Atomic Energy Agency, Vienna, p. 45.

Notermans, S., and Havelaar, A. H. (1980). Removal and inactivation of botulinum toxins during production of drinking water from surface water. *Antonie van Leeuwenhoek 46*: 511.

Ohye, D. F., and Scott, W. J. (1957). Studies in the physiology of *Clostridium botulinum* type E. *Austr. J. Biol. Sci. 10*: 85.

Orr, P. F. (1920). Studies in *Bacillus botulinus. J. Med. Res. 42*: 127.

Rose, S. A., Modi, N. K., Tranter, H. S., Bailey, N. E., Stringer, M. F., and Hambleton, P. (1988). Studies on the irradiation of toxins of *Clostridium botulinum* and *Staphylococcus aureus. J. Appl. Bacteriol. 65*: 223.

Schoenholz, P., and Meyer, K. F. (1924). Effect of direct sunlight, diffuse daylight and heat on potency of botulinus toxin in culture mediums and vegetable products. XXIV. *J. Infect. Dis. 35*: 361.

Scott, W. J. (1950). Thermal destruction of type A *Clostridium botulinum* toxin: The nature of the protective substances in canned vegetables. *Austr. J. Appl. Sci. 1*: 200.

Scott, W. J., and Stewart, D. F. (1950). The thermal destruction of *Clostridium botulinum* toxin in canned vegetables. *Austr. J. Appl. Sci. 1*: 188.

Skulberg, A. (1965). The resistance of *Clostridium botulinum* type E toxin to radiation. *J. Appl. Bact. 28*: 139.

Skulberg, A., and Coleby, B. (1960). The inactivation of the toxins of *Clostridium botulinum*, types A and B, by 4 MeV electrons. *Riso Report No. 16*: 59.

Smith, L. DS., and Sugiyama, H. (1988). *Botulism: The Organism, Its Toxins, the Disease*. 2nd ed. Charles C. Thomas, Springfield, IL.

Sommer, H., and Sommer, E. W. (1932). Botulinus toxin. VI. The destruction of botulinus toxin by heat. *J. Infect. Dis. 51*: 243.

Sonnabend, O., Sonnabend, W., Heinzle, R., Sigrist, T., Dirnhofer, R., and Krech, U. (1981). Isolation of *Clostridium botulinum* type G and identification of type G botulinal toxin in humans: Report of five sudden unexpected deaths. *J. Infect. Dis. 143*: 22.

Tranter, H. S., Modi, N. K., Hambleton, P., Melling, J., Rose, S., and Stringer, M. F. (1987). Food irradiation and bacterial toxins. *Lancet* (July 4): 48.

van Ermengem, E. (1897). Ueber einen neuen anaeroben Bacillus und seine Beziehungen zum Botulismus. *Ztschr. Hyg. Infekt. 26*: 1.

Wagenaar, R. O., and Dack, G. M. (1956). Effect in surface ripened cheese of irradiation on spores and toxin of *Clostridium botulinum* types A and B. *Food Res. 21*: 226.

Wagenaar, R. O., Dack, G. M., and Murrell, C. B. (1959). Studies on purified type A *Clostridium botulinum* toxin subjected to ultracentrifugation and irradiation. *Food Res. 24*: 57.

Wagenaar, R. O., and Dack, G. M. (1960). Studies on the inactivation of type A *Clostridium botulinum* toxin by irradiation with cobalt 60. *Food Res. 25*: 279.

Wagman, J., and Bateman, J. B. (1953). Botulinum type A toxin: Properties of a toxic dissociation product. *Arch. Biochem. Biophys. 45*: 375.

Woodburn, M. J., Schantz, E. J., and Rodriguez, J. (1979a). Thermal inactivation of botulinum toxins in canned salmon. *Home Econ. Res. J. 7*: 171.

Woodburn, M. J., Somers, E., Rodriguez, J., and Schantz, E. J. (1979b). Heat inactivation rates of botulinum toxins A, B, E and F in some foods and buffers. *J. Food Sci. 44*: 1658.

Woolford, A. L., Schantz, E. J., and Woodburn, M. J. (1978). Heat inactivation of botulinum toxin type A in some convenience foods after frozen storage. *J. Food Sci. 43*: 622.

Yao, M. G., Denny, C. B., and Bohrer, C. W. (1973). Effect of frozen storage time on heat inactivation of *Clostridium botulinum* type E toxin. *Appl. Microbiol. 25*: 503.

[illegible]

14

Predictive Modeling

David A. Baker

Nestec Research Center, Lausanne, Switzerland

Constantin Genigeorgis

Aristotelian University, Thessaloniki, Greece

I. INTRODUCTION

A model has been described as "a system of postulates, data, and inferences presented as a mathematical description of an entity, or state of affairs" (Webster, 1989). Predictive modeling offers food safety microbiologists a means to describe the survival, growth, and production of toxins by bacterial pathogens in substrates controlled by preservation factors, such as temperature, pH, water activity, oxidation-reduction potential (Eh), various additives, and their interactions. Acquisition of bacterial growth or toxigenesis data is labor-intensive, especially in the study of *Clostridium botulinum*. Thus, individual researchers have concentrated their studies on variables affecting specific media or food systems. The analysis of these separate studies has led to many modeling techniques, which may confuse readers, depending on their individual working knowledge of advanced statistics, calculus, solution of engineering problems, or use of computer applications. The future of food preservation science will most likely use models to reduce and correlate data from related factorial research designs. The best modeling approaches will incorporate terms that are consistent with the principles of microbial physiology, robust enough to be applicable to food product design, and offer a basis for risk assessment and the establishment of safety margins.

The objective of this chapter is to present the evolution and use of quantitative microbiology to assess the potential risk of *C. botulinum* growth in foods and to introduce design variables and general descriptions of the mathematical modeling methods being explored by various researchers. Some commentary will be made on the applicability of one approach or another, but this overview will not delve deeply into underlying assumptions or functional descriptions of the mathematics employed. Also, because of the disparity of experimental systems and analytical methodologies used, no in-depth attempt will be made to compare results. We hope that the reader will keep in mind that a prime objective of quantitative microbiology is to produce a model from intrinsic data, applicable to the prediction of outcome from an extrinsic, or real-world, situation. As Pflug (1987) points out, "heat preservation of food is not a scientific experiment." Rather, scientific models must lend themselves to use by food technologists and engineers. Thus, the responsibility and challenge of individual researchers should be to develop models that are simple enough to be widely usable.

II. PROBABILITY STATEMENTS FOR *C. BOTULINUM* THERMAL INACTIVATION

A. The 12-D Canning Concept for Low-Acid Canned Foods

Thermal destruction models for *C. botulinum* have been applied by the food industry since the early 1920s (Ball, 1923; Bigelow et al., 1920). Many of the lessons learned and problems experienced in their development also apply to the design and interpretation of contemporary experiments predicting the survival and growth of *C. botulinum* in multifaceted food-preservation systems.

The thermal destruction of bacteria or spores is best described by the time in minutes at a given temperature to achieve a decimal reduction (DR) of the initial viable number. Thus, the D-value describes a treatment that reduces the viable number by 90% (or conversely, to 10% of the original number). Research to be presented extrapolates this concept to describe inhibitory processes that reduce the number of *C. botulinum* spores able to initiate growth. The D-value can be obtained graphically by plotting the logarithm of the survivors against the time on an arithmetic scale (Fig. 1). Given a straight line response, the D-value can be read as the time for the curve to transverse one log cycle. The D-value is mathematically equal to the reciprocal of the exponential slope, or the velocity constant (k), of the survivor curve. The destruction of bacteria by moist heat is generally logarithmic. Therefore, equations for first-order kinetic reactions can also be used to describe survivor curves (Stumbo (1973):

$$\text{D-value} = \frac{t}{\log N_0 - \log N_t} \tag{1}$$

where t is the duration of heat treatment in minutes and N_0 and N_t are the initial and final numbers of viable cells or spores in samples of equal size. Pflug and Holcomb (1983) discussed methods utilizing this relationship in practical detail.

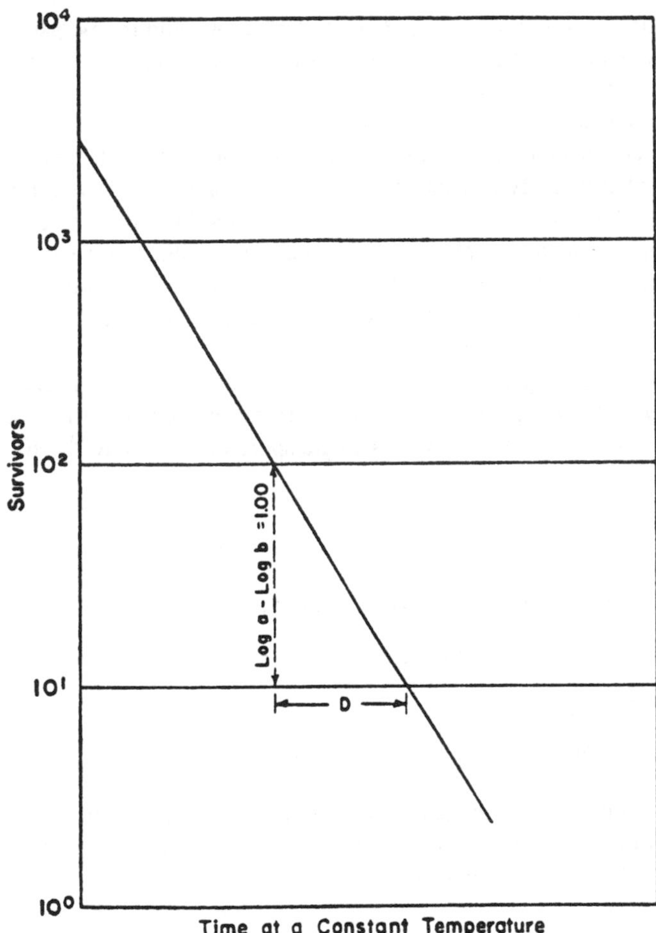

Figure 1 The logarithmic survivor curve.

The safety of commercial food canning evolved with the development of mathematical relations between the thermal process and the probability of bacterial spore survival. In Eq. 1, the volumes of samples N_0 and N_t must be equal. Increasing the number of sample units increases the initial number of viable cells (N_0), as well as the likelihood of at least one cell surviving (N_t). No time-temperature combination can assure sterility of an infinite number of containers, but if the initial volume or quantity of sample units is small, then N_t will become less than one, though it will never reach zero. A value of N_t less than one merely represents the probability of survival for the organism of interest within the sample unit. This reasoning has been extrapolated from a classic example (Example 1) to become the standard for safe processing of shelf-stable, low-acid foods, packaged in hermetically sealed containers. It was the first model to assess the risk of *C. botulinum* toxigenesis in a food product.

The precedent to develop a food safety standard for low-acid canned foods was based upon a predetermined reduction of bacterial spore numbers, regardless of the initial number actually present. Type A and B spores of *C. botulinum* were chosen because they were the most heat-resistant foodborne pathogens able to produce lethal toxins above pH 4.5. Esty and Meyer (1922) are widely credited with documenting the heat resistance of more than 100 strains of *C. botulinum* spores. The most resistant spores had a D-value of about 0.21 minutes at 121.1°C.

Given instantaneous heating and cooling, a 12-decimal reduction (12 DR) will result when *C. botulinum* spores are heated in neutral phosphate buffer for 2.52 minutes at 121.1°C. Thus, if there was a single spore in each of 10^{12} cans, then from Eq. 1:

Example 1:

$$2.52 = 0.21\,(\log 10^{12} - \log N_t)$$
$$12 = 12 - \log N_t$$
$$\log N_t = 0$$
$$N_t = 1$$

After thermal processing, one can expect only a single can of the original 10^{12} to contain a viable spore. Similarly, this process would result in a one in 10^{12} chance for the survival of a single, initial spore:

Example 2:

$$2.52 = 0.21\,(\log 1 - \log N_t)$$
$$\log N_t = -12$$
$$N_t = 10^{-12}$$

In the food industry, where direct enumeration is often difficult, the "fraction of negative units" method can be used to estimate the D-value (Pflug and Holcomb, 1983). In this procedure, a number of replicate units (r) are sampled after the heating process to determine the number of units with growth (p) and without growth (q) after incubation:

$$r = p + q \qquad (2)$$

Given the initial number of viable spores per replicate unit (N_0), the number of survivors per replicate unit (N_t) can be calculated:

$$N_t = \ln(r/q) = 2.303 \log_{10}(r/q) \quad (3)$$

The principles and practices of modern thermal processing (Hachigian, 1989; Hersom and Hulland, 1980; Pflug, 1987; Pflug and Holcomb, 1983; Pflug et al., 1985; Russel, 1982; Stumbo, 1973) are based upon increasingly intricate mathematics applied to the development of each specific commercially sterile product. Calculations must address the method by which heat is delivered, the thermal protection of pathogenic or spoilage organisms by media ingredients, the shape of the container being heated, and the rates of thermal distribution within a container or heat exchanger based on viscosity, particle size, flow characteristics, etc. The engineering aspects of modeling the thermal destruction of *C. botulinum* spores diverge from our topic, but the basis of thermal processing sets the framework for concepts in more recent attempts to quantify the safety of minimally processed foods.

III. PROBABILITY STATEMENTS OF *C. BOTULINUM* GROWTH INHIBITION

The microbial hurdles concept first proposed by Leistner and Rödel (1976) refers to physical, chemical, and/or biological effects that act as barriers or hurdles to microbial growth. A hurdle may act alone or interact additively or synergistically with others. Hurdles may be used to engineer new food products with predictable levels of safety and patterns of spoilage. Intrinsic and extrinsic factors and their possible interactions and synergies affect the survival of microorganisms in the food and the ability of the microbial environment to support growth. Extrinsic factors include heat, freezing, cold, radiation, and packaging, which affect the atmosphere and relative humidity surrounding the product. Intrinsic factors include pH and acidifying agents, water activity (a_w) and humectant, Eh, available nutrients, natural inhibitors of microbial growth, the presence of competitive microorganisms, and chemical additives. Comprehensive reviews of the individual effects of hurdles have been published (Genigeorgis, 1981; Leistner, 1985, 1987, 1988).

Quantitation of the effect of hurdles follows the 12 DR scheme devised for canning processes, where extreme heat and packaging impermeable to microorganisms virtually assure microbiological shelf stability. Extrapolating the concept of probability to the number of organisms required to overcome growth hurdles or, conversely, the number of effective decimal reductions of bacteria/spores due to such barriers provides a conceptual framework to model growth of pathogens and a common denominator for comparison of applicable qualitative research available in the literature.

A. Shelf-Stable, Canned, Cured Meats

Assessment of the safety of shelf-stable, canned, cured meats requires the incorporation of factors other than heat into the risk analysis of *C. botulinum* growth and toxigenesis. A brief review of the pertinent variables and some results of qualitative experiments will demonstrate the evolution of the first models describing interactions of preservatives and preservation processes in the prediction of food safety.

Shelf-stable canned luncheon meats were the first "modern" low-acid foods packaged in hermetically sealed packages that were not given the 12D thermal process. Many cured meats would be thoroughly inedible after even 1 minute at 121°C. They are commonly processed for the equivalent of only 0.4–0.6 minutes at 121°C (F_0 = 0.4–0.6). However, for comparable safety, the inhibition of *C. botulinum* spore outgrowth must be as effective in canned cured meats as the thermal spore destruction in low-acid canned products.

Inoculated pack studies have shown that the stability of canned luncheon meats depends on three factors: (1) a mild cook to kill vegetative cells and injure spores, (2) the action of nitrite, NaCl, ascorbate, and other curing agents to inhibit heat-damaged spores, and (3) a low initial concentration of *C. botulinum* spores in the raw product (Cerveny, 1980; Hauschild and Simonsen, 1985; Riemann, 1963; Silliker et al., 1958). The longstanding safety of shelf-stable canned cured meats cannot be attributed to any single factor such as nitrite, NaCl, pH, or thermal treatment used in commercial processes. Instances of botulism attributed to canned cured meat products resulting from process failures have been rare (Tompkin, 1980).

B. Inhibitory Action of Curing Ingredients

The observation that low numbers of spores do not cause spoilage in canned cured meat is explained by the inhibitory effect of curing salts on heat-injured spores. The curing action has been attributed to interactions between NaCl and nitrite. A multitude of inoculated pack studies has demonstrated limits to growth for product formulation, i.e., 10^6 type A and B spores produced toxin in chopped beef containing up to 7.8% NaCl but caused spoilage only at NaCl levels up to 6.7% (Greenberg et al., 1959).

C. Low Levels of *C. botulinum* Spore Contamination

The demonstration that canned cured meat preservation systems break down when many putrefactive anaerobic spores are present in raw meat stresses the importance of considering initial numbers in the experimental design process (Hauschild and Simonsen, 1985; Riemann, 1963, 1967). The growth requirements of spore-forming putrefactive anaerobic bacteria are similar to those of *C.*

botulinum. Therefore, it is generally assumed that conditions resulting in spoilage may also result in toxigenesis by *C. botulinum* (Pivnick et al., 1969). Riemann (1963) calculated the most probable numbers of putrefactive anaerobic spores as 0.02/g of cut beef trimmings, 0.54/g of pork, and 0.45/g of chopped ham, assuming a log normal distribution. However, individual samples contained up to 51 anaerobic spores/g. One pork luncheon meat sample contained 15 clostridial spores/g. Hauschild and Simonsen (1985) correlated *C. botulinum* spore prevalence data which suggest that the ratio of putrefactive anaerobe spores to those of *C. botulinum* in raw meats may be 10^4 to 1. This ratio should also be a valid frequency for spores in canned cured meats and other cured meat products. *C. botulinum* spores are not prevalent in fresh and semi-preserved meats, and their numbers have not been reported to exceed 4/kg (Genigeorgis, 1976). With margins of safety not well established, both Silliker et al. (1958) and Pivnick et al. (1969) demonstrated that the safety of preservation processes depends on a low initial *C. botulinum* spore contamination. Hauschild (1989) tabulated data on the incidence, level, and types of *C. botulinum* reported for raw and semi-preserved meats and fish.

Shorter times to *C. botulinum* toxigenesis have been attributed to increases in spore inoculum (Eklund, 1982; Eyles and Warth, 1981; Hauschild 1982; Lindsay, 1981). These observations are only as accurate as the sampling intervals used in each study. Other studies support the contention that, given a conducive environment, the spore inoculum will have little effect on the time to toxigenesis (Baker and Genigeorgis, 1990; Huss et al., 1979). However, when less favorable conditions are presented, increasing numbers of spores will help *C. botulinum* to overcome hurdles and allow earlier toxigenesis (Baker and Genigeorgis, 1990; Genigeorgis, 1981; Hauschild, 1982; Riemann, 1967).

D. Spore Number and Growth Probability

Pivnick and Petrasovits presented a rationale for the safety of canned, shelf-stable, cured meat as follows (see Hauschild and Simonsen, 1985):

$$\text{Pr} = \text{Ds} + \text{In} \qquad (4)$$

where Pr is preservation, Ds is spore destruction by heat, and In is spore inhibition by synergistic action of heat and curing agents.

The effectiveness of a preservation system to prevent growth of *C. botulinum* depends on the probability (P) that not even a single spore will be able to grow and produce toxin (Riemann, 1967). Presenting qualitative plus/minus data on a continuous probability scale has allowed quantitative comparisons to be made. Riemann (1967), Roberts et al. (1981a), and Hauschild (1982) were among the first investigators to estimate P of a single spore to initiate growth and toxigenesis as follows:

$$P = \frac{\text{MPN spores outgrowing}}{\text{MPN spores inoculated}} \tag{5}$$

where P relates the most probable number (MPN) of spores initially inoculated into each sample unit to the MPN of spores able to survive the heat process and outgrow despite the presence of hurdles. Hurley and Roscoe (1983) reviewed the origins of MPN methods and presented formulas for standard error and confidence interval for the MPN estimate. In studies where the inoculation was made at a single level, the MPN of outgrowing spores may be calculated by (Halvorson and Ziegler, 1933):

$$\text{MPN} = \ln (n/q) \tag{6a}$$

where n is the number of inoculated packs and q is the number of nontoxic packs after processing and incubation. This method assumes that the probability of a single spore to grow and initiate toxigenesis is not affected by the presence of other spores. P of toxin development from a single *C. botulinum* spore is determined by dividing the calculated MPN by the number of spores inoculated (s):

$$P = \frac{\ln (n/q)}{s} \tag{6b}$$

Multiplication of P by the natural prevalence of *C. botulinum* spores expected per sample (i) (Eq. 6c), yields a probability statement that incorporates experimental results with an estimate of product contamination, which can be used for further risk analysis (Hauschild et al., 1982; Hauschild, 1982):

$$P_i = \frac{\ln (n/q)}{s}\, i \tag{6c}$$

The expression log 1/P represents the number of spores or cells (in log units) required initially to enable one spore to germinate, outgrow, and give rise to toxin under given environmental conditions. Thus, conditions inhibitory to growth will require logarithmic increases in initial spore numbers for growth and toxigenesis, which is analogous to the decimal destruction of spores during heat treatment (Stumbo, 1973). For example, if 5 out of 100 sample units, i.e., cans, inoculated with 1000 *C. botulinum* spores demonstrate growth after processing and appropriate incubation, then:

Example 3

$$P = \frac{\ln (100/95)}{1000} = 5.1 \times 10^{-5}$$

Thus, each individual spore has a 1/510,000 chance for growth. The number of spores not able to grow for each capable one is expressed as the reciprocal of P(1/P). Thus, log 1/P = 4.3 is analogous to the D-value or decimal reduction (inhibition) of spores due to the above process (Example 4). Log 1/P is therefore equivalent to Pr = Ds + In.

Example 4:

> $P = 10^{-5}$ can be interpreted as equivalent to 5 DR, i.e., 100,000 spores/unit are reduced to one, or, starting with 10 spores/unit, there would be a 1/10,000 chance for one viable spore.

If we accept that it is not necessary to distinguish between spore destruction and inhibition, then log 1/P may be used conceptually for the design of quantitative experiments and employed for the reanalysis of existing inoculated pack data (Hauschild, 1982). Similarly, log $1/(P \times i)$ can express the decimal number of units produced for each unit that may be expected to become toxic, based on the expected prevalence of contaminating *C. botulinum* spores per unit (Hauschild and Simonsen, 1985). Inoculated pack studies demonstrate the possibility of botulinum toxin production in canned cured meats. Assessing the synergistic inhibition of *C. botulinum* due to heat, pH, and curing salts in terms of D-values allows product formulations and treatments to be compared quantitatively (Hauschild, 1982; Hauschild and Simonsen, 1985; Riemann, 1967).

Riemann (1963) used factorial design to study the interactions between thermal injury and curing agents on the stability of canned cured meats. The MPN of putrefactive anaerobes able to overcome the effects of thermal process, pH, NaCl, $NaNO_3$, and $NaNO_2$ (up to their regulated limits) was evaluated by inoculation in five decimal dilutions, with five replicate cans for each treatment level. The number of spores able to outgrow over a 6-month incubation period at 30°C was significantly affected by all factors, except $NaNO_3$, many in two- and three-way interactions. Quantitative analysis allowed the degree of improvement to be predicted: at $F_0 = 0.4$, 1% of inoculated spores outgrew, while only 0.1% outgrew at $F_0 = 1.0$; 150 ppm $NaNO_2$ was as effective as 50 ppm at $F_0 = 0.4$, but inhibited 10 times more spores at $F_0 = 1.0$. These experiments also demonstrated that the heat treatment served to extend the lag phase before growth.

Application of P calculations to experimental design allowed Riemann (1967) to quantitatively demonstrate the effect of spore number in overcoming intrinsic hurdles. With 4.5% NaCl there was approximately 5 DR of *C. botulinum* type E; increasing NaCl to 5.5% resulted in 6 DR. As the NaCl concentration increased from 0 to 4.67%, the pH required to inhibit the growth of 10^6 *C. botulinum* type E spores could be raised from 5 to 6.5.

Data on the growth of *C. botulinum* spores in canned luncheon meats (Pivnick et al., 1969) was reanalyzed (Table 1) in terms of DR(log 1/P) by Hauschild and Simonsen (1985). These data indicate the need for nitrite at low brine levels {brine% = NaCl% × 100/(NaCl% + H_2O%)}. With 3.6–4.6% brine and ≥75 ppm nitrite, the DR was 7–8. At brine concentrations of 5.0–5.8%, 8–9 DR were obtained without nitrite, and >9.5 DR with the addition of 75 ppm nitrite. The authors stressed that the heat resistance of the inoculated *C. botulinum* spores was low. A minimum of 5 DR resulted from heating to F_0 = 0.64 in neutral phosphate buffer. With more heat-resistant spores, the observed DR might have been significantly lower.

Similar assessments have been applied to the potential botulism hazard from a variety of other cured meats that rely upon refrigeration for part of their preservation. Here, too, qualitative inoculated pack studies concerning the safety of vacuum-packaged bacon and other nonfermented vacuum-packaged cured meats, including liver sausage, pasteurized canned meats, and vacuum-packaged fermented sausages, dominate the literature. Hauschild (1982) compared studies of *C. botulinum* toxigenesis in nonfermented, vacuum-packaged cured meats and liver sausage from which P could be calculated. This method

Table 1 Protection (Log 1/P) of Canned Luncheon Meats from *C. botulinum*[a]

NaCl (% brine)[b]	F_0	Nitrite (mg/kg) 0	75	150
3.6	0.64		7.6	7.6
4.1	0.64		7.5	7.8
4.6	0.64		7.6	7.8
4.8	0.68			8.7
4.9	0.57	<8.2	8.2	>9.5
5.0	0.57	8.1		
5.2	0.62	8.2	>8.2	8.0
5.2	0.62	8.5	9.1	>8.2
5.2	0.64		7.8	10.2
5.3	0.57			9.7
5.4	0.64		8.0	>8.8
5.5	0.57	8.3	8.6	>9.5
5.5	0.57			10.2
5.5	0.61			>10.2
5.8	0.57	8.8	>9.5	>9.5

[a]Log 1/P values calculated from data of Pivnick et al. 1969.
[b]% NaCl × 100/% NaCl + H_2O.
Source: Hauschild and Simonsen, 1985.

provided the first effort to bring together and relate previous qualitative studies through quantitative methods. The data of eight inoculated pack studies with pasteurized canned comminuted pork were related to one another in terms of P. An extra safety margin due to the presence of erythorbate could be quantified. When no erythorbate was added to the product with 50 ppm nitrite, P of toxigenesis was $>10^{-3}$. Addition of erythorbate decreased P to $<10^{-5}$, an additional 2 DR inhibition for *C. botulinum* outgrowth (Hauschild, 1982).

Figure 2 contains data on wieners, chicken frankfurters, turkey roll (ham), and liver sausage preserved with nitrite, NaCl, and sugar (Hauschild, 1982). Products with ≤4% brine were not very inhibitory to the outgrowth of *C. botulinum* in the presence of 50 or 100 ppm nitrite; wieners with 4.5–4.8% brine presented *C. botulinum* with a considerable hurdle, even without nitrite. Fermentable carbohydrate in wiener formulation may have augmented the hurdle offered from competing microorganisms. Turkey rolls are usually prepared without nitrite in the United States and Canada and are vacuum-packaged for retail sale, as are some wieners, liver sausages, bratwurst, and other products. Thus,

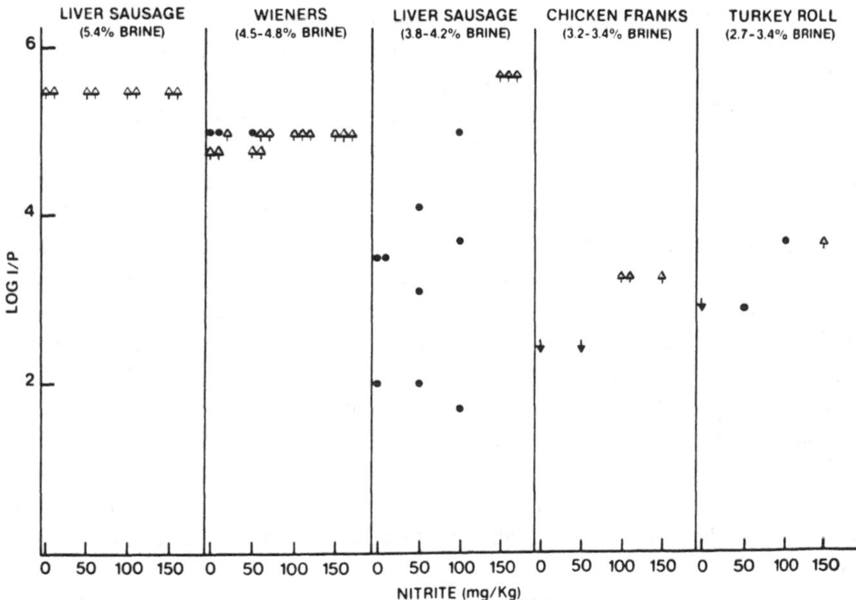

Figure 2 Vacuum-packaged meats and sausages: number of *C. botulinum* spores required for one spore to grow out and produce toxin (1/P) during temperature abuse (one week at 27°C). (↑), data with sign >; (↓), data with sign <; (•) finite values. (From Hauschild, 1982.)

fewer than 1200 *C. botulinum* spores were necessary to overcome the NaCl hurdle (2.7–3.4% brine) in turkey rolls. Overall, these products were much less inhibitory to *C. botulinum* growth than other products, such as vacuum-packaged bacon, which were shown by many researchers to have a much higher degree of safety (average $P < 10^{-6}$).

The estimation of P was more precise in liver sausage studies where the sausage was inoculated with decimal dilutions of *C. botulinum* types A and B (Hauschild et al., 1982). The protective effects of brine (approx. 4%) and nitrite concentrations between 0 and 150 ppm revealed that during temperature abuse toxigenesis was delayed with decreasing spore challenge and increasing nitrite input. Acceptable protection ($P < 2 \times 10^{-6}$) justified the consistent use of 150 ppm nitrite.

Lund and coworkers used P to summarize their findings in three separate studies (Lund and Wyatt, 1984; Lund et al., 1984, 1985). In the first two studies, P was calculated from the reciprocal of MPN estimates (Hurley and Roscoe, 1983; Meynell and Meynell, 1965) for the number of spores required to initiate growth under optimal anaerobic conditions divided by the estimated number required for outgrowth under test conditions. The Bayesian method (Mood et al., 1974) was used to calculate 95% confidence limits, based on the likelihood function of the $\log_{10}$ number of spores required to produce growth. In the third study, P was calculated according the method of Hauschild (1982).

P for *C. botulinum* growth under optimal conditions at Eh −400 millivolts (mV) was similar to media with Eh +60 mV when adjusted with air, while Eh levels above +122 mV produced 5 DR. The interaction of Eh and NaCl on P was quantified. Media with 3.25% NaCl at Eh −400 mV gave 2 DR, while raising the Eh to between +62 and +122 mV resulted in 6 DR (Lund and Wyatt, 1984). To develop further control of Eh in their experimental system, the second paper addressed the effect of the partial O_2 pressure (Lund et al., 1984). Partial pressures (pO_2) greater than 4.5×10^{-3} atm (approx. Eh +217 mV) affected P, reaching 4 to >5 DR at pO_2 $1.1–1.6 \times 10^{-2}$ atm (Eh +271–294 mV) (Lund et al., 1984). The third study assessed P of *C. botulinum* types A, B, nonproteolytic type B, and type E to germinate and initiate growth with toxin formation as related to the storage temperature (16, 12, 8°C) and pH (5.1, 4.8, 4.5) over an 8-week incubation period (Lund et al., 1985). Tabulated P values again allowed the reader to compare the effect of each combination; however, neither the relative contribution of each variable nor their interactions were accounted for in the final analysis.

Montville (1984) used the expression "plating efficiency" to evaluate the interactive effect of pH and NaCl concentration on the growth of *C. botulinum*:

$$\text{Plating efficiency} = \frac{\text{CFU/ml}}{\text{cells/ml}} \times 100 \qquad (7)$$

where CFU (colony forming units) refers to outgrowth enumerated and cells to inoculum level. A plating efficiency of 0.1% (10 CFU from an inoculum of 10^4 spores) was demonstrated on pH 6.5 media with 3% NaCl. This approach is synonymous with expressions of DR and P, which quantify the subpopulations of microorganisms in any given inoculum which may be more resistant to intrinsic (i.e., NaCl, pH) and/or extrinsic (i.e., heat, cold) factors in their growth environment (Ito and Chen, 1978; Riemann, 1967).

IV. MODELING THE PROBABILITY OF *C. BOTULINUM* GROWTH

The use of challenge studies to evaluate individual effects such as temperature, NaCl, nitrite, etc., have yielded a wealth of growth/no-growth information about various preservation systems. Qualitative studies illuminate trends but may not answer questions not included in the original experimental design. The ability to quantitatively compare research efforts on all foodborne pathogens (and food spoilage) is problematic. Ingram (1973) made the challenge: "What we need . . . is not more incubation pack experiments, but a rationale for interpreting them."

Microbiologists have long recognized the interaction of formulation, processing, and storage affects on the growth of pathogens. Many kinetic models have been used to describe the behavior of microbial populations. These models consider population growth and decline as a function of uniform environmental conditions, such as in controlled laboratory fermenters (Howell, 1983). Investigators have attempted to predict the shelf life or safety of foods with regard to *C. botulinum* by observing the fate of organisms inoculated in media or foods. Such challenge studies were limited to the discrete conditions encompassed in each test (Baird-Parker and Freame, 1967; Emodi and Lechowich, 1969; Ohye and Christian, 1967; Ohye et al., 1967; Pivnick et al., 1969; Segner et al., 1966). Recognition of the intricate relationships involving the intrinsic and extrinsic factors led to the proliferation of multifactorial experimentation.

A. Descriptive Analysis of Inhibition

Tables and graphic forms of descriptive analysis can convey the dynamic interactions of a limited number of conditions over a specific range on the growth of *C. botulinum*. However, with each variable added to a study, the number of combinations increases exponentially, while figures are limited to three dimensions. Figures and tables do not lend themselves to precise estimation between data points either, nor can experimental error or hidden functional relationships (Draper and Smith, 1981) be readily expressed.

Extensive tables may be utilized to condense results from factorial studies. For example, Baird-Parker and Freame (1969) had to use five tables to demonstrate the growth of *C. botulinum* (types A, B, or E) spores and vegetative cells affected by combinations of lowering a_w (0.997–0.890, with addition of NaCl or glycerol) in media with pH 5–7 during incubation at 20 or 30°C. Each of these tables demonstrated how the minimum a_w permitting germination or growth depended on the pH, storage temperature, and type of humectant incorporated for each *C. botulinum* type tested. Correlated sets of figures can also be used to relate the response of an organism to various environmental growth conditions. Siegel and Metzger (1980) used three such figures to relate the effects of gas sparging, temperature, and added glucose on growth and toxin production of *C. botulinum* type B under controlled conditions in a fermenter.

Bean (1983) presented three-dimensional surface response graphs by Braithwaite and Perigo (1971) (Fig. 3) to demonstrate the effects of pH and a_w on the survival and outgrowth of spores heated in a nutrient agar system. Results were reported as the most severe F_0 equivalent allowing spore outgrowth (assuming z = 10°C). The same data were presented in a more readable form using isopleths (Fig. 4).

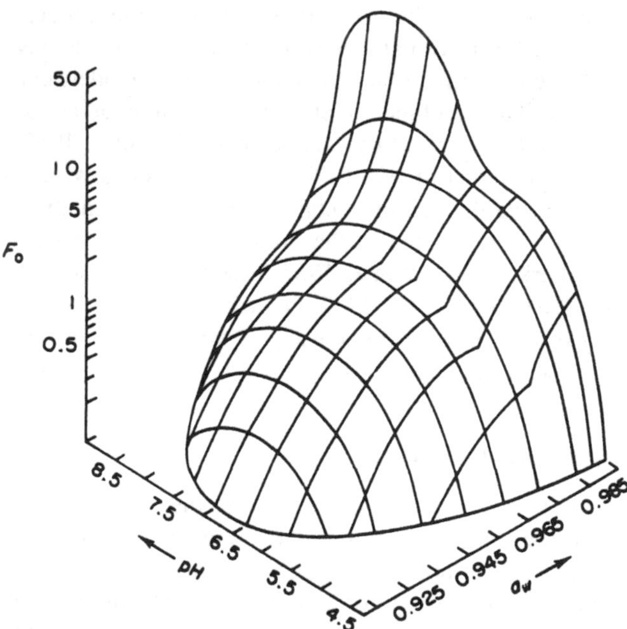

Figure 3 Isometric illustration of the combined effect of pH and a_w on the limiting F_0 value for combined suspension of culture collection strains of *Bacillus* spp. (From Braithwaite and Perigo, 1971.)

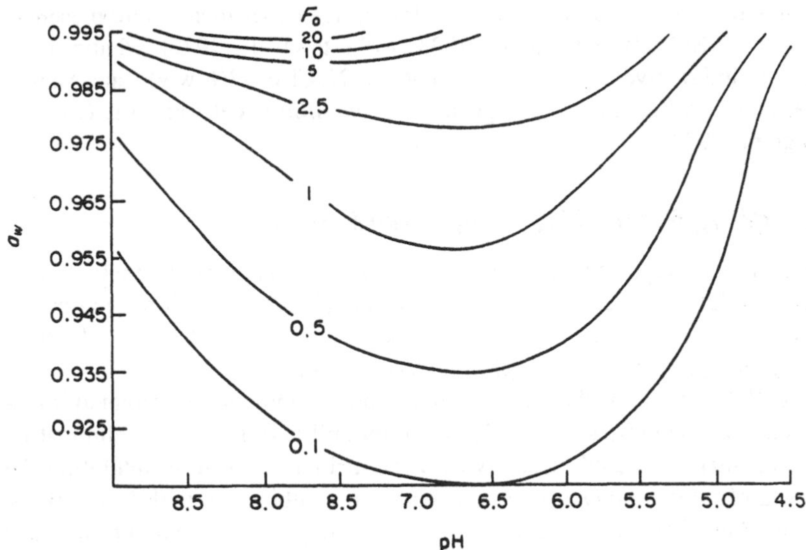

Figure 4 Combinations of pH and a_w that result in various limiting F_0 values for culture collection strains of *Bacillus* spp. (From Braithwaite and Perigo, 1971.)

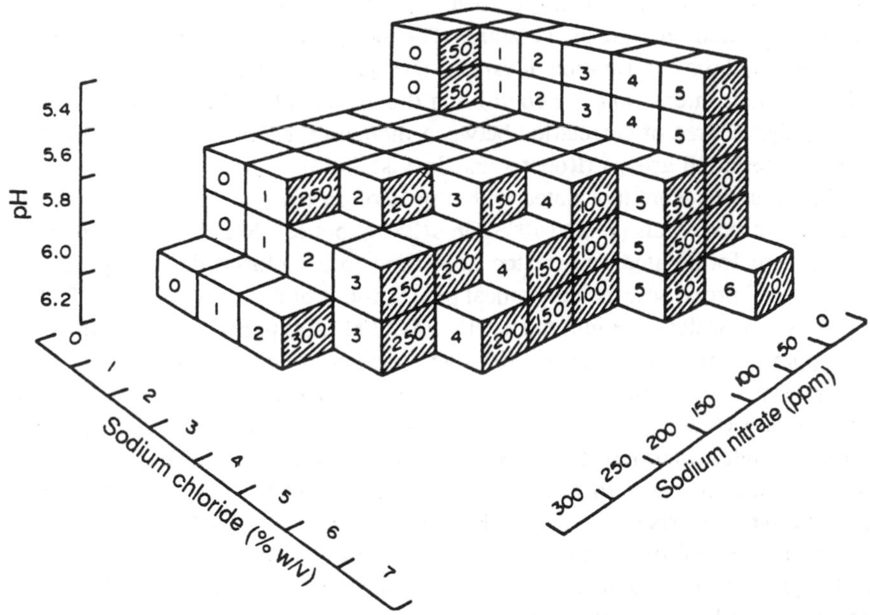

Figure 5 The effect of pH, NaCl, and $NaNO_2$ on growth of *C. botulinum* type A (NCTC 3806) at 35°C. (From Roberts and Ingram, 1973.)

Results from a very large series of collaborative experiments were presented as ecograms (Roberts and Ingram, 1973; Roberts et al., 1976). Figure 5 describes the interactive effect of pH (5.4–6.2), NaCl (0–7% w/v), and $NaNO_2$ (0–300 ppm) on the growth of vegetative *C. botulinum* cells at 35°C (Roberts and Ingram, 1973).

B. Modeling Probability Independent of Time

Perhaps the first reported use of factorial design to quantify interactions in a working model was presented by Adams et al. (1952). A multiple regression equation was used to relate the quality of chopped beef after storage to the thermal process, spore load, and level of added subtilin.

In the 1970s, concern about nitrosamines and carcinogenesis prompted investigators in many countries to investigate the feasibility of limiting or eliminating the use of nitrite in cured meats. Nitrite is a primary hurdle in inhibiting the outgrowth of *C. botulinum* spores, especially in combination with NaCl. Botulism control by nitrite and the controversy concerning its use in cured meats have been reviewed (Roberts et al., 1981a; Sofos et al., 1979).

A common experimental approach (Rhodes and Jarvis, 1976) was used by researchers at the British Meat Research Institute in Langford and the Leatherhead Food R.A. in Surry to determine the minimum effect nitrite concentration, based on its interactions with pH, NaCl, other selected preservatives, thermal processing, and storage temperature. The design and analysis quantitatively demonstrated the contribution of various intrinsic and extrinsic factors, i.e., hurdles and their interactions, in controlling the growth of *C. botulinum*, and there was a high degree of correlation between modeling results obtained by the two separate research groups (Roberts and Jarvis, 1983).

Pork slurries were inoculated with a pool of proteolytic *C. botulinum* A and B spores at two levels (10 and 10^3 spores/28-ml bottle). Slurries, five replicates each, were formulated and treated according to a factorial design of realistic commercial processes, including meat pH (5.5–6.3 or 6.3–6.8), NaCl (2.5, 3.5, 4.5% w/v in water), sodium nitrite (0, 100, 200, 300 ppm), sodium nitrate (0, 500 ppm), sodium isoascorbate (0, 1000, etc. ppm), polyphosphate (0, 0.3% w/v slurry), potassium sorbate (0, 0.26% w/v), heat process [0–70°C internal temperature (low), 70°C IT for 1 hr (high)], and storage at 15, 17.5, 20, and 35°C. Samples were tested for toxin upon observation of gas production, proteolysis, and change in meat coloration, or at the end of the six-month incubation period (Roberts et al., 1981a, b, c; Robinson et al., 1982).

The number of toxic replicates at each combination level tested was recorded as the proportion (p) of nontoxic to toxic samples. Data based on discrete counts are described by a binomial distribution (Draper and Smith, 1981; Jarvis, 1988; Sokal and Rohlf, 1981). Equality of variances, also referred to as homogeneity of variances or homoscedasticity, in a set of samples is an important precondition for several statistical tests based upon normal distributions. Using a Bartlett's test (Sokal and Rohlf, 1981), the error structure for the observed proportions of toxic samples was determined not to be homoscedastic (Roberts et al., 1981a). The heterogenous proportions were subsequently transformed to y, correcting them to a homogeneous normal error structure:

$$y = \frac{180}{\pi} \sin^{-1} p^{1/2} \tag{8}$$

Consequently, the analysis of variance was used initially to identify significant discrete (i.e., single-level) combinations of factors and interactions.

All individual factors significantly affected toxin production, although some levels were either not significant (e.g., the difference between zero and low-heat treatments) or were deleted from the reported analysis to make the overall experiments more closely related to naturally occurring levels (e.g., discontinued inoculation at the 10^3 spore level). Most interactions between two variables acted somewhat antagonistically, as their combined effect was greater than either factor acting alone but less than the sum of their individual effects. No synergistic interactions were found (Roberts et al., 1981a, b). Although analysis of variance indicated several three-factor interactions, these were difficult to justify and may have been as significant as other sources of experimental variation (Gibson et al., 1982).

Analysis of variance, and least significant difference tests, were used to fit linear models to identify factors and up to three interactions significantly affecting the transformed proportion of toxic samples:

$$y_{ijk\ldots rst} = \frac{180}{\pi} \sin^{-1} p^{1/2}{}_{ijk\ldots rst} \tag{9}$$

where (1) $ijk \ldots rst$ represents the level of discrete or continuous variables and (2) $y_{ijk\ldots rst}$ = mean + $(\text{nitrite})_i$ + $(\text{nitrate})_j$ + $(\text{polyphosphate})_k \ldots (\text{NaCl})_r$ + $(\text{isoascorbate})_s$ + $(\text{heat treatment})_t$ + $(\text{nitrite} \times \text{nitrate})_{ij} \ldots$ + $(\text{isoascorbate} \times \text{heat treatment})_{st} \ldots$ + $(\text{NaCl} \times \text{isoascorbate} \times \text{heat treatment})_{rst}$ + $\epsilon_{ijk\ldots rst}$. The expected proportion of toxic replicates was calculated by substitution into the inverse of the angular transformation used previously (Eq. 10). Comparison of observed and predicted proportions allowed preliminary screening of the *C. botulinum* inhibition due to individual treatment combinations.

$$p_{ijk\ldots rst} = \sin^2\left(\frac{\pi}{180} y_{ijk\ldots rst}\right) \tag{10}$$

Logistic regression (Plackett, 1974) was then used to model the binomial data (p) in these experiments directly (Roberts et al., 1981c, 1982; Robinson et al., 1982). The logistic model of probability has the following form:

$$P_{ijk\,\ldots\,rst} = \frac{1}{1 + e^{-\mu}} \qquad (11)$$

where $P_{ijk\ldots rst}$ is the probability of toxin production under the combination of treatments defined by the subscripts ijk...rst. The linear predictor, μ, is expressed by an equation that includes the individual terms of regression on factors and their interactions, which vary over the experimental range of each factor. Thus, combinations and interactions of intrinsic and extrinsic factors that minimize the probability of toxigenesis will be reflected by reducing the value of μ (Fig. 6). The logistic regression calculates predictor coefficients able to weight the overall effect of each continuous variable $_{ijk\ldots rst}$ (e.g., $\beta_{nitrite} \times$ quantity of nitrite), in terms of the status of the discrete variables $_{ijk\ldots rst}$ included (factors

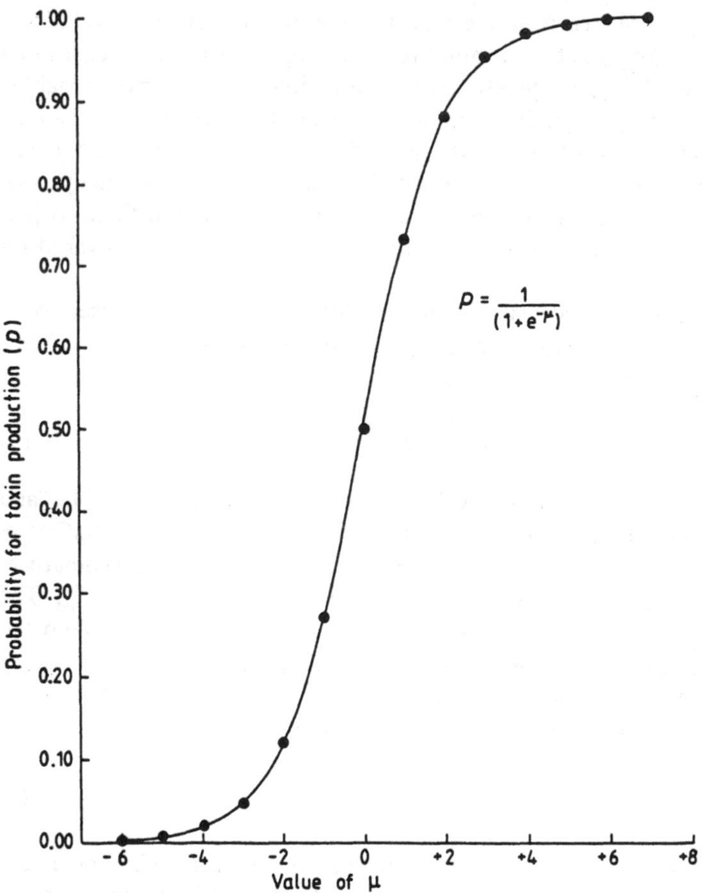

Figure 6 Relationship between the values of exponential function (μ) and the probability for toxin formation (p). (From Jarvis and Robinson, 1983.)

present at only two levels or either "absent" or "present"), which act together to affect the probability of toxin development (i.e., $P_{ijk\ldots rst}$). Figure 6 demonstrates that the probability of toxigenesis is an increasing function of the linear predictor (μ). Thus, the influence of any single factor on toxin production should be linear when plotted against μ (Roberts et al., 1982). Nonlinearity was demonstrated for storage temperature (Fig. 7) (Roberts et al., 1982). This observation and a similar, though nominal nonlinear effect due to NaCl led to the inclusion of quadratic transformations for both variables (i.e., $\beta_T \times T^2, \beta_{NaCl} \times NaCl^2$). The importance of adding these second-order variables to the logistic regression models was evident in their ability to decrease the amount of unexplained variation. Final models were presented for the effects of pig breed, cut and batch slurries (Gibson et al., 1982), effect of sorbate (Roberts et al., 1982), and "low"– and "high"–pH pork (Robinson et al., 1982). Equation 12 is an example of the predictive model to derive the linear predictor (μ) for slurries made with low–pH pork.

$$
\begin{array}{lll}
\mu = & -2.809 & \\
& -(3.364 \times N) & N = NaNO_2\ (\text{ppm} \times 10^2) \\
& +(0.8194 \times T) & T = \text{storage temp. } (^\circ C) \\
& -(1.926 \times S) & S = \text{NaCL (\% w/v on water)} \\
& +(1.915) & \text{if inoculum 1000 spores/bottle} \\
& +(2.360) & \text{if heat treatment (heat) LOW} \\
& -(0.08305) & \text{if heat HIGH} \\
& -(2.322) & \text{if 1000 ppm isoascorbate added} \\
& +(2.672) & \text{if 0.3\% w/v polyphosphate added} \\
& -(4.304) & \text{if 500 ppm } NaNO_3 \text{ added} \\
& -(4.329 \times N \times S) & \\
& -(0.01439 \times T^2) & \\
& -(0.05057 \times T) & \text{if heat LOW} \\
& -(0.02103 \times T) & \text{if heat HIGH} \\
& -(0.2765 \times S) & \text{if heat LOW} \\
& +(0.1413 \times S) & \text{if heat HIGH} \\
& -(0.5543) & \text{if heat LOW w/isoascorbate} \\
& +(0.05577) & \text{if heat HIGH w/isoascorbate} \\
& -(0.4831) & \text{if heat LOW w/polyphosphate} \\
& -(1.597) & \text{if heat HIGH w/polyphosphate} \\
& +(1.07) & \text{if heat LOW w/}NaNO_3 \\
& -(1.703) & \text{if heat HIGH w/}NaNO_3 \\
& +(1.275 \times N) & \text{if w/}NaNO_3 \\
& +(0.3755 \times N) & \text{if w/isoascorbate} \\
& +(0.3320 \times S) & \text{if w/isoascorbate} \\
& -(0.3745 \times S) & \text{if w/polyphosphate} \\
& +(0.5218 \times S) & \text{if w/}NaNO_3
\end{array}
\tag{12}
$$

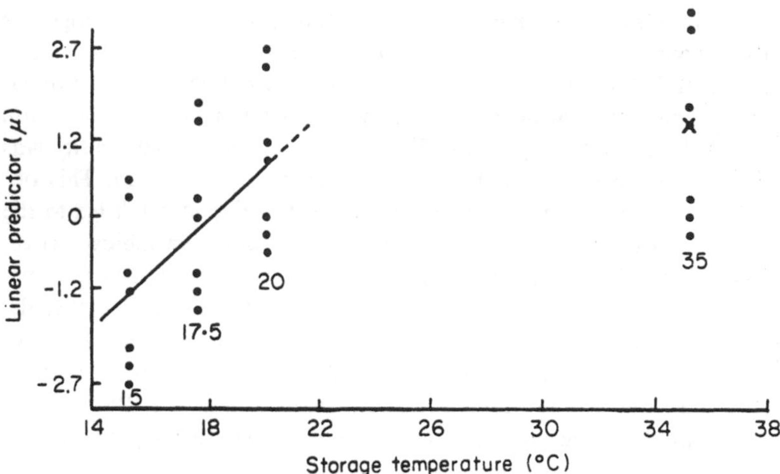

Figure 7 The effect of storage temperature on the linear predictor (μ). ×, Median point. (From Roberts et al., 1982.)

In previous studies heat treatment alone was not significant (Roberts et al., 1982), but it was included in Eq. 12 because it was involved in several significant interactions. Although several three-way interactions were significant, their overall influence on the probability of toxin production (i.e., contribution to explanation of variability) was minimal. Therefore, they were not included (Robinson et al., 1982). Once the model had been fitted to the experimental variables, the predicted estimates for P could be calculated within the range of each parameter tested. Extrapolation outside these limits may lead to considerable error.

The effects of treatments are reported as the probability (%) of toxin production by *C. botulinum* type A and B in pork slurries. Low-pH pork inoculated with 10^3 spores/bottle at the lowest NaCl level and given the high heat treatment required the highest level of $NaNO_2$ to keep the average P below 10%. The contributions of polyphosphate, sodium isoascorbate, and $NaNO_3$ were not evident. At the intermediate NaCl level, nitrite was still needed at its highest concentration, although the average P across all treatments and levels had decreased. The highest NaCl treatment further decreased P, as did increasing $NaNO_2$ levels and decreasing storage temperatures. P did not fall below 1% in this experiment (Robinson et al., 1982).

The contribution of individual hurdles and their interactions can be assessed statistically using an analysis of deviance. Similar to stepwise regression, the logistic model is fitted by first including the constant term, followed by including additional terms one by one, retaining the last term only if a significant reduction in the unexplained variation (residual deviance) occurs. The process is continued until no further reduction of unexplained variation is accomplished by further addition of independent variables or their interactions (Roberts et al., 1982). The magnitude of the deviance reflects the relative contribution of factors or interactions significantly affecting toxin production. One such analysis of a factorial experiment to evaluate the effect of potassium sorbate in a pork slurry system containing nitrite (40 ppm) showed that storage temperature was on average more than twice as important as any other single factor (18.9% of explainable variance) and was probably responsible for most of the significance of interactions, which included storage temperature. The pH accounted for 11.5%, and its interactions with storage temperature was the second most important hurdle, followed by potassium sorbate (8.1%) and NaCl (7.0%). Inoculum level (4.9%) also influenced the model significantly. Other interactions not related to storage temperature accounted for an additional 4.8%. Thus, the logistic model describing P, the portion of pork slurry samples expected to become toxic during 6 months of storage, accounted for 58.4% of the unexplained variation among observations (Roberts et al., 1982). As noted above, inclusion of quadratic terms for temperature, NaCl, etc. decreased the overall amount of unexplained variation in later analyses.

Companion research conducted at Leatherhead R.A. was initially analyzed by a risk factor assessment scheme (Jarvis et al., 1979; Farber, 1986). The authors acknowledged the imprecision of this approach and reanalyzed their data according to the methods used by the Meat Research Institute (Roberts et al., 1981 a,b,c,, 1982). Experimental designs, models based on logistic regression, and a comparison of results between Leatherhead R.A. and the Meat Research Institute have been summarized (Jarvis and Robinson, 1983; Roberts and Jarvis, 1983).

Data on the interactions between NaCl, sodium nitrite, and storage temperature on *C. botulinum* toxin development in pasteurized pork slurry (Jarvis and Patel, 1979) were subsequently fit with a growth-equilibrium model by So et al. (1987). Since the plots of % toxicity vs. temperature at the specific levels of NaCl and $NaNO_2$ tested were exponential in form, the following equation was used:

$$Y = \beta_0 \left[1 - e^{-\beta_1(T - T_0)}\right] \quad (13)$$

where Y is % toxicity, β_0, β_1, T_0 are regression coefficients, and T is the storage temperature. The final model expressed β_1 and T_0 as an exponential-type function, rather than a polynomial (So et al., 1987). Kempton et al. (1987) developed this mathematical procedure further, and discussed statistical tests, e.g., residual sum of squares (amount of variation explained), coefficients of correlation (dependents between variables), and examination of residuals (difference between observed and predicted), to evaluate the aptness of a fitted model.

Regression analysis was applied by Dymicky and Trenchard (1982) to develop a mathematical relationship describing the minimum inhibitory concentrations (MIC) of saturated aliphatic acids (C_1 to C_{20}) and *n*-alkyl formate, acetate, propionate, and butyrate esters (C_1 to C_{20}) needed to inhibit growth of *C. botulinum* 62A. Vegetative cells were exposed to seven concentrations of approximately 105 chemicals. The effect of carbon chain length (R) was a significant index of inhibition and could be expressed by an individual polynomial for each chemical group. Equation 14 demonstrates an example derived for the fumerates:

$$\ln \text{MIC} = 2.06 + 1.84\,\text{R} - 0.28\,\text{R}^2 + 0.01\,\text{R}^3 \tag{14}$$

Pflug et al. (1985) developed a model for $F_{121°C}$ as a function of pH between 4.6 and 6.0 by replotting D-value data comparing the effect of pH on *C. botulinum* type A in three food substrates (Xezones and Hutchings, 1965). A three-factor analysis of variance (ANOVA) was used to fit log $F_{250°F}$ to pH and temperature (T). The authors concluded that the lack of interactions between food and pH or temperature indicated that the function between log F_0 and temperature was linear, and the slope was independent of food type and pH level. Thus, a z-value of 17.8°F would be valid, allowing F-values to be predicted within the range of experimental data. A safety term was added to the final model to increase the $D_{250°F}$ at pH 6.0 from 0.12 to 0.25 minutes. The following model is able to predict a variable processing time yielding 12-D for *C. botulinum* at pH 6–7, given the generally accepted minimum botulinum process of 3.0 minutes (Pflug and Odlaug, 1978):

$$\begin{aligned} \log F_{250°F} = \log 12D &= 11.544 + 0.897(\text{pH}) - 0.0573(\text{pH}^2) \\ &\quad - 0.0586(\text{T}) + 0.264 \end{aligned} \tag{15}$$

where $4.6 < \text{pH} < 6.0$ and $220°\text{F} < \text{T} < 255°\text{F}$.

A table of F-values from 3.0 to 1.2 minutes for products ranging in pH from 7.0 to 4.6 was presented by Pflug et al. (1985) as a general guide for adjusting minimum safe thermal-processing standards. Application of Eq. 18 to other food products assumes that an F of 3.0 minutes is safe at pH 6.0–7.0 and that the acidulant used to adjust the pH and the relation between D-value and pH for the food in question are similar to the data of Xezones and Hutchings (1965).

V. PREDICTION OF *C. BOTULINUM* LAG PHASE AND GROWTH

Multifactorial design allows discriminate (classifications) and continuous (numerical) variables to covary within predetermined parameters. Various types of mathematical models incorporate significant independent variables into equations to predict the outcome of a dependent variable.

A. Probability as the Only Dependent Variable

The combined inhibitory effects of NaCl, disodium phosphate (Na_2HPO_4), moisture, and pH on *C. botulinum* were studied in pasteurized processed cheese spreads by Tanaka et al. (1986). The ability of types A and B spores (10^3/g sample) to produce toxin yielded growth/no growth data in 304 treatments, incubated at 30°C and sampled for up to 42 weeks. Logistic regression was used to model the effect of cheese spread formulation on the binary response variable (toxic or nontoxic). The logistic model was able to incorporate the contribution of main factors, their transformations (squares, etc.), and interactions to predict the proportion of positive samples at a specified time.

NaCl, Na_2HPO_4, pH, and moisture significantly affected *C. botulinum* toxin production. Statistical analysis demonstrated that the effects of NaCl and Na_2HPO_4 were additive at the levels studied, and these are represented in Figure 8, which plots the estimated boundary between inhibitory formulations and combinations supporting toxigenesis. The following mathematical basis for plotting curves predicting *C. botulinum* in cheese spreads were obtained from an unpublished report (Food Research Institute, University of Wisconsin, Madison,

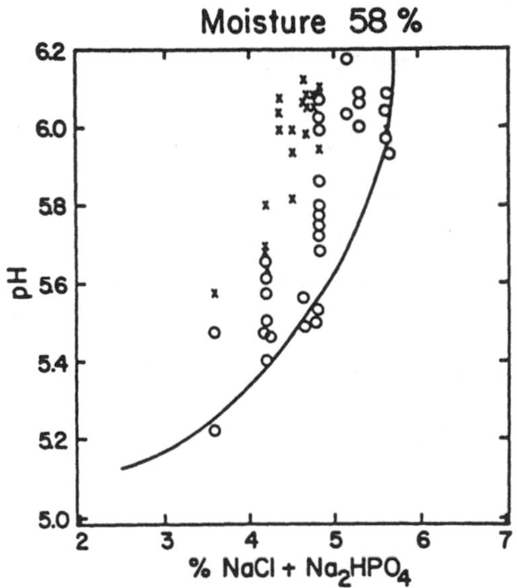

Figure 8 Formation of botulinum toxin in cheese spread incubated 42 weeks at 30°C. Moisture content, 59%. The solid line represents the boundary between the treatment combinations that would either inhibit or allow toxin production. A quadratic mathematical model was used to locate the line. (O), Batches without toxin; (×), batches with toxin. (From Tanaka et al., 1986.)

WI). A second-order model was used to express the most conservative estimate for toxicity development, when NaCl and phosphate were formulated as a percentage of the whole:

$$\begin{aligned}\ln(p/1 - p) = \ln(0.2/0.8) = \ln 0.25 = -1912 + 22.43\ W - 95.63\ N_2 \\ - 21.11\ P_2 + 449.5\ pH + 6.52\ T - 0.096\ W^2 - 3.80\ N_2{}^2 \\ - 28.88\ pH^2 - T^2 + 0.833\ W \cdot N_2 + 0.383\ W \cdot P_2 \\ - 2.11\ W \cdot pH - 2.88\ N_2 \cdot P_2 + 10.75\ N_2 \cdot pH \\ - 0.690\ N_2 \cdot T - 1.73\ P_2 \cdot T + 1.15\ pH \cdot T \qquad (16)\end{aligned}$$

where

p = probability of toxin development (i.e., $p = 1/5 = 0.2$)
T = ln(number of weeks to first toxicity)
W = % moisture
N_1 = % NaCl in water phase
N_2 = % NaCl in whole product
P_1 = % Na_2HPO_4 in water phase
P_2 = % Na_2HPO_4 in whole product

Treating the combined effects of NaCl and Na_2HPO_4 as a single factor reduces Eq. 6 to Eq. 7:

$$\begin{aligned}\ln(p/1 - p) = \ln 0.25 = -2134 + 29.20\ W - 91.97\ (N_2 + P_2) \\ + 484.6\ pH + 16.53\ T - 1.205\ T^2 - 29.17\ pH^2 \\ - 2.565\ (N_2 + P_2)^2 - 0.1342\ W^2 - 0.1174\ W \cdot T - 2.846\ W \\ \times pH + 1.084\ W \times (N_2 + P_2) + 7.650\ ph \times (N_2 + P_2) \\ - 0.5089\ T \times (N_2 + P_2) \qquad (17)\end{aligned}$$

Equation 16 may be rearranged to solve for T as time in weeks to first toxicity:

$$0 = a \times T^2 + b \times T + c$$

where

$a = 1.21$
$b = 6.52 - 0.690\ N_2 + 1.15\ pH$
c = sum of terms that do not involve T, which = ln 0.25 = $1912 + 22.43\ W - 95.63\ N_2 - 21.11\ P_2 + 449.5\ pH - 0.096\ W^2 - 3.80\ N_2{}^2 - 28.88\ pH^2 + 0.833\ W \times N_2 + 0.383\ W \times P_2 - 2.11\ W \times pH - 2.88\ N_2 \times P_2 + 10.75\ N_2 \times pH$

Thus,

$$T = \ln(\text{time to ontset of toxicity in weeks})$$
$$= \frac{-b + (b^2 - 4 \times a \times c)}{2a(\text{number of weeks})} = e^T$$

The effects of pH, undissociated sorbic acid, and storage temperature on P have also been modeled in a series of three papers (Lund et al., 1987a, b, 1990). Inoculation of vegetative *C. botulinum* cells was intended to maximize their probability for growth as compared to spores that would have to germinate first. The MPN method was used to enumerate growth over time at 30°C in nutrient media, and probability (P) of growth of a single bacterium (Hauschild, 1982) was used as the dependent variable for modeling.

Even meticulous experimental design, especially with regard to maintenance of strict anaerobic conditions (redox potential of −300 mV or lower, at pH 7), led to significant variation for P over time between replicate experiments. Random-effects regression models (Racine and Moppert, 1984; Stiratelli et al., 1984) were chosen to relate P to pH, incubation time, and temperature, etc. These models were fit in two stages. One stage modeled the kinetics of the

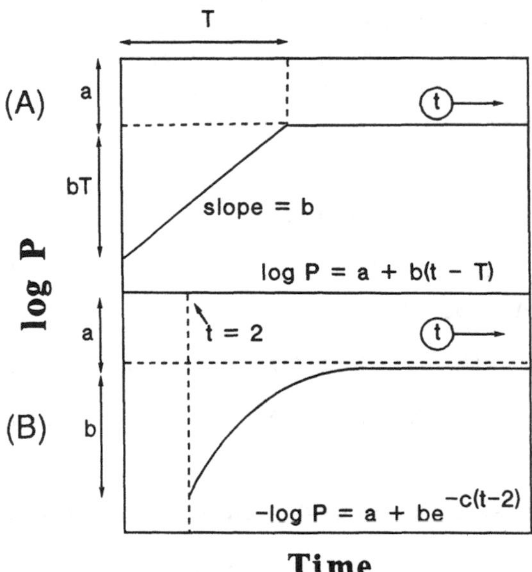

Figure 9 (A) Probability model on linear increase, reaching an asymptote at T, the time of maximum P (based on Eq. 19, Lund et al., 1987). (B) Probability model assuming a smooth (exponential curve) increase until becoming asymptotic to maximum P (based on Eq. 20, Lund et al., 1987b).

increase in P as a function of the maximum P achieved (A), rate of increase of P (B), and incubation time (T) to reach maximum P. The second stage utilized A, B, and T as the dependent variables in individual regression analyses to model the combined effects of pH, undissociated acid, incubation temperature, etc. and their interactions.

In the first stage, observed values for log P and incubation time (t) were plotted for each treatment, i.e., at each pH level (Lund et al., 1987a), or concentration of undissociated sorbic acid (Lund et al., 1987b). In subsequent studies, observed data were fitted to segmented regression lines by a grid search method (Lerman, 1980), from which A, B, and T could be estimated mathematically (Lund et al., 1990).

The observed log P in the first paper (Lund et al., 1987a) increased linearly until a maximum at T, when it became asymptotic to t. A segmented linear model was used to model P (Eq. 19). The model used (Lund et al., 1987b) is different in that it assumes a smooth curve, rather than a point of transition from the linear part to the asymptote (Eq. 20). Figure 9 demonstrates the differences between the models presented in Lund et al. (1987a,b). The slopes of the two models were a function of B and of both B and C, respectively. It is of interest that the authors did not find the logistic models for P used by Roberts et al. (1981c) or log P (%) used by Lindroth and Genigeorgis (1986) adequate. Lund et al. (1990) utilized the segmented model (Eq. 21a, b), as the data appeared to fit a segmented straight line model better than the curved model in Lund et al. (1987b) (Fig. 10). The authors also felt that the parameters of the segmented model are more easily interpreted.

$$\log P = A + B(T - t) \quad \text{for } 1 \le t \le T, \tag{19}$$
$$\log P = A \text{ (otherwise)} \quad \text{where } t > T$$

$$-\log P = A + Be^{-c(t-2)} \quad \text{or,} \tag{20}$$
$$\log P = -A - Be^{-c(t-2)}$$

$$\log P = A - s(X - t) \quad \text{for } t \le X, \text{ equivalet to:} \tag{21a}$$

$$\log P = A + s(t - X) \quad \text{for } t \le X \tag{21b}$$
$$\log P = -A \quad \text{for } t > X$$

Note: Since (T = X) and (B = s), equations used in Lund et al. (1987a) and Lund et al. (1990) are equal.

In the second stage, regression analysis was used to fit models relating the range of independent experimental variables to the random effects of variables, A, B, and T. In some cases, better numerical stability was achieved by performing the regression analysis on a transformation of the dependent variable; examples are shown in Eqs. 22a, b, and c (Lund et al., 1990). Either linear

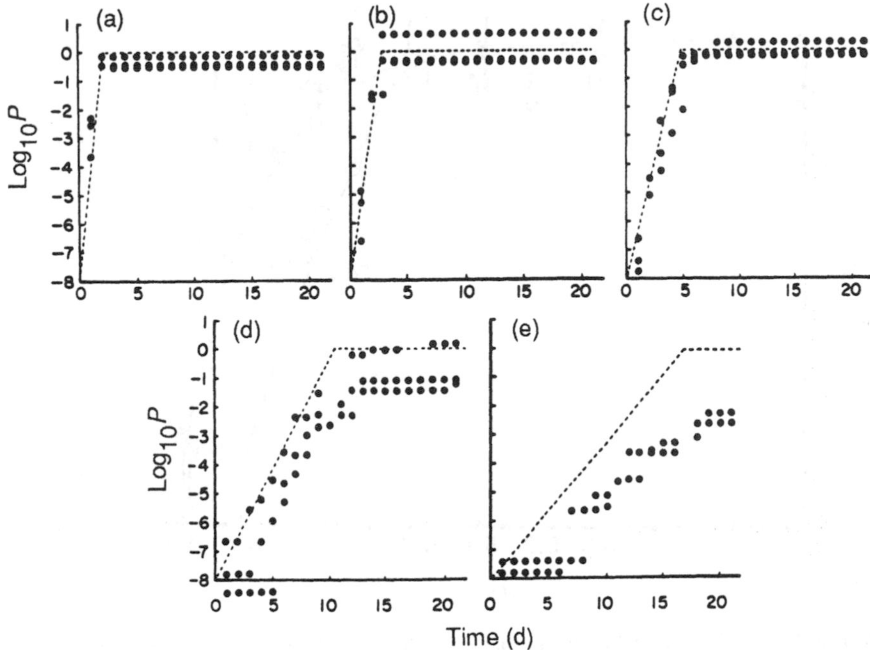

Figure 10 Effect of incubation time on the probability of growth from a single bacterium of nonproteolytic, type B *C. botulinum* at pH 6.8–7.0 and temperatures of (a) 20°C, (b) 16°C, (c) 12°C, (d) 8°C, and (e) 6°C. Dots indicate experimental results; lines, results calculated from the combined model (Eqs. 21a and b). Values for a, T, and b calculated from Eqs. 22a, b, and c for the specific conditions of temperature, pH, and undissociated sorbic acid. 95% confidence limits of each value of $\log_{10} P = \pm 0.72$. (From Lund et al., 1990.)

regression models were employed to find predictive equations for log A, B, and T, or a nonlinear model was used to relate A, B, and T to the exponent of a linear regression of the independent variables. In both cases, the best fit equations were formulated using the significant independent variables (temperature, pH, etc.) and their squares, cubes, and cross-product interactions as needed.

The models to predict the effect of pH alone (Lund et al., 1987a) required six fitted parameters (two parameters for each random variable, A, B, or T) generated from the second stage of analyses. Substitution of pH values into these equations allowed log P to be predicted for incubation times greater than 2 days, at pH ≥4.6, and otherwise optimal conditions. Figure 11 demonstrates the correlation between observed and predicted values after 3 and 14 days of incubation at 30°C.

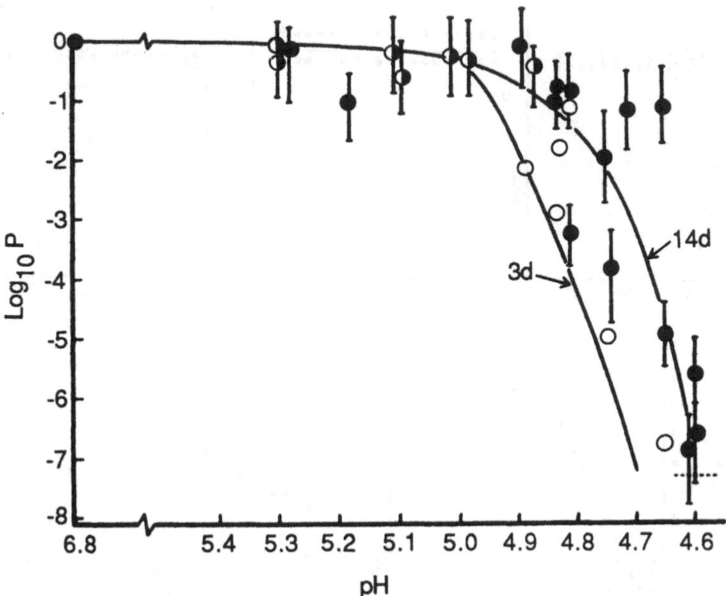

Figure 11 Effect of pH on the probability (P) of growth of *C. botulinum* within 3 days and 14 days at 30°C. Continuous lines predicted from the model based on results for strain ZK3 after incubation for 3 days and 14 days. Symbols represent experimental results. Open circles, after incubation for 3 days; closed circles, after incubation for 14 days. Bars represent 95% confidence limits of $\log_{10}$ P after 14 days. The confidence limits for $\log_{10}$ P 3 days were similar and have been omitted for clarity. (From Lund et al., 1987a.)

Predictive equations generated for the full model based on temperature, pH, and undissociated sorbic acid for A, B, and T are as follow:

$$A = -\exp[-10.894 + 1.502s - 5.979p - 8.229t - 3.610pt + 1.522t^2] \tag{22a}$$

$$T = \exp[1.150 + 1.045s - 1.290p - 1.183t - 0.952st + 1.692pt + 0.753t^2 - 2.374p^2 - 1.150p^3 - 0.134s^2 + 0.314s^2t + 0.825p^2t] \tag{22b}$$

$$B = \exp[1.005 - 2.023s + 3.017p + 0.991t + 0.555st - 0.820pt - 1.359t^2 + 5.101p^2 + 2.231p^3 + 1.588s^2 - 0.459s^3] \tag{22c}$$

where

p = pH 7
t = (temp. − 15)/10
s = calculated concentration of undissociated sorbic acid/100

Figure 10 demonstrates the predictive capabilities of this modeling method, where the random variables (A, B, T) regenerated from regression equations (polynomials) for temperatures between 6 and 20°C. The predicted results in Figure 10d are conservative in that they generally predict higher Ps than those observed (Lund et al., 1990). The authors chose to calculate the slope (B) using Eq. 23, which originated at time zero with P = log −8, rather than modeling the slope and extrapolating to P at time zero.

$$B = \frac{A + 8}{T} \tag{23}$$

Thus, P would be predicted to constantly increase with time. When growth was observed, the model would predict the minimum number of *C. botulinum* cells necessary to initiate growth or the number to overcome the lag period. However, the model was not able to directly handle a lag period, as demonstrated in Figure 10e, where storage at 6°C required approximately 5 days before growth was observed.

B. Fitting Growth Data to Sigmoidal Curve Equations

Plotting the logarithm of bacterial growth (L(t)) against time typically results in a sigmoidal curve (Fig. 12) (Monod, 1949). Zwietering et al. (1990) evaluated several sigmoidal functions for use in modeling bacterial growth curves. Three-parameter models based upon the maximum specific growth rate defined as the tangent to the inflection point, the lag time defined as the x-axis intercept of this tangent, and the asymptote to the maximal cell number achieved were described. These were compared to a four-parameter model, which included an additional term for shape. They concluded that microbial growth data generated by their laboratory and others on a variety of bacteria were fit better by the Gompertz model than by logistic, linear, quadratic, t^{th} power, and exponential models. Three-parameter models were recommended for the following reasons: (1) these parameters could be mathematically redefined to relate biological significance; (2) they were simpler, easier to use, and, since they were less correlated, they were more stable; and (3) estimates from three-parameter models yielded more degrees of freedom, allowing curves to be fitted with fewer observations.

The Gompertz function (Buchanan and Phillips, 1990; Gibson et al., 1987; Jeffries and Brain, 1984) is described by the following:

$$L(t) = A + C \exp[-\exp(-B(t - M))] \tag{24}$$

where

L(t) = log count of bacteria at time (t)
A = asymptotic log count of bacteria as t decreases indefinitely

C = asymptotic log count of bacteria as t increases indefinitely
M = time at which the absolute growth rate is maximal
B = relative growth rate at M (log[cfu/ml]/hour)
e = $\log_2$

The parameters of the Gompertz equation can be used to characterize bacterial growth as follow:

$$\text{lag time} = M - (1/B)$$
$$\text{exponential growth rate} = BC/e$$
$$\text{generation time} = \log(2)e/BC$$

The logistic equation

$$L(t) = \frac{A + C}{1 + \exp(-B(t - M))} \quad (25)$$

is symmetric around M, making it a special case of the Gompertz function, where M can be modeled at any point between asymptotes of increasing and decreasing values of t. Growth kinetics equations can be derived for the logistic

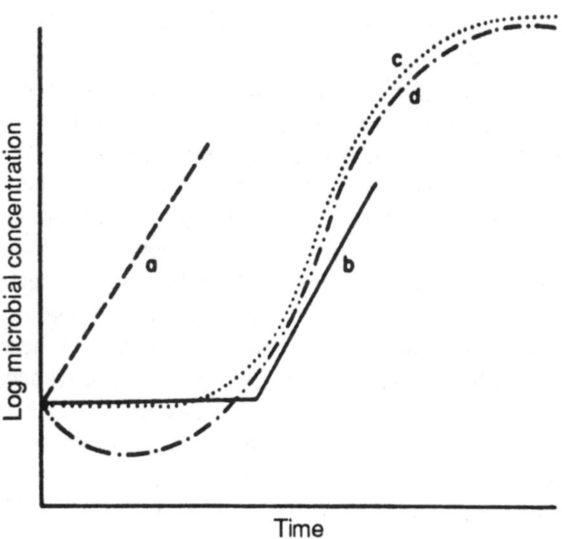

Figure 12 Bacterial growth curves: (a) simple description for constant microbial growth without a lag phase; b) simple description for microbial growth at a constant rate preceded by a lag phase; c) growth curve showing a lag phase followed by growth to a maximum population density; d) growth curve showing an initial drop in viable numbers followed by growth to a maximum population density. (From Baird-Parker and Kilsby, 1987.)

equation in a manner similar to the Gompertz equation (Gibson et al., 1987). Researchers have reported that the specific growth rate of bacteria in culture is in fact not constant throughout the exponential phase, but rather increases to a maximum and then declines (Broughall and Brown, 1984; Gibson et al., 1987; Janson, 1983). The ability of the Gompertz equation to incorporate the time of optimal growth enabled several research groups to produce models applicable to the prediction of bacterial growth (Gibson et al., 1988; Buchanan and Phillips, 1990; Buchanan et al., 1989).

Application of sigmoidal growth curve models to factorial experimental designs is also a two-stage process. Individual growth curves are characterized by parameters for each experimental system in the first stage. Regression analysis is used in the second stage to relate growth curve parameters from individual experiments (as dependent variables) to the independent variables of the given system (i.e., pH, acidulant, inoculum level, etc.). Gibson et al. (1987) attempted to utilize this approach to develop a predictive expression for the growth of *C. botulinum* in pasteurized meat slurry media. Forty-three growth curves were generated from 24 combinations of NaCl (1.5, 2.5, 3.5, 4.5% w/v), four levels of mild heat treatments, and storage temperatures (15, 20, or 27°C) for up to 211 days. Growth was measured by laborious direct agar plating methods, combined with confirmation by detection of botulinum toxin, which precluded replicating each experimental treatment.

Figure 13 demonstrates the reproducibility between replicate growth curves fit to experimental data using a maximum likelihood, nonlinear regression program (Gibson et al., 1987). Widely differing germination rates of individual spores (Roberts and Thomas, 1982) added to the variability of enumerating *C. botulinum* cells in this food system. Relatively few data points could be collected for a number of experimental treatments. If the lag time was very short and the growth rapid, few observations during these periods led to B being infinite and M being "indefinite." Conversely, when the lag time was long and the growth rate was slow, the stationary phase was not attained, and excessive estimates for M and C were possible. In either case, an adequate number of data points, evenly spaced to clearly delineate the shape of the growth curve, is necessary for each treatment to enable accurate modeling. Roberts (1989) suggested that 10–15 observations are necessary to accurately fit an equation to a set of growth data. The Gompertz curve is more sensitive to unevenly weighted data because the curve is asymmetrical about M, while the logistic curve is symmetrical and can be generated from data collected either before or after the time when the absolute growth rate is maximal. The error encountered in fitting growth curves to some treatments precluded accurate estimation of parameters for the Gompertz function, which in turn would have interfered with development of separate predictive polynomial models for the response variables B, M, C, and A as functions of pH, NaCl, temperature, and their interactions.

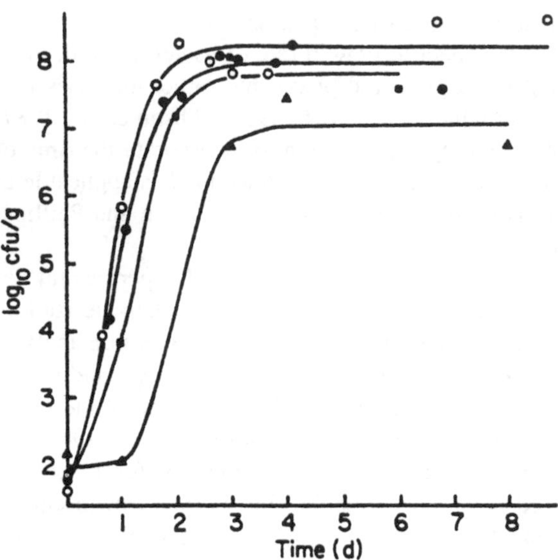

Figure 13 Replicate growth curves of *C. botulinum* type A in pasteurized pork slurry containing 1.5% NaCl, heated for 2 hr at 70°C, and stored at 27°C. Symbols represent separate data sets. (From Gibson et al., 1987.)

Examples of successful applications of the preceding technique are being used with interactive programs on personal computers (Buchanan and Philips, 1990; Gibson et al., 1988). The programs query for pathogen type (*Salmonella* or *Listeria monocytogenes*), intrinsic and extrinsic parameters of the growth environment, and the initial number of cells. In order for the program to determine the lag time, generation time, maximum population, etc., it must estimate the parameters B, M, A, and C. The natural logarithm of each estimate is calculated from the derived polynomial model, based upon the values of pH, temperature, atmosphere, etc. Modeling based upon the natural logarithm transformation ensures that predicted parameters for B, M, A, and C are positive and has decreased variability.

C. Kinetic Growth Models Based Upon the Arrhenius Equation

Derivations of the linear Arrhenius equation have been successfully applied to describe the primary effect of temperature on the growth rate of bacteria (McMeekin et al., 1988), and other effects have been included with increased levels of mathematical complexity. Growth rate is not a constant function of temperature (Mohr and Krawiec, 1980). The relation between the logarithm of lag time and temperature was shown to be nonlinear for *C. botulinum* (Ohye and Scott,

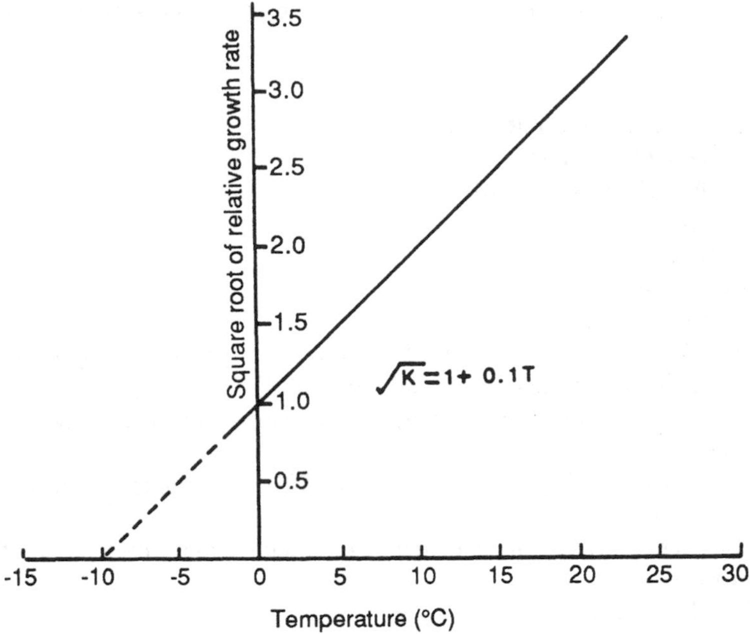

Figure 14 Square root of relative growth rate as a function of temperature for psychrotrophic spoilage bacteria (T_{min} = −10°C). (From McMeekin and Olley, 1986.)

1953). Schoolfield et al. (1981) proposed a nonlinear regression of the Arrhenius equation, while Ratkowsky et al. (1982, 1983) related the square root of growth rate (k) to temperature (Fig. 14).

Neither the Schoolfield nor the Ratkowsky bacterial growth models readily incorporated the inhibitory effects of parameters, other than temperature, or their interactions. Broughall and Brown (1984) were able to use the Schoolfield equation to model the secondary effects of a_w and pH on the growth of *Staphylococcus* only by successive reexpression of these variables in terms of temperature.

A few researchers have applied Arrhenius-based equations to relate temperature to *C. botulinum* growth kinetics. Good agreement between growth of *C. botulinum* at reduced water activities (data from Ohye and Christian, 1967) and predictions calculated with the square root model have been reported (McMeekin et al., 1988; Olley et al., 1987).

Adair et al. (1989) conducted a series of inoculated pack studies, including proteolytic *C. botulinum* in cooked turkey. Growth parameters for lag and generation time were fitted for a minimum of five temperatures. The data were modeled by an alternative form of the Schoolfield equation (Eq. 26) and the square root model (Eq. 27) (Ratkowsky et al., 1983). Both models were written to cover the entire biokinetic temperature range.

$$\ln(K) = A + (B/T) - \ln T + \ln\{1 + \exp(F + D/T) + \exp(G + H/T) \quad (26)$$

where

K = lag or generation time in days
T = temperature in kelvin
A = $\ln 298 - (HA/R \times 298)) - \ln \Theta(25)$
B = HA/R
D = $-HL/R$
F = $HL/(T1/2L \times R)$
$\Theta(25)$ = growth rate at 24°C
HA = constant describing the enthalpy of activation for microbial growth
R = universal gas constant (1.987 cal mol^{-1} k^{-1})
HL = constant describing the enthalpy of low temperature inactivation of growth
T1/2L = constant describing the temperature for 50% low-temperature inactivation of growth
HH = constant describing the enthalpy of high-temperature inactivation of growth
G = $HH/R \times T1/2L$
H = $-HH/R$

$$\sqrt{r} = b(T - T_{min}) \left\{1 - \exp[c(T - T_{max})]\right\} \quad (27)$$

where

r = growth rate constant (or 1/lag or 1/generation time)
b = regression coefficient
T = temperature in kelvin
T_0 = conceptual temperature of no metabolic significance (temperature below which the rate of growth is zero or lag time is infinite)
T_{min}, T_{max} = minimum and maximum temperatures at which the rate of growth is zero or lag is infinite
c = additional parameter to enable model to fit data for temperatures above the optimal temperature

Table 2 compares the observed data with values predicted from both the Schoolfield and square root models. The mean square error (MSE) (Eq. 28) was calculated to quantify differences between observed and predicted data. Smaller MSEs for the Schoolfield equation were demonstrated for each bacterium and medium tested, as compared to the square root model, indicating an overall better fit. The authors recommended use of the Schoolfield model, especially for its greater accuracy at lower temperatures.

Table 2 Comparison of Observed Data from Growth of *C. botulinum* (Proteolytic) in Raw Minced Turkey with Predicitons from the Square Root and Schoolfield Models

	Lag Time (hr)		
Temperature (°C)	Observed data	Schoolfield prediction	Square root prediction
15–0	672–0	622–89	387–81
17–5	154–08	184–42	148–01
25–0	22–22	23–25	32–12
30–0	21–65	21–30	17–47
Mean square error for the fit		683–71	16185–6
Number of observations = 5			

Source: Adair et al., 1989.

$$\text{MSE} = \sum \frac{\{(\text{obs} - \text{pred})^2\}}{\text{n}} \tag{28}$$

where obs is the observed value of the lag or generation time in hours, pred is the predicted value of the lag or generation time in hours, and n is the number of observations in the data set for predicting bacterial growth in foods.

Baird-Parker and Kilsby (1987) also found the Schoolfield equation better than the square root model for predicting growth of *C. botulinum* at low temperatures in minced meat (Table 3).

Table 3 Comparison of Ratkowsky (Rat.) and Schoolfield (Sch.) Models for Growth of *C. botulinum* in Vacuum-Pack Minced Meat

	Lag time (hr)			Generation time (hr)		
Temp. (°C) prediction	Observed data	Sch. prediction	Rat. prediction	Observed	Sch. data	Rat. prediction
15	672–0	622–89	387–81	72–0	75–34	37–73
17–5	154–08	184–42	148–01	20–16	17–67	17–38
20	74–16	64–92	77–47	6–05	7–16	9–95
25	22–22	23–25	32–12	4–54	3–92	4–51
30	21–65	21–30	17–47	2–78	2–97	2–56
Mean square error for fit		1–187	28–100		0–0066	0–4157

Source: Baird-Parker and Kilsby, 1987.

D. Prediction of Both Probability and Lag Time

The initial goal of Genigeorgis and coworkers was to develop a technique to assess the safety of modified atmosphere packaged fresh fish. Their methodology was broken into two basic steps. First, multiple linear regression was used to model the lag time (LT) prior to *C. botulinum* growth or toxigenesis in media, muscle tissue microsystems, or inoculated packs. The effects of spore inoculum level (I), incubation temperature (T), and modified atmosphere of storage (MA) were primarily used to describe LT, the dependent variable. Models for LT were then incorporated into secondary models based upon logistic regression, which related the probability (P) of one *C. botulinum* spore initiating toxigenesis, to the storage time (ST), T, MA, etc.

Experimental Design

Initial studies were conducted in media (Ikawa et al., 1986; Jensen et al., 1987). Factorial design experiments and MPN methodology were used to evaluate the probability (P) of growth initiation of 17 individual proteolytic (A, B, F) and nonproteolytic (B, E, F) *C. botulinum* strains (spores and cells) in a model broth as affected by temperature (4–47°C) and time of incubation (up to 28 days). These modeling techniques produced good correlations between predicted and observed P values in broth studies.

Experiments using fish as the growth substrate were also arranged in a factorial fashion, which included seven 10-fold inoculum levels of 10^4–10^{-2} spores from pools made of individual or multiple nonproteolytic types B, E and F *C. botulinum*. Other continuous variables included up to seven incubation temperatures between 4 and 30°C, the initial level of microbial contamination, which was kept to a minimum by aseptic filleting of very fresh fish, and multiple sampling times during storage for up to 60 days. The discrete variables included two spore pool inocula (equal levels of nonproteolytic B, E and F pools, or an E pool), three MA (vacuum, 100% CO_2, or 70% CO_2 and 30% N_2), three fish species (rockfish, salmon, or sole), and two tissue preparations (muscle homogenate or fillets). Muscle homogenate was inoculated in microsystems (MS) in which approximately 2.5 g of homogenate was loaded into 24-well tissue culture plates. Fish were homogenized together to make composite muscle samples, eliminating "between-fish" sources of variation. Wells were inoculated in triplicate with 20µl of inoculum of 10^4–10^{-2} spores/well (Baker et al., 1990; Garcia and Genigeorgis, 1987; Garcia et al., 1987; Ikawa and Genigeorgis, 1987; Lindroth and Genigeorgis, 1986). Similar methodologies were used to study the safety of vacuum-packaged, processed turkey roll (Genigeorgis et al., 1991).

Calculation of Probability of Growth

The MPN of spores initiating growth and toxigenesis in the highest dilution was calculated from the number of toxigenic samples among the 21 inoculated, using a modification of MPN tables for seven 10-fold × three samples (Fisher and

Yates 1957; Lindroth and Genigeorgis, 1986). The number of spores initiating growth at 30°C in the fish medium was accepted as the baseline inoculum level for statistical purposes. The probability of one spore to exhibit growth and toxigenesis (P) was defined as:

$$P = \frac{\text{final MPN}}{\text{initial inoculum}} \tag{29}$$

When none of 21 samples were toxic, the probability was defined as $10^{-3}\%$. Thus, P was related to spore growth inhibition, expressed as a number of functional decimal reductions (DR) by:

$$P = \frac{1}{\text{antilog of DR}} \tag{30}$$

Statistical Analysis

A linear regression program, BMDP1R (Dixon and Brown, 1983), was used to construct the best fit model for predicting the LT (log LT = a + β_1(parameter 1) + . . . + β_n(parameter n)). Lag time is defined as the time (in days) of incubation to the sampling period prior to the first toxic samples, a was the y-intercept, and β values were linear predictors for each variable. A single model was constructed for each MA, or MA was specified by a dummy variable (0 = vacuum, 1 = 100% CO_2, etc.). The relative effect of each parameter (or hurdle) was described by the magnitude and sign of β. In one example, the β for vacuum and 100% CO_2 would decrease the predicted LT by 0.85 or 0.86 days, respectively. Thus, the mechanics of the linear equation allowed one to calculate the effect of a hurdle upon the outcome of the dependent variable.

Logistic regression was performed by the BMDPLR program to describe P as it increased in relation to the decreasing spore inocula resulting in toxigenesis over the incubation period. The final models predicting P were combinations of linear regression and logistic regression equations (Baker et al., 1990; Garcia and Genigeorgis, 1987; Garcia et al., 1987; Genigeorgis et al., 1991; Ikawa and Genigeorgis, 1987; Lindroth and Genigeorgis, 1986). Logistic regression incorporated all experimental variables to predict P, or the ability of one spore to initiate toxigenesis as related to significant storage variables. The logistic regression used is equivalent to Eq. 11, though expressed in the form:

$$\frac{e^y}{1 + e^y} \tag{31}$$

where

$$y = a + b_1X_1 + b_2X_2 + b_3X_3 \ldots b_iX_i$$
$$a = \text{intercept}$$
$$b_1, b_2 \ldots b_i = \text{regression coefficient of predictive variables}$$
$$X_1, X_2 \ldots X_i = \text{predictive variables (T, ST} - \text{LP, MA)}$$

The time element (ST − LT) refers to the time from the predicted LT through the peak of the exponential phase of toxin production. P values that declined after reaching a maximum were discarded, because toxin had probably deteriorated (Huss et al., 1979).

In many cases there was good agreement between the observed and predicted P values. The ability to predict P of one spore to initiate growth allows one to ascertain how many spores would be needed to initiate toxigenesis in a food product stored under defined conditions. The rate at which P increases expresses the growth rate for *C. botulinum*. The steeper the slope of the logistic regression, the shorter the time needed to attain a detectable level of toxin. However, even low probabilities of *C. botulinum* toxicity in food products are undesirable, since any observation of toxin production exceeds acceptable risks for this pathogen.

Defining LT as the sampling period prior to toxin detection makes LT a conservative estimation of the true lag phase and thus a good estimate for safe storage time. Baker and Genigeorgis (1990) pooled LT data from 927 experimental combinations, accounting for approximately 18,700 inoculated fish samples. The BMDP2V program for analysis of variance (Dixon and Brown, 1983) was used to determine significance of differences in LT in relation to the continuous and discrete variables described above. The optimum relation between design variables with respect to LT predictions was obtained using the GLM program

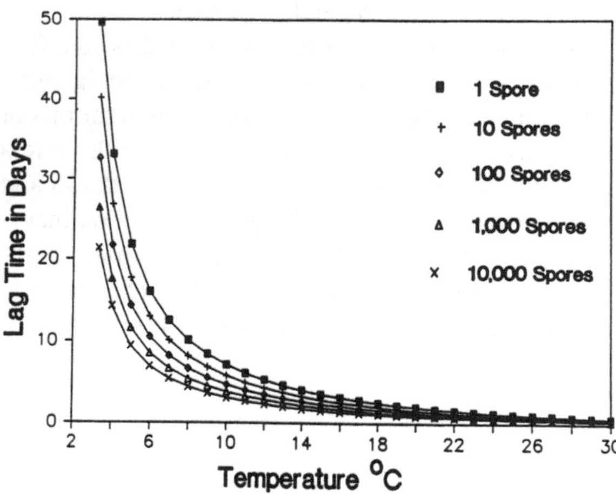

Figure 15 Predicted lag time in rockfish, inoculated with nonproteolytic *C. botulinum* types B, E, and F, as affected by spore inoculum and storage temperature when stored under vacuum. (From Baker and Genigeorgis, 1990.)

for multiple linear regression (SAS Institute, Inc., 1985). The discrete factors, fish species, and spore pool (type of *C. botulinum* inoculum), were highly significant ($P < 0.0001$), while MA (vacuum or CO_2/N_2 gas atmospheres) was less significant ($P < 0.02$). Preparation of fish tissue as a fillet or as a homogenate was insignificant ($P > 0.05$). The continuous factors, temperature and spore inoculum, were also highly significant ($P < 0.0001$). The reciprocal of T (1/T) transformation was highly significant and was included in formulas to add biological realism by forcing the graph of LT to become asymptotic as T approaches the lower limits for nonproteolytic *C. botulinum* growth (Fig. 15). Though some two-way interactions were shown to exert a significant effect on LT, the amount of variability explained by these interactions in the predictive models deemed them to be unnecessary components.

Based on the analysis of the compiled experiments with fish, a general formula of the most conservative model for prediction of LT was constructed. Equation 32 yielded calculated LT values that were very close to observed data and LTs predicted from models constructed from data of individual experiments, but which were shorter in most cases.

$$\begin{aligned}\log \text{LT} = 0.974 - 0.042(\text{temp.°C}) + (2.74/\text{temp.°C}) - \\ 0.091(\log \text{spore inoculum}) + 0.035(\text{initial log APC})\end{aligned} \tag{32}$$

where APC is the aerobic plate count. This general formula, used to plot Figure 15, yields predictions for the safest storage of MA packaged fresh fish within the limits of these experimental designs. Figure 15 also demonstrates that increased spore loads have the greatest effect on shortening LT at the lower refrigeration temperatures, i.e., below 8°C. As temperature increases, the effect of spore inoculum diminishes, and P of growth from a single *C. botulinum* spore increases.

The total variation within a factorial data set explained by each single factor, or interaction, presents a more concise view of the relative contribution of each factor upon the dependent variable. Expression of the proportion of explained experimental variation due to each variable as a relative height allows quantitation and pictorial presentation of the "Hurdle" principle (Fig. 16). Temperature alone accounted for 74.6% of the total explained variation ($r^2 = 0.883$). The initial spore inoculum accounted for 7.4%. Differences between spore type, fish species, and MA composition together accounted for only a 2.3% increase in r^2. Thus, it became apparent that temperature was the only substantial growth hurdle in these studies, whereas MA, fish species, etc., merely refined the predictive capabilities of the LT models.

A similar approach to multifactorial experimental design based upon MPN methodology was applied to studies of *C. botulinum* toxigenesis as affected by a_w, pH, and storage time at 25°C in vacuum-packaged, mashed potatoes (Dodds, 1989). Analysis of variance and multiple regression yielded predictive

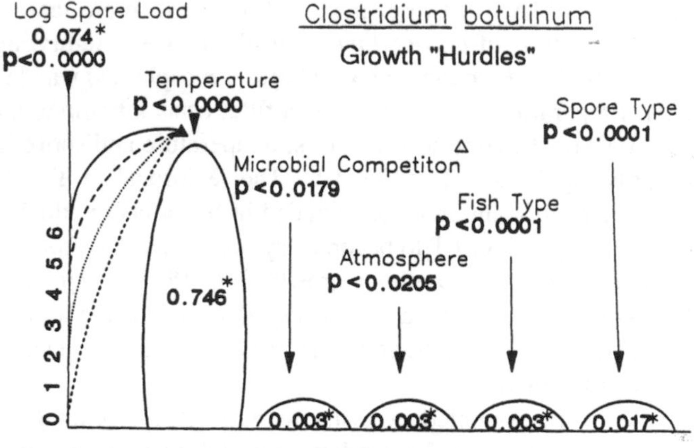

Figure 16 Pictorial representation of the effects of six variables (temperature, spore load, microbial competition, modified atmosphere, fish species, and spore type) on the length of lag time of *C. botulinum* inoculated in fresh fish. (From Baker and Genigeorgis, 1990.)

models for the lag time (LT) until observation of the first toxic sample, (Eq. 33) and P (Eq. 34).

$$\log_{10} \text{lag time} = 17{,}870 - 35{,}570(a_w) + 232.4(\text{pH}) - 17{,}712(a_w{}^2) + 227.5(a_w)(\text{pH}) + 1.1(\text{pH})^2 \qquad r^2 = 0.98 \tag{33}$$

$$\log P = -577.62 - 6.96(\text{days}) + 1399.49(a_w) - 41.97\,\text{pH} - 840.64(a_w{}^2) + 44.64(a_w)(\text{pH}) + 7.23(\text{days})(a_w) \qquad r^2 = 0.71 \tag{34}$$

Storage time, a_w, pH, and several interactions affected P ($p < 0.0001$). Descriptive analysis provided by plots of log P against a_w were curvilinear for each pH level (5.25–5.75) at 45 days of storage, indicating the benefit of adding a quadratic term for a_w ($p < 0.0001$) in the regression equation for prediction of log P.

When LT was treated as the dependent variable, ANOVA and descriptive analysis identified significant factors, transformations, and interactions to be included in an analysis of multiple regression. The resulting model (Eq. 33) was able to explain 98% of the observed variation, an indication of how well the data could be intrinsically predicted. This model predicted shorter lag times in all but a single case, when compared to results of a subsequent experiment using the

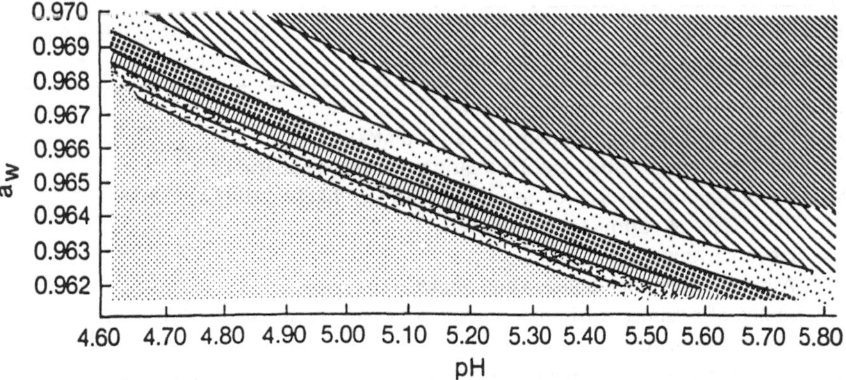

Figure 17 Contour plot of a_w versus pH in which the contours represent the predicted lag time until toxin production by *C. botulinum* in cooked, vacuum-packed potatoes set to initial a_w of 0.960–0.970 and initial pH 4.75–5.75 and incubated at 25°C. All predicted values greater than 100 were set equal to 100 to show the contour lines over the range of days studied. Contour patterns from lower left corner to the upper right are: >100; 80.9–93.6; 69.1–80.9; 55.4–69.1; 42.6–55.4; 29.9–42.6; 17.1–29.9; and 10.8–17.1 days, respectively. (From Dodds, 1989.)

same design factors. Figure 17 presents contour plots for the predicted lag time range for *C. botulinum* in vacuum-packaged potatoes for combinations of a_w (adjusted with NaCl) and pH (adjusted with ascorbic acid) when stored at 25°C. Although the use of this model should not be extrapolated to predict the safety of other potato products, it may be of value in the design of quantitative or qualitative food safety experiments with similar pH and a_w combinations but different formulations.

VI. PROBLEMS ASSOCIATED WITH MODELING STRATEGIES

A. Experimental Limitations

Besides the main factors, i.e., temperature, pH, and a_w, there are many other variables that complicate and confound experimental design and analysis. Roberts and Jarvis (1983) listed many such factors (Table 4) that make it impossible to simultaneously test all treatment combinations. These authors suggested that researchers ''design a range of experimental foods containing differing proportions of carbohydrates, proteins and fats, and other components as necessary,'' in order to focus experimental designs on formulations that broadly represent an important cross-section of potentially hazardous products. Such an approach

Table 4 Factors to Be Taken into Account in Laboratory Studies of Microbial Growth Responses

1. The pH value and buffering capacity of the system
2. The water activity and choice of humectant(s)
3. The temperature, time, and fluctuations during incubation
4. The nature and severity of any prior treatment likely to damage cells
5. The effects of antimicrobial additives
6. Substrate limitation
7. (Anti)microbial metabolites and competition
8. Choice of strain, e.g., from a culture collection or isolation from food
9. Work with single or mixed cultures
10. Initial numbers of microbes and, if more than one type, the initial proportions
11. Experimental variation
12. Laboratory to laboratory variation also important
13. Data formatting and the most appropriate statistical/mathematical analysis of the results

Source: Roberts and Jarvis, 1983.

would facilitate the exchange of comparable food safety data collected using more standardized procedures, and which might be consolidated in a widely accessible databank (Roberts and Jarvis, 1983).

B. Reproducibility

Many of the factors affecting the reproducibility of *C. botulinum* thermal destruction studies are also relevant to food safety models. Stumbo (1973) summarized observations that may account for deviations from the logarithmic order of death in survivor curves.

C. Heat Activation of Spore Germination

The ability of sublethal heat treatments to help induce some spores to germinate was referred to as heat activation by Curran and Evans (1945). Russell (1982) reviewed studies on the mechanism of spore heat activation. Modeling studies are difficult to compare when spore inocula are given different pretreatments. For example, Adair et al. (1989) and Tanaka et al. (1986) heat-shocked *C. botulinum* spores prior to use, while others either did not use thermally activated spores (Baker and Genigeorgis, 1990; Dodds, 1989; Lund et al., 1982) or tested exposure to mild heat processes as a variable (Roberts et al., 1982a, b, c).

The activation of a spore population with heat depends on: (1) the species or strain, (2) the medium in which the spores are produced, (3) the medium in which the spores are stored, (4) the storage temperature of the spores, (5) the age of the spores, (6) the media in which the spores are heated prior to inoculation, and (7) the temperature of heating (Stumbo, 1973) (see Chapter 6). These factors are equally likely to affect the ability of a spore to germinate and grow in an experimental medium or food system, regardless of whether the spores are heat treated or not. These ramifications may also affect the ability of *C. botulinum* spores to outgrow at suboptimal temperatures.

As a consequence of thermal activation, the germinated spore is then subject to thermal inactivation during subsequent heating. Both processes obey first-order kinetics as demonstrated by Shull et al. (1963). Working with heat-resistant dormant spores, the following equations were derived to predict the number of viable spores after a mild heat treatment:

$$A_t = A_t e^{-kt} + \frac{aN_0}{k - a}(e^{-at} - e^{-kt}) \tag{35a}$$

$$L_t = N_t + A_t \tag{35b}$$

where

L_0, L_t = numbers of viable spores
A_0, A_t = numbers of activated spores
N_0, N_t = numbers of nonactivated spores at times 0 and t, respectively (i.e., $N_0 + A_0 = L_0$)
k = inactivation and activation rate constants, respectively

Similar reasoning may be applied to the inoculation of activated or germinated spores into suboptimal growth media, whereas growth from a dormant spore may provide an acclimation period more favorable for subsequent growth.

D. Spore Variability

For models to be valid, the bacteria and spores used in experiments should be physiologically similar to naturally occurring microbes in real food systems. Bean (1983) used *Clostridium* and *Bacillus* spores from natural sources to repeat experiments similar to those conducted by Braithwaite and Perigo (1971) (see Figs. 3 and 4). The effect of a_w, pH, and heating on 13 clostridial species isolated from spoiled canned foods are presented in Figure 18. Data from 10,342 individual experiments indicated that the spores isolated from nature had a wider range of limiting parameters (a_w, pH, and heat) than the laboratory strains used by Braithwaite and Perigo (1971). Figure 19 demonstrates that the two experiments had similar response surface shapes. However, the portion of the upper surface attributable to natural spores shows a resistance to combination levels of low a_w, pH extremes, and increased F_0.

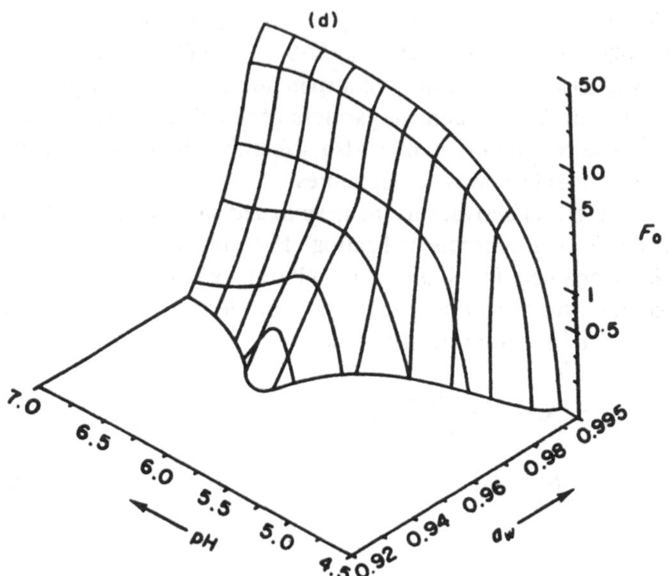

Figure 18 Isometric illustration of combined effect of pH and a_w on limiting F_0 of spoilage *Clostridium* spp. The heating/culturing medium was Reinforced Clostridial Agar (Oxoid). (From Bean, 1983.)

E. Inoculation Number

The preceding discussion draws attention to studies based on the probability of one *C. botulinum* spore out of a population to germinate and grow. This approach allows the intrinsic and extrinsic factors to be characterized by their combined ability to cause "functional" decimal reductions of the initial number of spores able to outgrow. Other studies have been based on inoculum level similar to those naturally occurring in raw food ingredients (as observed thus far) and their ability to outgrow (Hauschild, 1989). Thus, an experimental design dilemma is reached. The greater the number of spores inoculated, the more conservative a model or challenge study becomes, perhaps even to the point of making a commercial food process appear to be excessively conducive to *C. botulinum* growth. The experimental logistics of incorporating many inoculum levels along with other important treatment variables can rapidly overwhelm labor, laboratory, and financial capabilities.

Cerf (1977) reviewed the tailing survival curves observed in the destruction of bacterial spores. Investigators should consider the fact that individuals in a population are not identical. If resistance characteristics within microbial populations are normally distributed, then a small proportion may possess abilities

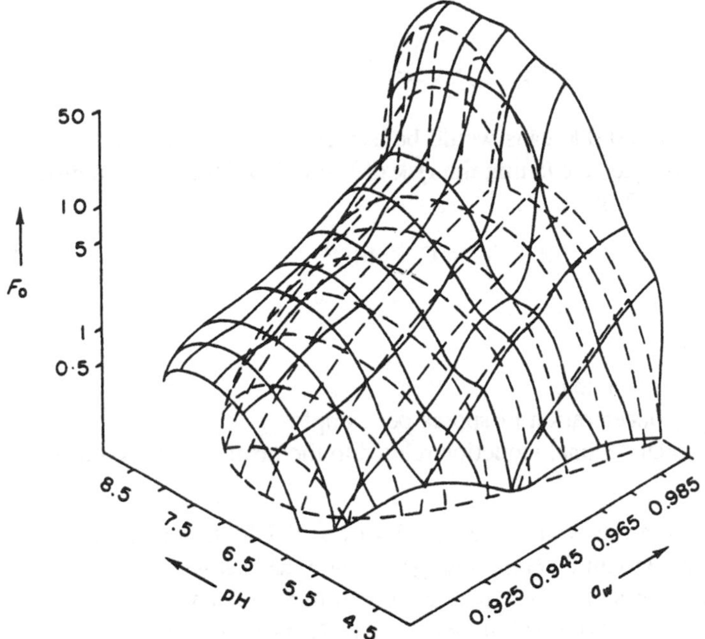

Figure 19 Comparison of the original and new composite isometric illustrations of pH and a_w effects on limiting F_0 values. Dashed line, original data from *Bacillus* spp. spores in nutrient agar; solid line, combined data from naturally occurring spores, *Clostridium* spp. and *Bacillus* spp. spores. (From Bean, 1983.)

to overcome the food system hurdles of combined germination and/or growth inhibitors. Thus, typical survivor curves should be considered sigmoidal or to have an upward concave tendency (Cerf, 1977). Since heat is only one aspect of preservation based upon the hurdle concept (Leistner and Rödel, 1976), only one side of the argument concerning possible genetic heterogeneity of spore populations is presented. Ultimately, the upper inoculation level necessary will be based upon some form of risk analysis.

To estimate the risk associated with the safety of canned, cured meats, the number of *C. botulinum* spores expected to be present in the product must be considered:

$$\text{risk} = \frac{\log 1}{\text{Pxi}} \tag{36}$$

where i is the incidence of *C. botulinum* per unit sample. Thus, if the incidence of *C. botulinum* spores is 0.1 per can, and P is from Example 3:

Example 5

$$\text{risk} = \frac{\log 1}{5.1 \times 10^{-5} \times 0.1 = 5.3}$$

or approximately 1/200,000 cans would be expected to produce toxin.

Another approach to calculating the risk of one unit volume being nonsterile is given by (Cerf, 1977):

$$R = V \times C \times 10^{-DR} \tag{37}$$

where

V = the volume of sample
C = the initial concentration of spores per sample
DR = the functional decimal reduction related to the media and its storage over time

If DR increases, the risk will decrease. However, if DR decreases due to the presence of more resistant *C. botulinum* types, then the risk will be greater than predicted. On the basis of viable spore counts from Scandinavian waters where mud samples contained 50–60 *C. botulinum* type E spores/g, Cann et al. (1967) determined that inoculated pack studies should be conducted with 10^4 spores/g fish to allow for an adequate safety margin. Spore numbers building up in processing plant microenvironments may be sources of *C. botulinum* (Hobbs and Shewan, 1969). Baker and Genigeorgis (1990) reasoned that if the highest number of *C. botulinum* spores ever observed in a raw food (5.3 spores/g of farm-raised trout) (Huss et al., 1974) is approximately doubled to 10 spores/g, and that a maximal size for a processed food portion would be 1 kg, then challenge studies conducted with 10^4 *C. botulinum* spores per pack (1000 g or less) would provide a convenient experimental unit to account for the worst-case scenario of naturally contaminating spores and a margin of safety allowing for spore population diversity.

F. Work in Media Vs. Food Systems

Experimentation with sterile media is often more controllable and reproducible than similar experimentation in either a sterilized or "fresh" food preparation. Roberts and Jarvis (1983) point out that the "Perigo" effect observed in media is much greater than in meat products. Furthermore, the role of nitrite as an antimicrobial agent in media or foods is poorly understood (Roberts et al., 1981a; Benedict, 1980). The influence of the pH of pork meat (range 5.34–6.36 or 6.3–6.72) used in experimental slurries was significant (Roberts et al., 1981a, b, c), which led to examination of pig breed, cut, and batch for additional explanation of experimental variance (Gibson et al., 1982). Similarly, fish type was a significant variable in lag time models, whereas differences between salmon, sole, or rockfish and between batches of each fish type only accounted for 0.3% of the overall variation (Baker and Genigeorgis, 1990).

Growth studies in microbiologically pure systems are far easier than in natural food systems. Recently, models for growth of *C. botulinum* (Gibson et al., 1987), *Salmonella* (Gibson et al., 1988), and *Listeria monocytogenes* (Buchanan and Philips, 1990; Buchanan et al., 1989) in media have been completed. The objective was to develop a database to aid in the design of subsequent challenge studies in foods. Results from a limited challenge study that agree well with the predicted outcome from well-replicated media studies may be more easily accepted as representing the behavior, safety, shelf life, etc. of the food system under ''real-world'' conditions.

G. Method of Inoculation

Clumping of spores affects the accuracy of the initial inoculation number, whether enumerated by direct counts, colony counts, or MPN methods. Linear thermal survivor curves are achieved when the number of viable spores per clump is reduced to one. Similarly, deviations in probability studies based on dilution of *C. botulinum* spores to less than one per inoculated pack can be attributed to clumping.

H. Anaerobiosis

In general, the more adverse a growth environment is to an obligate anaerobe, the more imperative it will become to maintain strict anaerobiosis in the medium throughout incubation. This factor may be especially important during extended incubation of samples packaged in plastic films.

I. Sampling Plans

Models based on actual counts or observance of toxicity are limited by the feasibility of extended sampling periods. Most probability studies strive to minimize the sampling interval between nontoxic samples and the first observation of toxicity. Lag time models based on the sampling period prior to toxicity are conservative (erring toward safety) and only approximate the growth of *C. botulinum* (Baker and Genigeorgis, 1990).

Modeling the logistic growth curve, whether using actual increases in bacterial counts (e.g., Gompertz function) or probability of toxicity (e.g., logistic regression), is subject to data acquisition problems that directly affect their mathematical basis. In general these procedures are most appropriate when data are evenly distributed along the sigmoid curve, i.e., in the lag phase, the periods of accelerating and decelerating change, and the stationary period. Bratchell et al. (1989) demonstrated the risk of developing "erroneous models" for *Salmonella* growth when data points were systematically removed or night and weekend observations were excluded. The fewer points available for curve fitting, the greater the influence each point has. Outlying points can severely affect the curve fit, especially if not offset by a highly conforming data set. Collection of adequate data during periods of rapid change, i.e., between lag and log phase, is particularly important. Baker et al. (1990) noted that reliability of the logistic regression could be improved if more samples were concentrated during inflection periods of growth. This kind of data is difficult to obtain without a well-documented growth system (Siegel, 1981).

J. Fluctuating Temperatures During Storage

During distribution, retail sale, and possession by the end user, foods are not always held under constant refrigeration. This fact raises the question of how shifts between proper storage temperature and thermal abuses affect P of *C. botulinum* growth and toxigenesis. Hauschild (1982) points out that thermal abuse following refrigerated storage may (1) facilitate outgrowth as residual nitrite decreases, or (2) inhibit outgrowth, as psychrotrophic microorganisms will have likely increased in numbers sufficient to compete with *C. botulinum* and lowered the pH of the food environment.

Small thermal fluctuations may lead to spore germination, followed by death if outgrowth is suboptimal. In studies with liver sausage inoculated with *C. botulinum* spores, P decreased marginally during storage at 8°C for 6 weeks (Hauschild et al., 1982), whereas storage of wieners at 7°C for 3 weeks had little effect on P (Hustad et al., 1973). Toxigenesis was more likely in canned cured comminuted pork when thermally abused after refrigerated storage (Tompkin et al., 1978, 1979). Smelt (1980) observed that the lag time of proteolytic *C. botulinum* spores in cooked meat medium at 20°C was not affected by prior incubation at 10°C.

K. Effects of Competitive Microorganisms

Consideration of the competitive aspects of resident microorganisms on the outgrowth of *C. botulinum* will present many new modeling opportunities in the future and perhaps stimulate interest in the application of modeling techniques used by population ecologists. *C. botulinum* models may be affected by sporicidal, bacteriocidal, sporostatic, or bacteriostatic microbial byproducts, such as bacteriocins (Anastasio et al., 1971; Beerens and Tahon, 1967; Ellison and Kautter, 1970; Kautter et al., 1966; Lau et al., 1974), bacteriophages (Eklund and Poysky, 1981; Hariharan and Mitchell, 1976; Iida and Oguma, 1981; Oguma et al., 1976), and other agents (Moberg and Sugiyama, 1979; Okereke and Montville, 1991; Smith, 1975; Wentz et al., 1967).

Although microbial competition should be addressed as a possible hurdle (Genigeorgis and Riemann, 1979), few studies have attempted to quantify this as a study variable. Garcia and Genigeorgis (1987) incorporated three levels of initial bacterial contamination (IMF) in factorial designs to model the toxigenesis of nonproteolytic *C. botulinum* in modified atmosphere packaged fresh fish. Increasing levels of IMF ($10^{2.5}$, 10^5, $10^{7.5}$) significantly ($P < 0.005$) affected P and interacted with storage temperature (IMF × T), but not with storage time ($P > 0.02$). IMF was included in a formula to predict lag time as a continuous variable (Eq. 38). Predictions for lag time were subsequently used to model P, as described above.

$$\log_{10} \text{LT} = -2.595 + 0.398\text{T} - 0.015\text{T}^2 + 12.698(1/\text{T}) - 0.141\text{I} + 0.026\text{IMF} \qquad r^2 = 0.93 \tag{38}$$

where

LT = lag time prior to *C. botulinum* toxigenesis in salmon homogenate packaged in 70% CO_2, 30% N_2
T = temperature (°C)
I = log *C. botulinum* spore inoculum
IMF = log initial microbial contamination

Equation 38 demonstrates that the $\log_{10}$ of LT increased by 0.026 units, equal to 1.06 days LT extension for every $\log_{10}$ increase in the initial numbers of microorganisms on the fish muscle. The authors demonstrated that the effect of IMF in their study was probably due to other mechanisms, since the pH of homogenate did not fluctuate more than 0.1 unit. Although the naturally occurring IMF did not completely inhibit *C. botulinum* growth in the preceding study, increased initial populations of other strains could (Okereke and Montville, 1991).

VII. INTEGRATION OF PREDICTIVE MODELING INTO HAZARD ANALYSIS OF CRITICAL CONTROL POINTS (HACCP) PROGRAMS

Requirements of food-preservation systems have been rapidly changing in the past few years to meet the consumers' perceived need for fresh, minimally processed foods with few or no chemical additives. Food processors responding to the market demand are faced with the logistical constraints of distribution and adequate shelf life for their products. Innovative packaging concepts and marketing strategies are being introduced to retail and food service customers, which include the use of vacuum and other modified atmospheres (MA).

Modified-atmosphere packaging extends the shelf life of refrigerated fresh foods largely by reducing the surface Eh. The normal microbial spoilage pattern shifts from one dominated by psychrotrophic aerobic bacteria responsible for slime, rancidity, and discolorations to one dominated by the slower-growing facultative anaerobes including lactic acid bacteria. The greatest advantage of MA packaging is achieved by its early application when the microbial load is low.

Temperature abuse of MA-packaged foods leads to increased growth rates of both spoilage and pathogenic bacteria. The initial number of microorganisms may influence the duration of the lag phase, especially when growth conditions are not optimal. The growth rate is independent of inoculum size.

Nonproteolytic *C. botulinum* (types B, E, and F) may germinate and grow under refrigeration. Schmidt et al. (1961) demonstrated growth of *C. botulinum* type E in sterile beef stew at 3.3°C, while growth occurred in raw, vacuum-packaged herring at 5°C (Cann et al., 1965). Production of toxin has been reported in refrigerated MA-packaged food prior to the onset of overt organoleptic spoilage (Eklund, 1982; Garcia et al., 1987; Post et al., 1985). The lowest temperature allowing growth of a pathogen in a food varies with intrinsic and extrinsic factors. Incorporation of one or more hurdles or barriers, in addition to refrigeration, will increase the minimum growth temperature. Additional safeguards to inhibit the psychrotrophic growth of pathogens during storage at refrigeration temperatures and during accidental temperature abuse are particularly important in food products with extended shelf lives.

The risk of *C. botulinum* growth and toxigenesis in MA-packaged foods that are not stabilized by a low pH, low a_w, or chemical additives should be of prime concern in the development of new products and marketing strategies. Inclusion of inoculated pack studies as part of a comprehensive HACCP (ICMSF, 1988) quality assurance program will allow processors to establish maximum time/temperature storage conditions for their products, based on worst-case scenarios with respect to *C. botulinum* spore numbers, either in the raw product or resulting from in-plant contamination and potential consumer abuse.

A. Marketing Strategies Based on *C. botulinum* Growth

The HACCP approach could use worst-case scenarios of this type to establish which product formulations are conducive to *C. botulinum* growth. Thus, *C. botulinum* inoculated pack studies help to ensure that marketing strategies are consistent with safe time/temperature storage schemes. Baker and Genigeorgis (1990) accepted the model of *C. botulinum* lag time in raw fish (Baker and Genigeorgis, 1989) as being the worst-case scenario in the absence of other baseline data. Challenge tests were conducted with 51 complex food products. Most categories of cooked foods were represented in these inoculated pack studies, including fish, beef, chicken, lamb, and pork (with and without seasoning or gravies), pasta, rice, potatoes, vegetables, soups, sauces, and stews. Figure 20 presents the results of these studies. In general, the baseline LT model de-

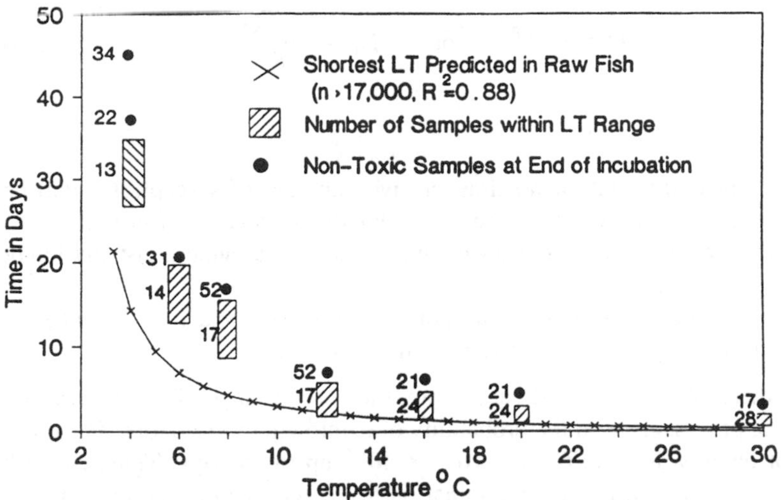

Figure 20 Comparison between lag time for *C. botulinum* inoculated into 69 *Sous Vide*–type food products (10^4 spores/pack) and the "general formula" for *C. botulinum* lag time in fresh fish reported by Baker and Genigeorgis (1990).

rived from fresh fish studies was an accurate guide upon which to base the safe refrigerated shelf life of sensitive products; it also highlighted their sensitivity to storage temperatures above 4°C. Figure 20 also demonstrates the need to develop less conservative baseline safety models more appropriate to other product groups and applicable to longer storage periods at higher temperatures. If marketing strategies were based on the time/temperature storage set by the LT model in a HACCP plan, then products allowing faster growth of *C. botulinum* would be considered "out of control." Figure 20 demonstrates one such product, which was rejected after challenge testing and subsequently reformulated, so that its ability to support growth of *C. botulinum* conformed to the HACCP plan.

B. Application of Risk Analysis

In general, if definitive critical control points are not incorporated to eliminate *C. botulinum* spores, then a HACCP plan can only address the risks associated with the product's raw materials, processing, distribution, marketing, and use by the consumer. Smelt (1980) applied a general model for risk analysis introduced by Mossel and Drion (1979) to estimate numbers of pathogens acceptable in foods, based on their minimal infective dose (MID). Smelt (1980) substituted the minimum number of *C. botulinum* needed to initiate growth in a food for the MID and presented an analysis of the hazard from *C. botulinum* in jellied milk with fruit stored under refrigeration. Risk was assessed in the following terms:

$$Q = e^{-N_f R} \qquad \text{or} \qquad N_f = \frac{-\ln Q}{R} \tag{39}$$

where

- Q = the probability that, at no time, a given number of susceptibles will be exposed to a minimum number of *C. botulinum* capable of growth
- N_f = number of spores able to grow per unit, assuming a random distribution of *C. botulinum* spores
- R = V × I, where V = size of the population at risk, and I = number of units annually produced, given a limited storage life

N_f must therefore be estimated based upon the normal range of contamination, lethality of processing, and the probability of outgrowth by survivors. The probability of outgrowth at 20°C was $<0.4 \times 10^{-6}$ up to 24 days. Thus, the risk analysis was based on initial contamination levels, chance of exposure to temperatures >10°C, and the associated probability of outgrowth after processing. Consequently, this risk analysis estimated that an equivalent of less than 10^{-11} spores per unit could be expected in jellied milk with fruit, given that the intended refrigerated storage is ≤24 days (Smelt, 1980).

Risk has been related to the observed safety of individual commercial products on a volume basis over the time of their marketing history (Hauschild and Simonsen, 1982). Recalling that log 1/(Pxi) represents the expected rate at which a unit would become toxic based on an assigned prevalence of *C. botulinum* per unit, then safety units (SU) could be an industrial interpretation relating the decimal number of units marketed per number of units causing illness (e.g., if $10^{8.5}$ units were marketed safely, then SU ≥ 8.5. Hauschild and Simonsen (1982) applied production data for self-stable, canned, cured meats to estimate minimum safety in terms of SU. The margins of safety for luncheon meat, ham, and shoulder and sausages ranged from >6.6 to >9.5 SU, depending on the source and volume of production data available. The authors caution that SU values and log (Pxi) are not strictly correlated, since (1) experimental periods are defined, while commercial turnover of each unit may be much faster, (2) most toxic experimental units are detected, while only a fraction of commercial units that turn toxic would be expected to cause illness, as the rest would not be consumed due to signs of organoleptic deterioration, and (3) reporting of all botulism cases and their etiological agent may not always be complete. Thus, estimates of S are less conservative than log (Pxi) values.

C. Time/Temperature Indicators

Time/temperature indicators (TTI) are an emerging technology, which may be included in food distribution systems to provide a warning mechanism of product temperature abuse for the food processor, retailer, and/or consumer (Mistry and Kosikowski, 1983; Wells and Singh, 1988). The expiration indicated by these devices relates to the effect of temperature on the rate of their biochemical or physical kinetics. The potential usefulness of time/temperature indicators will require a strong correlation with the lag time prior to *C. botulinum* toxigenesis. Ultimately, TTIs will be developed with different thermal sensitivities to match the abilities of different food groups to support *C. botulinum* toxigenesis during refrigerated storage or abuse, providing a final critical control point in future HACCP systems.

VIII. CONCLUSION

Development of standard methodologies to predict *C. botulinum* lag phase and P of toxigenesis in packaged foods should be employed in the food industry as part of product development. This is especially true for the refrigerated foods industry. Sharing of pertinent information would allow a database to be developed in which foods could be classified by their generic formulation and ability to support *C. botulinum* growth. Analysis of a large body of data on food products would help the industry and consumers to (1) help develop an understanding of the relative risk of particular products, (2) teach greater awareness of the time/temperature combinations that may allow *C. botulinum* to produce toxin in a contaminated product without other signs of spoilage, and (3) focus upon the development of new hurdles for food safety.

Figure 21 presents a comparison between results of the initial toxicities observed in MA-packaged fish inoculated with *C. botulinum* spores reported in the international literature and the shortest *C. botulinum* lag times in fish predicted from a multiple liner regression model (Baker and Genigeorgis, 1990). The model is in close agreement with the other independent observations, yet in most cases the predicted onset of risk is more conservative. Thus, the database underlying this model demonstrates the feasibility of developing similar databases for risk analysis of other food systems and the intrinsic and extrinsic factors contributing to their preservation.

Modeling strategies should aim towards consolidation of mathematical and experimental approaches, thus providing the food industry with rational guidelines for design and routine safety testing. Standardization of factorial and other design experimentation would allow the refrigerated foods industry to develop a *C. botulinum* growth (LT, P) database. Manufacturers would then be able to design general formulas including nutritional, physical, and chemical factors pertinent to their specific product. Access to such a database would greatly reduce the cost of conducting product challenge studies by providing *C. botulinum* growth expectations. It would also improve the reliability of challenge studies through tighter bracketing of sampling times and temperatures. Jarvis (1983) found support among the leading food scientists surveyed for the future usefulness of data banks to relate composition and physical data of selected food

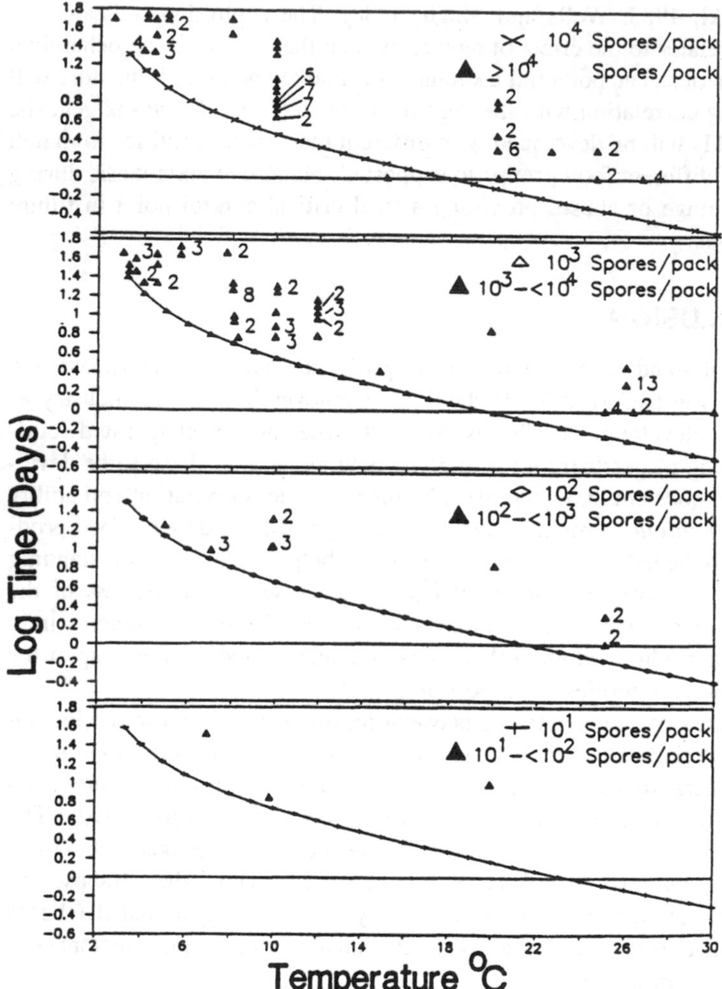

Figure 21 Comparison of first-time toxicities (days) in MA-packaged fish inoculated with *C. botulinum* spores reported in the literature with the shortest *C. botulinum* lag times in fish predicted from a multiple linear regression model. (From Baker and Genigeorgis, 1990.)

groups to critical inhibition limits for growth or survival of pathogenic bacteria and the use of such information in risk hazard analysis.

Methods of analysis must be suited to the accuracy of the experimental data and yield predictions within a range of precision upon which management decisions can be based with confidence. Researchers should strive for simplicity to fulfill these goals, in order to facilitate transfer of improved food safety technology to all sectors of the food-processing industry.

REFERENCES

Adair, C., Kilsby, D. C., and Whittall, P. T. (1989). Comparison of the Schoolfield (non-linear Arrhenius) model and the square root model for predicting bacterial growth in foods. *Food Microbiol. 6*: 7.

Adams, A. T., Ayres, J. C., Tischer, R. G., and Ostle, B. (1952). Effect of subtilin on spoilage of thermal processed beef. *Food Technol. 11*: 421.

Anastasio, K. L., Souchek, J. A., and Sugiyama, H. (1971). Boticinogeny and actions of the bacteriocin. *J. Bacteriol. 107*: 143.

Baird-Parker, A. C., and Freame, B. (1967). Combined effects of water activity, pH and temperature on the growth of *Clostridium botulinum* from spore and vegetative cell inocula. *J. Appl. Bacteriol. 30*: 420.

Baird-Parker, A. C., and Kilsby, D. C. (1987). Principles of predictive food microbiology. *J. Appl. Bacteriol.* 43S.

Baker, D. A., and Genigeorgis, C. (1989). Microbiological risk assessment of refrigerated foods: The safety of "Sous-Vide" products. In *Proceedings of the Xth International Symposium World Association Vet. Food Hygienists*, Stockholm, Sweden, pp. 132–141.

Baker, D. A., and Genigeorgis, C. (1990). Predicting the safe storage of fresh fish under modified atmospheres with respect to *Clostridium botulinum* toxigenesis by modeling length of the lag phase of growth. *J. Food Prot. 53*: 131.

Baker, D. A., Genigeorgis, C., Glover, J., and Razavilar, V. (1990). Growth and toxigenesis of *C. botulinum* type E in fishes packaged under modified atmospheres. *Intl. J. Food Microbiol. 10*: 269.

Ball, C. O. (1923). Determining, by methods of calculation, the time necessary to process canned foods. Bull. 7. National Res. Council, Washington, DC.

Bean, P. G. (1983). Developments in heat treatment processes for shelf-stable products. In *Food Microbiology: Advances and Prospects* (T. A. Roberts and F. A. Skinner, eds.), Academic Press, New York, p. 97.

Beerens, H., and Tahon, M. (1967). Production of bacteriocins by different types of *Clostridium botulinum*. In *Botulism 1966*, Proceedings of the 5th International Symposium on Food Microbioly (M. Ingram and T. A. Roberts, eds.), Chapman and Hall, Ltd., London, p. 424.

Benedict, R. C. (1980). Biochemical basis for nitrite-inhibition of *Clostridium botulinum* in cured meat. *J. Food Prot. 43*: 877.

Bigelow, W. D., Bohart, G. S., Richardson, A. C., and Ball, C. O. (1920). Heat penetration in processing canned foods. Bull. No. 16-L., Res. Lab. National Canners Association, Washington, DC.

Braithwaite, P. J., and Perigo, J. A. (1971). The influence of pH, water activity and recovery temperature on the heat resistance and outgrowth of *Bacillus* spores. In *Spore Research 1971* (A. M. Barker, G. W. Gould, and J. Wolf, eds.), Academic Press, New York, p. 289.

Bratchell, N., Gibson, A. M., Truman, M., Kelly, T. M., and Roberts, T. A. (1989). Predicting microbial growth: The consequences of quantity of data. *Intl. J. Food. Microbiol. 8*: 47.

Broughall, J. M., and Brown, C. (1984). Hazard analysis applied to microbial growth in foods: Development and application of three-dimensional models to predict bacterial growth. *Food Microbiol. 1*: 13.

Buchanan, R. L., and Phillips, J. G. (1990). Response surface model for predicting the effects of temperature, pH, sodium chloride content, sodium nitrite concentration and atmosphere on the growth of *Listeria monocytogenes*. *J. Food Prot. 53*: 370.

Buchanan, R. L., Stahl, H. G., and Whiting, R. C. (1989). Effects and interactions of temperature, pH, atmosphere, sodium chloride, and sodium nitrite on the growth of *Listeria monocytogenes*. *J. Food Prot. 52*: 844.

Cann, D. C., Wilson, B. B., Hobbs, G., and Shewan, J. M. (1965). The growth and toxin production of *Clostridium botulinum* type E in certain vacuum-packaged fish. *J. Appl. Bacteriol. 28*: 431.

Cann, D. C., Wilson, B. B., Hobbs, G., and Shewan, J. M. (1967). Toxin production by *C. botulinum* type E in vacuum packaged fish. In *Botulism 1966*, Proceedings of the 5th International Symposium on Food Microbioly (M. Ingram and T. A. Roberts, eds.), Chapman and Hall, Ltd., London, England, p. 202.

Cerf, O. (1977). Tailing of survival curves of bacterial spores: A review. *J. Appl. Bacteriol. 42*: 1.

Cerveny, J. G. (1980). Effects of changes in the production and marketing of cured meats on the risk of botulism. *Food Technol. 34*: 240.

Curran, H. R., and Evans, F. R. (1945). Heat activation inducing germination in the spores of thermotolerant and thermophilic aerobic bacteria. *J. Bacteriol. 49*: 335.

Dixon, W. J., and Brown, M. B. (1983). *Biomedical Computer Programs P-Series*. University of California Press, Berkeley, CA.

Dodds, K. L. (1989). Combined effect of water activity and pH on inhibition of toxin production by *Clostridium botulinum* in cooked, vacuum-packed potatoes. *Appl. Environ. Microbiol. 55*: 656.

Draper, N. R., and Smith, H. (1981). *Applied Regression Analysis*. John Riley & Sons, Inc., Toronto.

Dymicky, M., and Trenchard, H. (1982). Inhibition of *Clostridium botulinum* 62A by saturated n-aliphatic acids, n-alkyl formates, acetates, propionates and butyrates. *J. Food Prot. 45*: 1117.

Eklund, M. W. (1982). Significance of *Clostridium botulinum* in fishery products preserved short of sterilization. *Food Technol. 36*(12): 107.

Eklund, M. W., and Poysky, F. T. (1981). Relationship of bacteriophages to the toxigenicity of *Clostridium botulinum* and closely related organisms. In *Biomedical Aspects of Botulism* (G. E. Lewis, Jr., ed.), Academic Press, New York, p. 93.

Ellison, J. S., and Kautter, D. A. (1970). Purification and some properties of two botulicins. *J. Bacteriol. 104*: 19.

Emodi, A. S., and Lechowich, R. V. (1969). Low temperature growth of type E *Clostridium botulinum* spores. I. Effects of sodium chloride, sodium nitrite and pH. *J. Food Sci. 34*: 78.

Esty, J. R., and Meyer, K. F. (1922). The heat resistance of spores of *B. botulinus* and allied anaerobes. *J. Infect. Dis. 31*: 650.

Eyles, M. J., and Warth, A. D. (1981). Assessment of the risk of botulism from vacuum-packaged raw fish: A review. *Food Technol. Aust. 33*: 574.

Farber, J. M. (1986). Predictive modeling of food deterioration and safety. In *Foodborne Microorganisms and Their Toxins—Developing Methodology* (N. J. Stern and M. D. Pearson, eds.), Marcel Dekker, Inc., New York, p. 57.

Fisher, R. A., and Yates, F. (1957). *Statistical Tables for Biological, Agricultural and Medical Research*, 5th ed. Oliver and Boyd, London.

Garcia, G., and Genigeorgis, C. (1987). Quantitative evaluation of *Clostridium botulinum* nonproteolytic types B, E, and F growth risk in fresh salmon tissue homogenates stored under modified atmospheres at low temperatures. *J. Food Prot. 50*: 390.

Garcia, G., Genigeorgis, C., and Lindroth, S. (1987). Risk of *Clostridium botulinum* non-proteolytic types B, E and F growth and toxin production in salmon fillets stored under modified atmospheres at low and abused temperatures. *J. Food Prot. 50*: 330.

Genigeorgis, C. (1976). Quality control for fermented meats. *J. Am. Vet. Med. Assoc. 11*: 1220.

Genigeorgis, C. (1981). Factors affecting the probability of growth of pathogenic microorganisms in foods. *J. Am. Vet. Med. Assoc. 179*: 1410.

Genigeorgis, C., and Riemann, H. (1979). Food processing and hygiene. In *Foodborne Infections and Intoxications*, 2nd ed. (H. Riemann and F. L. Brian, eds.), Academic Press, New York, p. 613.

Genigeorgis, C., Meng, J., and Baker, D. A. (1991). Behavior of nonproteolytic *Clostridium botulinum* type B and E spores in cooked turkey and modeling lag phase and probability of toxigenesis. *J. Food Sci. 56*: 373.

Gibson, A. M., Bratchell, N., and Roberts, T. A. (1988). Predicting microbial growth: Growth responses of salmonellae in a laboratory medium as affected by pH, sodium chloride and storage temperature. *Intl. J. Food Microbiol. 6*: 155.

Gibson, A. M., Bratchell, N., and Roberts, T. A. (1987). The effect of sodium chloride and temperature on the rate and extent of growth of *Clostridium botulinum* type A in pasteurized pork slurry. *J. Appl. Bacteriol. 62*: 479.

Gibson, A. M., Roberts, T. A., and Robinson, A. (1987). Factors controlling the growth of *Clostridium botulinum* type A in pasteurized, cured meats. IV. The effect of pig breed, cut and batch of pork. *J. Food Technol. 17*: 471.

Greenberg, R. A., Silliker, J. H., and Flatta, L. D. (1959). Influence of sodium chloride on toxin production and organoleptic breakdown in perishable cured meat inoculated with *Clostridium botulinum*. *Food Technol. 139*: 509.

Hachigian, J. (1989). An experimental design for determination of D-values describing inactivation kinetics of bacterial spores; design parameters selected using computer simulation. *J. Food Sci. 54*: 720.

Halvorson, H. O., and Ziegler, N. R. (1933). Application of statistics to problems in bacteriology. *J. Bacteriol. 25*: 101.

Hariharan, H., and Mitchell, W. R. (1976). Observations on bacteriophages of *Clostridium botulinum* type C isolated from different sources and the role of certain phages in toxigenicity. *Appl. Environ. Microbiol. 32*: 145.

Hauschild, A. H. W. (1989). *Clostridium botulinum*. In *Foodborne Bacterial Pathogens* (M. P. Doyle, ed.), Marcel Dekker, New York, p. 111.

Hauschild, A. H. W. (1982). Assessment of botulism hazards from cured meat products. *Food Technol. 36*(12): 95.

Hauschild, A. H. W., and Simonsen, B. (1985). Safety of shelf-stable canned cured meats. *J. Food Prot. 48*: 997.

Hauschild, A. H. W., Hilsheimer, R., Jarvis, G., and Raymond, D. P. (1982). Contribution of nitrite to the control of *Clostridium botulinum* in liver sausage. *J. Food Prot. 45*: 500.

Hersom, A. C., and Hulland, E. D. (1980). *Canned Foods: Thermal Processing and Microbiology*, 7th ed., Churchill Livingstone, New York, p. 367.

Hobbs, G., and Shewan, J. M. (1969). Present status of radiation preservation of fish and fishery products in Europe. In *Freezing and Irradiation of Fish* (R. Kreuzer, ed.), Fishing News Books, Ltd., London, p. 488.

Howell, J. A. (1983). Mathematical models in microbiology: Mathematical tool-kit. In *Mathematics in Microbiology* (M. Bazin, ed.), Academic Press, New York, p. 37.

Hurley, M. A., and Roscoe, M. E. (1983). Automated statistical analysis of microbial enumeration by dilution series. *J. Appl. Bacteriol. 55*: 159.

Huss, H. H., Schaeffer, I., Rye Petersen, E., and Cann, D. C. (1979). Toxin production by *Clostridium botulinum* type E in fresh herring in relation to the measured oxidation-reduction potential (Eh). *Nord. Vet. Med. 31*: 81.

Huss, H. H., Pederson, A., and Cann, D. C. (1974). The incidence of *Clostridium botulinum* in Danish trout farms. 1. Distribution in fish and their environment. *J. Food Technol. 9*: 445.

Hustad, G. O., Cerveny, J. G., Trenk, H., Deibel, R. H., Kautter, D. A., Fazio, T., Johnson, R. W., and Kolari, O. E. (1973). Effect of sodium nitrite and sodium nitrate on botulinal toxin production and nitrosamine formation in weiners. *Appl. Microbiol. 17*: 542.

Iida, H., and Oguma, K. (1981). Toxin production and phage in *Clostridium botulinum* types C and D. In *Biomedical Aspects of Botulism* (G. E. Lewis, Jr., ed.), Academic Press, New York, p. 109.

Ikawa, J. Y., and Genigeorgis, C. (1987). Probability of growth and toxin production by nonproteolytic *Clostridium botulinum* in rockfish fillets stored under modified atmospheres. *Intl. J. Food Microbiol. 4*: 167.

Ikawa, J. Y., Genigeorgis, C., and Lindroth, S. (1986). Temperature and time effect on the probability of *Clostridium botulinum* growth in a model broth. *Proceedings of the 2nd World Congress on Foodborne Infection and Intoxication*, Robert von Ostertag Institute, West Berlin, pp. 370–374.

International Commission of Microbiological Specifications for Foods (ICMSF). (1988). *Microorganisms in Foods 4. Application of the Hazard Analysis Critical Control Point (HACCP) System to Ensure Microbiological Safety and Quality*, Blackwell Scientific Publications, London.

Ingram, M. (1973). The microbiological effects of nitrite. *Proceedings of the 2nd International Symposium on Nitrite in Meat Products*, Zeist. Pudoc, Wageningen, The Netherlands, pp. 67–75.

Ito, K. A., and Chen, J. K. (1978). Effect of pH on growth of *Clostridium botulinum* in foods. *Food Technol. 32*: 71.

Janson, A. C. (1983). A deterministic model for monophasic-growth of batch cultures of bacteria. *Antonie van Leeuvenhoek 49*: 513.

Jarvis, B. (1983). Food microbiology into the twenty-first century: A delphi forecast. In *Food Microbiology: Advances and Prospects* (T. A. Roberts and F. A. Skinner, eds.), Academic Press, New York, p. 334.

Jarvis, B. (1988). *Statistical Aspects of the Microbiological Analysis of Foods. Progress in Industrial Microbiology*, Vol. 21, Elsevier, New York.

Jarvis, B., and Patel, M. (1979). The occurrence and control of *Clostridium botulinum* toxin production in model cured meat systems. Research report #414. Leatherhead Food Manufacturing Industries Research Association, Surrey, United Kingdom.

Jarvis, B., Rhodes, A. C., and Patel, M. (1979). Microbiological safety of pasteurized cured meats: Inhibition of *Clostridium botulinum* by curing salts and other additives. In *Food Microbiology and Technology* (B. Jarvis, J. H. B. Christian, and H. D. Michener, eds.), Medicina Viva, Italy, p. 251.

Jeffries, C. J., and Brain, P. (1984). A mathematical model of pollen tube penetration in apple styles. *Planta. 160*: 52.

Jensen, M. J., Genigeorgis, C., and Lindroth, S. (1987). Probability of growth of *Clostridium botulinum* as affected by strain, cell and serologic type, inoculum size and temperature and time of incubation in a model broth system. *J. Food Safety 8*: 109.

Kautter, D. A., Harmon, S. M., Lynt, R. K., and Lilly, T. (1966). Antagonistic effect of *Clostridium botulinum* type E by an organism resembling it. *Appl. Microbiol. 14*: 616.

Kempton, A. G., So, K., and Moneib, N. A. (1987). A procedure for mathematical modelling the production of a metabolic marker of *Clostridium sporogenes* grown under variable nutritional and environmental conditions. *Food Microbiol. 4*: 51.

Lau, A. H. S., Hawirko, R. Z., and Chow, C. T. (1974). Purification and properties of boticin P produced by *Clostridium botulinum. Can. J. Microbiol. 20*: 385.

Leistner, L. (1985). Hurdle technology applied to meat products of the shelf stable product and intermediate moisture food types. In *Properties of Water in Foods* (D. Simatos and J. L. Multon, eds.), Martinus Nijfoff Publishers, The Netherlands, p. 309.

Leistner, L. (1987). Shelf-stable products and intermediate moisture foods based on meat. In *Water Activity: Theory and Applications to Foods* (L. B. Rockland and L. R. Beuchat, eds.), Marcel Dekker, New York, p. 295.

Leistner, L. (1988). Shelf-stable edible meat. In *Proceedings of the 34th International Congress Meat Science Technology*, Brisbane, Australia, pp. 78–82.

Leistner, L., and Rodel, W. (1976). The stability of intermediate moisture foods with respect to micro-organisms. In *Intermediate Moisture Foods* (R. Davies, G. G. Birch, and K. J. Parker, eds.), Applied Science Publishers, Ltd., London, p. 120.

Lerman, P. M. (1980). Fitting segmented regression lines by grid search. *Appl. Statistics 29*: 77.

Lindroth, S. E., and Genigeorgis, C. A. (1986). Probability of growth and toxin production by nonproteolytic *Clostridium botulinum* in rockfish stored under modified atmosphere. *Intl. J. Food Microbiol. 3*: 167.

Lindsay, R. (1981). Modified atmosphere packaging systems for refrigerated fresh fish providing shelflife extension and safety from *Clostridium botulinum* toxigenesis. *Proceedings of the 1st National Conference on Modified and Controlled Atmosphere Packaging of Seafood Products* (R. E. Martin, ed.), National Fisheries Institute, Washington, DC, pp. 30–51.

Lund, B. M., and Wyatt, G. M. (1984). The effect of redox potential, and its interaction with sodium chloride concentration, on the probability of growth of *Clostridium botulinum* type E from spore inocula. *Food Microbiol. 1*: 49.

Lund, B. M., Graham, A. F., George, S. M., and Brown, D. (1990). The combined effect of incubation temperature, pH and sorbic acid on the probability of growth of non-proteolytic, type B *Clostridium botulinum*. *J. Appl. Bacteriol. 69*: 481.

Lund, B. M., Graham, A. F., and Franklin, J. G. (1987a). The effect of acid pH on the probability of growth of proteolytic strains of *Clostridium botulinum*. *Intl. J. Food Microbiol. 4*: 215.

Lund, B. M., Graham, A. F., and Franklin, J. G. (1987b). Inhibition of type A and type B (proteolytic) *Clostridium botulinum* by sorbic acid. *Appl. Environ. Microbiol. 53*: 935.

Lund, B. M., Wyatt, G. M., and Graham, A. F. (1985). The combined effect of low temperature and low pH on survival of, and growth and toxin formation from, spores of *Clostridium botulinum*. *Food Microbiol. 2*: 135.

Lund, B. M., Knox, M. R., and Sims, A. P. (1984). The effect of oxygen and redox potential on the growth of *C. botulinum* type E from a spore inoculum. *Food Microbiol. 1*: 277.

McMeekin, T. A., and Olley, J. (1986). Predictive microbiology. *Food Technol. Australia 38*: 331.

McMeekin, T. A., Olley, J., and Ratkowsky, D. A. (1988). Temperature effects on bacterial growth rates. In *Physiological Models in Microbiology*, Vol. I (M. J. Bazin and J. I. Prosser, eds.), CRC Press, Boca Raton, FL, p. 75.

Meynell, G. G., and Meynell, E. (1965). *Theory and Practice in Experimental Bacteriology*, Cambridge University Press, Cambridge, United Kingdom.

Mistry, V. V., and Kosikowski, F. V. (1983). Use of time-temperature indicators as quality control devices for market milk. *J. Food Prot. 46*: 52.

Moberg, L. J., and Sugiyama, H. (1979). Microbial ecological basis of infant botulism as studied with germfree mice. *Infect. Immun. 25*: 653.

Mohr, P. W., and Krawiec, S. (1980). Temperature characteristics and Arrhenius plots for nominal psychrophiles, mesophiles and thermophiles. *J. Gen. Microbiol. 121*: 311.

Monod, J. (1949). The growth of bacterial cultures. *Annual Reviews Microbiology*, Vol. III (C. E. Clifton, S. Raffel, and H. A. Barker, eds.), Annual Reviews, Inc., Stanford, CA, p. 371.

Montville, T. J. (1984). Quantitation of ph- and NaCl-tolerant subpopulations from *Clostridium botulinum*. *Appl. Environ. Microbiol. 47*: 28.

Mood, A. M., Graybill, F. A., and Boes, D. C. (1974). *Introduction to the Theory of Statistics*, 3rd ed., McGraw-Hill, Tokyo.

Mossell, D. A. A., and Drion, E. F. (1979). Risk analysis. Its application to the protection of the consumer against food-transmitted diseases of microbial aetiology. *Antonie van Leeuwenhoek 45*: 321.

Oguma, K., Iida, H., and Shiozaki, M. (1976). Phage conversion to hemagglutinin production in *Clostridium botulinum* types C and D. *Infect. Immun. 14*: 597.

Ohye, D. F., and Christian, J. H. B. (1967). Combined effects of temperature, pH and water activity on growth and toxin production by *Clostridium botulinum* types A, B and E. In *Botulism 1966*, Proceedings of the 5th International Symposium on Food Microbioly (M. Ingram and T. A. Roberts, eds.), Moscow, July 1966, Chapman & Hall, London, pp. 217–223.

Ohye, D. F., and Scott, W. J. (1953). The temperature relations of *Clostridium botulinum*, types A and B. *Aust. J. Biol. Sci. 6*: 178.

Ohye, D. F., Christian, J. H. B., and Scott, W. J. (1967). Influence of temperature on the water relation of growth of *Clostridium botulinum* type E. In *Botulism 1966*, Proceedings of the 5th International Symposium on Food Microbioly (M. Ingram and T. A. Roberts, eds.), Moscow, July 1966, Chapman & Hall, London, pp. 136–143.

Okereke, A., and Montville, T. J. (1991). Bacteriocin inhibition of *Clostridium botulinum* spores by lactic acid bacteria. *J. Food Prot. 54*: 349.

Olley, J., Doe, P. E., and Heruwati, E. S. (1987). The influence of drying and smoking on the nutritional properties of fish; an introductory overview. In *The Effect of Smoking and Drying on the Nutritional Properties of Fish* (J. R. Burt, ed.), Min. Agric. Fish and Food, Torry Res. Stn., Aberdeen, Scotland.

Pflug, I. J. (1987). Using the straight-line semilogarithmic microbial destruction model as an engineering design model for determining the F-value for heat processes. *J. Food Prot. 50*: 342.

Pflug, I. J., and Holcomb, R. G. (1983). Principles of thermal destruction of microorganisms. In *Disinfection, Sterilization and Preservation*, 3rd ed. (S. S. Block, ed.), Lea and Febiger, Philadelphia, p. 751.

Pflug, I. J., and Odlaug, T. E. (1978). A review of z- and F-values used to ensure the safety of low-acid canned foods. *Food Technol. 32*(6): 63.

Pflug, I. J., Odlaug, T. E., and Christensen, R. (1985). Computing a minimum public health sterilizing value for food with pH values from 4.6 to 6.0. *J. Food Prot. 48*: 848.

Pivnick, H., Barnett, H. W., Nordin, H. R., and Rubin, L. J. (1969). Factors affecting the safety of canned, cured, shelf-stable luncheon meat inoculated with *Clostridium botulinum*. *Can. Inst. Food Sci. Technol. J. 2*: 141.

Plackett, R. L. (1974). *The Analysis of Categorical Data*, Monograph 35, (A. Stewart, ed.), Charles Griffin & Co., Ltd., London.

Post, L. S., Lee, D. A., Solberg, M., Furgang, D., Specchio, J., and Graham, C. (1985). Development of botulinal toxin and sensory deterioration during storage of vacuum and modified atmosphere packaged fish fillets. *J. Food Sci. 50*: 990.

Racine, A., and Moppert, J. (1984). Concentration-effect relationship of oxyprenol in health volunteers: proposal of a new mathematical model. *Br. J. Pharmacokinet. 19*: 143.

Ratkowsky, D. A., Lowry, R. K., McMeekin, T. A., Stokes, A. N., and Chandler, R. E. (1983). Model for bacterial culture growth rate throughout the entire biokinetic temperature range. *J. Bacteriol. 154*: 1222.

Ratkowsky, D. A., Olley, J., McMeekin, T. A., and Ball, A. (1982). Relationship between temperature and growth rate of bacterial cultures. *J. Bacteriol. 149*: 1.

Rhodes, A. C., and Jarvis, B. (1976). A pork slurry system for studying inhibition of *Clostridium botulinum* by curing salts. *J. Food Technol. 11*: 13.

Riemann, H. (1963). Safe heat processing of canned cured meats with regard to bacterial spores. *Food Technol. 1*: 39.

Riemann, H. (1967). The effect of the number of spores on growth and toxin formation by *C. botulinum* type E in inhibitory environments. In *Botulism 1966*, Proceedings of the 5th International Symposium on Food Microbioly (M. Ingram and T. A. Roberts, eds.), Moscow, July 1966, Chapman & Hall, London, pp. 150–168.

Roberts, T. A. (1989). Combination of antimicrobials and processing methods. *Food Technol. 43*: 156.

Roberts, T. A., and Ingram, M. (1973). Inhibition of growth of *Clostridium botulinum* at different pH values by sodium chloride and sodium nitrite. *J. Food Technol. 8*: 467.

Roberts, T. A., and Jarvis, B. (1983). Predictive modelling of food safety with particular references to *Clostridium botulinum* in model cured meat systems. In *Food Microbiology: Advances and Prospects* (T. A. Roberts and F. A. Skinner, eds.), Academic Press, New York, p. 87.

Roberts, T. A., and Thomas, J. A. (1982). Germination and outgrowth of single spores of *Clostridium botulinum* and putrefactive anaerobes. *J. Appl. Bacteriol. 53*: 317.

Roberts, T. A., Gibson, A. M., and Robinson, A. (1981a). Factors controlling the growth of *Clostridium botulinum* types A and B in pasteurized, cured meats. I. Growth in pork slurries prepared from 'low' pH meat (pH range 5.5–6.3). *J. Food Technol. 16*: 239.

Roberts, T. A., Gibson, A. M., and Robinson, A. (1981b). Factors controlling the growth of *Clostridium botulinum* types A and B in pasteurized, cured meats. II. Growth in pork slurries prepared from 'high' pH meat (pH range 6.3–6.8). *J. Food Technol. 16*: 267.

Roberts, T. A., Gibson, A. M., and Robinson, A. (1981c). Prediction of toxin production by *Clostridium botulinum* in pasteurized pork slurry. *J. Food Technol. 16*: 337.

Roberts, T. A., Gibson, A. M., and Robinson, A. (1982). Factors controlling the growth of *Clostridium botulinum* types A and B in pasteurized, cured meats. III. The effect of potassium sorbate. *J. Food Technol. 17*: 307.

Roberts, T. A., Jarvis, B., and Rhodes, A. C. (1976). Inhibition of *Clostridium botulinum* by curing salts in pasteurized pork slurry. *J. Food Technol. 11*: 25.

Robinson, A., Gibson, A. M., and Roberts, T. A. (1982). Factors controlling the growth of *Clostridium botulinum* types A and B in pasteurized, cured meats. V. Prediction of toxin production: Non-linear effects of storage temperature and NaCl concentration. *J. Food Technol. 17*: 727.

Russel, A. D. (1982). *The Destruction of Bacterial Spores*, Academic Press, New York.

SAS Institute Inc. (1985). *SAS User's Guide: Statistics*, Version 5, Cary, NC.

Schmidt, C. F., Lechowich, R. V., and Folinazzo, J. F. (1961). Growth and toxin production by type E *Clostridium botulinum* below 40°F. *J. Food Sci. 26*: 616.

Schoolfield, R. M., Sharpe, P. J. H., and Magnuson, C. E. (1981). Non-linear regression of biological temperature-dependent rate models based on absolute reaction-rate theory. *J. Theor. Biol. 88*: 719.

Segner, W. P., Schmidt, C. F., and Boltz, J. K. (1966). Effect of sodium chloride and pH on the outgrowth of spores of type E *Clostridium botulinum* at optimal and suboptimal temperatures. *Appl. Microbiol. 14*: 49.

Shull, J. J., Cargo, G. T., and Ernst, R. R. (1963). Kinetics of heat activation and thermal death of bacterial spores. *Appl. Microbiol. 11*: 485.

Siegel, L. S. (1981). Fermentation kinetics of botulinum toxin production (types A, B and E). In *Biomedical Aspects of Botulism* (G. E. Lewis, Jr., ed.), Academic Press, New York, p. 121.

Siegel, L. S., and Metzger, J. F. (1980). Effect of fermentation conditions on toxin production by *Clostridium botulinum* type B. *Appl. Environ. Microbiol. 40*: 1023.

Silliker, J. H., Greenberg, R. A., and Schack, W. R. (1958). Effect of individual curing ingredients on the shelf stability of canned comminuted meats. *Food Technol. 12*(10): 551.

Smelt, J. P. (1980). Heat resistance of *Clostridium botulinum* in acid ingredients and its significance for the safety of chilled foods. Thesis, University of Utrecht, The Netherlands.

Smith, L. DS. (1975). The inhibition of *Clostridium botulinum* by strains of *Clostridium perfringens* isolated from soil. *Appl. Microbiol. 30*: 319.

So, K., Moneib, N. A., and Kempton, A. G. (1987). Conversion of ecogram data to a mathematical equation of four dimension. *Food Microbiol. 4*: 67.

Sofos, J. N., Busta, F. F., and Allen, C. E. (1979). Botulism control by nitrite and sorbate in cured meats: A review. *J. Food Prot. 42*: 739.

Sokal, R. R., and Rohlf, F. J. (1981). *Biometry: The Principles and Practices of Statistics in Biological Research*, 2nd ed., W. H. Freeman & Co., New York.

Stiratelli, R., Laird, N., and Ware, J. H. (1984). Random-effects models for serial observations with binary response. *Biometrics 40*: 961.

Stumbo, R. C. (1973). *Thermobacteriology in Food Processing*, 2nd ed., Academic Press, New York.

Tanaka, N., Traisman, E., Plantinga, P., Finn, L., Flom, W., Meske, L., and Guggisberg, J. (1986). Evaluation of factors involved in antibotulinal properties of pasteurized process cheese spreads. *J. Food Prot. 49*: 526.

Tompkin, R. B. (1980). Botulism from meat and poultry products—a historical perspective. *Food Technol. 34*: 229.

Tompkin, R. B., Christiansen, L. N., and Shaparis, A. B. (1978). Effect of prior refrigeration on botulinal outgrowth in perishable canned cured meat when temperature abused. *Appl. Environ. Microbiol. 35*: 863.

Tompkin, R. B., Christiansen, L. N., and Shaparis, A. B. (1979). Isoascorbate level and botulinal inhibition in perishable canned cured meat. *J. Food Sci. 44*: 1147.

Webster's Ninth Collegiate Dictionary. (1989). Merriam-Webster, Inc., Springfield, MA.

Wells, J. H., and Singh, R. P. (1988). Application of time-temperature indicators in monitoring changes in quality attributes of perishable and semipreserved foods. *J. Food Sci. 53*: 148.

Wentz, M. W., Scott, R. A., and Vennes, J. W. (1967). *Clostridium botulinum* type F: seasonal inhibition by *Bacillus licheniformis*. *Science 155*: 89.

Xezones, J., and Hutchings, I. J. (1965). Thermal resistance of *Clostridium botulinum* (62A) spores as affected by fundamental food constituents. *Food Technol. 19*: 113.
Zwietering, M. H., Jongenburger, I., Rombouts, F. M., and van't Riet, K. (1990). Modeling of the bacterial growth curve. *Appl. Environ. Microbiol. 56*: 8175.

Index